AF411430

Chemical Engineering and Technology in Mineral Processing and Extractive Metallurgy

Chemical Engineering and Technology in Mineral Processing and Extractive Metallurgy

Editors

Shuai Wang
Xingjie Wang
Jia Yang

MDPI • Basel • Beijing • Wuhan • Barcelona • Belgrade • Manchester • Tokyo • Cluj • Tianjin

Editors

Shuai Wang
College of Chemistry and
Chemical Engineering
Central South University
Changsha
China

Xingjie Wang
Institute of Geological Survey
China University of
Geosciences
Wuhan
China

Jia Yang
National Engineering
Research Center for Vacuum
Metallurgy
Kunming University of
Science and Technology
Kunming
China

Editorial Office
MDPI
St. Alban-Anlage 66
4052 Basel, Switzerland

This is a reprint of articles from the Special Issue published online in the open access journal *Minerals* (ISSN 2075-163X) (available at: www.mdpi.com/journal/minerals/special_issues/MPEM).

For citation purposes, cite each article independently as indicated on the article page online and as indicated below:

LastName, A.A.; LastName, B.B.; LastName, C.C. Article Title. *Journal Name* **Year**, *Volume Number*, Page Range.

ISBN 978-3-0365-5426-6 (Hbk)
ISBN 978-3-0365-5425-9 (PDF)

Contents

Jinhong Zhang and Wei Zhang
AFM Image Analysis of the Adsorption of Xanthate and Dialkyl Dithiophosphate on Chalcocite
Reprinted from: *Minerals* **2022**, *12*, 1018, doi:10.3390/min12081018 1

Jian Peng, Wei Sun, Le Xie, Haisheng Han and Yao Xiao
An Experimental Study of Pressure Drop Characteristics and Flow Resistance Coefficient in a
Fluidized Bed for Coal Particle Fluidization
Reprinted from: *Minerals* **2022**, *12*, 289, doi:10.3390/min12030289 35

Jinhong Zhang
An Investigation of the Adsorption of Xanthate on Bornite in Aqueous Solutions Using an
Atomic Force Microscope
Reprinted from: *Minerals* **2021**, *11*, 906, doi:10.3390/min11080906 51

**Tomasz Gawenda, Agata Stempkowska, Daniel Saramak, Dariusz Foszcz, Aldona
Krawczykowska and Agnieszka Surowiak**
Assessment of Operational Effectiveness of Innovative Circuit for Production of Crushed
Regular Aggregates in Particle Size Fraction 8–16 mm
Reprinted from: *Minerals* **2022**, *12*, 634, doi:10.3390/min12050634 75

Qinghong Yang, Fengyang Chen, Lin Tian, Jianguo Wang, Ni Yang and Yanqing Hou et al.
Boron Impurity Deposition on a Si(100) Surface in a SiHCl$_3$-BCl$_3$-H$_2$ System for
Electronic-Grade Polysilicon Production
Reprinted from: *Minerals* **2022**, *12*, 651, doi:10.3390/min12050651 91

Meng Li, Junfan Yuan, Bingbing Liu, Hao Du, David Dreisinger and Yijun Cao et al.
Detoxification of Arsenic-Containing Copper Smelting Dust by Electrochemical Advanced
Oxidation Technology
Reprinted from: *Minerals* **2021**, *11*, 1311, doi:10.3390/min11121311 105

Kui Li, Wei Zhang, Menglong Fu, Chengzhi Li and Zhengliang Xue
Discussion on Criterion of Determination of the Kinetic Parameters of the Linear Heating
Reactions
Reprinted from: *Minerals* **2022**, *12*, 81, doi:10.3390/min12010081 119

Renlin Zhu, Jianli Li, Jiajun Jiang, Yue Yu and Hangyu Zhu
Effect of Basicity on the Sulfur Precipitation and Occurrence State in Kambara Reactor
Desulfurization Slag
Reprinted from: *Minerals* **2021**, *11*, 977, doi:10.3390/min11090977 133

**Gloria I. Dávila-Pulido, Adrián A. González-Ibarra, Mitzué Garza-García and Danay A.
Charles**
Effect of Magnesium on the Hydrophobicity of Sphalerite
Reprinted from: *Minerals* **2021**, *11*, 1359, doi:10.3390/min11121359 145

**Yuki Semoto, Gde Pandhe Wisnu Suyantara, Hajime Miki, Keiko Sasaki, Tsuyoshi Hirajima
and Yoshiyuki Tanaka et al.**
Effect of Sodium Metabisulfite on Selective Flotation of Chalcopyrite and Molybdenite
Reprinted from: *Minerals* **2021**, *11*, 1377, doi:10.3390/min11121377 157

Hao Peng, Bing Li, Wenbing Shi and Zuohua Liu
Efficient Recovery of Vanadium from High-Chromium Vanadium Slag with Calcium-Roasting
Acidic Leaching
Reprinted from: *Minerals* **2022**, *12*, 160, doi:10.3390/min12020160 . . . . **171**

Tülay Türk, Zeynep Üçerler, Fırat Burat, Gülay Bulut and Murat Olgaç Kangal
Extraction of Potassium from Feldspar by Roasting with $CaCl_2$ Obtained from the Acidic
Leaching of Wollastonite-Calcite Ore
Reprinted from: *Minerals* **2021**, *11*, 1369, doi:10.3390/min11121369 . . . . **183**

**Martina Laubertová, Alexandra Kollová, Jarmila Trpčevská, Beatrice Plešingerová and
Jaroslav Briančin**
Hydrometallurgical Treatment of Converter Dust from Secondary Copper Production: A Study
of the Lead Cementation from Acetate Solution
Reprinted from: *Minerals* **2021**, *11*, 1326, doi:10.3390/min11121326 . . . . **197**

**Norman Toro, Carlos Moraga, David Torres, Manuel Saldaña, Kevin Pérez and Edelmira
Gálvez**
Leaching Chalcocite in Chloride Media—A Review
Reprinted from: *Minerals* **2021**, *11*, 1197, doi:10.3390/min11111197 . . . . **213**

Youming Yang, Xiaolin Zhang, Kaizhong Li, Li Wang, Fei Niu and Donghui Liu et al.
Maintenance of the Metastable State and Induced Precipitation of Dissolved Neodymium (III)
in an Na_2CO_3 Solution
Reprinted from: *Minerals* **2021**, *11*, 952, doi:10.3390/min11090952 . . . . **229**

Bo Yang, Ruzhen Peng, Dan Zhao, Ni Yang, Yanqing Hou and Gang Xie
Performance of TiB_2 Wettable Cathode Coating
Reprinted from: *Minerals* **2021**, *12*, 27, doi:10.3390/min12010027 . . . . **243**

Wei Wang, Shuai Wang, Jia Yang, Chengsong Cao, Kanwen Hou and Lixin Xia et al.
Preparation of Antimony Sulfide and Enrichment of Gold by Sulfuration–Volatilization from
Electrodeposited Antimony
Reprinted from: *Minerals* **2022**, *12*, 264, doi:10.3390/min12020264 . . . . **255**

Katarzyna Ochromowicz, Kurt Aasly and Przemyslaw B. Kowalczuk
Recent Advancements in Metallurgical Processing of Marine Minerals
Reprinted from: *Minerals* **2021**, *11*, 1437, doi:10.3390/min11121437 . . . . **269**

Changxin Li, Xiang Li, Qingwu Zhang, Li Li and Shuai Wang
The Alkaline Fusion-Hydrothermal Synthesis of Blast Furnace Slag-Based Zeolite (BFSZ): Effect
of Crystallization Time
Reprinted from: *Minerals* **2021**, *11*, 1314, doi:10.3390/min11121314 . . . . **299**

Mohammed El Khalloufi, Olivier Drevelle and Gervais Soucy
Titanium: An Overview of Resources and Production Methods
Reprinted from: *Minerals* **2021**, *11*, 1425, doi:10.3390/min11121425 . . . . **311**

 *minerals*

Article

AFM Image Analysis of the Adsorption of Xanthate and Dialkyl Dithiophosphate on Chalcocite

Jinhong Zhang * and Wei Zhang

Department of Mining and Geological Engineering, The University of Arizona, Tucson, AZ 85721, USA
* Correspondence: jhzhang@email.arizona.edu; Tel.: +1-520-626-9656

Abstract: Atomic force microscopy (AFM) has been applied to study the adsorption morphology of various collectors, i.e., potassium ethyl xanthate (KEX) and potassium amyl xanthate (PAX) and Cytec Aerofloat 238 (sodium dibutyl dithiophosphate), on chalcocite in situ in aqueous solutions. The AFM images show that all these collectors adsorb strongly on chalcocite. Xanthate adsorbs mainly in the form of insoluble cuprous xanthate (CuX), which binds strongly with the mineral surface without being removed by flushing with ethanol alcohol. This xanthate/chalcocite adsorption mechanism is very similar to the one obtained with the xanthate/bornite system; while it is different from the one of the xanthate/chalcopyrite systems, for which oily dixanthogen is the main adsorption product on chalcopyrite surface. On the other hand, dibutyl dithiophosphate adsorbs on chalcocite in the form of hydrophobic patches, which can be removed by rinsing with ethanol alcohol. AFM images show that the adsorption of collectors increases with increasing adsorption time and collectors' concentration. In addition, increasing the solution pH to 10 does not prevent the adsorption of xanthate and Aerofloat 238 on chalcocite and the result is in line with the fact that chalcocite floats well in a wide pH range up to 12 with xanthate and dialkyl dithiophosphate being used as collectors. The blending collectors study shows that xanthate and dialkyl dithiophosphate can co-adsorb with both insoluble cuprous xanthate and oily $Cu(DTP)_2$ (Cu dibutyl dithiophosphate) on a chalcocite surface. The present study helps to clarify the flotation mechanism of chalcocite in industry practice using xanthate and dialkyl dithiophosphate as collectors.

Keywords: flotation; chalcocite; xanthate; dialkyl dithiophosphate; cuprous xanthate; AFM

Citation: Zhang, J.; Zhang, W. AFM Image Analysis of the Adsorption of Xanthate and Dialkyl Dithiophosphate on Chalcocite. *Minerals* **2022**, *12*, 1018. https://doi.org/10.3390/min12081018

Academic Editors: Jan Zawala and Jean-François Blais

Received: 30 June 2022
Accepted: 11 August 2022
Published: 13 August 2022

Publisher's Note: MDPI stays neutral with regard to jurisdictional claims in published maps and institutional affiliations.

1. Introduction

In copper sulfide extractive metallurgy, the most efficient and widely applied separation technique is froth flotation, for which the adsorption of the collector on the mineral surface is vital to achieving satisfactory flotation efficiency and metal recovery. By now, some works have been carried out to clarify the adsorption mechanism of collectors on copper sulfides. [1–6]. For the flotation of chalcocite, xanthate and dialkyl dithiophosphate have been applied as collectors in flotation practice.

For the case of the adsorption of xanthate on chalcocite surface in an aqueous solution, the work has attracted the interest of many researchers. For example, Gaudin and Schuhmann [1] studied the solubility of potassium ethyl xanthate and its reaction products in organic solvents and firstly proposed there was initially chemisorbed xanthate (X^-) on chalcocite surface followed by insoluble cuprous xanthate (CuX). Allison et al. [3] studied the reaction products of various sulfide minerals with xanthate solutions and reported that the measured rest potential of chalcocite in 6.25×10^{-4} M KEX solution at pH 7 was +60 mV and the reaction product of PAX on chalcocite was cuprous xanthate. Mielczarski and Suoninen [7,8] applied X-ray photoelectron spectroscopy (XPS) and studied the adsorption of potassium ethyl xanthate on cuprous sulfide. It was reported that there was a relatively rapid formation of a well-oriented monolayer of xanthate ions followed by the slow growth of disordered cuprous xanthate molecules on top of this layer. Richardson et al. [9] carried out an electrochemical study of chalcocite flotation by using ethyl xanthate as a collector. It was proposed that cuprous xanthate and possibly cupric xanthate were adsorbed hydrophobic species on chalcocite surfaces due to the exchange and charge transfer oxidation reactions. Leppinen et al. [10] applied an in situ Fourier-transform infrared spectroscopy (FTIR) study of ethyl xanthate adsorption on sulfide minerals under conditions of controlled potential and they concluded that, for chalcocite, there were three different potential regions of xanthate adsorption, including chemisorption, copper ethyl xanthate formation on the surface and multilayer formation.

Some studies have been carried out to study the adsorption of dialkyl dithiophosphate on the chalcocite surface and therefore its impact on chalcocite flotation. Goold and Finkelstein [11] reported that only $Cu(DTP)_2$ was present on copper sulfide. Solozhenkin and Koptsia [12] found cuprous diethyl dithiophosphate (CuDTP) and cupric diethyl dithiophosphate ($Cu(DTP)_2$). Chander and Fuerstenau [13] studied the effect of potassium diethyl dithiophosphate (DTP) on the electrochemical properties of copper sulfide in aqueous solutions. It was reported that multilayers of reaction products, i.e., CuDTP and $Cu(DTP)_2$ were formed when copper sulfide was immersed in reagent solutions of dithiophosphate.

Zhang et al. [14] studied the hydrophobic flocculation of marmatite fines in aqueous suspensions by the addition of butyl xanthate (KBX) and ammonium dibutyl dithiophosphate (ADD) using laser particle size analysis, microscopy analysis, electrophoretic light scattering, contact angle measurement and infrared spectroscopy. It was claimed that the chemisorption of butyl xanthate ions or dibutyl dithiophosphate ions on marmatite resulted in hydrophobic flocculation, and therefore a greatly improved flotation response of marmatite because of the formation of flocs.

Recently, Dhar et al. [15] investigated the interaction of dithiophosphate and a mixture of xanthate and dithiophosphate collector on copper ore sample using zeta potential, quantitative adsorption, FTIR studies and Hallimond tube flotation. It was reported that using this mixture of collectors can improve both the grade and recovery of copper flotation concentrate.

All the studies reviewed above have revealed a lot of information, such as the reaction, product and mechanism, of the adsorption of xanthate and dithiophosphate collectors on chalcocite surface. It is also of great interest to directly obtain the image of collectors on mineral surfaces changing with pulp chemistry, such as pH and chemical dosage. For example, the AFM imaging technique has been widely used in the surface characterization of various materials, and recently the technique has been successfully applied to study in situ the adsorption of chemicals on the surface of minerals [16–22]. The novel analysis method has greatly expanded the understanding of the impact of solution chemistry on the collectors' adsorption on the mineral surface and the flotation mechanism as well.

In the present investigation, an AFM image analysis technique has been applied to obtain the surface morphology of chalcocite in various collectors' solutions, i.e., KEX, PAX and dialkyl dithiophosphate, at different pHs. By comparing the AFM images obtained under different conditions, such as the collector's type, dosage, and contact time, one can study the impact of water chemistry on the adsorption of collectors on chalcocite. The image analysis results will help answer some of the questions, for example, what is the morphology of the adsorbate on the chalcocite surface? What is the impact of the collector's dosage, the contact time and solutions' pH on the adsorption of the collector on the chalcocite surface? What is the binding strength between the adsorbate and chalcocite? What is the adsorption morphology on chalcocite by using a mixed collectors scheme of xanthate and dialkyl dithiophosphate? All this information will help clarify the reaction and adsorption mechanisms of xanthate and dialkyl dithiophosphate on chalcocite changing with solutions' chemistry, and therefore its impact on chalcocite flotation in industry practice.

2. Materials and Methods

2.1. Materials

Research-grade chalcocite (Cu_2S) was obtained from Wards Natural Science Establishment Inc. (Rochester, NY, USA). Mineral samples were finely polished by consecutively using #800, #1200 and #2400 sandpaper, and then a diamond-polishing paste of 10, 5, 2.5 and 1 microns. (MTI Inc., Richmond, CA, USA) Mineral samples were further cleaned by rinsing thoroughly with ethanol and water. A 1.2 cm × 1.2 cm sample was used for the AFM surface image analysis. The DI (deionized) water used in the present work had a conductivity of 18.2 $M\Omega \cdot cm^{-1}$ at 22 °C and surface tension of 72.8 mN/m at 22 °C. Potassium amyl xanthate (PAX, >98%), potassium ethyl xanthate (KEX, >98%) and NaOH (>99%) were obtained from Alfa Aesar and used without further purification. Cytec Aerofloat 238 (sodium di-butyl dithiophosphate) was obtained from Cytec (Tempe, AZ, USA). Chemical solutions were freshly prepared at various concentrations and pH levels as needed each time right before an experiment was carried out.

2.2. AFM Surface Image Analysis

AFM surface image measurements were carried out with a Digital Instrument Nanoscope IIID (Veeco, San Jose, CA, USA) AFM using the contact mode at room temperature (22 ± 1 °C). SNL cantilevers were obtained from Veeco, San Jose, CA, USA. Triangular Si_3N_4 cantilevers with a nominal spring constant of 0.12~0.58 N/m were used for both AFM imaging and force measurements.

To study the mineral surface in water, surface image measurements were carried out after 5 mL of DI water was gently injected into an AFM fluid cell. Extreme care was taken to avoid the entrapment of air in the cell. After surface images were collected in water, a 10 mL solution of a specific chemical's concentration was flushed through the liquid cell, and the cell was left undisturbed for the adsorption of chemicals on the mineral surface. AFM image analysis measurement was commenced after the exposure of the mineral plate to the chemical solution for a specific time. The AFM images as reported in this study, which were processed by no image modification other than being flattened, include both height and deflection images obtained in the contact mode. The same silicon nitride probe was applied to obtain the AFM image of the mineral plate in the solutions at different conditions.

3. Results

3.1. AFM Image of Mineral Surface in Xanthate Solutions

Figure 1 shows the surface images of a bare chalcocite surface that has been covered by nanopure water in an AFM liquid cell for 10 min. Figure 1A is the 10 μm × 10 μm height image, from which one can see that the solid surface is still largely smooth with very few adsorbates on the sample surface detected by the AFM probe. Figure 1B is the 3-D image of Figure 1A. Figure 1C is the section analysis of Figure 1A. Figure 1D is the deflection image of Figure 1A with a 10 nm data scale.

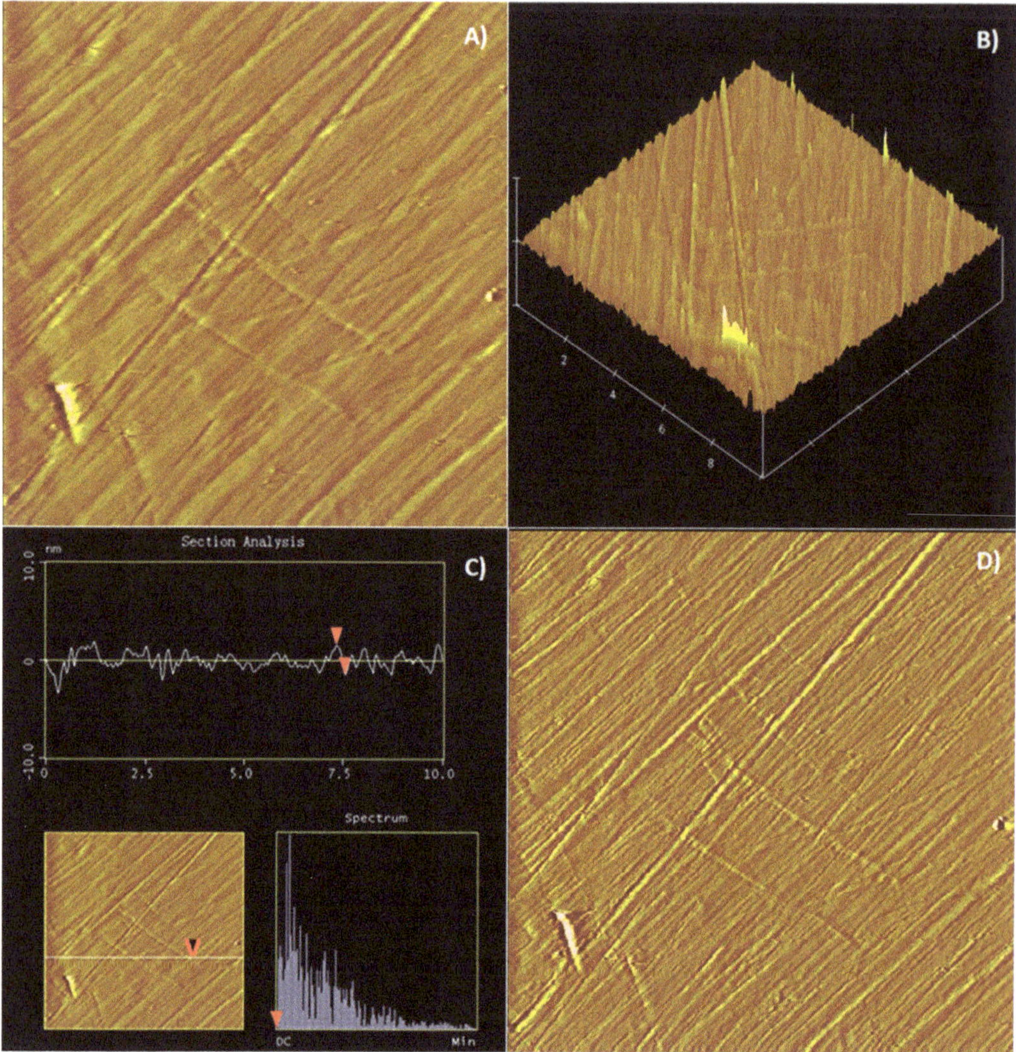

Figure 1. AFM images of a chalcocite surface in water for 10 min. (**A**) The 10 μm × 10 μm height image with a data scale of 20 nm; (**B**) the 3-D image; (**C**) the section analysis (the red arrows indicate the top and the bottom of an average asperity); and (**D**) the deflection image with a data scale of 10 nm.

Figure 2A is the height image of a chalcocite surface in 5×10^{-5} M KEX solution at pH 6 for 10 min. Compared to Figure 1A, a lot of adsorbate shows up on chalcocite when the mineral surface contacts the xanthate solution for 10 min. Figure 2B is the 3-D image of Figure 2A. Figure 2C is the section analysis of Figure 2A and the surface roughness clearly increases due to the adsorption. Figure 2D is the deflection image of Figure 2A with a 10 nm data scale.

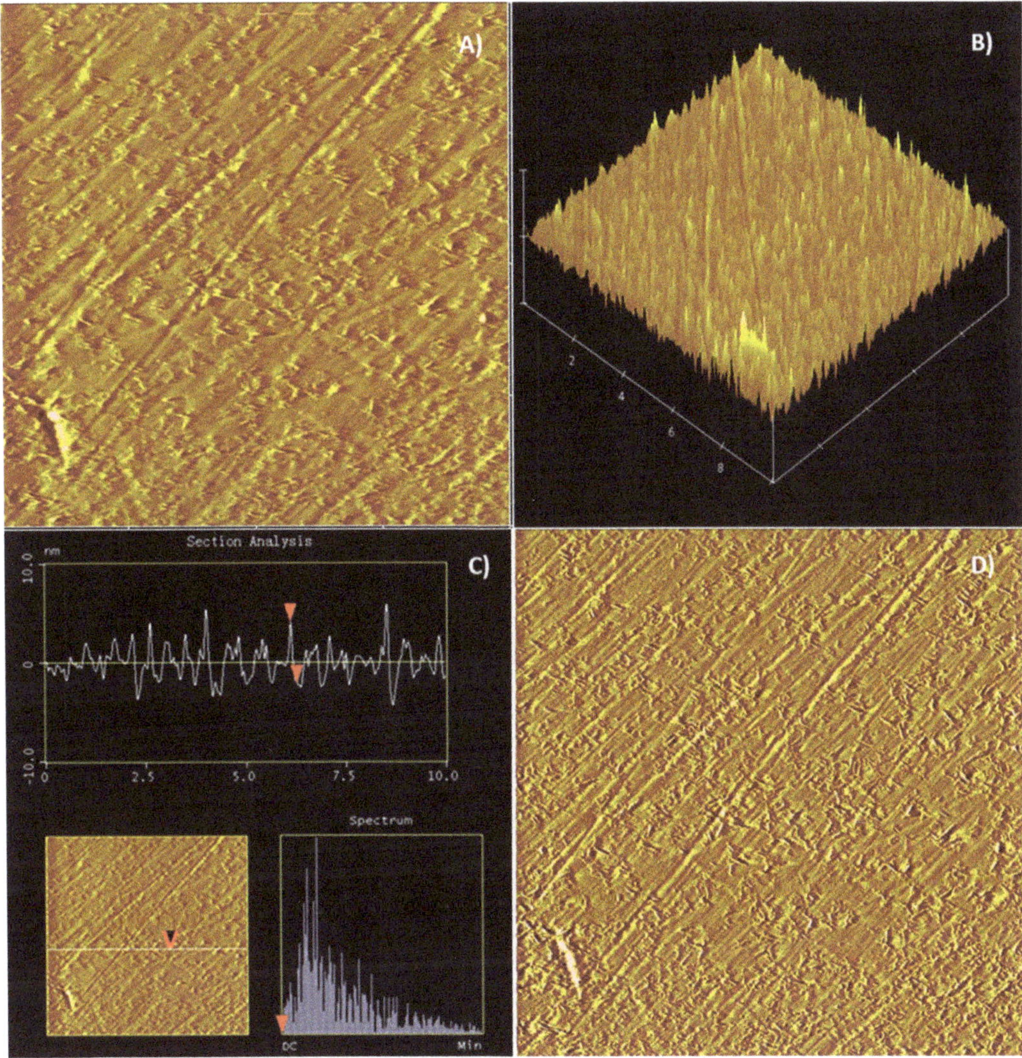

Figure 2. AFM images of a chalcocite surface in 5×10^{-5} M KEX solution at pH 6 for 10 min. (**A**) The $5\ \mu m \times 5\ \mu m$ height image with a data scale of 20 nm; (**B**) the 3-D image; (**C**) the section analysis (the red arrows indicate the top and the bottom of an average asperity); and (**D**) the deflection image with a data scale of 10 nm.

Figure 3A is the height image of a chalcocite surface in 1×10^{-4} M KEX solution at pH 6 for 10 min. Compared to Figure 1A, a lot of adsorbate exists on chalcocite when the mineral surface contacts 1×10^{-4} M KEX solution for 10 min. In addition, by comparing it to Figure 2A, one can find that the mineral surface becomes rougher, suggesting more precipitates are forming at the solid/liquid interface, when the KEX's concentration increases from 5×10^{-5} M to 1×10^{-4} M. Figure 3B is the 3-D image of Figure 3A. Figure 3C is the section analysis of Figure 3A and it confirms that the surface roughness increases due to the adsorption of the collector. Figure 3D is the deflection image of Figure 3A with a 10 nm data scale.

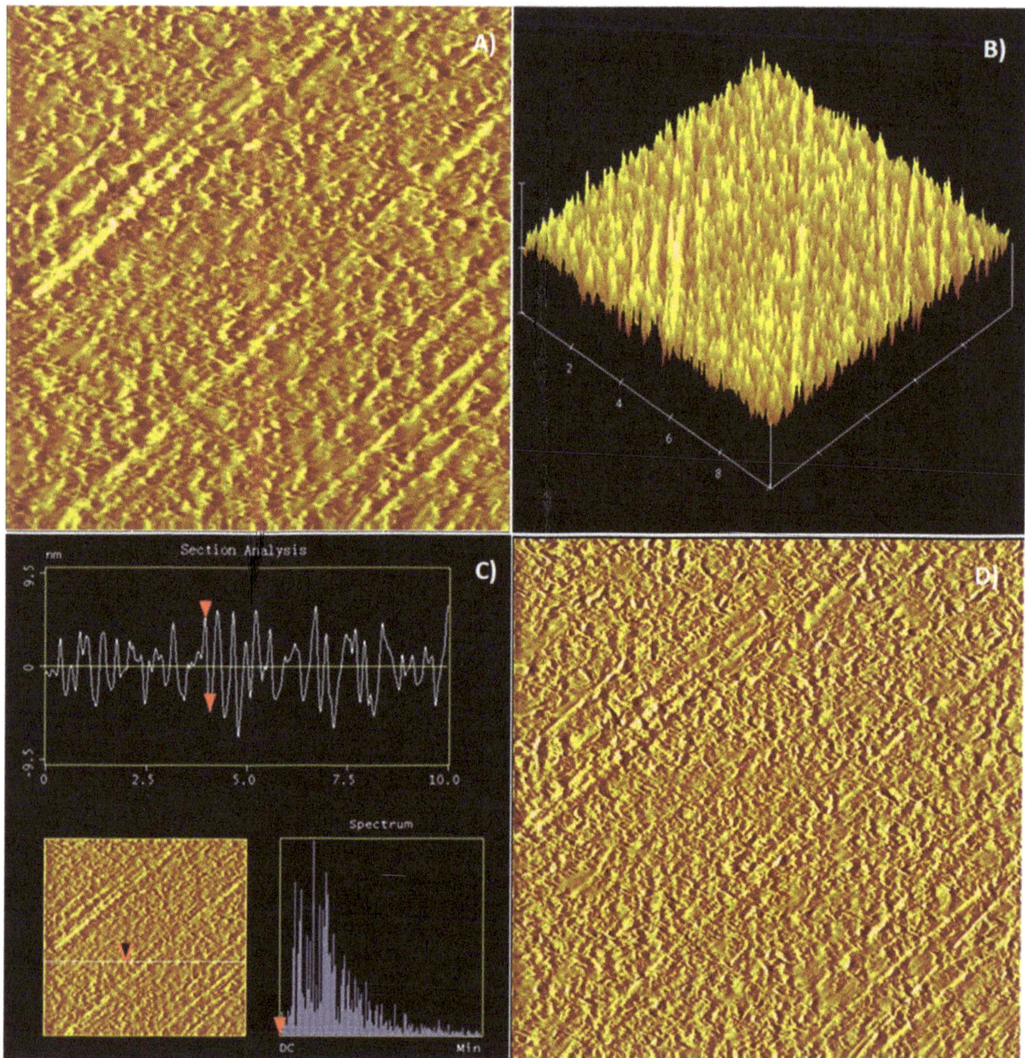

Figure 3. AFM images of a chalcocite surface in 1×10^{-4} M KEX solution at pH 6 for 10 min. (**A**) The 5 μm × 5 μm height image with a data scale of 20 nm; (**B**) the 3-D image; (**C**) the section analysis (the red arrows indicate the top and the bottom of an average asperity); and (**D**) the deflection image with a data scale of 10 nm.

Figure 4 shows the AFM images of a chalcocite surface in 5×10^{-4} M KEX solution at pH 6 for 10 min. By comparing Figure 4A, the height image, to Figure 1A, one can observe a lot of adsorbate showing up on chalcocite after the mineral surface contacts the xanthate solution. In addition, by comparing it to Figures 2A and 3A, one can find that, when the KEX's concentration increases, more precipitates are forming at the solid/liquid interface and the mineral surface becomes much rougher. Figure 4B is the 3-D image of Figure 4A. Figure 4C, the section analysis of Figure 4A, confirms that the surface roughness increases greatly when the chalcocite surface contacts a high concentration of KEX. Figure 4D is the deflection image of Figure 4A with a 10 nm data scale.

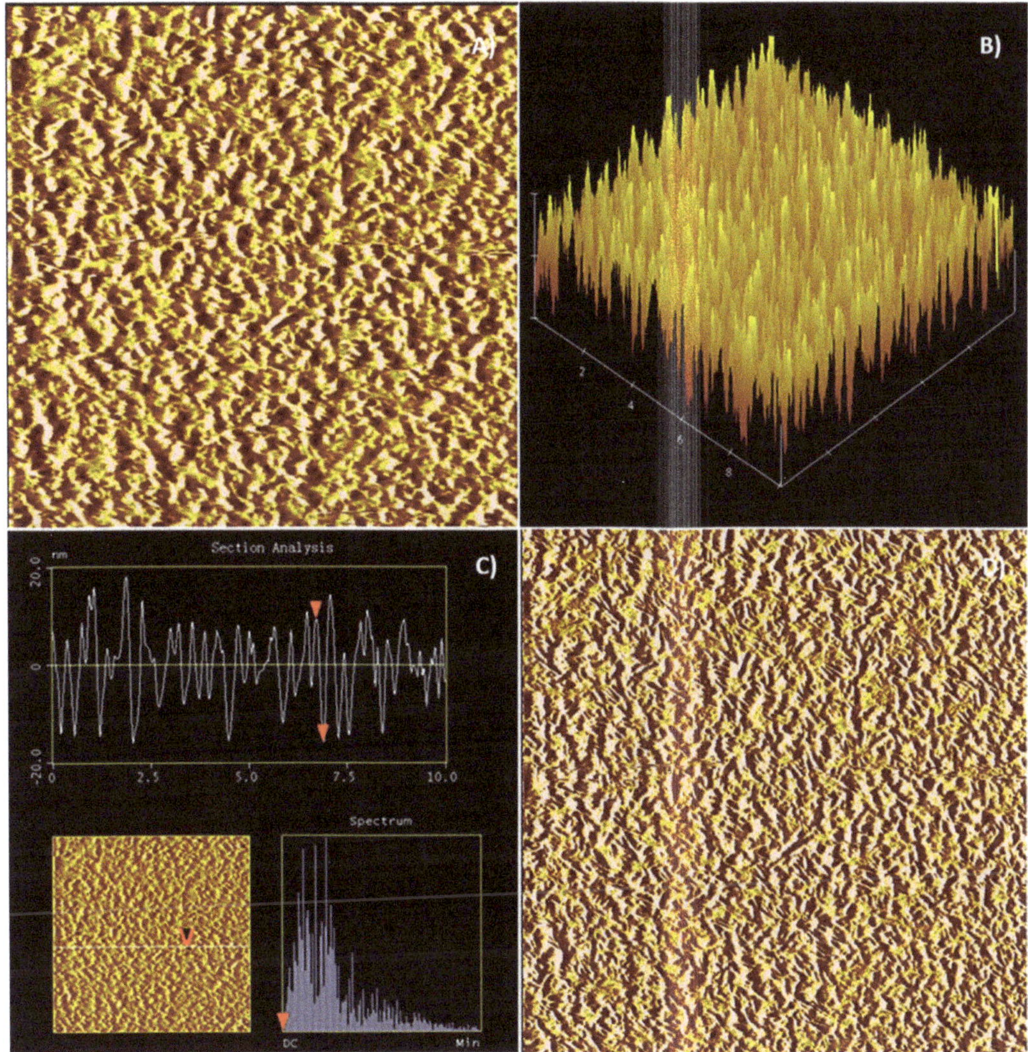

Figure 4. AFM images of a chalcocite surface in 5×10^{-4} M KEX solution at pH 6 for 10 min. (**A**) The 10 μm × 10 μm height image with a data scale of 20 nm; (**B**) the 3-D image; (**C**) the section analysis (the red arrows indicate the top and the bottom of an average asperity); and (**D**) the deflection image with a data scale of 10 nm.

Figure 5A is the height image of a chalcocite surface in 5×10^{-4} M KEX solution at pH 6 for 20 min and further rinsed with 10 mL ethanol and 5 mL water consecutively. The images are finally obtained when the mineral sample is placed in water. Compared to Figure 4A, one can see that the precipitates as observed in Figure 4A still exist on the chalcocite surface without being dissolved and rinsed off from the mineral surface. Figure 5B is the 3-D image of Figure 5A. Figure 5C is the section analysis of Figure 5A. Figure 5D is the deflection image of Figure 5A with a 10 nm data scale.

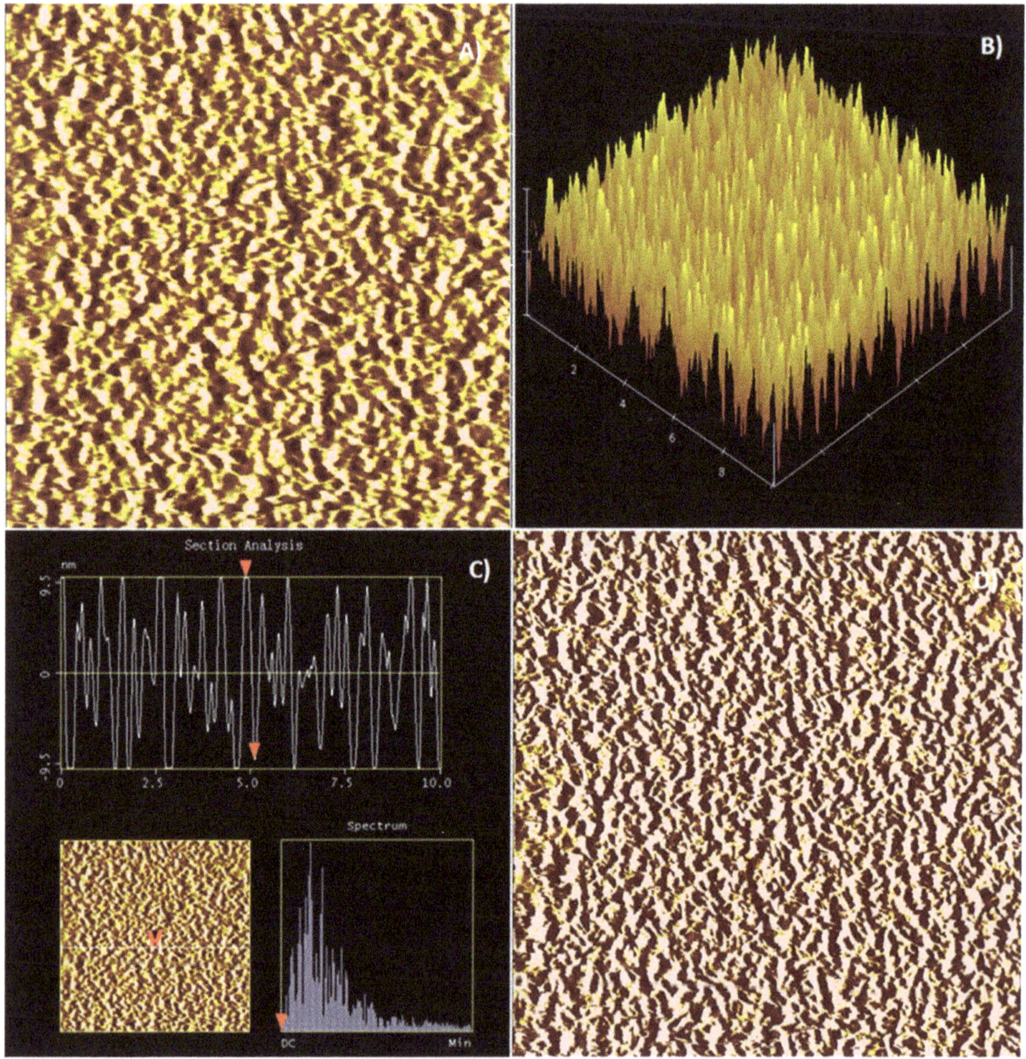

Figure 5. AFM images of a chalcocite surface in 5×10^{-4} M KEX solution at pH 6 for 30 min and further rinsed with ethanol and water. (**A**) The 10 μm × 10 μm height image with a data scale of 20 nm; (**B**) the 3-D image; (**C**) the section analysis (the red arrows indicate the top and the bottom of an average asperity); and (**D**) the deflection image with a data scale of 10 nm.

Figure 6 shows the AFM images of a chalcocite surface in 1×10^{-5} M PAX solution at pH 6 for 10 min. By comparing Figure 6A, the height image, to Figure 1A, one can find that there is a little adsorbate showing up on chalcocite when the mineral surface contacts PAX solution at pH 6 for 10 min. Figure 6B is the 3-D image of Figure 6A. Figure 6C, the section analysis of Figure 6A, shows that the surface roughness increases a little bit due to the adsorption of PAX. Figure 6D is the deflection image of Figure 6A with a 10 nm data scale.

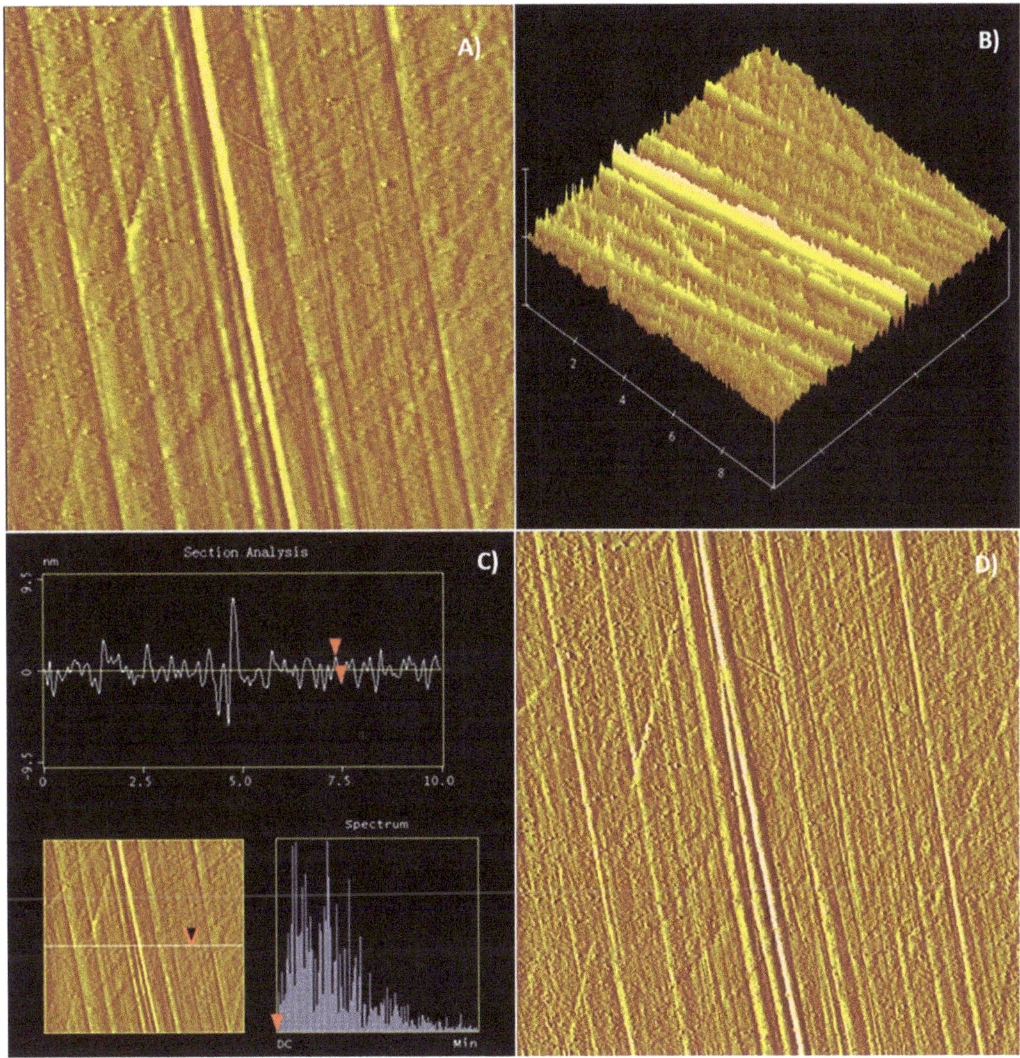

Figure 6. AFM images of a chalcocite surface in 1×10^{-5} M PAX solution at pH 6 for 10 min. (**A**) The 10 μm × 10 μm height image with a data scale of 20 nm; (**B**) the 3-D image; (**C**) the section analysis (the red arrows indicate the top and the bottom of an average asperity); and (**D**) the deflection image with a data scale of 10 nm.

Figure 7 shows the AFM images of a chalcocite surface in 5×10^{-5} M PAX solution at pH 6 for 10 min. By comparing Figure 7A, the height image, to Figure 1A, one can observe that there is a lot of adsorbate showing up on chalcocite when the mineral surface contacts 5×10^{-5} M PAX solution for 10 min. By comparing Figure 7A to Figure 6A, one can also find that when the PAX's concentration increases from 1×10^{-5} M to 5×10^{-5} M, the mineral surface becomes much rougher with more precipitates forming at the solid/liquid interface. Figure 7B is the 3-D image of Figure 7A. Figure 7C is the section analysis of Figure 7A and it confirms that the surface roughness increases with increasing the PAX's concentration. Figure 7D is the deflection image of Figure 7A with a 10 nm data scale.

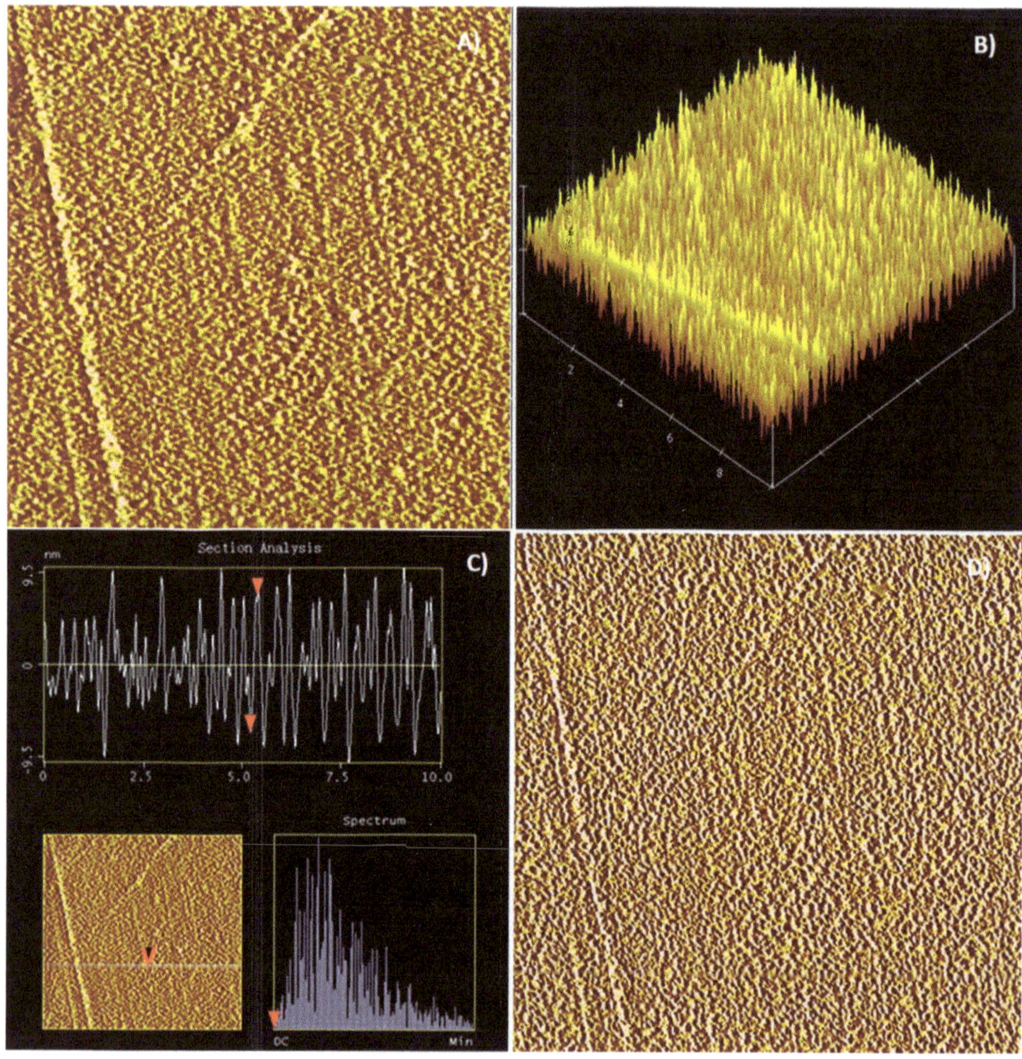

Figure 7. AFM images of a chalcocite surface in 5×10^{-5} M PAX solution at pH 6 for 10 min. (**A**) The 10 μm × 10 μm height image with a data scale of 20 nm; (**B**) the 3-D image; (**C**) the section analysis (the red arrows indicate the top and the bottom of an average asperity); and (**D**) the deflection image with a data scale of 10 nm.

Figure 8 shows the AFM images of a chalcocite surface in 1×10^{-4} M PAX solution at pH 6 for 10 min. By comparing Figure 8A, the height image, to Figure 1A, one can observe that there is a lot of adsorbate showing up on chalcocite when the mineral surface contacts 1×10^{-4} M PAX solution for 10 min. By comparing Figure 8A to Figures 6A and 7A, one can observe that when the PAX's concentration increases, the mineral surface becomes much rougher with more precipitates forming at the solid/liquid interface. Figure 8B is the 3-D image of Figure 8A. Figure 8C, the section analysis of Figure 8A, shows that the surface roughness increases with increasing the PAX's concentration. Figure 8D is the deflection image of Figure 8A with a 10 nm data scale.

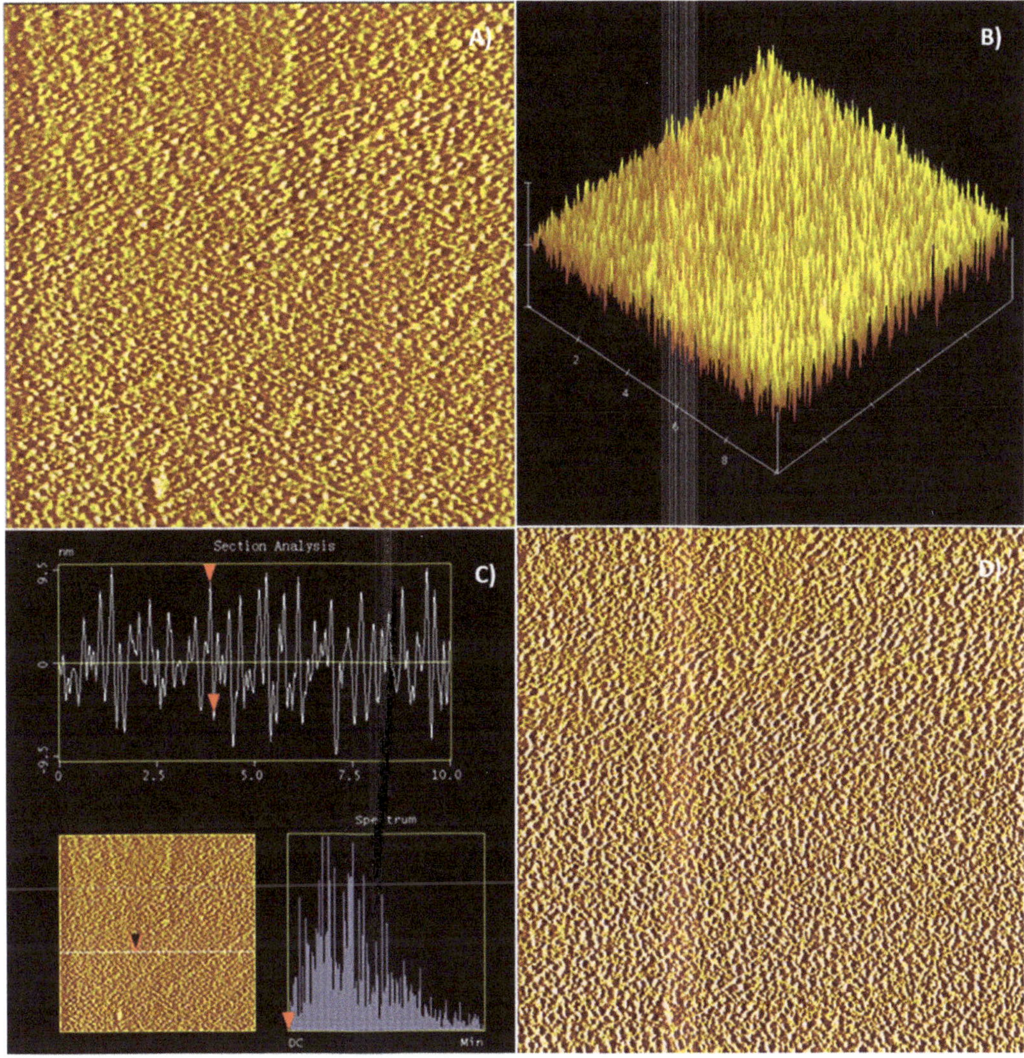

Figure 8. AFM images of a chalcocite surface in 1×10^{-4} M PAX solution at pH 6 for 10 min. (**A**) The 10 μm × 10 μm height image with a data scale of 20 nm; (**B**) the 3-D image; (**C**) the section analysis (the red arrows indicate the top and the bottom of an average asperity); and (**D**) the deflection image with a data scale of 10 nm.

To verify the adsorption of the PAX on chalcocite in solution, after Figure 8 is obtained, a 10 μm × 5 μm area is scanned for one time with a much larger scan force being applied to intentionally remove the adsorbate. Further, the same position is scanned again in a 13 μm × 13 μm area with a normal scan force being applied and the result is shown in Figure 9. One can see from Figure 9A, the height image, that a 10 μm × 5 μm blank 'window' is shown in the center of the image due to the removal of the adsorbate from the mineral surface under the previously applied large scan force. That is, the 'window' in the center is the bare chalcocite surface and the surrounding area is the mineral surface still covered by the adsorbate without being disturbed by the applied large scan force. By comparing Figure 9B, the 3-D image of Figure 9A, to Figure 8B, one can easily observe a pit existing on the mineral surface with adsorbate covering the surrounding area. Figure 9C is the section analysis of Figure 9A and it shows the height difference between the blank 'window' (as shown by the green markers) and the surrounding area being covered with adsorbate (as shown by the red markers). Figure 9D is the deflection image of Figure 9A with a 10 nm data scale.

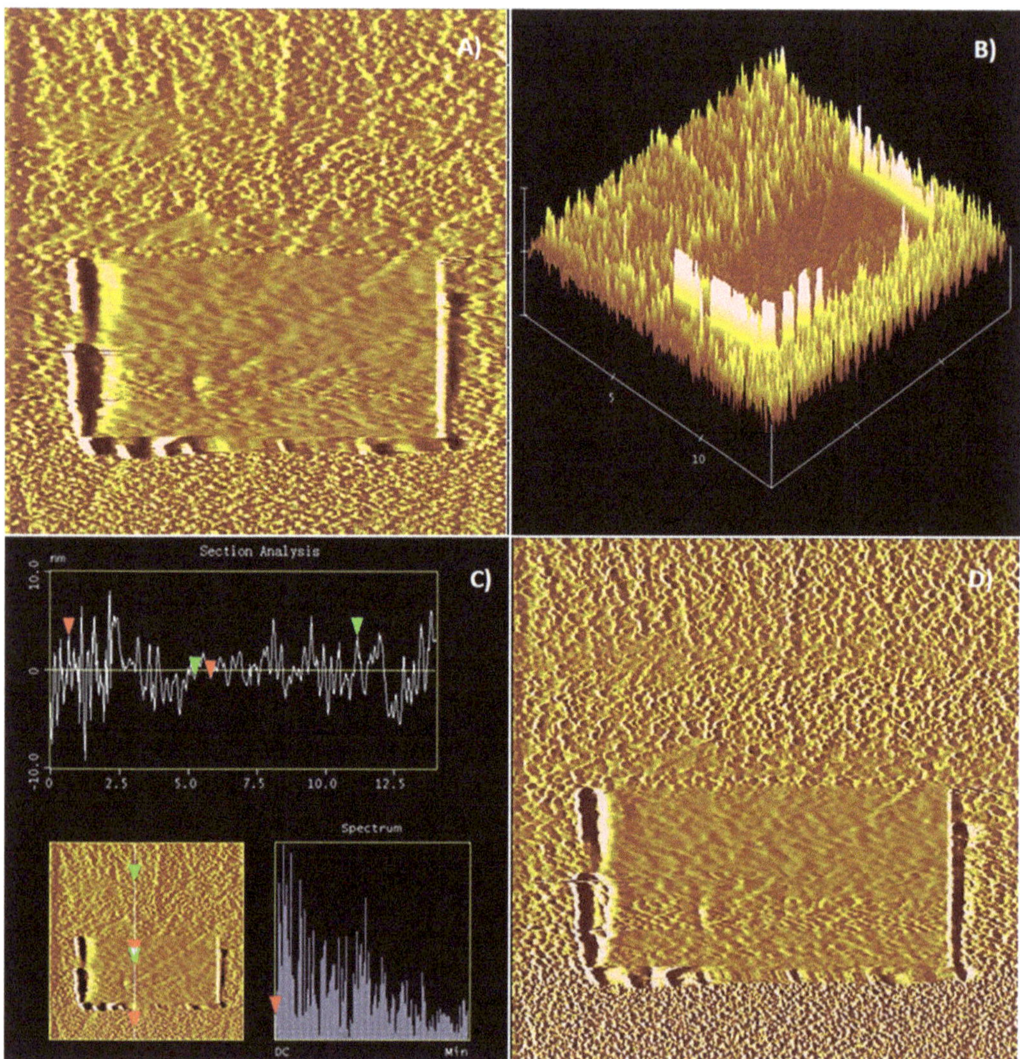

Figure 9. AFM images of a chalcocite surface in 1×10^{-4} M PAX solution at pH 6 for 20 min. (**A**) The 13 μm $\times$ 13 μm height image with a data scale of 20 nm; (**B**) the 3-D image; (**C**) the section analysis (the arrows indicate the top and the bottom of average asperities of different zones); and (**D**) the deflection image with a data scale of 10 nm.

Figure 10 shows the AFM images of a chalcocite surface in 5×10^{-5} M KEX solution at pH 10 for 10 min. By comparing Figure 10A, the height image to Figure 1A, one can find a lot of adsorbate showing up on chalcocite when the mineral surface contacts the

KEX solution. Figure 10B is the 3-D image of Figure 10A. Figure 10C is the section analysis of Figure 10A and the surface roughness clearly increases due to the adsorption of KEX. Figure 10D is the deflection image of Figure 10A with a 10 nm data scale.

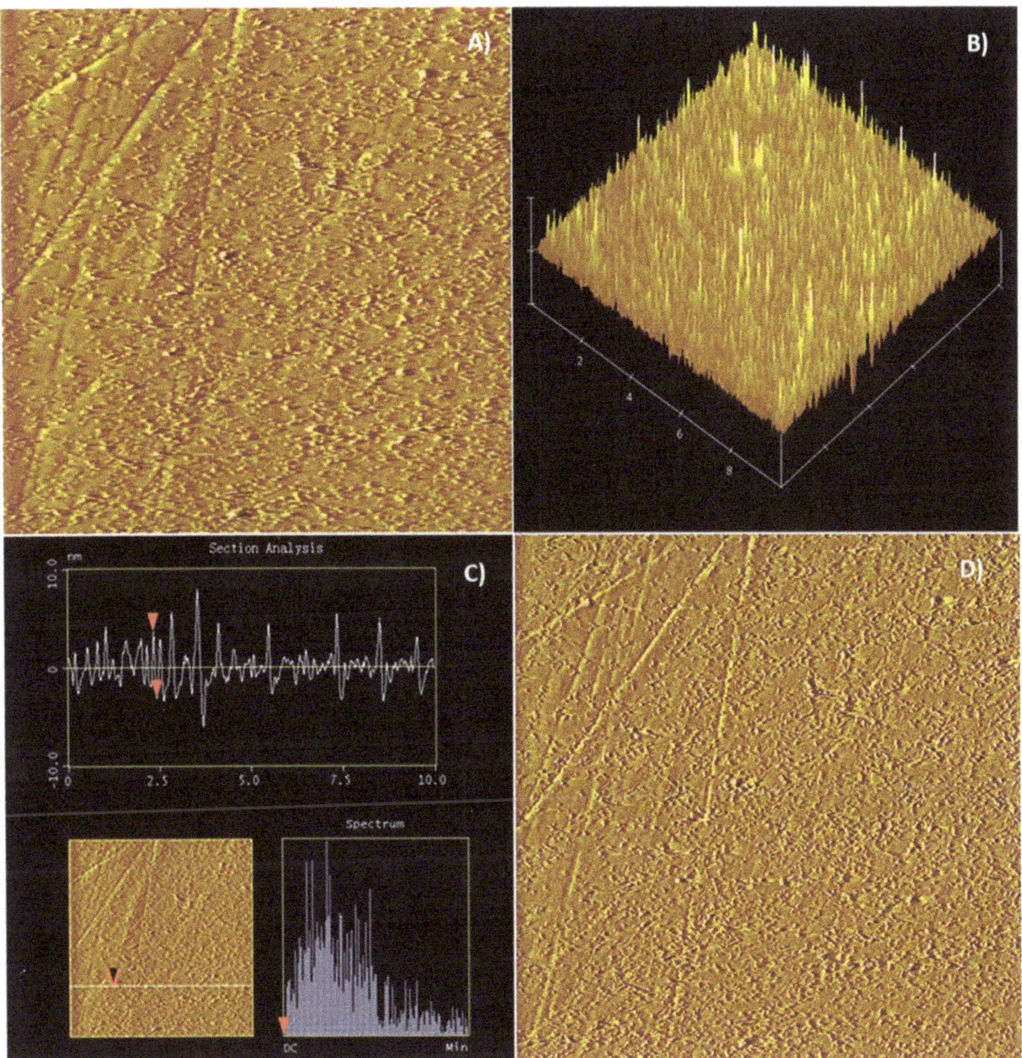

Figure 10. AFM images of a chalcocite surface in 5×10^{-5} M KEX solution at pH 10 for 10 min. (**A**) The 10 μm × 10 μm height image with a data scale of 20 nm; (**B**) the 3-D image; (**C**) the section analysis (the red arrows indicate the top and the bottom of an average asperity); and (**D**) the deflection image with a data scale of 10 nm.

Figure 11 shows the AFM images of a chalcocite surface in 1×10^{-4} M KEX solution at pH 10 for 10 min. Figure 11A is the height image, from which one can see a lot of adsorbate showing up on chalcocite. In addition, by comparing it to Figure 10A, one can find that the mineral surface becomes much rougher, suggesting more precipitates are forming at the solid/liquid interface, when the KEX's concentration increases from 5×10^{-5} M to 1×10^{-4} M at pH 10. Figure 11B is the 3-D image of Figure 11A. Figure 11C, the section analysis of Figure 11A, confirms that the surface roughness increases with increasing the concentration of KEX. Figure 11D is the deflection image of Figure 11A with a 10 nm data scale.

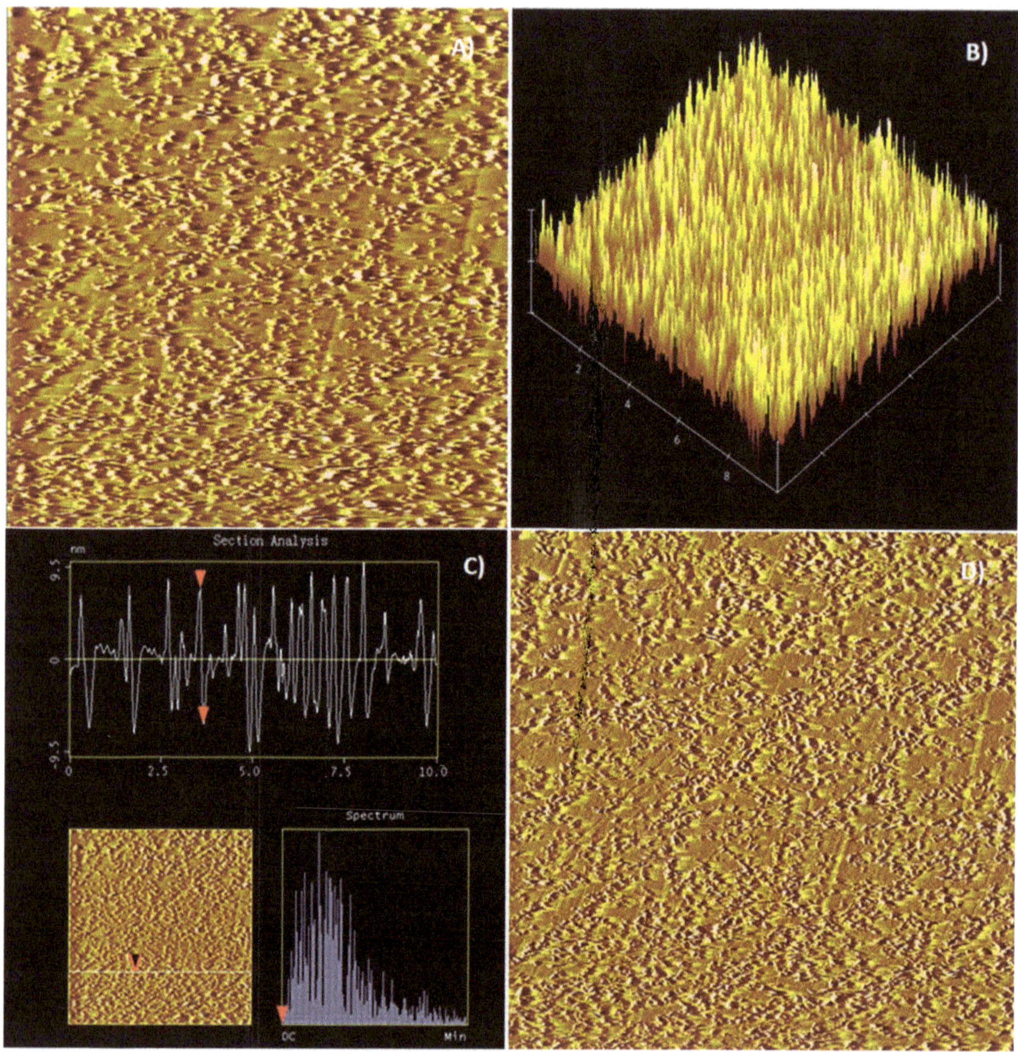

Figure 11. AFM images of a chalcocite surface in 1×10^{-4} M KEX solution at pH 10 for 10 min. (**A**) The 10 μm × 10 μm height image with a data scale of 20 nm; (**B**) the 3-D image; (**C**) the section analysis (the red arrows indicate the top and the bottom of an average asperity); and (**D**) the deflection image with a data scale of 10 nm.

Figure 12 shows the AFM images of a chalcocite surface in 5×10^{-4} M KEX solution at pH 10 for 10 min. By comparing Figure 12A, the height image, to Figure 1A, one can notice a lot of adsorbate showing up on chalcocite after the mineral surface contacts the xanthate solution. In addition, by comparing it to Figures 10A and 11A, one can find that, when the KEX's concentration increases, more precipitates are forming at the solid/liquid interface and the mineral surface becomes much rougher. Figure 12B is the 3-D image of Figure 12A. Figure 12C, the section analysis of Figure 12A, confirms that the surface roughness increases greatly when the chalcocite surface contacts a high concentration of KEX. Figure 12D is the deflection image of Figure 12A with a 10 nm data scale.

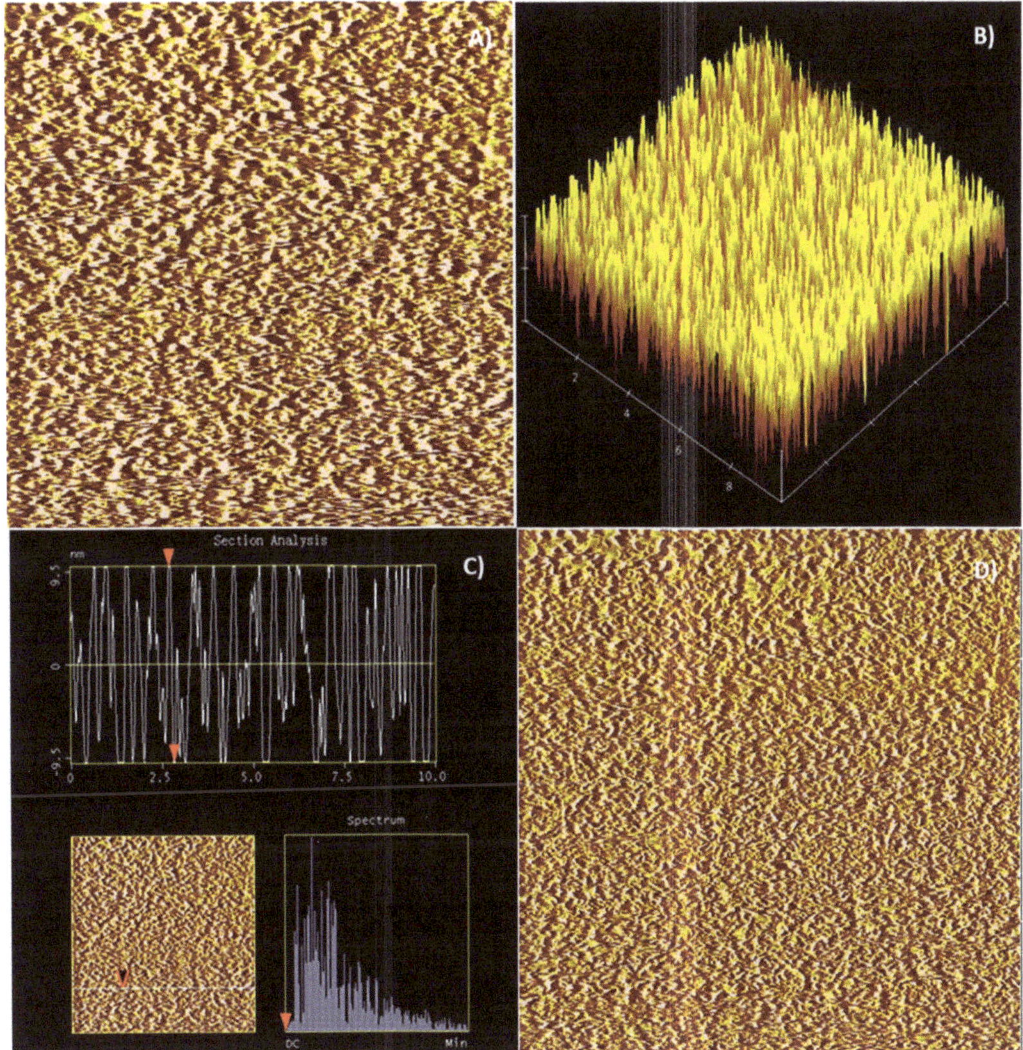

Figure 12. AFM images of a chalcocite surface in 5×10^{-4} M KEX solution at pH 10 for 10 min. (**A**) The 10 μm × 10 μm height image with a data scale of 20 nm; (**B**) the 3-D image; (**C**) the section analysis (the red arrows indicate the top and the bottom of an average asperity); and (**D**) the deflection image with a data scale of 10 nm.

Similar to as obtained in Figure 9, to verify the adsorption of PAX on chalcocite at pH 10, after Figure 12 is obtained, a 5 μm × 5 μm area is scanned with a much larger scan force being applied to intentionally remove the adsorbate. Further, the same position is scanned again in a 10 μm × 10 μm area with a normal scan force being applied and the result is shown in Figure 13. One can observe from Figure 13A, the height image, that a 5 μm × 5 μm 'window' is shown in the center of the image due to the removal of the adsorbate from the mineral surface under the previously applied large scan force. That is, the 'window' in the center is the bare chalcocite surface and the surrounding area is the mineral surface still covered by the adsorbate without being disturbed by the applied large scan force. By comparing Figure 13B, the 3-D image of Figure 13A, to Figure 12B, one can easily observe a pit existing on the mineral surface with adsorbate covering the surrounding area. Figure 13C is the section analysis of Figure 13A and it shows the height difference between the 'window' (as shown by the green markers) and the surrounding area being covered with adsorbate (as shown by the red markers). Figure 13D is the deflection image of Figure 13A with a 10 nm data scale.

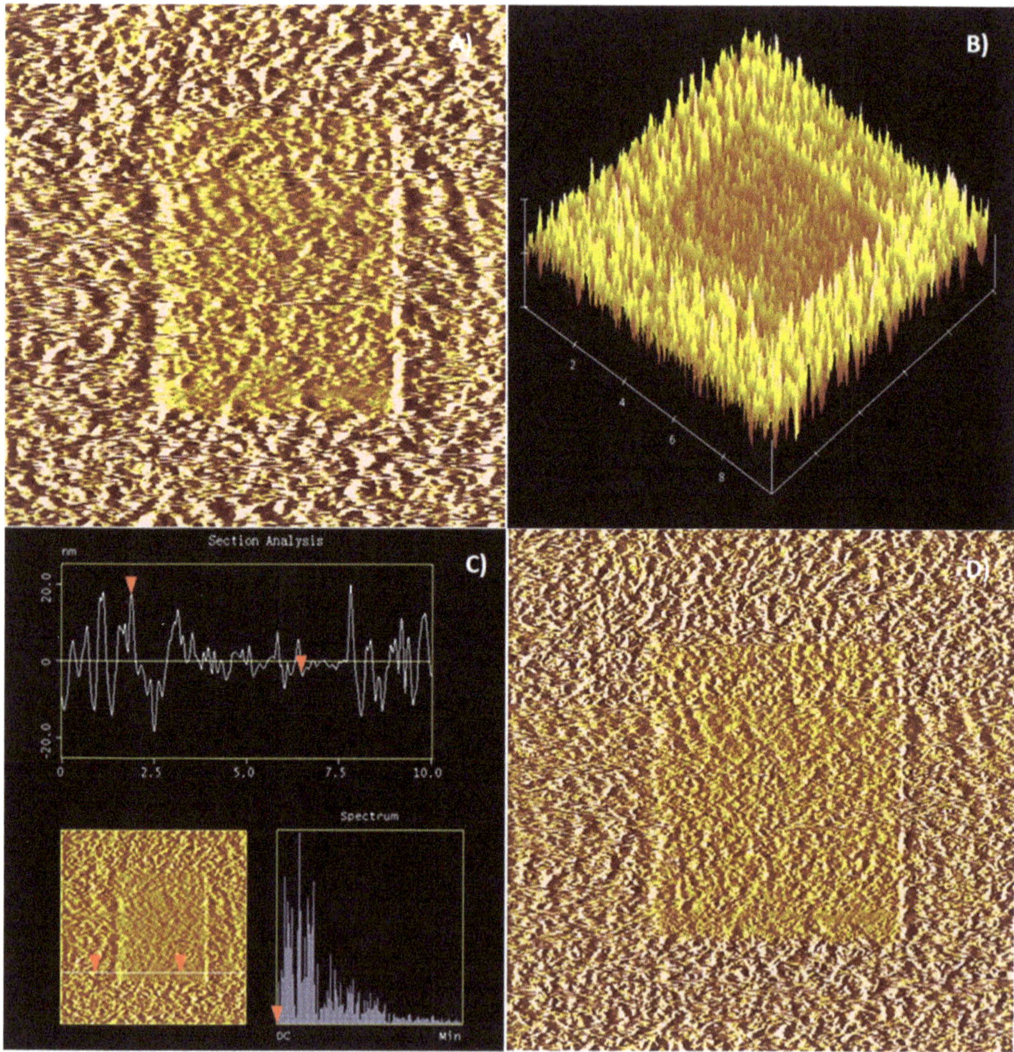

Figure 13. AFM images of a chalcocite surface in 5×10^{-4} M KEX solution at pH 10 for 20 min. (**A**) The 10 μm × 10 μm height image with a data scale of 20 nm; (**B**) the 3-D image; (**C**) the section analysis (the red arrows indicate the top and the bottom of an average asperity); and (**D**) the deflection image with a data scale of 10 nm. The 5 μm × 5 μm 'window' in the center of the image is due to the removal of the adsorbate from the mineral surface under the intentionally applied large scan force.

Figure 14A is the height image of a chalcocite surface in 5×10^{-4} M KEX solution at pH 10 for 30 min and further rinsed with 10 mL ethanol and 5 mL water consecutively. The images are finally obtained when the mineral sample is placed in water. By comparing it to Figure 12A, one can observe that the precipitates as observed in Figure 12A still exist on the chalcocite surface without being dissolved and rinsed off from the mineral surface. Figure 14B is the 3-D image of Figure 14A. Figure 14C is the section analysis of Figure 14A. Figure 14D is the deflection image of Figure 14A with a 10 nm data scale.

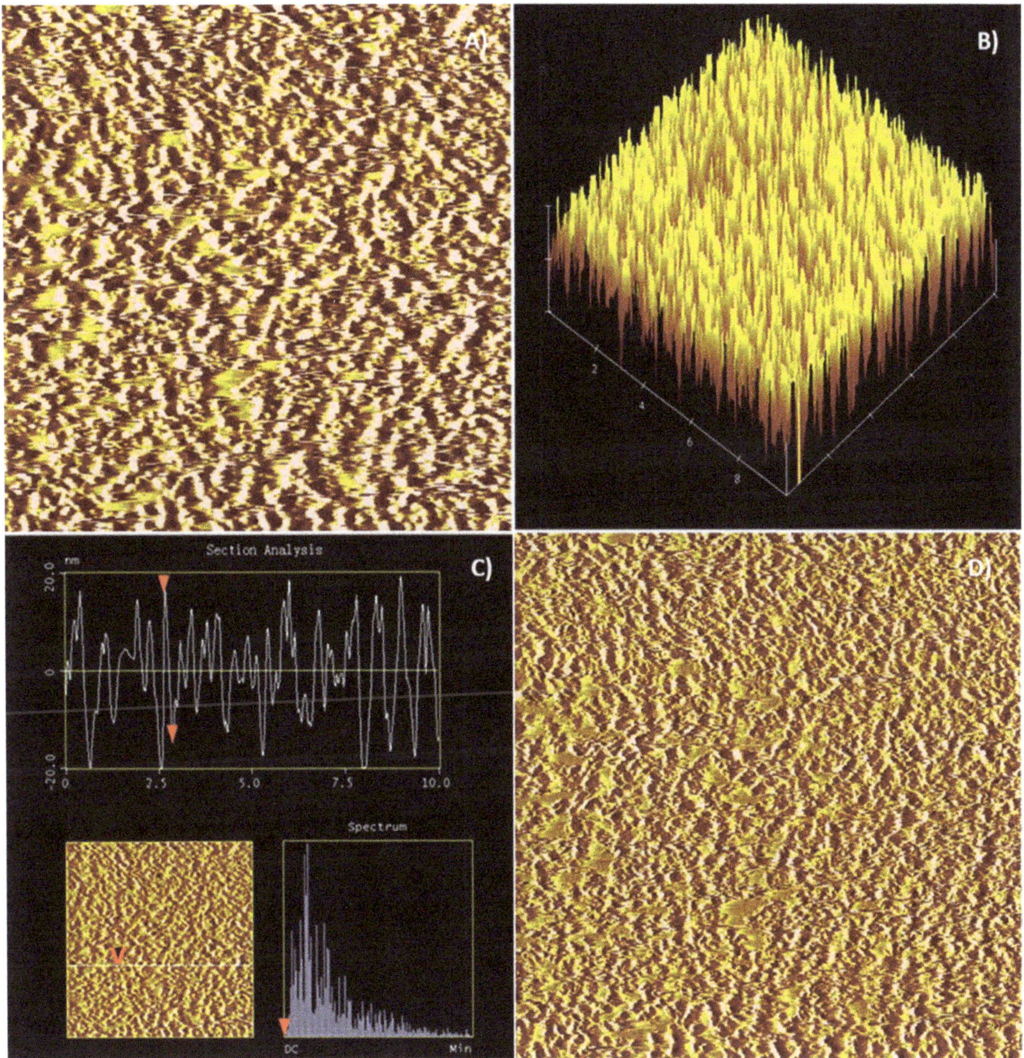

Figure 14. AFM images of a chalcocite surface in 5×10^{-4} M KEX solution at pH 10 for 30 min, and further rinsed with 10 mL ethanol and 5 mL water consecutively. (**A**) The 10 μm × 10 μm height image with a data scale of 20 nm; (**B**) the 3-D image; (**C**) the section analysis (the red arrows indicate the top and the bottom of an average asperity); and (**D**) the deflection image with a data scale of 10 nm.

Figure 15 shows the AFM images of a chalcocite surface in 1×10^{-5} M PAX solution at pH 10 for 10 min. From Figure 15A, the height image, one can find that there is a little adsorbate showing up on chalcocite when the mineral surface contacts PAX solution at pH 10 for 10 min. Figure 15B is the 3-D image of Figure 15A. Figure 15C, the section analysis of Figure 15A, shows that the surface roughness increases due to the adsorption of PAX. Figure 15D is the deflection image of Figure 15A with a 10 nm data scale.

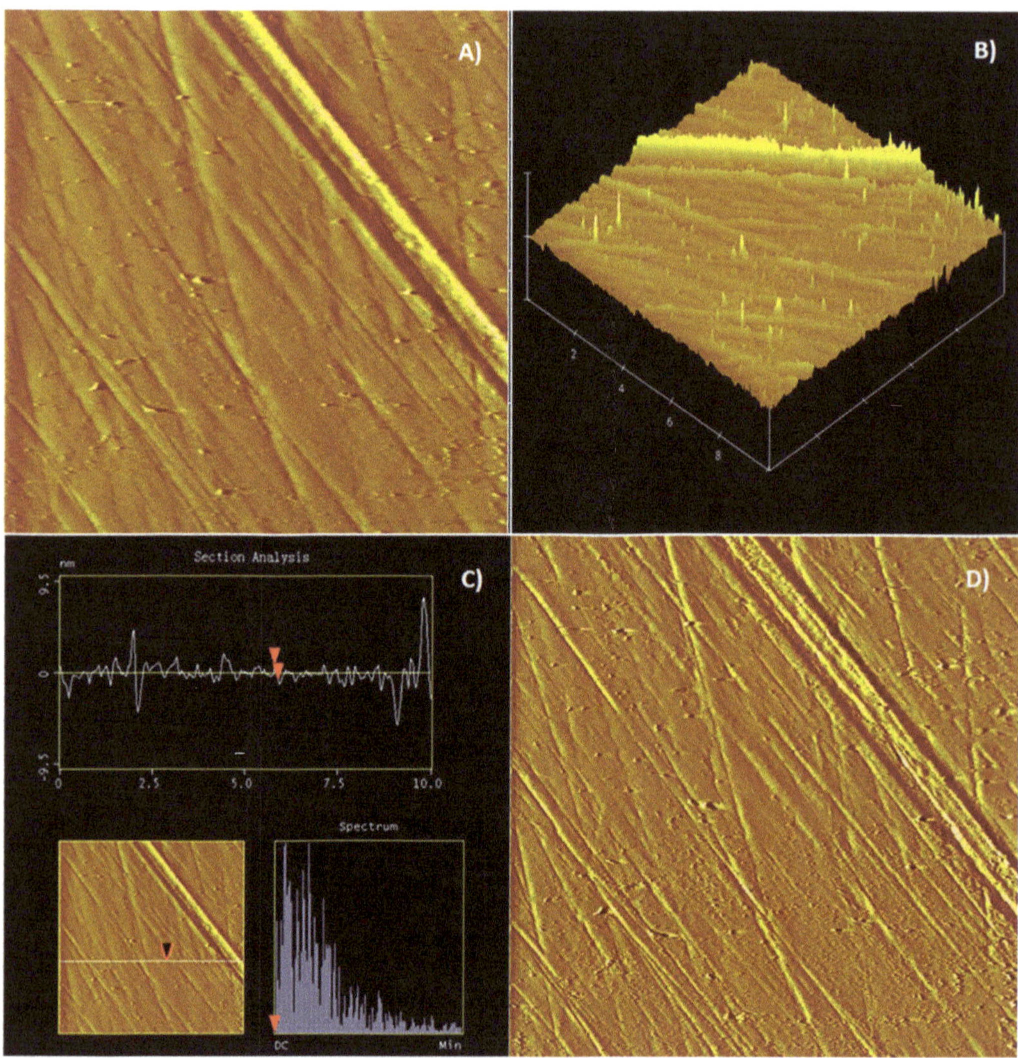

Figure 15. AFM images of a chalcocite surface in 1×10^{-5} M PAX solution at pH 10 for 10 min. (**A**) The 10 μm × 10 μm height image with a data scale of 20 nm; (**B**) the 3-D image; (**C**) the section analysis (the red arrows indicate the top and the bottom of an average asperity); and (**D**) the deflection image with a data scale of 10 nm.

Figure 16 shows the AFM images of a chalcocite surface in 5×10^{-5} M PAX solution at pH 10 for 10 min. Figure 16A is the height image, from which one can see a lot of adsorbate showing up on chalcocite. In addition, by comparing it to Figure 15A, one can find that the mineral surface becomes much rougher, suggesting more precipitates are forming at the solid/liquid interface, when the PAX's concentration increases from 1×10^{-5} M to 5×10^{-5} M at pH 10. Figure 16B is the 3-D image of Figure 16A. Figure 16C, the section analysis of Figure 16A, confirms that the surface roughness increases with increasing the concentration of PAX. Figure 16D is the deflection image of Figure 16A with a 10 nm data scale.

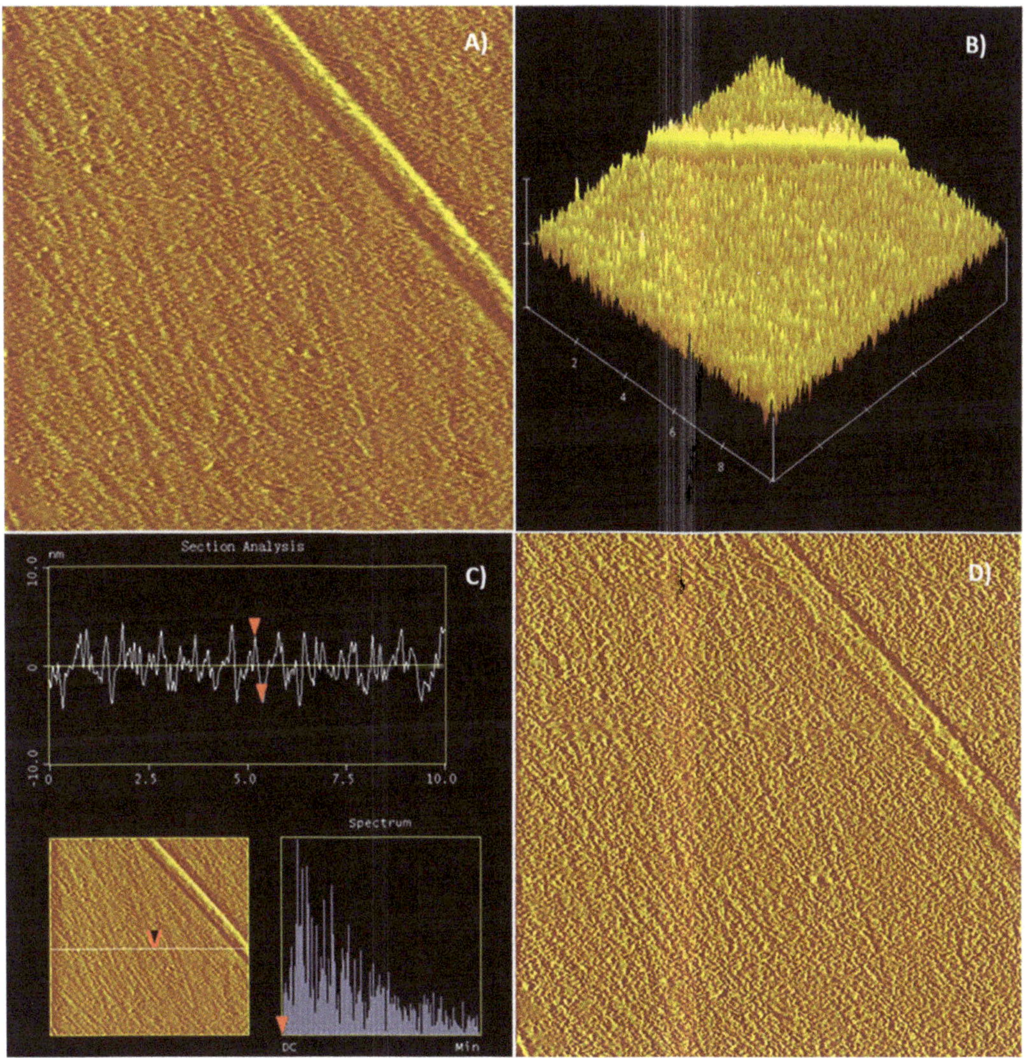

Figure 16. AFM images of a chalcocite surface in 5×10^{-5} M PAX solution at pH 10 for 10 min. (**A**) The 10 µm × 10 µm height image with a data scale of 20 nm; (**B**) the 3-D image; (**C**) the section analysis (the red arrows indicate the top and the bottom of an average asperity); and (**D**) the deflection image with a data scale of 10 nm.

Figure 17 shows the AFM images of a chalcocite surface in 1×10^{-4} M PAX solution at pH 10 for 10 min. By comparing Figure 17A, the height image, to Figure 1A, one can notice a lot of adsorbate showing up on chalcocite after the mineral surface contacts the xanthate solution. In addition, by comparing it to Figures 15A and 16A, one can find that, when the PAX's concentration increases, more precipitates are forming at the solid/liquid interface and the mineral surface becomes much rougher. Figure 17B is the 3-D image of Figure 17A. Figure 17C, the section analysis of Figure 17A, confirms that the surface roughness increases greatly when the chalcocite surface contacts a high concentration of PAX. Figure 17D is the deflection image of Figure 17A with a 10 nm data scale.

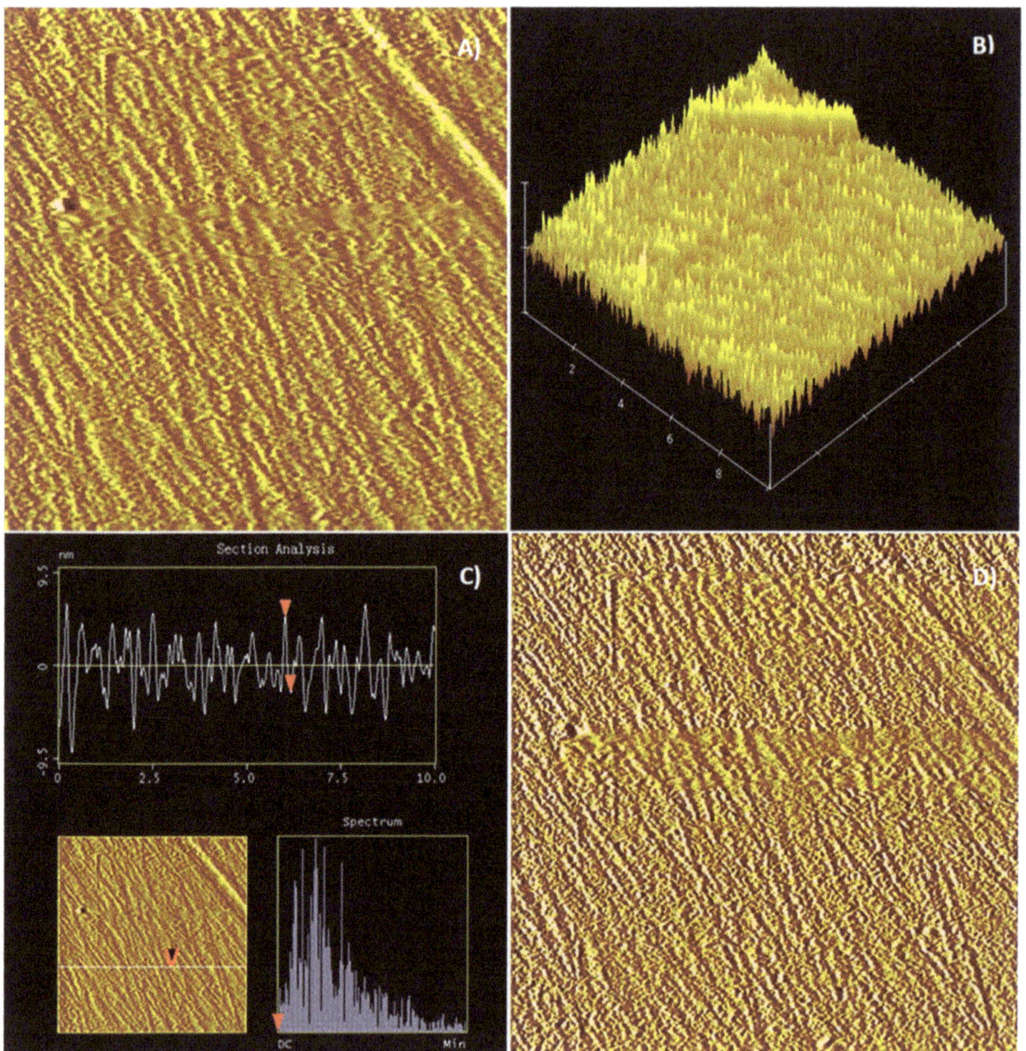

Figure 17. AFM images of a chalcocite surface in 1×10^{-4} M PAX solution at pH 10 for 10 min. (**A**) The 10 μm × 10 μm height image with a data scale of 20 nm; (**B**) the 3-D image; (**C**) the section analysis (the red arrows indicate the top and the bottom of an average asperity); and (**D**) the deflection image with a data scale of 10 nm.

3.2. AFM Image of Mineral Surface in Cytec Aerofloat 238 Solutions

Figure 18 shows the AFM images of a chalcocite surface in 200 ppm Aerofloat 238 solution at pH 6 for 10 min. From Figure 18A, the height image, one can observe that there is a lot of adsorbate showing up on chalcocite when the mineral surface contacts the Aerofloat 238 solution at pH 6 for 10 min. The figure also shows that the morphology of the adsorbate, which shows round and smooth edges, differs greatly from those obtained with xanthate, for which the adsorbate is generally precipitates with irregular sharp edges. Figure 18B is the 3-D image of Figure 18A. Figure 18C is the section analysis of Figure 18A, and Figure 18D is the deflection image of Figure 18A with a 10 nm data scale.

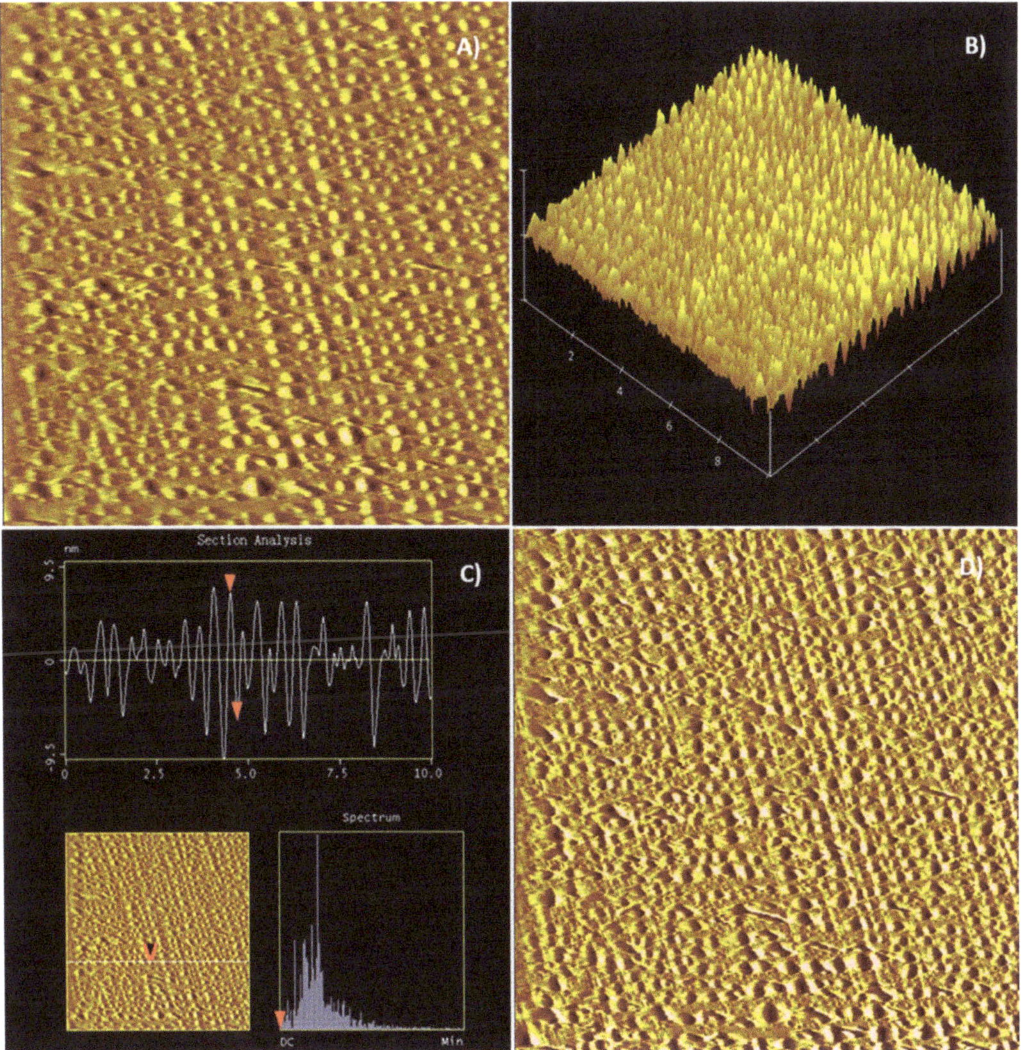

Figure 18. AFM images of a chalcocite surface in 200 ppm Aerofloat 238 solution at pH 6 for 10 min. (**A**) The 10 μm × 10 μm height image with a data scale of 20 nm; (**B**) the 3-D image; (**C**) the section analysis (the red arrows indicate the top and the bottom of an average asperity); and (**D**) the deflection image with a data scale of 10 nm.

Figure 19 shows the AFM images of a chalcocite surface in 200 ppm Aerofloat 238 solution at pH 6 for 20 min. By comparing Figure 19A, the height image, to Figure 18A, one can see that the patches adsorbed on chalcocite become larger in size when the adsorption time increases from 10 min to 20 min. Again, the figure also shows that the adsorbate has round and smooth edges and it is greatly different from those obtained with xanthate, the adsorbate of which is generally precipitates with irregular sharp edges. Figure 19B is the 3-D image of Figure 19A. Figure 19C is the section analysis of Figure 19A, showing the surface roughness increases with adsorption time increases. Figure 19D is the deflection image of Figure 19A with a 10 nm data scale.

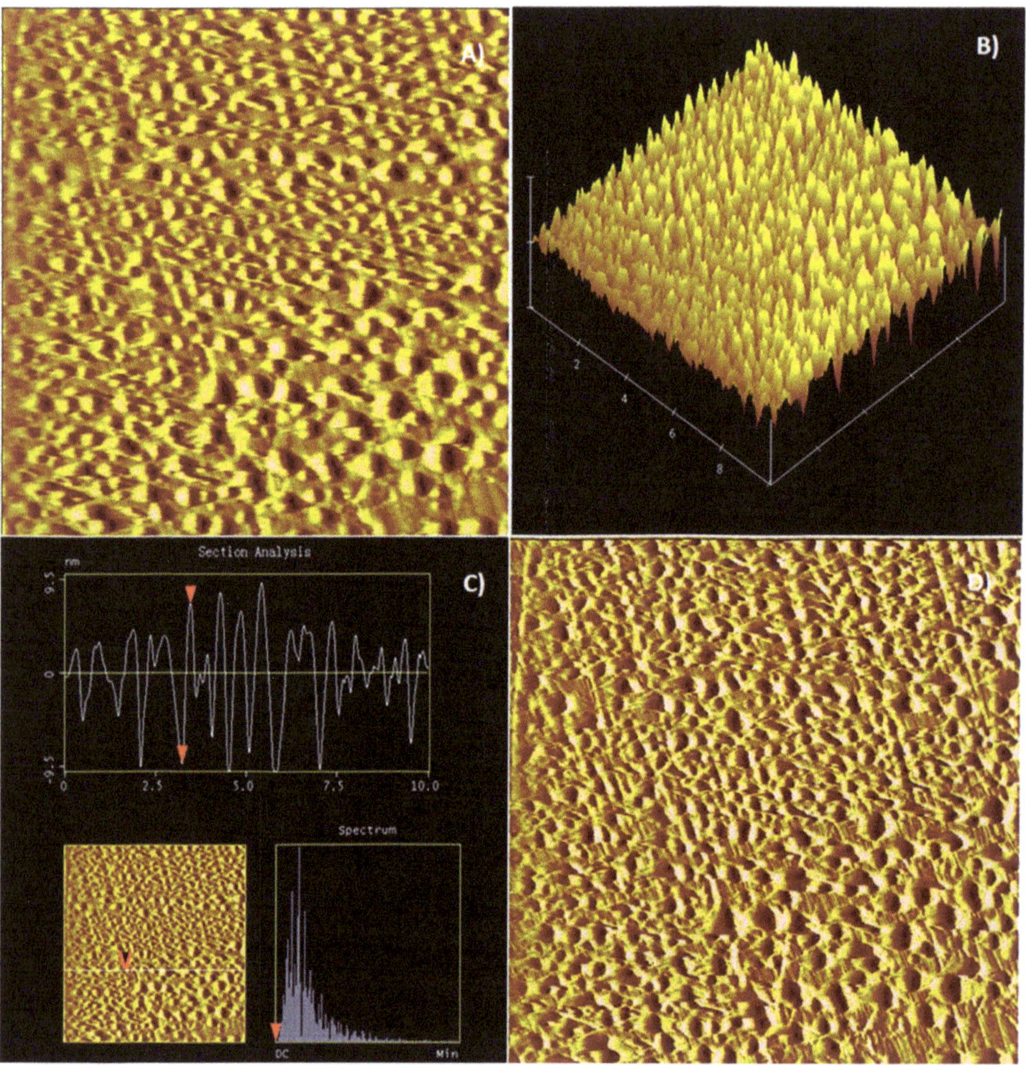

Figure 19. AFM images of a chalcocite surface in 200 ppm Aerofloat 238 solution at pH 6 for 20 min. (**A**) The 10 μm × 10 μm height image with a data scale of 20 nm; (**B**) the 3-D image; (**C**) the section analysis (the red arrows indicate the top and the bottom of an average asperity); and (**D**) the deflection image with a data scale of 10 nm.

Similar to as obtained in Figure 13, to verify the adsorption of Aerofloat 238 on chalcocite, after Figure 19 is obtained, a 10 μm × 5 μm area is scanned with a much larger scan force being applied to intentionally remove the adsorbate. Further, the same position is scanned again in a 10 μm × 10 μm area with a normal scan force being applied and the result is shown in Figure 20. One can find from Figure 20A, the height image, that a 10 μm × 5 μm 'window' is shown in the upper image due to the removal of the adsorbate from the mineral surface under the previously applied large scan force. That is, the upper 'window' is the bare chalcocite surface and the lower section is the mineral surface still covered by the adsorbate without being disturbed by the applied large scan force. Figure 20B, Figure 20C, and Figure 20D, are, respectively, the 3-D image, the section analysis, and the deflection image of Figure 20A with a 10 nm data scale.

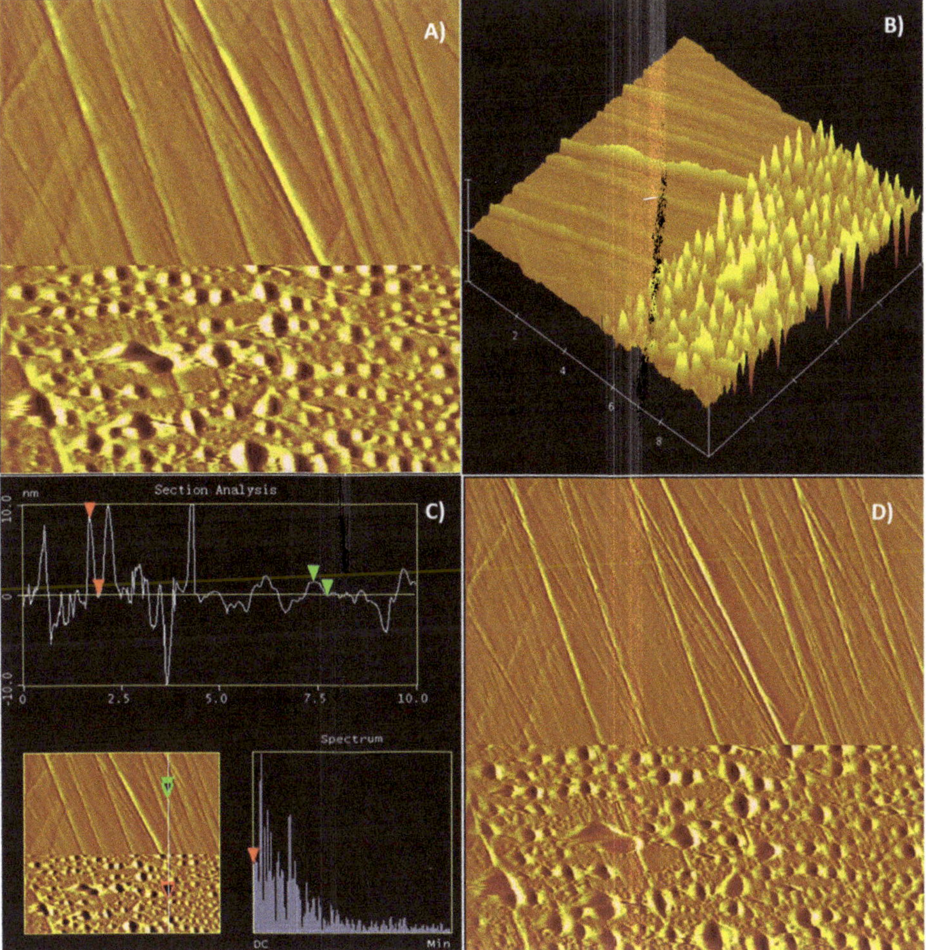

Figure 20. AFM images of a chalcocite surface in 200 ppm Aerofloat 238 solution at pH 6 for 20 min. (**A**) The 10 μm × 10 μm height image with a data scale of 20 nm; (**B**) the 3-D image; (**C**) the section analysis (the red arrows indicate the top and the bottom of an average asperity in a zone of adsorbate and the green arrows indicate those in a zone with adsorbate being removed); and (**D**) the deflection image with a data scale of 10 nm. The 10 μm × 5 μm 'window' in the upper part is due to the removal of the adsorbate from the mineral under the intentionally applied large scan force.

Figure 21 shows the AFM images of a chalcocite surface in 200 ppm Aerofloat 238 solution at pH 10 for 10 min. From Figure 21A, the height image, one can find that there are many patches being adsorbed on chalcocite when the mineral contacts the Aerofloat 238 solution at pH 10. Again, the figure also shows that the adsorbate has round and smooth edges, which is greatly different from those obtained with xanthate. Figure 21B, Figure 21C, and Figure 21D, are, respectively, the 3-D image, the section analysis, and the deflection image of Figure 21A with a 10 nm data scale.

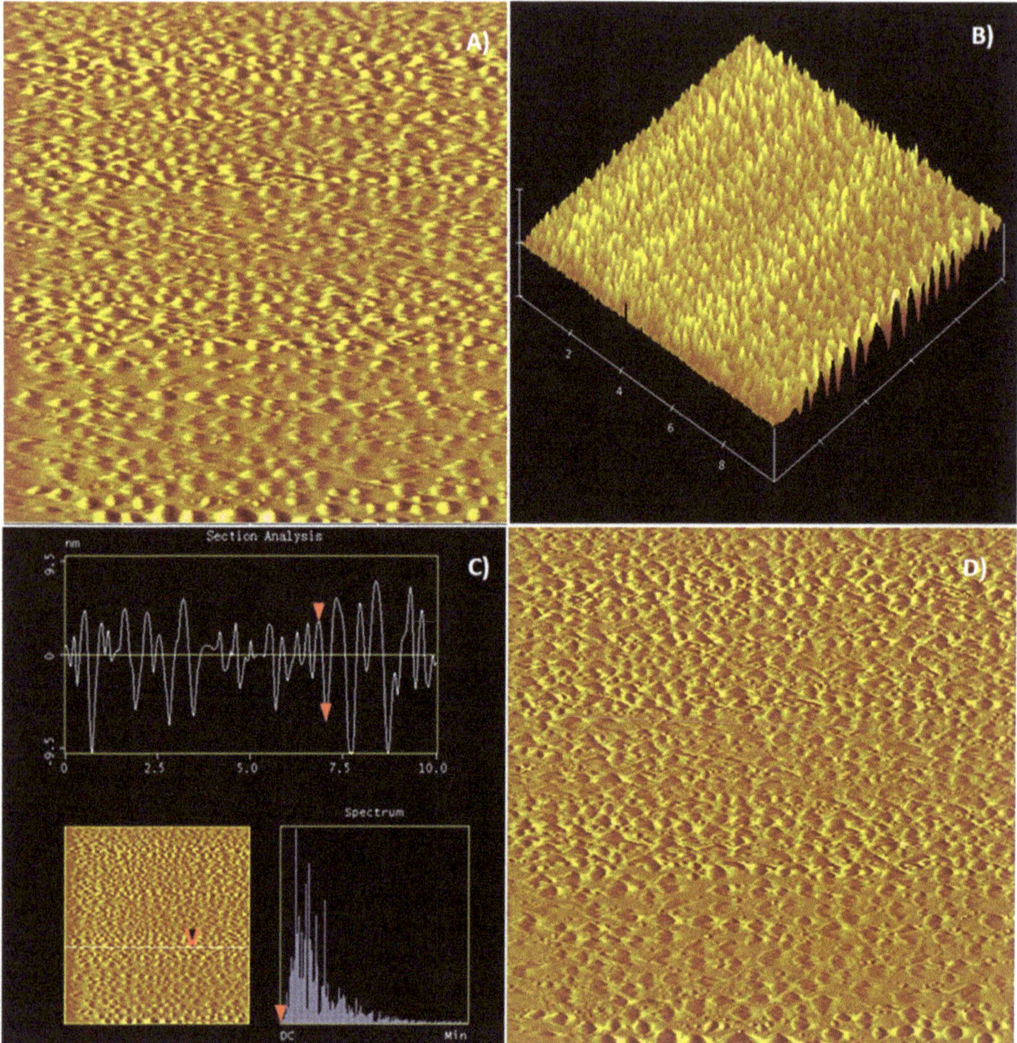

Figure 21. AFM images of a chalcocite surface in 200 ppm Aerofloat 238 solution at pH 10 for 10 min. (**A**) The 10 μm × 10 μm height image with a data scale of 20 nm; (**B**) the 3-D image; (**C**) the section analysis (the red arrows indicate the top and the bottom of an average asperity); and (**D**) the deflection image with a data scale of 10 nm.

Figure 22 shows the AFM images of a chalcocite surface in 200 ppm Aerofloat 238 solution at pH 10 for 20 min. By comparing Figure 22A, the height image, to Figure 21A, one can find that when adsorption time increases, there are more patches being adsorbed on chalcocite when the mineral contacts the Aerofloat 238 solution at pH 10. Figures 22B, 22C and 22D are, respectively, the 3-D image, the section analysis, and the deflection image of Figure 22A with a 10 nm data scale.

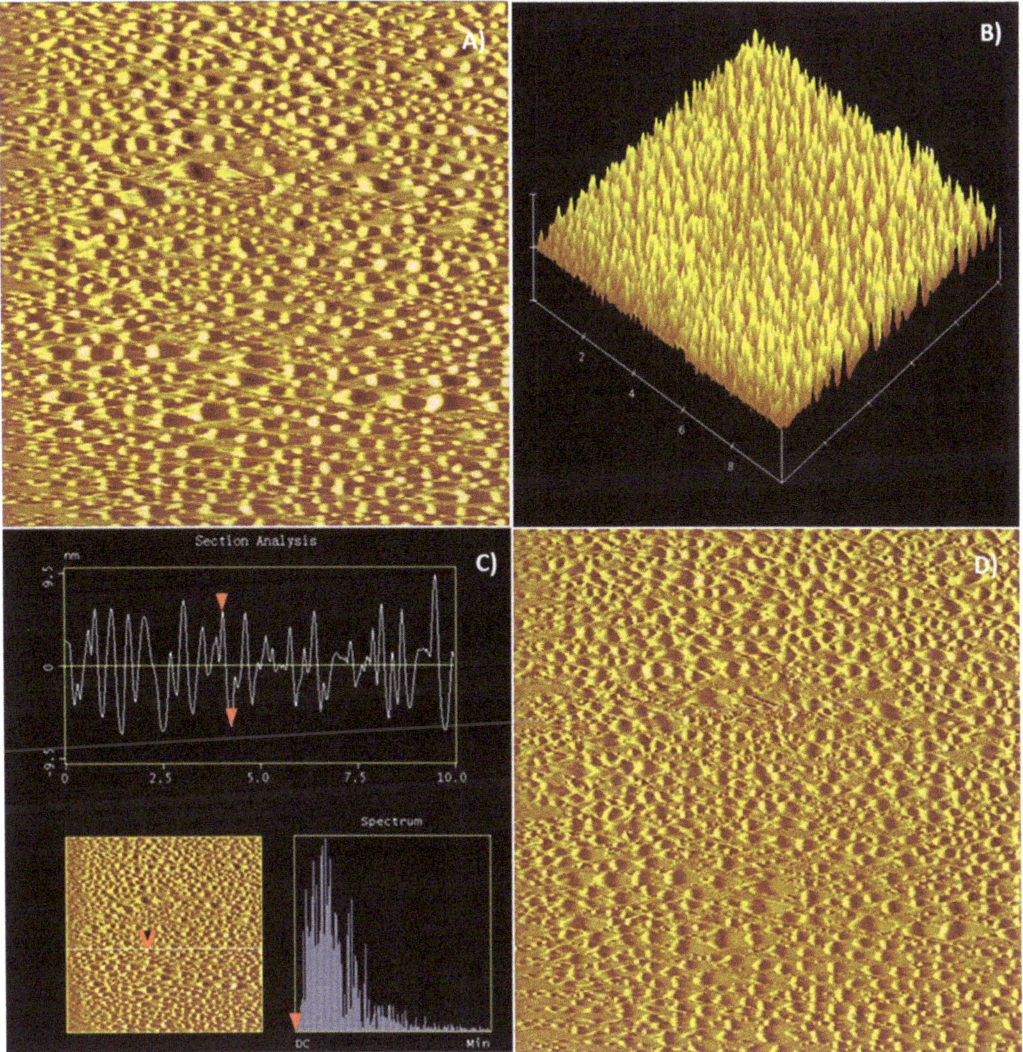

Figure 22. AFM images of a chalcocite surface in 200 ppm Aerofloat 238 solution at pH 10 for 20 min. (**A**) The 10 μm × 10 μm height image with a data scale of 20 nm; (**B**) the 3-D image; (**C**) the section analysis (the red arrows indicate the top and the bottom of an average asperity); and (**D**) the deflection image with a data scale of 10 nm.

Figure 23A is the height image of a chalcocite surface in 200 ppm Aerofloat 238 solution at pH 10 for 20 min, and further rinsed with 10 mL ethanol and 5 mL water consecutively. The images are finally obtained when the mineral sample is in water. By comparing it to Figure 22A, one can tell that the adsorbate as observed in Figure 23A does not exist on the chalcocite surface anymore because it has been dissolved and rinsed off from the mineral surface by flushing with ethanol. Figure 23B, Figure 23C, and Figure 23D, are, respectively, the 3-D image, the section analysis, and the deflection image of Figure 23A with a 10 nm data scale.

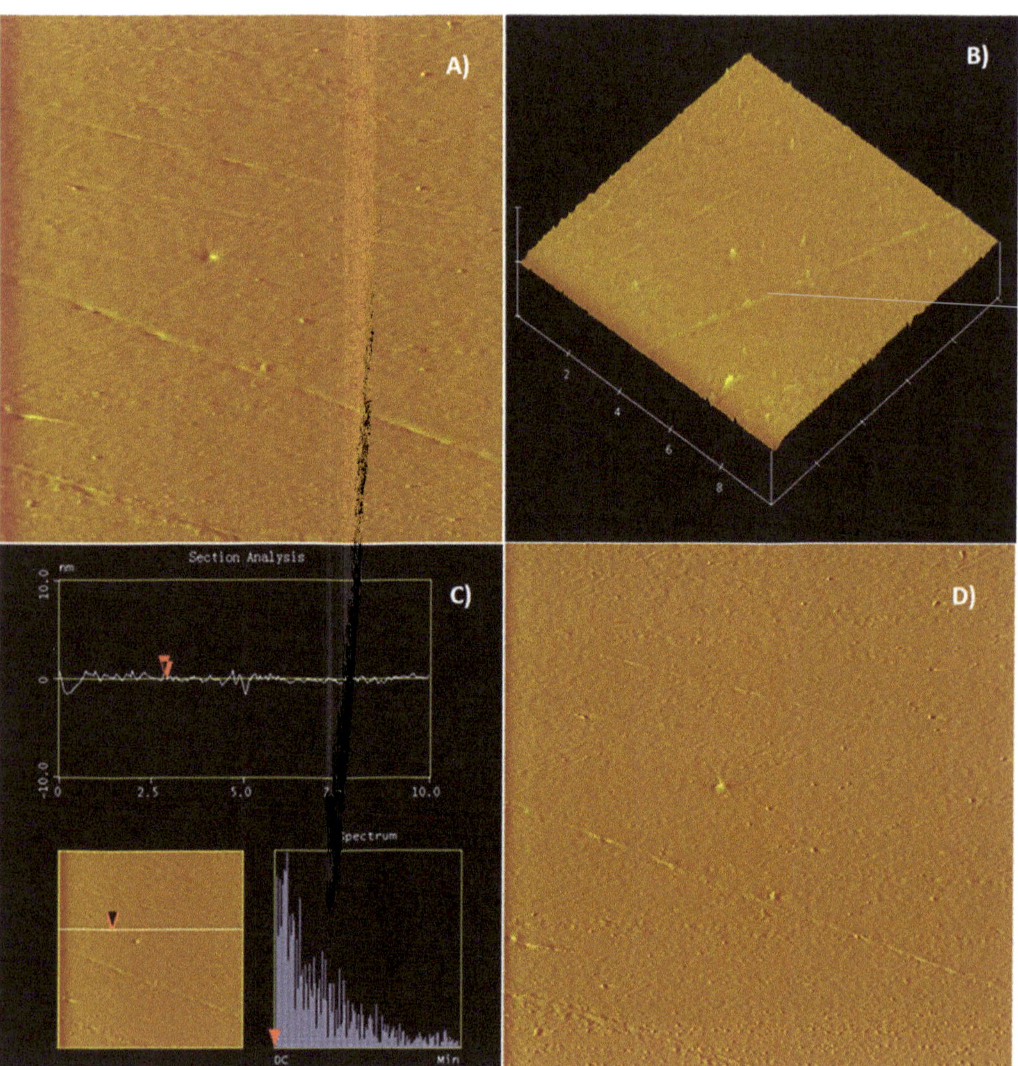

Figure 23. AFM images of a chalcocite surface in 200 ppm Aerofloat 238 solution at pH 10 for 20 min, and further rinsed with 10 mL ethanol and 5 mL water consecutively. (**A**) The 10 μm × 10 μm height image with a data scale of 20 nm; (**B**) the 3-D image; (**C**) the section analysis (the red arrows indicate the top and the bottom of an average asperity); and (**D**) the deflection image with a data scale of 10 nm.

Figure 24A is the height image of a chalcocite surface in the mixture of 1×10^{-4} M PAX and 300 ppm Aerofloat 238 solution (1:1 vol. ratio) at pH 6. One can clearly see that there are two different types of adsorption morphology, i.e., as indicated by the "Red" arrow, one is the morphology of patches and it is similar to those as obtained with the addition of Aerofloat 238; as indicated by the "Blue" arrow, the one is the morphology of precipitates, and it is similar to those as obtained with the addition of xanthate. Figure 24B, Figure 24C, and Figure 24D, are, respectively, the 3-D image, the section analysis, and the deflection image of Figure 24A with a 10 nm data scale.

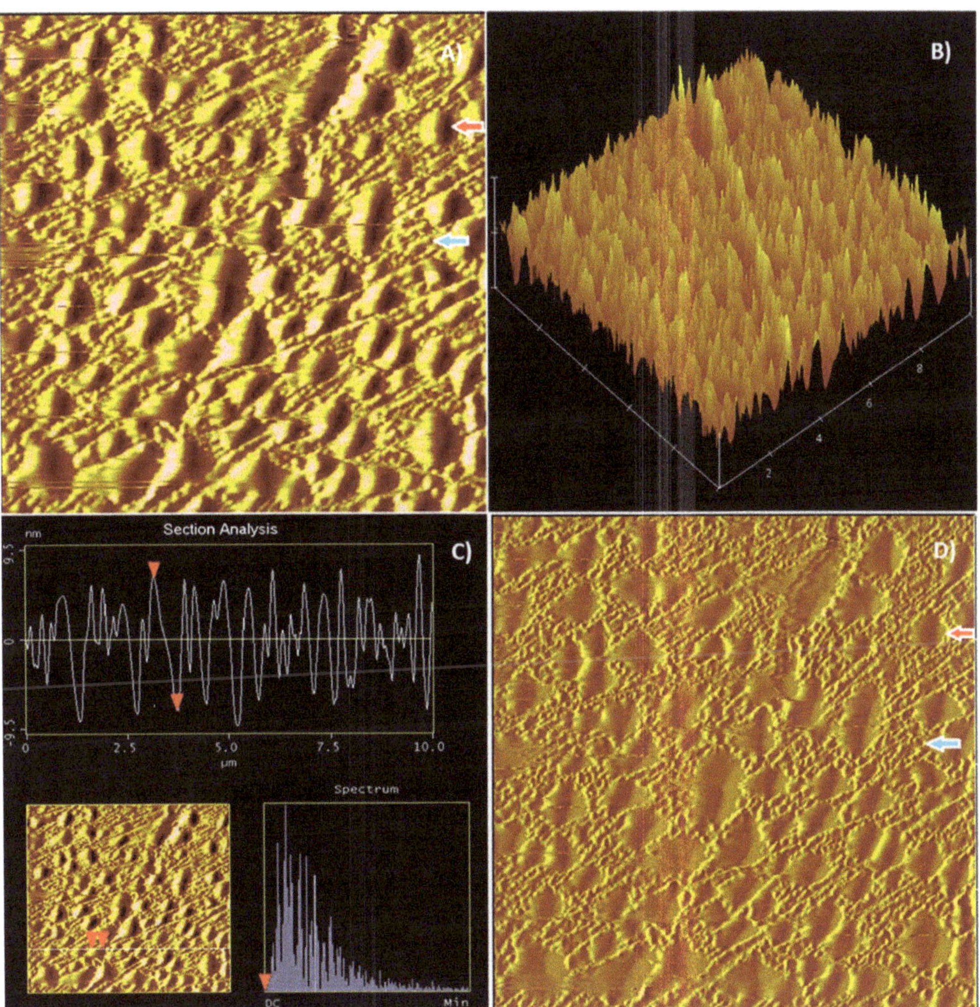

Figure 24. AFM images of a chalcocite surface in the mixture of 1×10^{-4} M PAX and 300 ppm Aerofloat 238 solution (1:1 volume ratio) at pH 6 for 20 min. (**A**) The 10 µm × 10 µm height image with a data scale of 20 nm; (**B**) the 3-D image; (**C**) the section analysis (the red arrows indicate the top and the bottom of an average asperity); and (**D**) the deflection image with a data scale of 10 nm. The "Red" arrow indicates the patches due to the adsorption of Aerofloat 238 and the "Blue" arrow indicates the precipitates due to the adsorption of PAX.

4. Discussion

4.1. Adsorption of Xanthate on Chalcocite Surface

Figures 2–17 show that a significant amount of adsorbate can be observed on the chalcocite surface when it contacts xanthate solutions for a specific time. The adsorbate changes the surface morphology of chalcocite accordingly to changing solution chemistry such as collector's concentration, solution pH and adsorption time. The roughness analysis of the AFM images of a chalcocite surface in collector solutions is summarized and listed in Table 1 as follows.

Table 1. Roughness analysis of the AFM images of a chalcocite surface in xanthate solutions.

Image Source	Collector	pH	Concentration (M)	Ra(Sa) (nm) *	Rms(Sq) (nm) **
Figure 1	H_2O	6	0	0.651	0.925
Figure 2	KEX	6	5×10^{-5} M	1.110	1.538
Figure 3	KEX	6	1×10^{-4} M	1.752	2.293
Figure 4	KEX	6	5×10^{-4} M	5.242	6.695
Figure 5	KEX	6	5×10^{-4} M	5.787	7.269
Figure 6	PAX	6	1×10^{-5} M	1.044	1.442
Figure 7	PAX	6	5×10^{-5} M	2.633	3.308
Figure 8	PAX	6	1×10^{-4} M	2.641	3.345
Figure 10	KEX	10	5×10^{-5} M	1.349	2.070
Figure 11	KEX	10	1×10^{-4} M	3.097	4.601
Figure 12	KEX	10	5×10^{-4} M	7.400	9.553
Figure 14	KEX	10	5×10^{-4} M	7.758	9.920
Figure 15	PAX	10	1×10^{-5} M	0.506	0.722
Figure 16	PAX	10	5×10^{-5} M	1.364	1.735
Figure 17	PAX	10	1×10^{-4} M	1.789	2.283
Figure 18	Aerofloat238	6	200 ppm	1.857	2.605
Figure 19	Aerofloat238	6	200 ppm	2.691	3.663
Figure 20	Aerofloat238	10	200 ppm	1.578	2.203
Figure 22	Aerofloat238	10	200 ppm	2.677	3.537
Figure 23	ethanol	6	0	0.295	0.425
Figure 24	PAX + Aerofloat238	6	1:1	3.589	4.484

Note: * Ra(Sa): arithmetic average of the absolute values of the surface height deviations. ** Rms(Sq): root mean square average of height deviations taken from the mean image data plane.

Table 1 shows that, during the same timeframe, i.e., 10 min, surface roughness in general increases with increasing the concentration of xanthate. For example, as shown by Figures 2–4, at pH 6, when the concentration of KEX increases from 5×10^{-5} M to 1×10^{-4} M, the roughness Ra value increases from 1.110 nm to 1.752 nm, and the value increases to 5.242 nm when the concentration is further increased to 5×10^{-4} M. A similar trend is also observed for the case of PAX, although the change in values is not as significant as the one obtained with KEX. At pH 10, the same conclusion is also applicable. For example, as shown by Figures 10–12, when the concentration of KEX increases from 5×10^{-5} M to 1×10^{-4} M, the roughness Ra value increases from 1.349 nm to 3.097 nm, and the value increases to 7.400 nm when the concentration is further increased to 5×10^{-4} M. A similar trend is also observed for the case of PAX as well.

In the above section, as shown by Figures 2–17, when chalcocite contacts xanthate solutions for a specific time, a lot of adsorbates will show up on the mineral surface and the adsorption changes the surface morphology. Clearly this change in surface morphology of chalcocite cannot be attributed to the reaction of the mineral with water because the AFM images obtained with the addition of various xanthate solutions are totally different from Figure 1, which is captured during the same time frame. Therefore, the adsorbate shown in Figures 2–17 must be due to the adsorption of xanthate at the mineral/liquid interface. The same conclusion can also be drawn for the cases of Aerofloat 238, as shown by Figures 18–24. That is, both these collectors can effectively adsorb on the chalcocite surface in a short time frame as applied for the AFM imaging analysis.

According to Leja [5] and Woods [6], generally, the main mechanisms for the increase in hydrophobicity of sulfide minerals in flotation by the addition of collectors include (1) the adsorption of metal xanthate with low solubility and (2) the oxidation of xanthate into dixanthogen on a sulfide mineral surface in an aqueous solution. Previous AFM studies with chalcopyrite and pyrite [16–18] have shown that, because under ambient conditions, i.e., room temperature and normal pressure, dialkyl dixanthogen is usually in a liquid form with a low melting point [23], the adsorbed dixanthogen on sulfides in an aqueous solution demonstrates patches with smooth and round edges, which fits well with the fact that oily dixanthogen is generally insoluble in water and that the circular boundary is the direct result of the high interfacial tension between hydrophobic dixanthogen and water. In the present investigation, as shown by Figures 2–17, the adsorbate shows no evidence of smooth and round edges, and the surface morphology of chalcocite after adsorption is similar to what has been obtained with the bornite/xanthate system instead [22]. In addition, Figures 8 and 13 clearly show that the adsorbate is not soft at all, because even when a large scan force is applied to remove the adsorbate and open a 'window' in the center of the AFM image, there is still some residual precipitates left on the mineral surface. Further, as shown in Figure 14, the precipitate persists on the chalcocite surface after the mineral surface is rinsed with copious ethanol alcohol, suggesting that the adsorbate is insoluble in ethanol, which is clearly not an essential property of nonpolar oily dixanthogen. As such, it is reasonable to rule out the possibility that the observed adsorbate on the chalcocite in xanthate solutions is dixanthogen.

It has been proposed that the semiconductor type of sulfide minerals determines the final adsorption products on sulfides. For example, on n-type minerals, dixanthogen is usually formed, while on p-type minerals metal xanthate is observed. Chalcocite is "known as a fairly good but variable conductor", and it is classified as "a consistently p-type mineral" [24], which favors the formation of metal xanthate. As reviewed in the previous section, the adsorption of xanthate on the chalcocite surface in an aqueous solution is mainly due to the initially chemisorbed xanthate (X^-) on the chalcocite surface followed by insoluble cuprous xanthate (CuX) [1,3,7–10]. Clearly, the findings obtained from the current AFM imaging analysis results are in line with what has been previously reported.

In general, the 'chemisorption' of xanthate on chalcocite occurs via: [10]

$$X^- = X_{ads} + e^- \tag{1}$$

The adsorption of xanthate on chalcocite is due to an anodic reaction simplified as follows introducing insoluble cuprous xanthate on the chalcocite surface [10,25]:

$$Cu_2S + nX^- = nCuX + Cu_{2-n}S + ne^- \tag{2}$$

The section analysis of the obtained AFM images also suggests that the irregular adsorbate cannot be attributed to the chemisorbed xanthate, of which the maximum adsorption peaks at a monolayer surface coverage [1,10] because the height of the adsorbate is generally above 5 nm and the value is dramatically larger than a monolayer length of ethyl xanthate and amyl xanthate as used in the present study. In the present study, solution potential is not intentionally controlled, and it suggests that the chalcocite surface will undertake an oxidation reaction to some extent when it contacts water.

However, the claim that the observed irregular adsorbate is cuprous xanthate does not rule out the existence of chemisorbed xanthate on the chalcocite surface. In the present study, chemisorbed xanthate may co-exist with cuprous xanthate on the chalcocite surface; however, it is technically too difficult to observe the substance from the obtained AFM images. The difficulty of detecting chemisorbed xanthate using an AFM lies in two facts. Firstly, the hydrocarbon chain length of KEX and PAX as used in the present study is very short, i.e., less than 1 nm, and this makes it difficult to identify such a small change in surface morphology of a very soft substance, i.e., xanthate, with high confidence. Secondly, the chalcocite mineral sample as studied in the present work is prepared by polishing the mineral surface, and the polishing process results in some degree of surface roughness. As shown in Figure 1, there are some scratch lines existing on the polished chalcocite surface and the preexisting surface roughness makes it unrealistic to study the surface morphology change in angstrom. Therefore, it is possible that xanthate chemisorbs on chalcocite surface in a very low profile without being successfully detected by an AFM.

In order to infer the binding strength of the adsorbate and chalcocite mineral surface in quality, as shown in Figure 9, a 10 μm × 5 μm area is scanned for one time by applying a much larger scan force to intentionally remove the adsorbate, and therefore a 10 μm × 5 μm 'window' is shown in the image due to the partial removal of the adsorbate from the mineral surface under the intentionally applied large scan force. Similarly, Figure 13 is obtained in the same methodology. Both Figures 9 and 13 show that there is still some residual adsorbate adsorbing in the "window" area although a large force is applied. In addition, Figure 14 shows that rinsing with copious ethanol does not remove the adsorbate. This conclusion coincides well with those reported by Allison et al. [3], quote, "no product of reaction with the methyl and ethyl homologues could be detected, although both reacted very extensively with the surface". As such, xanthate adsorbs on chalcocite intensively with the mineral surface being covered quickly by adsorbate. The binding of the adsorbate, i.e., cuprous ethyl xanthate, and chalcocite is very strong, and the adsorbate cannot be completely removed by applying a large scan force. Rinsing with ethanol alcohol cannot extract the adsorbate from the chalcocite surface.

4.2. Effect of the Rank of Xanthate

In froth flotation, the rank of a collector is an important parameter determining the collectivity and the selectivity of the collector. In general, a xanthate collector with a low rank has a low collectivity and therefore a high selectivity and vice versa. In the present work, the effect of the collector's rank on the adsorption of xanthate on chalcocite has been studied by using both KEX and PAX as collectors. By comparing Figures 2–5 to Figures 6–8 and Figures 10–14 to Figures 15–17, one can see that when the collector's concentration is the same, PAX adsorbs on the chalcocite surface in much higher surface coverage and a more uniform layer structure. The formation of cuprous amyl xanthate occurs at a lower surfactant concentration than that of cuprous ethyl xanthate, and it is due to a lower solubility product for the former. For example, it was reported that the solubility product of cuprous amyl xanthate is 8.0×10^{-22} and that for cuprous ethyl xanthate is 5.2×10^{-20} [26,27]. Allison et al. [3] also reported that the percentage of reacted xanthate increased by one fold with the carbon number of xanthate increasing from 2 to 5 at the same surfactant concentration. Therefore, the fact that cuprous amyl xanthate forms a more uniform layer than cuprous ethyl xanthate does is due to the longer hydrocarbon chain of the former and therefore an increased lateral hydrophobic attraction between hydrocarbon chains. Therefore, a high-rank xanthate, i.e., PAX, is more powerful than a low-rank xanthate, i.e., KEX, for the adsorption on chalcocite surface; and therefore, a highly enhanced flotation collectivity. The same conclusion is also applicable to the case of the xanthate/bornite system [22].

4.3. Effect of the Concentration of Xanthate

Equation (2) shows that increasing the concentration of xanthate, i.e., the reactant, facilitates the reaction to proceed rightward, resulting in more reaction product, i.e., CuX, on the mineral surface. The AFM images obtained in the present work clearly show such a concentration effect of xanthate. For example, by comparing Figures 2, 3 and 5, one can easily observe that the surface coverage and the height of the adsorbate (surface roughness) increase, suggesting that at pH 6 the amount of adsorbate increases greatly when the concentration of KEX increases from 5×10^{-5} M to 5×10^{-4} M. The same conclusion can also be drawn for the case of the PAX/chalcocite system. For example, by comparing Figures 7–9, one can observe that when the concentration of PAX increases from 1×10^{-5} M to 1×10^{-4} M, the amount of adsorbate increases with the same contact time frame. The same concentration effect of xanthate (KEX and PAX) is also observed at pH 10. All these results show that during a wide pH range in flotation practice, increasing the concentration of xanthate is beneficial for a high surface coverage of cuprous xanthate, and therefore a high flotation recovery.

4.4. Effect of Adsorption Time of Xanthate

By comparing Figure 4 to Figure 5, Figure 8 to Figure 9, and Figure 12 to Figure 13, one can see that when the adsorption time increases from 10 min to 20 min in xanthate solution, the surface coverage and the height of the adsorbate increase as well as. The same trend is applicable for both pH 6 and pH 10. It is then concluded from these AFM images that the adsorption of xanthate on chalcocite increases with the adsorption time. The finding is in line with a common industrial practice of copper ore beneficiation by adding collectors in a mill, the objective of which includes increasing the adsorption time and promoting the adsorption of the collector on the freshly exposed mineral surface.

4.5. Adsorption of Dialkyl Dithiophosphate on Chalcocite Surface

The AFM images of a chalcocite surface in Aerofloat 238 solution at pH 6 for different contact times are shown in Figures 18–20, from which one can see that dialkyl dithiophosphate adsorbs intensively on chalcocite at pH 6. In addition, increasing contact time from 10 min to 20 min increases the amount of adsorbate on the chalcocite surface by increasing both surface coverage and surface roughness as shown in Table 1. Figures 21–23 are the AFM images of a chalcocite surface in Aerofloat 238 solution at pH 10 for different contact times. The figures show that at pH 10 dialkyl dithiophosphate can adsorb intensively on chalcocite as well, suggesting a wide pulp pH range for the flotation of chalcocite using Aerofloat 238 as a collector.

Figure 18 shows that there is a lot of adsorbate existing on chalcocite when the mineral surface contacts the Aerofloat 238 solution and the morphology of the adsorbate, which shows round and smooth edges, differs greatly from those as obtained with xanthate, for which the adsorbate is generally precipitates with irregular sharp edges. In addition, Figure 19 shows that when the mineral surface is scanned with a much larger scan force being applied to intentionally remove the adsorbate, a 10 μm × 5 μm 'window' is shown in the upper image due to the removal of the adsorbate from the mineral surface under the applied large scan force. As the 'window' area is close to the bare chalcocite surface, it suggests that the binding strength between the adsorbate and chalcocite is weak, and the finding is different from the one obtained in Figure 9, which shows a strong binding strength between the precipitate of cuprous xanthate and chalcocite.

Additionally, as shown in Figure 23, rinsing the chalcocite surface with 10 mL ethanol after it contacts the Aerofloat 238 solution for 20 min removes the adsorbate almost completely from chalcocite, and it suggests that the adsorbate is generally an oily substance of high solubility in ethanol, but not in water.

As to the adsorbate on chalcocite, there have been some studies with different conclusions [11–13]. For example, Goold and Finkelstein [11] reported that only $Cu(DTP)_2$ was present on copper sulfide. Solozhenkin and Koptsia [12] reported that the products were both CuDTP and $Cu(DTP)_2$. Chander and Fuerstenau [13] proposed that the oxidation of DTP^- at copper sulfide was a two-step process consisting of the discharge of the DTP^- ion to form a free $DTP^{\cdot}$ radical ($DTP^- = DTP^{\cdot} + e^-$) followed by the subsequent reaction of $DTP^{\cdot}$. They also claimed that the most probable reaction involving $DTP^{\cdot}$ at copper sulfide were:

$$Cu_2S + (2-y)DTP^{\cdot} + yDTP^- = 2CuDTP + S + ye^- \tag{3}$$

$$Cu_2S + (4-m)DTP^{\cdot} + mDTP^- = 2Cu(DTP)_2 + S + me^- \tag{4}$$

Equations (3) and (4) show that CuDTP and $Cu(DTP)_2$ are formed when copper sulfide is immersed in reagent solutions of dithiophosphate.

Zinc dialkyl dithiophosphate ($Zn(DTP)_2$) has been widely applied as anti-wear additives in lubricants. It is characterized as a nonpolar oily liquid at room temperature, which is insoluble in water. As cupric dialkyl dithiophosphate ($Cu(DTP)_2$) has a similar molecular structure as that of zinc dialkyl dithiophosphate, $Cu(DTP)_2$ is an oily liquid substance as well at room temperature. Figures 18–23 clearly show that the adsorbate is generally an oily substance of high solubility in ethanol, and the finding confirms well with the claim that $Cu(DTP)_2$ is the adsorbate formed when copper sulfide is immersed in reagent solutions of dithiophosphate. As such, $Cu(DTP)_2$ is the main adsorbate on the chalcocite surface in dialkyl dithiophosphate solutions. Of course, CuDTP can also co-adsorb together with $Cu(DTP)_2$ on chalcocite, but its amount is too sparse to be observed from AFM images when compared to $Cu(DTP)_2$.

4.6. Adsorption of the Mixture of Xanthate and Dialkyl Dithiophosphate on Chalcocite Surface and Its Implication for Chalcocite Flotation

In froth flotation, specific chemicals, i.e., collectors or promoters, are added to pulp to increase the surface hydrophobicity of a target mineral. This will result in the increase of both attractive hydrophobic force and adhesion force between mineral particles and bubbles. The former can facilitate a particle/bubble attachment and the latter can retard a particle/bubble detachment, which are both beneficial for froth flotation.

AFM force measurements [17,21,22] have shown that, compared to insoluble metal xanthate, an oily substance such as dixanthogen adsorbs on the mineral surface resulting in a large adhesion force both in magnitude and range between an AFM tip and the mineral surface, suggesting a strong adhesion between a bubble and a mineral surface, which is beneficial for better flotation. That is, on the mineral surface, an oily adsorbate is better than insoluble precipitates when the surface coverage of adsorbate is the same.

The AFM images as obtained in the present work show that xanthate adsorbs on chalcocite in the form of insoluble cuprous xanthate, while dialkyl dithiophosphate adsorbs on chalcocite mainly in the form of oily $Cu(DTP)_2$. Figure 24 shows the AFM images of chalcocite in the mixture of 1×10^{-4} M PAX and 300 ppm Aerofloat 238 solution at pH 6. One can clearly see that there are two very different types of adsorption morphology. Firstly, as indicated by the "Red" arrow, the morphology is of oily patches, i.e., $Cu(DTP)_2$, due to adsorption of dialkyl dithiophosphate; Secondly, as indicated by the "Blue" arrow, the morphology is of precipitates, i.e., cuprous xanthate, due to the adsorption of xanthate. Figure 24 shows that xanthate and dialkyl dithiophosphate can co-adsorb with both insoluble cuprous xanthate and oily $Cu(DTP)_2$ on the chalcocite surface. The former increases the surface coverage of the collector on the mineral surface, while the latter increases the adhesion between a bubble and chalcocite. Clearly, this mixed collectors scheme is beneficial for an improved flotation of chalcocite. The conclusions obtained from the present AFM images analysis explain a recent flotation practice of chalcocite, which claims that using a mixture of xanthate and dithiophosphate collector can largely improve both grade and recovery of copper flotation concentrate [15].

5. Conclusions

AFM image analysis has been applied to study the adsorption morphology of KEX, PAX and Aerofloat 238 on chalcocite in situ in aqueous solutions changing with solutions' chemistry such as collector type and dosage, and pH. AFM images show that: (1) All these collectors, i.e., xanthate and dialkyl dithiophosphate, adsorb strongly on chalcocite but in different manners. (2) Xanthate adsorbs mainly in the form of insoluble cuprous xanthate (CuX), which binds strongly with the mineral surface without being removed by flushing with ethanol alcohol. This xanthate/chalcocite adsorption mechanism is very similar to the one obtained with the xanthate/bornite system; while it is different from the one of the xanthate/chalcopyrite system. (3) On the other hand, dibutyl dithiophosphate adsorbs on chalcocite in the form of hydrophobic patches, which can be removed by rinsing with ethanol alcohol. (4) AFM images show that the adsorption of collectors increases with increasing adsorption time and collectors' concentration. (5) Using the mixture of xanthate and dithiophosphate as a collector can help with the adsorption of collectors on chalcocite surface, as shown by the co-adsorption images of xanthate and dithiophosphate on chalcocite, and therefore an improved flotation. The present AFM image analysis work clarifies the flotation mechanism of chalcocite in industry practice using xanthate and dialkyl dithiophosphate as collectors.

Author Contributions: Conceptualization, J.Z.; methodology, J.Z. and W.Z.; formal analysis, J.Z.; investigation, J.Z. and W.Z.; resources, J.Z.; data curation, J.Z. and W.Z.; writing—original draft preparation, J.Z.; writing—review and editing, J.Z.; supervision, J.Z.; project administration, J.Z. All authors have read and agreed to the published version of the manuscript.

Funding: This research received no external funding.

Data Availability Statement: The data presented in this study are available on request from the corresponding author. The data are not publicly available due to ethical.

Conflicts of Interest: The authors declare no conflict of interest.

References

1. Gaudin, A.M.; Schuhmann, R., Jr. The action of potassium n-amyl xanthate on chalcocite. *J. Phys. Chem.* **1936**, *40*, 257–275. [CrossRef]
2. Poling, G.W.; Leja, J. Infrared Study of Xanthate Adsorption on Vacuum Deposited Films of Lead Sulfide and Metallic Copper under Conditions of Contolled Oxidation. *J. Phys. Chem.* **1963**, *67*, 2121–2126. [CrossRef]
3. Allison, S.A.; Goold, L.A.; Nicol, M.J.; Granville, A.D. A determination of the products of reaction between various sulfide minerals and aqueous xanthate solution, and a correlation of the products with electrode rest potentials. *Metall. Trans.* **1972**, *3*, 2613–2618. [CrossRef]
4. Fuerstenau, M.C. Sulphide Mineral Flotation: In Principles of Flotation. *SAIMM Monogr. Ser.* **1982**, *3*, 159–182.
5. Leja, J. *Surface Chemistry of Flotation*; Plenum Press: New York, NY, USA, 1982.
6. Woods, R. Electrochemistry of sulphide flotation. In *Principles of Mineral Flotation, The Wark Symposium*; Jones, M.H., Woodcock, J.T., Eds.; AIMM: Melbourne, Australia, 1984; pp. 91–115.
7. Mielczarski, J.; Suoninen, E. XPS Study of Ethyl Xanthate Adsorbed onto Cuprous Sulfide. *Surf. Interface Anal.* **1984**, *6*, 34–39. [CrossRef]
8. Mielczarski, J. XPS study of ethyl xanthate adsorption on oxidized surface of cuprous sulfide. *J. Colloid Interface Sci.* **1987**, *120*, 201–209. [CrossRef]
9. Richardson, P.E.; Stout, J.V., III; Proctor, C.L.; Walker, G.W. Electrochemical flotation of sulfides: Chalcocite-ethyl xanthate interactions. *Int. J. Miner. Process.* **1984**, *12*, 73–93. [CrossRef]
10. Leppinen, J.O.; Basilio, C.I.; Yoon, R.H. In-Situ FTIR study of ethyl xanthate adsorption on sulfide minerals under conditions of controlled potential. *Int. J. Miner. Process.* **1989**, *26*, 259–274. [CrossRef]
11. Solozhenkin, P.; Koptsia, N. *International Mineral Processing Congrress, Leningrad, 1968*; Institut Mekhanobr: St Petersburg, Russia, 1968; Volume VIII.
12. Goold, U.; Finkelstein, N. *The Reaction of Sulphide Minerals with Thiol Compounds*; National Institute of Metallurgy: Johannesburg, South Africa, 1972.
13. Chander, S.; Fuerstenau, D. The effect of potassium diethyl dithiophosphate on the electrochemical properties of platinum, copper and copper sulfide in aqueous solutions. *Electroanal. Chem. Interfacial Electrochem.* **1974**, *56*, 217–247. [CrossRef]
14. Zhang, T.; Qin, W.; Yang, C.; Huang, S. Floc flotation of marmatite fines in aqueous suspensions induced by butyl xanthate and ammonium dibutyl dithiophosphate. *Trans. Nonferrous Met. Soc. China* **2014**, *24*, 1578–1586. [CrossRef]
15. Dhar, P.; Thornhill, M.; Rao Kota, H. Investigation of Copper Recovery from a New Copper Ore Deposit (Nussir) in Northern Norway: Dithiophosphates and Xanthate-Dithiophosphate Blend as Collectors. *Minerals* **2019**, *9*, 146. [CrossRef]
16. Zhang, J.; Zhang, W. An AFM Study of Chalcopyrite Surface in Aqueous Solution. In Proceedings of the SME Annual Meeting and Exhibit, Phoenix, AZ, USA, 28 February–3 March 2010; pp. 21–24.
17. Zhang, J.; Zhang, W. *The Adsorption of Collectors on Chalcopyrite Surface Studied by an AFM in 'Separation Technologies'*; Young, C., Luttrell, G., Eds.; SME: Englewood, CO, USA, 2012; pp. 65–73. ISBN 978-0-87335-339-7.
18. Zhang, J.; Zhang, W. An Atomic Force Microscopy Study of the Adsorption of Collectors on Chalcopyrite. In *Microscopy: Advances in Scientific Research and Education*; Méndez-Vilas, A., Ed.; Formatex Research Center: Badajoz, Spain, 2014; Volume 2, pp. 967–973. ISBN 978-84-942134-4-1.
19. Zhang, J.; Zhang, W. An AFM Study of the Adsorption of Collector on Chalcocite. In Proceedings of the 2015 SME Annual Conference and Expo and CMA 117th National Western Mining Conference-Mining: Navigating the Global Waters, Denver, CO, USA, 15–18 February 2015.

20. Zhang, J.; An, D.; Withers, J. A Micro-Scale Investigation of the Adsorption of Collectors on Bastnaesite. *Min. Metall. Explor.* **2019**, *36*, 957.
21. An, D.; Zhang, J. A Study of Temperature Effect on the Xanthate's Performance during Chalcopyrite Flotation. *Minerals* **2020**, *10*, 426. [CrossRef]
22. Zhang, J. An Investigation of the Adsorption of Xanthate on Bornite in Aqueous Solutions Using an Atomic Force Microscope. *Minerals* **2021**, *11*, 906. [CrossRef]
23. Rao, S.R. *Xanthate and Related Compounds*; Dekker: New York, NY, USA, 1971.
24. Shuey, R.T. *Semiconducting Ore Minerals*; Elsevier: Amsterdam, The Netherlands, 1975.
25. Woods, R.; Richardson, P.E. The flotation of sulfide minerals: Electrochemical aspects. In *Advances in Mineral Processing*; Somasundran, P., Ed.; SME: Littleton, CO, USA, 1986; pp. 154–170.
26. Kakovsky, I.A. Physiochemical properties of some flotation reagents and their salts with ions of heavy non-ferrous metals. In Proceedings of the Second International Congress of Surface Activity, London, UK, 1 January 1957; pp. 225–237.
27. Ackerman, P.K.; Harris, G.H.; Klimpel, R.R.; Aplan, F.F. Evaluation of flotation collectors for copper sulfides and pyrite, III. Effect of xanthate chain length and branching. *Int. J. Miner. Process.* **1987**, *21*, 141–156. [CrossRef]

Article

An Experimental Study of Pressure Drop Characteristics and Flow Resistance Coefficient in a Fluidized Bed for Coal Particle Fluidization

Jian Peng [1,2], Wei Sun [1,2], Le Xie [3,*], Haisheng Han [1,2,*] and Yao Xiao [1,2]

[1] School of Minerals Processing and Bioengineering, Central South University, Changsha 410083, China; peng.jian@csu.edu.cn (J.P.); sunmenghu@csu.edu.cn (W.S.); 215601040@csu.edu.cn (Y.X.)
[2] Key Laboratory of Hunan Province for Clean and Efficient Utilization of Strategic Calcium-Containing Mineral Resources, Central South University, Changsha 410083, China
[3] College of Chemistry and Chemical Engineering, Central South University, Changsha 410083, China
* Correspondence: xiele2018@csu.edu.cn (L.X.); hanhai5086@csu.edu.cn (H.H.); Tel./Fax: +86-731-8883-6873 (L.X.); +86-731-8887-9616 (H.H.)

Abstract: Liquid–solid fluidized beds have a wide range of applications in metallurgical processing, mineral processing, extraction, and wastewater treatment. Great interest on their flow stability and heterogeneous fluidization behaviors has been aroused in research. In this study, various fluidization experiments were performed by adjusting the operating conditions of particle size, particle density, and liquid superficial velocity. For each case, the steady state of liquid–solid fluidization was obtained, and the bed expansion height and pressure drop characteristics were analyzed. The time evolution of pressure drop at different bed heights can truly reflect the liquid–solid heterogeneous fluidization behaviors that are determined by operating conditions. With the increase in superficial liquid velocity, three typical fluidization stages were observed. Accordingly, the flow resistance coefficient was obtained based on the experimental data of bed expansion height and pressure drop. The flow resistance coefficient experiences a decrease with the increase in the modified particle Reynolds number and densimetric Froude number.

Keywords: pressure drop; liquid–solid; fluidized bed; flow resistance coefficient; fluidization experiments

Citation: Peng, J.; Sun, W.; Xie, L.; Han, H.; Xiao, Y. An Experimental Study of Pressure Drop Characteristics and Flow Resistance Coefficient in a Fluidized Bed for Coal Particle Fluidization. *Minerals* **2022**, *12*, 289. https://doi.org/10.3390/min12030289

Academic Editor: Luis A. Cisternas

Received: 10 January 2022
Accepted: 23 February 2022
Published: 25 February 2022

Publisher's Note: MDPI stays neutral with regard to jurisdictional claims in published maps and institutional affiliations.

1. Introduction

Liquid–solid fluidized beds have been widely used in metallurgical processing, mineral processing, extraction, and wastewater treatment [1–9] for their excellent fluid–solid coupling as well as heat and mass transfer performance. Heterogeneous fluidization behaviors have recently aroused great interest for the purpose of operation stability and optimized operating conditions [10,11]. The transition from the homogeneous to heterogeneous regime occurs at some critical operating conditions. The intensity of turbulence is much higher in the heterogeneous regime, resulting in higher rates of heat, mass, and momentum transfer and better mixing. Such conditions are preferable in many applications. However, too much turbulence may be detrimental to the performance in operations such as flotation, causing higher rates of detachment of particles from the bubble surface. Therefore, an insight into the heterogeneous flow structures of liquid–solid fluidized systems is critical.

Generally, liquid–solid fluidization systems are homogeneous when compared with gas–solid systems. As a result, liquid–solid fluidization has received little attention despite its wide applications. Liquid–solid fluidized beds have unique characteristics, such as low solid–fluid density ratio, and thus, homogeneous fluidization phenomena have been frequently observed [12,13]. For small and light particles, particle entrainment occurs easily in the fluidized state. Severe heterogeneous liquid–solid fluidization phenomena (i.e., particle clustering, bubbling, and velocity fluctuations) occur only under some extreme conditions [14–22].

In the past few decades, various experiments and numerical simulations have been carried out to study liquid–solid flow characteristics. The earliest research on the liquid–solid heterogeneous fluidization phenomenon can be traced back to 1948 when Wilhelm and Kwauk performed the fluidization of lead particles in water [23]. Later, Di Felice [13] found that solid–fluid density ratio, particle morphology, particle size, and size distribution were responsible for heterogeneity. In recent years, experimental techniques have improved greatly for visualizing multiphase flow fields. Razzak et al. [24–26] measured the liquid phase holdup and velocity distribution using resistance tomography, whereas the solid phase holdup was measured using a pressure sensor. The solid phase holdup decreases with the increase in liquid superficial velocity. Similarly, Zbib et al. [27,28] measured the particle phase holdup by electrical resistance tomography in a liquid–solid fluidized bed. Reddy et al. [29] performed flow visualization experiments using particle image velocimetry. They found that a liquid–solid fluidized bed was a homogenous system for particle Reynolds number (Re_t) in the range of 51–759.

Additionally, computational fluid dynamics (CFD) simulations have great advantages in investigating the hydrodynamic characteristics of liquid–solid fluidized beds, particularly for the purpose of scaling up. The accuracy of a CFD model can be validated by experimental data. An entire flow field can be captured using a CFD simulation, including turbulence intensity, particle concentration, particle velocity, and particle clusters, which are beneficial to clarify the fluidization mechanism. Some studies consider the local heterogeneity of liquid–solid flow in CFB when calculating the drag force. Liu et al. [30] showed that compared with experimental data, the multi-scale resistance coefficient model predicted better particle concentration distribution by taking into account the effect of mesoscale cluster structure. In addition, Xie et al. proposed an effective resistance closure method for fluidization of coarse coal particles in transition liquid–solid two-phase flow by clarifying the relationship between resistance coefficient and Reynolds number, Froude number and Stokes number. The output prediction is in good agreement with the experimental data [11].

The effects of operating conditions, such as liquid superficial velocity, particle properties, and fluidized bed structures on fluidization behaviors have been widely investigated in recent years [31–35]. It is also generally accepted that liquid–particle and particle–particle interactions play an important role in determining liquid–solid fluidization characteristics [36–40]. However, there is still limited research on heterogeneous fluidization phenomena [41,42]. Moreover, the flow stability and flow resistance in a liquid–solid fluidized bed have received limited attention [28,43,44].

This study aims to investigate the heterogeneous fluidization and flow resistance in a square liquid–solid fluidized bed where coarse coal particles and water are the solid and liquid phases, respectively. Various fluidization experiments were carried out by adjusting the operating conditions such as particle size, particle density, and liquid superficial velocity. The bed-expansion height and pressure drop under different conditions were measured and used to analyze the flow stability and resistance. According to the measured experimental data, relationships between the flow resistance coefficient and the modified particle Reynolds number (Re_s) and densimetric Froude number (Fr_p) are obtained. To the best of our knowledge, the experimental data of the bed expansion height and pressure drop obtained can provide insight into the fluidization mechanism of a liquid–solid fluidized bed.

2. Experimental Design and Procedures

The lab-grade fluidized bed unit shown in Figure 1 is constructed of clear plexiglass of 100 mm in length, 20 mm in width, and 500 mm in height. A rectangular distributor is arranged at the bottom of the fluidized bed. The distributor has an opening ratio of 40% and a pore size of 1 mm. The coal particle samples for the solid phase were purchased from Pingdingshan Coal Industry Group Co. Ltd., located in Henan, China. Raw coal was crushed by a laboratory jaw crusher and then sieved to produce two large size fractions

(0.7 ± 0.1 and 1.25 ± 0.25 mm). Two methods were employed to determine the particle size of each sample: classical sieve analysis and microscopic static image analysis. These were further subdivided according to different densities using the $ZnCl_2$ heavy liquid method. Five density fractions of coal samples (1500 ± 50 and 1700 ± 50 kg/m^3) were obtained. As a result, four different coal particle samples were used in this study. The physical properties of the coal particles are given in Table 1.

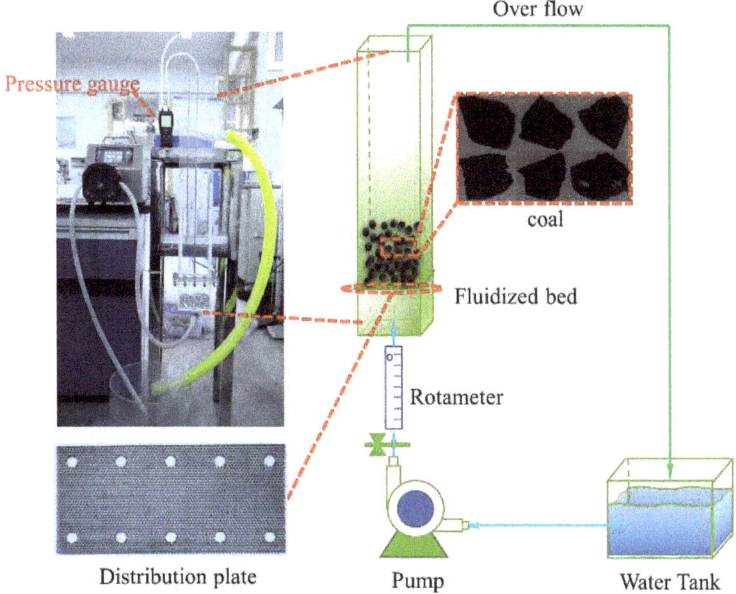

Figure 1. Schematic of the liquid–solid fluidized bed experimental setup.

Table 1. Operating conditions for fluidization experiments.

Scheme	Particle Density/kg/m^3	Particle Size/mm	Condition/State
1#	1500 ± 50	0.7 ± 0.1	Fluidized bed
2#	1700 ± 50	0.7 ± 0.1	100 mm × 20 mm × 500 mm
3#	1500 ± 50	1.25 ± 0.25	Initial bed height: 100 mm
4#	1700 ± 50	1.25 ± 0.25	Superficial velocity: 0.008–0.04 m/s

The fluidization experiment was carried out as follows. The prepared coal sample is initially placed on the distribution plate with an initial filling height of 100 mm. Water is delivered to the bottom of the fluidized bed by a peristaltic pump (WT600-3J) manufactured by Longer Precision Pump Co., Ltd. (Baoding, China). The flow error of the peristaltic pump is less than ±0.5%. The inlet speed at the bottom of the fluidized bed is controlled by the speed of the peristaltic pump. When the superficial velocity of the water exceeds the minimum fluidization velocity, the coal particles leave the distribution plate and are suspended in the fluidized bed. Then, a ruler is used to record the expansion height of the bed. When the particle fluidization reaches a steady state, the bed expansion height is measured three times, and the average value is taken as the final experimental measurement value.

Then, the bed expansion ratio can be calculated as follows:

$$\chi = \frac{H - H_0}{H_0} \times 100\%$$

(1)

where χ is the bed expansion ratio, H_0 is the initial bed height, and H is the bed height after fluidization.

Additionally, the pressure drop continues to fluctuate with flow time, even though the bed height is stable. In this study, four sections, namely, $z = 50$ mm, $z = 100$ mm, $z = 150$ mm, and $z = 200$ mm are selected, and the time evolutions of the pressure drop are recorded. The differential pressure gauge is composed of two pressure-measuring copper tubes, digital display pressure gauge (Xima AS510) manufactured by Dongguan Wanchuang Electronic Products Co., Ltd. (Dongguan, China), and rubber hose. The accuracy of the pressure gauge is ±3 Pa and the resolution is ±1 Pa. When measuring the pressure drop, the pressure drop at the end of two copper tubes is directly read by the digital differential pressure gauge. The total duration of pressure drop measurement is 30 s, and a data point is recorded every 0.5 s. Distilled water from the laboratory was employed as the fluidizing medium. The density and viscosity of water are 998.2 kg/m^3 and 0.001 Pa, respectively. The experiments were performed at a temperature of 20 °C under atmospheric conditions.

In this study, 31 groups of fluidization experiments (see Figure 2) were performed based on the developed fluidized bed equipment. Figure 2 shows the bed expansion ratio at different superficial velocities for four different coal particles. Two different fluidization characteristics by particle size are observed. For small coal particles (0.7 $\pm$ 0.1 mm), the bed expansion ratio almost increases linearly with the superficial velocity. Generally, small particles achieve homogeneous fluidization easier, and the bed height increases uniformly and does not change abruptly. For large coal particles with sizes of 1.25 $\pm$ 0.25 mm, however, the linear relationship between bed expansion ratio and superficial velocity is only observed at low superficial velocity (<0.025 m/s). When the superficial velocity exceeds minimum fluidization velocity, the bed expansion ratio meets a significant increase, deviating from the linear growth trend. We guess that this trend may also be applicable to small particles if the superficial velocity is high enough. As shown in Figure 2, the growth trend of the bed expansion ratio is accelerated for small particles when the superficial velocity is higher than 0.02 m/s. Additionally, the bed expansion ratio decreases with the increase in particle density and particle size. When compared with particle density, the effect of particle size is more significant. When the superficial velocity is 0.0208 m/s, the bed expansion ratio decreases from 190% to 37.5% as the particle size increases from 0.7 $\pm$ 0.1 to 1.25 $\pm$ 0.25 mm. For both particle sizes, particle density has a limited effect on the bed expansion ratio at low superficial velocity. The particle density effect can only be highlighted at high superficial velocity. It should be noted that the local solid holdup is still fluctuating even though the whole bed height is stable. This fluctuation is reflected by the pressure drop of the cross section and will be analyzed in Section 3.1.

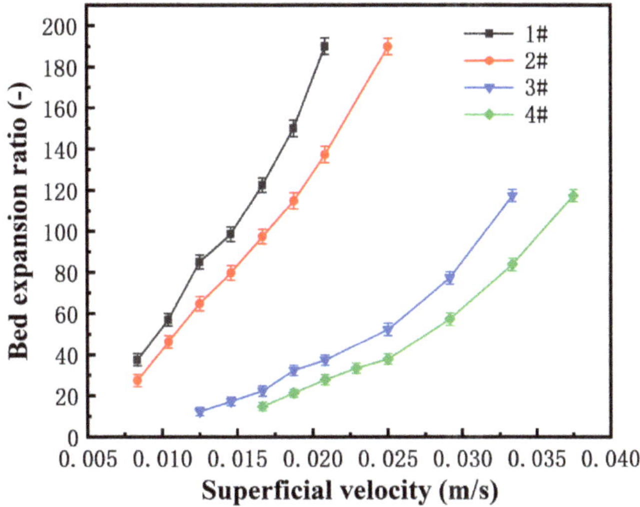

Figure 2. The bed expansion ratio at different superficial velocities for the various samples 1#, 2#, 3#, and 4#.

3. Results and Discussion

3.1. Pressure Drop Characteristics

In this section, the pressure drop was measured at different bed heights when the steady state of liquid–solid fluidization was obtained, for example, when the average bed expansion height remained constant. Generally, the pressure drop continues to fluctuate even though the bed height is stable. Therefore, the pressure drop can be written as

$$\Delta p = \overline{\Delta p} + \Delta p' \tag{2}$$

$$\overline{\Delta p} = \frac{1}{t_1} \int_0^{t_1} \Delta p \, dt \tag{3}$$

Then, the fluctuation intensity of the pressure drop can be defined as:

$$I = \frac{\sqrt{\left(\sum_1^N \Delta p'^2 \right) / N}}{\overline{\Delta p}} \tag{4}$$

where $\Delta p'$ is the fluctuating pressure drop and $\overline{\Delta p}$ is the time-averaged pressure drop.

Figure 3 shows the time evolution of the pressure drop at different bed heights and superficial velocities when the particle size is 0.7 ± 0.1 mm and particle density is 1500 kg/m^3. The fluctuation of the pressure drop is very small at low flow rates (see Figure 3A). The calculated fluctuation intensities of the pressure drops are 0.097, 0.049, 0.029, and 0.031, respectively, indicating that the particle fluidization is homogeneous. With the increase in superficial velocity, however, the heterogeneous fluidization phenomenon was first observed at the top of the bed, accompanied by significant fluctuations in the pressure drop (see Figure 3B,C). The possible reason is liquid void motion. Nijssen and Kramer et al. also observed the formation of voids through experiments and numerical simulation in the study of liquid–solid heterogeneous fluidization behavior [10]. Small liquid voids are generated at the bottom of the fluidized bed for low superficial liquid velocity. These liquid voids will grow in the process of rising up, and finally break up at the top of the fluidized bed. The liquid void motion and breakage lead to changes in the local solid phase holdup, which is reflected by fluctuations in the pressure drop. When the superficial liquid velocity further increases to 0.0188 m/s (see Figure 3D), the fluctuation in the pressure drop mainly appears at the bottom of the fluidized bed. Of course, this is exactly the opposite phenomenon when compared with that at low superficial liquid velocity. High superficial liquid velocity is often accompanied by high bed expansion height. However, high superficial liquid velocity may only influence particle fluidization at the bottom of the fluidized bed, and homogeneous fluidization is observed at the top of the fluidized bed.

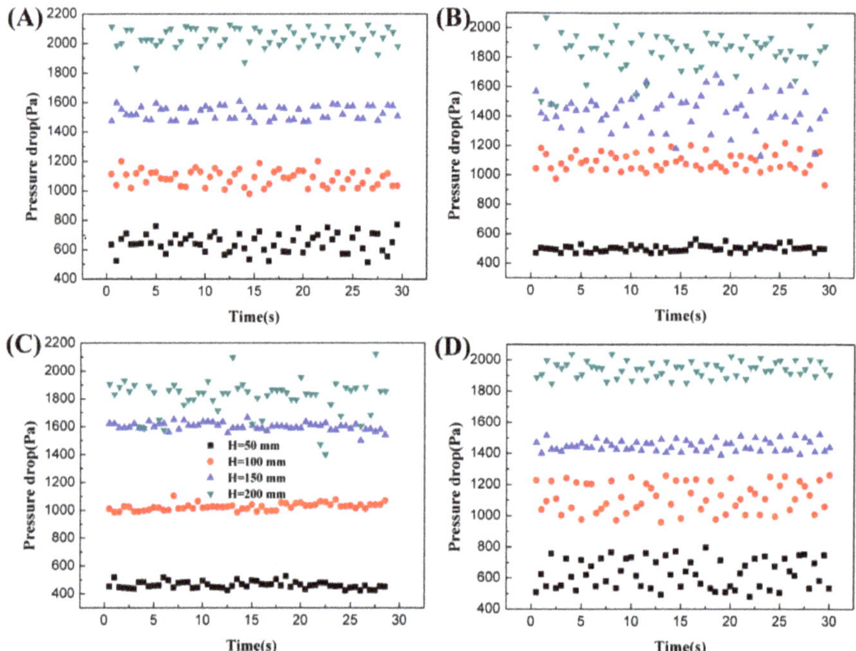

Figure 3. The time evolution of the pressure drop at different bed heights and superficial velocities when the particle size is 0.7 ± 0.1 mm and particle density is 1500 kg/m^3: (**A**) $v_s = 0.0125$ m/s; (**B**) $v_s = 0.0146$ m/s; (**C**) $v_s = 0.0167$ m/s; (**D**) $v_s = 0.0188$ m/s.

When the particle density increases to 1700 kg/m^3, the fluidization becomes more stable at the same superficial liquid velocity (see Figure 4). The fluctuation intensity of the pressure drop is less than 0.065 when the superficial liquid velocity increases from 0.0125 m/s to 0.0167 m/s. That is to say, heavy particle fluidization is insensitive to the superficial liquid velocity. In this study, we observed a large fluctuation in the pressure drop only when the superficial liquid velocity increases to 0.0208 m/s. The pressure drop shows a periodic trend throughout the fluidized bed, indicating that liquid–solid fluidization reaches a steady state.

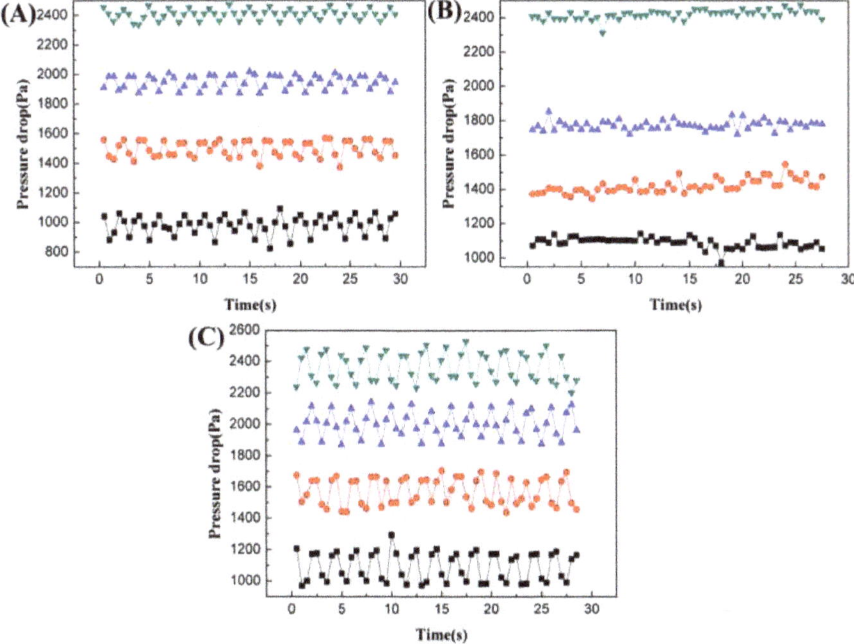

Figure 4. The time evolution of the pressure drop at different bed heights and superficial velocities when the particle size is 0.7 ± 0.1 mm and particle density is 1700 kg/m^3: (**A**) $v_s = 0.0125$ m/s; (**B**) $v_s = 0.0167$ m/s; (**C**) $v_s = 0.0208$ m/s.

Figure 5 shows the time evolution of the pressure drop at different bed heights and superficial velocities when the particle size further increases to 1.25 ± 0.25 mm. In determining fluidization behaviors, gravity dominates for large and heavy particles, while drag force dominates for small and light particles. With the increase in particle size, the fluidized bed also becomes more stable at the same superficial liquid velocity (less than 0.025 m/s), although a significant decrease in bed expansion height is observed. Similarly, when the superficial liquid velocity increases to 0.0292 m/s, a very chaotic pressure drop trend is monitored, indicating that the local particle concentration varies greatly.

From Figures 3–5, it can be concluded that liquid–solid fluidization behavior is highly dependent on the superficial liquid velocity, particle size, and particle density. In most cases, the liquid–solid fluidized bed is in a homogeneous fluidization state. With the increase in particle size and particle density, the fluidized bed will become more stable. As the superficial liquid velocity increases, however, the homogeneous fluidization will change into heterogeneous fluidization. Particle fluidization behavior is controlled by the drag force, gravity, and the heterogeneous flow structure. For small and light particles, the drag force is much larger than gravity. As superficial liquid velocity continues to increase, heterogeneous flow structures, such as liquid voids, may evolve and determine the particle fluidization behaviors. For the large and heavy particles, gravity is greater than the drag force at low superficial liquid velocity. As a result, the particles are inclined to stay at the bottom of the fluidized bed. When the drag force increases to the same level as gravity, the liquid–solid flow is unstable and will produce oscillations. In the following, we will provide explanations by analyzing the relative magnitudes of the drag force, inertia force, and gravity.

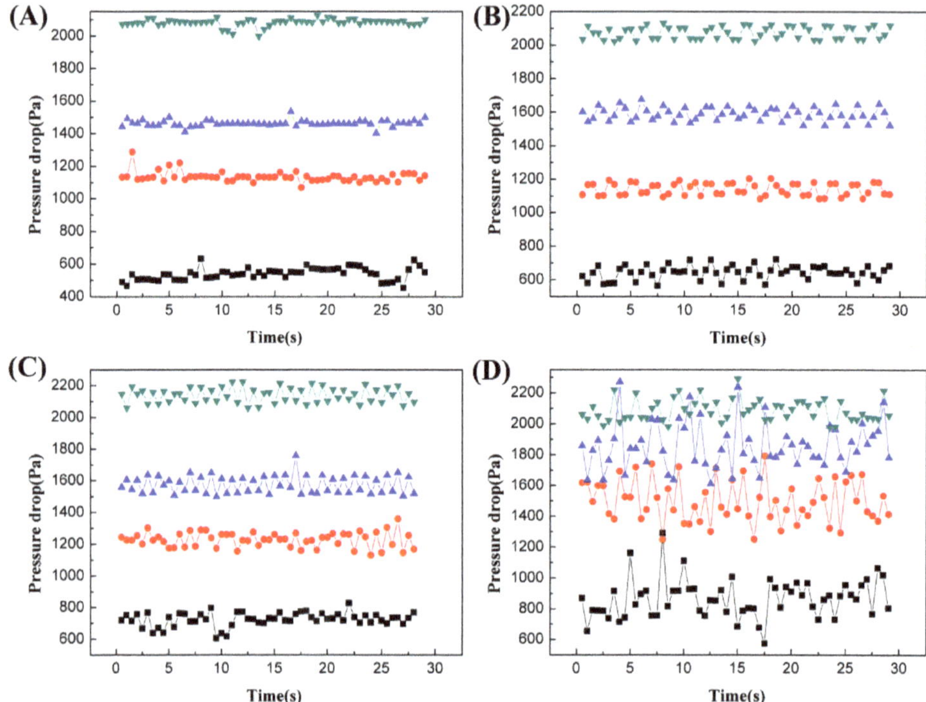

Figure 5. The time evolution of the pressure drop at different bed heights and superficial velocities when the particle size is 1.25 ± 0.25 mm and particle density is 1500 kg/m^3: (**A**) v_s = 0.0167 m/s; (**B**) v_s = 0.0208 m/s; (**C**) v_s = 0.025 m/s; (**D**) v_s = 0.0292 m/s.

3.2. Drag Force Coefficient

In fluidized beds, the fluid will be accompanied by a pressure drop when passing through a granular bed due to frictional loss and inertia. Even though the fluidization reaches steady state, for example, when the bed height is nearly constant, the pressure drop may still fluctuate because of local changes in the bed voidage. As recommended by Ergun [45], the pressure drop in a fluidized bed is defined as:

$$\Delta p = f \frac{H v_0^2 \rho_l (1 - \varepsilon)}{d_s \varepsilon^3} \tag{5}$$

where Δp, f, v_0, ρ_l, H, d_s, and ε are the pressure drop, drag coefficient, flow velocity, fluid density, bed height, particle size, and bed voidage, respectively. For a fluidized bed, the pressure drop can be calculated as follows [46–48]:

$$\Delta p = (\rho_s - \rho_l) g (1 - \varepsilon) H \tag{6}$$

As shown in Equation (6), the pressure drop is determined by the bed voidage and bed expansion height. Generally, the bed voidage is also one of the most important evaluation indexes for the fluidization performance. The Richardson–Zaki relation has been widely used to calculate the bed voidage [48]:

$$\varepsilon^n = \frac{v_s}{v_t} \tag{7}$$

where v_s, is the superficial velocity, which can be determined by the flow rate and pipe diameter; v_t is the terminal velocity of an isolated particle in an unbounded fluid. Stokes has derived the famous expression for the settling velocity of a sphere, now known as Stokes' settling velocity v_{ts}.

$$v_{ts} = \frac{g}{18}(\rho_s - \rho_l)d_s{}^2 \tag{8}$$

Phillip P. Brown and Desmond F. Lawler [49] obtained a new settlement velocity expression based on 480 data points, which can predict settlement velocity very accurately when the terminal Reynolds number is less than 4000.

$$v_t = v_{ts}\frac{3.13 + Re_t{}^{0.682}}{3.13 + Re_t{}^{0.682} + 0.409Re_t{}^{0.682+1/3} + 0.00985Re_t{}^{0.682+2/3}} \tag{9}$$

In Equation (7), n is the Richardson–Zaki coefficient, which is highly dependent on the Reynolds number. A number of researchers have modified this correlation. In the literature, a collection of equations is given to estimate the Richardson–Zaki index n of which the most popular are presented in Table 2.

Table 2. Richardson–Zaki index equations from literature.

	References	Equation	
Classical Richardson–Zaki equation	[48]	$n = \begin{cases} 4.65 & Re_t < 0.2 \\ 4.4Re_t^{-0.03} & 0.2 \leq Re_t < 1 \\ 4.4Re_t^{-0.1} & 1 \leq Re_t < 500 \\ 2.4 & Re_t \geq 500 \end{cases}$	(10)
General expression	[50]	$n = c_1\,Re_t^{c_2}$	(11)
Garside and *Al*-Dibouni equation	[51,52]	$\frac{n_L-n}{n-n_T} = \alpha\,Re_t^{\beta}$	(12)
Khan and Richardson	[53]	$\frac{n_L-n}{n-n_T} = \alpha Ar^{\beta}$	(13)

In Table 2, the particle Reynolds number Re_t under terminal settling conditions and the Archimedes number Ar can be determined as follows:

$$Re_t = \frac{d_s v_t \rho_l}{\mu_l} \tag{14}$$

$$Ar = \frac{g d_s{}^3 \rho_l(\rho_s - \rho_l)}{\mu_l{}^2} \tag{15}$$

Kramer et al. [54] obtained a new expression for n based on more ideal fluidization experiments:

$$\frac{4.8 - n}{n - 2.4} = 0.043Re_t^{0.75} \tag{16}$$

$$\frac{4.8 - n}{n - 2.4} = 0.015Ar^{0.5} \tag{17}$$

which improves the accuracy deviation of the Richardson–Zakie equation from 15% to 3%.

Figure 6 shows the relationship between bed voidage and superficial velocity for the four different coal particle samples. With the increase in superficial velocity, the bed voidage significantly increases, especially for the small and light particles. The maximum bed voidage is 0.74 when the particle size is 0.7 ± 0.1 mm, the particle density is 1500 kg/m^3, and the superficial velocity is 0.0208 m/s. For the large and heavy particles, the bed voidage is 0.65 even though the superficial velocity increases to 0.0375 m/s. As discussed above, the linear relationship between bed expansion ratio and superficial velocity was only observed at low superficial velocities. It can be seen in Figure 6 that there is a power function relationship between bed voidage and superficial velocity throughout the entire range of velocities. The goodness of fit (R^2) is higher than 0.988. The reliable and straightforward voidage prediction model obtained by Kramer et al. [55] in a fluidization system is also a power function relationship.

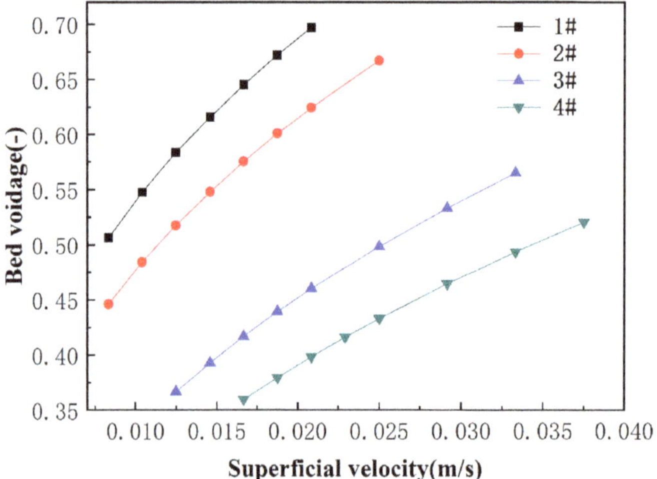

Figure 6. The relationship between bed voidage and superficial velocity.

The dimensionless drag force coefficient can be determined as follows:

$$f = \frac{(\rho_s - \rho_l)}{\rho_l} \frac{g d_s}{v_0^2} \varepsilon^3 \tag{18}$$

In addition to the modified particle Reynolds number (Re_s), the densimetric Froude number (Fr_p) is also commonly used in liquid–solid fluidization. These values are defined as follows:

$$\mathrm{Re}_s = \frac{\rho_l d_s v_0}{\mu_l} \frac{1}{1 - \varepsilon} \tag{19}$$

$$Fr_p = \frac{v_s}{\sqrt{(\rho_s/\rho_l - 1)g d_s}} \tag{20}$$

By adjusting particle size, particle density, and apparent velocity, 31 fluidization experimental datasets (see Figures 2 and 6) were obtained. Figure 7 shows the relationship between the drag force coefficient and the modified particle Reynolds number. The drag force coefficient decreases as the particle Reynolds number increases. In this study, the least squares method is used to fit the experimental data. As reported in the literature, the relationship between drag coefficient and Reynolds number satisfies a power function [45,56–59]. Therefore, a power function is considered as the target fitting function in this study. The following drag coefficient functional relationship is obtained:

$$f = 44.56\mathrm{Re}_s^{-0.674} \quad R^2 = 0.9909 \tag{21}$$

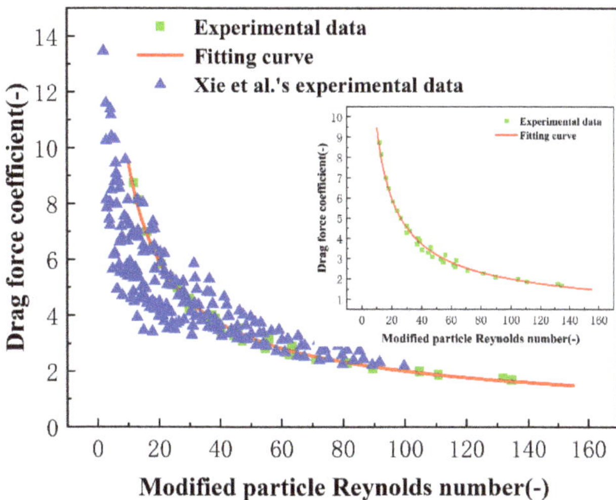

Figure 7. The relationship between drag force coefficient and modified particle Reynolds number [11].

As shown, the goodness of fit (R^2) is higher than 0.99. Additionally, our experimental data are compared with those reported by Xie et al., who investigated the drag coefficient correlation for coarse coal particle fluidization in a transitional flow regime [11]. Although the experimental conditions are different, the calculated dimensionless drag coefficients agree well with the reported data, thus validating the rationality of our experimental data. In addition, the regularity of our experimental data is better. Therefore, the proposed drag force coefficient correlation may have wider application prospects.

The densimetric Froude number is widely used to evaluate the fluidization characteristics of a liquid–solid fluidization system. As early as 1948, Wilhelm and Kwauk proposed that the two fluidization states, namely particulate fluidization and aggregative fluidization, can be distinguished according to the Froude number [17]. The Froude number expresses the influence of gravity on flow; its physical meaning is the ratio of inertial force to gravity. A plot of the drag force coefficient versus Froude number is shown in Figure 8. In this study, the Froude number ranges from 0.12 to 0.43. This means that the fluidization of coal particles does not belong to particulate fluidization or aggregative fluidization but is in the transition stage between the two. Generally, the drag force coefficient decreases with the increase in Froude number. The relationship between drag coefficient and Froude number also satisfies a power function. However, these functional relationships are related to particle density and particle size.

As can be seen from Equations (18) and (20), both particle size and density have an impact on drag force coefficient and Froude number. In the case of a constant apparent flow rate, the drag force coefficient increases with the increase in particle size and density, while Froude number decreases with the increase of particle size and density.

In Figure 8, the experimental comparison of 1# and 2# shows that the particle sizes are both 0.7 mm, and the particle density increases from 1500 to 1700 m^3/kg. Compared with curve 1#, curve 2# deviates to the lower left, indicating that the influence of particle density on Froude number is greater than the drag force coefficient. The comparison of test 1# and 3# shows that the particle density is the same, and the particle size increases from 0.7 mm to 1.25 mm. Compared with curve 1#, curve 3# moves to the lower left with a larger amplitude, indicating that the influence of particle size on Froude number is greater than the drag force coefficient.

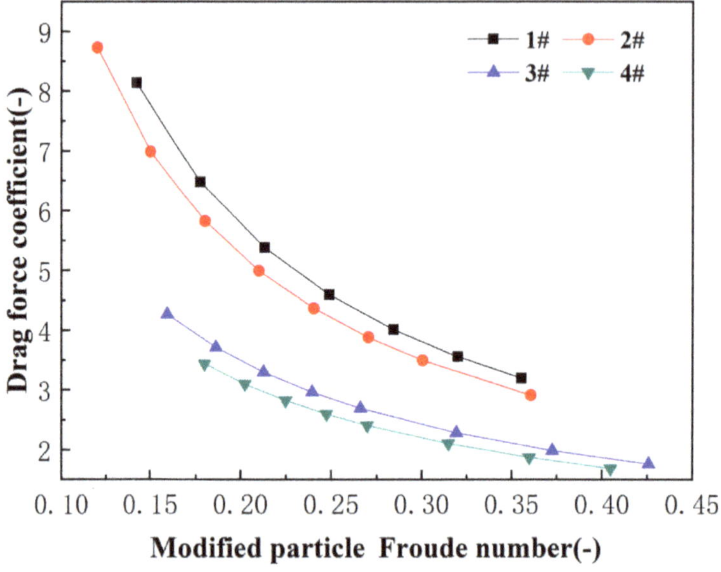

Figure 8. The relationship between drag force coefficient and modified particle Froude number.

4. Conclusions

In this study, the heterogeneous fluidization and flow resistance in a square liquid–solid fluidized bed were investigated. Various fluidization experiments were performed by adjusting the operating conditions of particle size, particle density, and liquid superficial velocity. For each case, the bed expansion height and the pressure drop were measured and used to analyze the flow stability and flow resistance.

In most cases, the liquid–solid fluidized bed is in a homogeneous fluidization state. With the increase in particle size and particle density, the fluidized bed will become more stable. As the superficial liquid velocity increases, however, the homogeneous fluidization will change to heterogeneous fluidization, which is reflected by fluctuations in the pressure drop.

The relationships between the flow resistance coefficient and the Reynolds number and Froude number both satisfy a power function. The relationship between the drag coefficient and Froude number is dependent on particle density and particle size, while the relationship between the drag coefficient and Reynolds number is independent of particle density and particle size. The obtained experimental data provide insight into the fluidization characteristics for liquid–solid fluidized beds. In future work, attention will focus on the fluidization mechanism of liquid–solid systems.

Author Contributions: Conceptualization, J.P.; Formal analysis, J.P., Y.X.; Investigation, J.P.; Methodology, J.P.; Project administration, W.S.; Software, J.P.; Supervision, H.H. and L.X.; Validation, J.P.; Visualization, J.P.; Writing—Original draft preparation, J.P.; Writing—Review and Editing, H.H. and L.X. All authors have read and agreed to the published version of the manuscript.

Funding: This research received no external funding.

Data Availability Statement: The data presented in this study are available on request from the corresponding author.

Acknowledgments: This research was supported by the National Natural Science Foundation of China (NSFC) (No. 52122406), Hunan High-tech Industry Technology Innovation Leading plan (No. 2022GK4056), the Fundamental Research Funds for the Central Universities of Central South University (No.2020zzts204), and the Hunan Provincial Innovation Foundation for Postgraduate (No. CX20200249).

Conflicts of Interest: The authors declare no conflict of interest.

Nomenclature

Ar	Archimedes number, [-]
d_s	Diameter of particles, [mm]
f	Drag force coefficient, [-]
Fr	Froude number, [-]
Fr_p	Densimetric or particle Froude number, [-]
g	Gravitational acceleration, [m s^{-2}]
H	Bed expansion height, [mm]
H_0	Initial bed expansion height, [mm]
I	Fluctuation intensity of the pressure drop, [-]
n	Richardson–Zaki coefficient, [-]
p	Pressure, [Pa]
Re_t	Particle Reynolds number, [-]
Re_s	Modified particle Reynolds number, [-]
t	Time, [s]
v_s	Superficial velocity, [m s^{-1}]
v_t	Terminal velocity, [m s^{-1}]
v_{ts}	Stokes' settling velocity, [m s^{-1}]
ρ_s	Solid density, [kg m^{-3}]
ρ_l	Liquid density, [kg m^{-3}]
μ	Viscosity, [Pa s]
ε	Voidage, [-]
χ	Bed expansion ratio, [-]

References

1. Galvin, K.P.; Zhou, J.; van Netten, K. Dense medium separation in an inverted fluidised bed system. *Miner. Eng.* **2018**, *126*, 101–104. [CrossRef]
2. Jameson, G.J.; Cooper, L.; Tang, K.K.; Emer, C. Flotation of coarse coal particles in a fluidized bed: The effect of clusters. *Miner. Eng.* **2020**, *146*, 106099. [CrossRef]
3. Lu, L.Q.; Yoo, K.; Benyahia, S. Coarse-grained-particle method for simulation of liquid-solids reacting flows. *Ind. Eng. Chem. Res.* **2016**, *55*, 10477–10491. [CrossRef]
4. Sowmeyan, R.; Swaminathan, G. Performance of inverse anaerobic fluidized bed reactor for treating high strength organic wastewater during start-up phase. *Bioresour. Technol.* **2008**, *99*, 6280–6284. [CrossRef] [PubMed]
5. Tripathy, A.; Bagchi, S.; Biswal, S.K.; Meikap, B.C. Study of particle hydrodynamics and misplacement in liquid-solid fluidized bed separator. *Chem. Eng. Res. Des.* **2017**, *117*, 520–532. [CrossRef]
6. Liu, B.; Li, X.; Li, Z.; Sui, H.; Li, H. Fluidized countercurrent solvent extraction of oil pollutants from contaminated soil. Part 1: Fluid mechanics. *Chem. Eng. Res. Des.* **2015**, *94*, 501–507. [CrossRef]
7. Yang, W.-C. Handbook of fluidization and fluid-particle systems. *China Particuology* **2003**, *1*, 137. [CrossRef]
8. Gibilaro, L.G. Fluidization-dynamics, the formulation and applications of a predictive theory for the fluidized state. In *Fluidization-Dynamics, the Formulation and Applications of a Predictive Theory for the Fluidized State*; Gibilaro, L.G., Ed.; Butterworth-Heinemann: Oxford, UK, 2001; pp. 1–7.
9. Epstein, N. Applications of liquid-solid fluidization. *Int. J. Chem. React. Eng.* **2002**, *1*. [CrossRef]
10. Nijssen, T.M.J.; Kramer, O.J.I.; de Moel, P.J.; Rahman, J.; Kroon, J.P.; Berhanu, P.; Boek, E.S.; Buist, K.A.; van der Hoek, J.P.; Padding, J.T.; et al. Experimental and numerical insights into heterogeneous liquid-solid behaviour in drinking water softening reactors. *Chem. Eng. Sci. X* **2021**, *11*, 100100. [CrossRef]
11. Xie, L.; Wang, D.; Wang, H. Effective drag coefficient correlation for coarse coal particle fluidization in transitional flow regime. *Chem. Eng. Res. Des.* **2021**, *172*, 109–119. [CrossRef]
12. Wang, S.Y.; Li, X.Q.; Wu, Y.B.; Li, X.; Dong, Q.; Yao, C.H. Simulation of flow behavior of particles in a liquid-solid fluidized bed. *Ind. Eng. Chem. Res.* **2010**, *49*, 10116–10124. [CrossRef]
13. Di Felice, R. Hydrodynamics of liquid fluidisation. *Chem. Eng. Sci.* **1995**, *50*, 1213–1245. [CrossRef]
14. Cheng, Y.; Zhu, J. Hydrodynamics and scale-up of liquid-solid circulating fluidized beds: Similitude method vs. Cfd. *Chem. Eng. Sci.* **2008**, *63*, 3201–3211. [CrossRef]
15. Esteghamatian, A.; Bernard, M.; Lanced, M.; Hammouti, A.; Wachs, A. Micro/meso simulation of a fluidized bed in a homogeneous bubbling regime. *Int. J. Multiph. Flow* **2017**, *92*, 93–111. [CrossRef]

16. Esteghamatian, A.; Euzenat, F.; Hammouti, A.; Lance, M.; Wachs, A. A stochastic formulation for the drag force based on multiscale numerical simulation of fluidized beds. *Int. J. Multiph. Flow* **2018**, *99*, 363–382. [CrossRef]
17. Kasat, G.R.; Khopkar, A.R.; Ranade, V.V.; Pandita, A.B. Cfd simulation of liquid-phase mixing in solid-liquid stirred reactor. *Chem. Eng. Sci.* **2008**, *63*, 3877–3885. [CrossRef]
18. Kramer, O.J.I.; Padding, J.T.; van Vugt, W.H.; de Moel, P.J.; Baars, E.T.; Boek, E.S.; van der Hoek, J.P. Improvement of voidage prediction in liquid-solid fluidized beds by inclusion of the froude number in effective drag relations. *Int. J. Multiph. Flow* **2020**, *127*, 103261. [CrossRef]
19. Sardeshpande, M.V.; Juvekar, V.A.; Ranade, V.V. Hysteresis in cloud heights during solid suspension in stirred tank reactor: Experiments and cfd simulations. *Aiche J.* **2010**, *56*, 2795–2804. [CrossRef]
20. Sardeshpande, M.V.; Sagi, A.R.; Juvekar, V.A.; Ranade, V.V. Solid suspension and liquid phase mixing in solid-liquid stirred tanks. *Ind. Eng. Chem. Res.* **2009**, *48*, 9713–9722. [CrossRef]
21. Song, Y.F.; Sun, Z.N.; Zhang, C.; Zhu, J.; Lu, X.F. Numerical study on liquid-solid flow characteristics in inverse circulating fluidized beds. *Adv. Powder Technol.* **2019**, *30*, 317–329. [CrossRef]
22. Song, Y.; Zhu, J.; Zhang, C.; Sun, Z.N.; Lu, X.F. Comparison of liquid-solid flow characteristics in upward and downward circulating fluidized beds by cfd approach. *Chem. Eng. Sci.* **2019**, *196*, 501–513. [CrossRef]
23. Wilhelm, R.; Kwauk, M. Fluidization of solid particles. *Chem. Engng. Prog.* **1948**, *44*, 201–208.
24. Razzak, S.A.; Barghi, S.; Zhu, J.X. Electrical resistance tomography for flow characterization of a gas-liquid-solid three-phase circulating fluidized bed. *Chem. Eng. Sci.* **2007**, *62*, 7253–7263. [CrossRef]
25. Razzak, S.A.; Barghi, S.; Zhu, J.X. Axial hydrodynamic studies in a gas-liquid-solid circulating fluidized bed riser. *Powder Technol.* **2010**, *199*, 77–86. [CrossRef]
26. Razzak, S.A.; Barghi, S.; Zhu, J.X.; Mi, Y. Phase holdup measurement in a gas-liquid-solid circulating fluidized bed (glscfb) riser using electrical resistance tomography and optical fibre probe. *Chem. Eng. J.* **2009**, *147*, 210–218. [CrossRef]
27. Zbib, H.; Ebrahimi, M.; Ein-Mozaffari, F.; Lohi, A. Comprehensive analysis of fluid-particle and particle-particle interactions in a liquid-solid fluidized bed via cfd-dem coupling and tomography. *Powder Technol.* **2018**, *340*, 116–130. [CrossRef]
28. Zbib, H.; Ebrahimi, M.; Ein-Mozaffari, F.; Lohi, A. Hydrodynamic behavior of a 3-D liquid-solid fluidized bed operating in the intermediate flow regime-application of stability analysis, coupled cfd-dem, and tomography. *Ind. Eng. Chem. Res.* **2018**, *57*, 16944–16957. [CrossRef]
29. Reddy, R.K.; Sathe, M.J.; Joshi, J.B.; Nandakumar, K.; Evans, G.M. Recent developments in experimental (piv) and numerical (dns) investigation of solid-liquid fluidized beds. *Chem. Eng. Sci.* **2013**, *92*, 1–12. [CrossRef]
30. Liu, G.; Wang, P.; Wang, S.; Sun, L.; Yang, Y.; Xu, P. Numerical simulation of flow behavior of liquid and particles in liquid–solid risers with multi scale interfacial drag method. *Adv. Powder Technol.* **2013**, *24*, 537–548. [CrossRef]
31. Cornelissen, J.T.; Taghipour, F.; Escudie, R.; Ellis, N.; Grace, J.R. Cfd modelling of a liquid-solid fluidized bed. *Chem. Eng. Sci.* **2007**, *62*, 6334–6348. [CrossRef]
32. Dadashi, A.; Zhu, J.X.; Zhang, C. A computational fluid dynamics study on the flow field in a liquid-solid circulating fluidized bed riser. *Powder Technol.* **2014**, *260*, 52–58. [CrossRef]
33. Hua, L.N.; Lu, L.Q.; Yang, N. Effects of liquid property on onset velocity of circulating fluidization in liquid-solid systems: A cfd-dem simulation. *Powder Technol.* **2020**, *364*, 622–634. [CrossRef]
34. Wang, S.Y.; Guo, S.; Gao, J.S.; Lan, X.Y.; Dong, Q.; Li, X.Q. Simulation of flow behavior of liquid and particles in a liquid-solid fluidized bed. *Powder Technol.* **2012**, *224*, 365–373. [CrossRef]
35. Ye, X.; Chu, D.Y.; Lou, Y.Y.; Ye, Z.L.; Wang, M.K.; Chen, S.H. Numerical simulation of flow hydrodynamics of struvite pellets in a liquid-solid fluidized bed. *J. Environ. Sci.* **2017**, *57*, 391–401. [CrossRef]
36. Luo, H.; Zhang, C.; Zhu, J. Development of a numerical model for the hydrodynamics simulation of liquid-solid circulating fluidized beds. *Powder Technol.* **2019**, *348*, 93–104. [CrossRef]
37. Koerich, D.M.; Lopes, G.C.; Rosa, L.M. Investigation of phases interactions and modification of drag models for liquid-solid fluidized bed tapered bioreactors. *Powder Technol.* **2018**, *339*, 90–101. [CrossRef]
38. Wu, Y.C.; Yang, B. An overview of numerical methods for incompressible viscous flow with moving particles. *Arch. Comput. Methods Eng.* **2019**, *26*, 1255–1282. [CrossRef]
39. Xie, L.; Luo, Z.H. Modeling and simulation of the influences of particle-particle interactions on dense solid-liquid suspensions in stirred vessels. *Chem. Eng. Sci.* **2018**, *176*, 439–453. [CrossRef]
40. Liu, G.D.; Yu, F.; Lu, H.L.; Wang, S.; Liao, P.W.; Hao, Z.H. Cfd-dem simulation of liquid-solid fluidized bed with dynamic restitution coefficient. *Powder Technol.* **2016**, *304*, 186–197. [CrossRef]
41. Fan, L.; Grace, J.R.; Epstein, N. Investigation of nonuniformity in a liquid-solid fluidized bed with identical parallel channels. *Aiche J.* **2010**, *56*, 92–101. [CrossRef]
42. Mazzei, L.; Lettieri, P. Cfd simulations of expanding/contracting homogeneous fluidized beds and their transition to bubbling. *Chem. Eng. Sci.* **2008**, *63*, 5831–5847. [CrossRef]
43. Ghatage, S.V.; Peng, Z.B.; Sathe, M.J.; Doroodchi, E.; Padhiyar, N.; Moghtaderi, B.; Joshi, J.B.; Evans, G.M. Stability analysis in solid-liquid fluidized beds: Experimental and computational. *Chem. Eng. J.* **2014**, *256*, 169–186. [CrossRef]
44. Palkar, R.R.; Patnaikuni, V.S.; Shilapuram, V. Step by step methodology of designing a liquid-solid circulating fluidized bed using computational fluid dynamic approach. *Chem. Eng. Res. Des.* **2018**, *138*, 260–279. [CrossRef]

45. Ergun, S. Fluid flow through packed column. *Chem. Eng. Prog.* **1952**, *48*, 89–94.
46. Chen, X.Z.; Shi, D.P.; Gao, X.; Luo, Z.H. A fundamental cfd study of the gas-solid flow field in fluidized bed polymerization reactors. *Powder Technol.* **2011**, *205*, 276–288. [CrossRef]
47. Islam, M.T.; Nguyen, A.V. Effect of particle size and shape on liquid-solid fluidization in a hydrofloat cell. *Powder Technol.* **2021**, *379*, 560–575. [CrossRef]
48. Richardson, J.F.; Zaki, W.N. Sedimentation and fluidization: Part i. *Chem. Eng. Res. Des.* **1997**, *75*, S82–S100. [CrossRef]
49. Brown Phillip, P.; Lawler Desmond, F. Sphere drag and settling velocity revisited. *J. Environ. Eng.* **2003**, *129*, 222–231. [CrossRef]
50. Siwiec, T. The experimental verification of Richardson-Zaki law on example of selected beds used in water treatment. *Electron. J. Pol. Agric. Univ.* **2007**, *10*. Available online: http://www.ejpau.media.pl/volume10/issue2/art-05.html (accessed on 1 January 2022).
51. Garside, J.; Al-Dibouni, M.R. Velocity-voidage relationships for fluidization and sedimentation in solid-liquid systems. *Ind. Eng. Chem. Process Des. Dev.* **1977**, *16*, 206–214. [CrossRef]
52. Rowe, P.N. A convenient empirical equation for estimation of the Richardson-Zaki exponent. *Chem. Eng. Sci.* **1987**, *42*, 2795–2796. [CrossRef]
53. Khan, A.R.; Richardson, J.F. The resistance to motion of a solid sphere in a fluid. *Chem. Eng. Commun.* **1987**, *62*, 135–150. [CrossRef]
54. Kramer, O.J.I.; de Moel, P.J.; Baars, E.T.; van Vugt, W.H.; Padding, J.T.; van der Hoek, J.P. Improvement of the richardson-zaki liquid-solid fluidisation model on the basis of hydraulics. *Powder Technol.* **2019**, *343*, 465–478. [CrossRef]
55. Kramer, O.J.I.; de Moel, P.J.; Padding, J.T.; Baars, E.T.; Hasadi, Y.M.F.E.; Boek, E.S.; van der Hoek, J.P. Accurate voidage prediction in fluidisation systems for full-scale drinking water pellet softening reactors using data driven models. *J. Water Process Eng.* **2020**, *37*, 101481. [CrossRef]
56. Carman, P.C. Fluid flow through granular beds. *Trans. Inst. Chem. Eng.* **1937**, *15*, 150–166. [CrossRef]
57. Dallavalle, J.M. *Micromeritics: The Technology of Fine Particles*, 2nd ed.; Pitman: London, UK, 1948.
58. Schiller, V.L. Uber die grundlegenden berechnungen bei der schwerkraftaufbereitung. *Z. Vernes Dtsch. Inge* **1933**, *77*, 318–320.
59. Dijk, J.; Wilms, D.A. Water treatment without waste material; fundamentals and state of the art of pellet softening. *Aqua* **1991**, *40*, 263–280.

Article

An Investigation of the Adsorption of Xanthate on Bornite in Aqueous Solutions Using an Atomic Force Microscope

Jinhong Zhang

Department of Mining and Geological Engineering, The University of Arizona, Tucson, AZ 85721, USA; jhzhang@email.arizona.edu

Abstract: An atomic force microscope (AFM) was applied to study of the adsorption of xanthate on bornite surfaces in situ in aqueous solutions. AFM images showed that xanthate, i.e., potassium ethyl xanthate (KEX) and potassium amyl xanthate (PAX), adsorbed strongly on bornite, and the adsorbate bound strongly with the mineral surface without being removed by flushing with ethanol alcohol. The AFM images also showed that the adsorption increased with the increased collector concentration and contact time. Xanthate adsorbed on bornite in a similar manner when the solution pH changed to pH 10. The AFM force measurement results showed that the probe–substrate adhesion increased due to the adsorption of xanthate on bornite. The sharp "jump-in" and "jump-off" points on force curve suggest that the adsorbate is not "soft" in nature, ruling out the existence of dixanthogen, an oily substance. Finally, the ATR-FTIR (attenuated total reflection-Fourier-transform infrared) result confirms that the adsorbate on bornite in xanthate solutions is mainly in the form of insoluble cuprous xanthate (CuX) instead of dixanthogen. This xanthate/bornite adsorption mechanism is very similar to what is obtained with the xanthate/chalcocite system, while it is different from the xanthate/chalcopyrite system, for which oily dixanthogen is the main adsorption product on the chalcopyrite surface. The present study helps clarify the flotation mechanism of bornite in industry practice using xanthate as a collector.

Keywords: flotation; xanthate; adsorption; bornite; cuprous xanthate; AFM; FTIR

Citation: Zhang, J. An Investigation of the Adsorption of Xanthate on Bornite in Aqueous Solutions Using an Atomic Force Microscope. *Minerals* **2021**, *11*, 906. https://doi.org/10.3390/min11080906

Academic Editors: Shuai Wang, Xingjie Wang, Jia Yang and Lev Filippov

Received: 22 July 2021
Accepted: 18 August 2021
Published: 21 August 2021

Publisher's Note: MDPI stays neutral with regard to jurisdictional claims in published maps and institutional affiliations.

1. Introduction

Flotation has been widely studied as the most efficient separation technique in the copper extraction industry. The adsorption of the collector on the mineral surface is vital for a successful flotation process to achieve a recovery. Historically, many works have been carried out to clarify the adsorption mechanism of collectors on sulfides [1–6].

Compared to other copper minerals, such as chalcopyrite and chalcocite, the adsorption of collector on bornite has been rarely studied. Allison et al. [3] studied the reaction products of various sulfide minerals with xanthate solutions. The authors reported that the measured rest potential of bornite in 6.25×10^{-4} M KEX solution at pH 7 was +60 mV, and the reaction product of PAX on bornite was cuprous alkyl xanthate. Mielczarski and Suoninen [7,8] applied XPS and studied the adsorption of potassium ethyl xanthate on cuprous sulfide. The authors reported that there was a relatively rapid formation of a well-oriented monolayer of xanthate ions followed by a slow growth of disordered cuprous xanthate molecules on top of this layer. Buckley et al. [9] investigated the surface oxidation of bornite by linear potential sweep voltammetry and X-ray Photoelectron Spectroscopy, and proposed the adsorption products on bornite depended on the solution potential. Zachwieja et al. [10] studied the electrochemical flotation of the bornite-ethylxanthate system and reported that KEX reacted with bornite through an electrochemical oxidation reaction, forming cuprous xanthate between -0.4 v and -0.2 v (SCE, saturated calomel electrode). Hangone et al. [11] studied the flotation of a bornite-rich copper sulfide ore using thio collectors and their mixtures, and reported that the highest copper recoveries were obtained with the diethyl dithiophosphate (di C_2-DTP). Recently, Dhar et al. [12] investigated the improvement of the copper recovery from Nussir Copper Ore Deposit in Northern Norway, using the blend of xanthate and dithiophosphate as collectors.

These previous studies have revealed a significant amount of information, such as the possible reaction product and reaction mechanism of the adsorption of collectors on the bornite surface. Technically, it is also of great interest to directly obtain the image of the collector on the bornite surface changing with the pulp chemistry, such as the type and dosage of chemicals, solution pH and adsorption time. An autoradiography technique was first applied by Polkin et al. [13] and Plaksin et al. [14] to obtain the images of xanthate radioactive isotopes absorbed on sulfide minerals. The authors reported that there was a mosaic distribution of xanthate collectors on the sulfide mineral surface. Later, scanning tunneling microscopy (STM) was applied by Kim et al. [15] and Smart et al. [16] to collect surface images for the study of the reaction, i.e., oxidation, reaction and absorption, of the galena surface under flotation-related conditions. The reported STM images showed that the pulp chemistry, such as the pH and chemical dosage, impacted the reaction and its products on the galena surface. Recently, the AFM imaging technique has also been successfully applied for the in situ study of the adsorption of chemicals on various mineral surfaces [17–22]. The novel analysis method has greatly expanded the understanding of the impact of solution chemistry on the collectors' adsorption on mineral surface and the flotation mechanism.

In the present investigation, an AFM was applied to obtain the surface morphology of bornite in KEX and PAX solutions. By comparing the AFM images obtained under different conditions, such as the collector's type, dosage and contacting time, we studied the impact of water chemistry on the adsorption of collectors on bornite. The results will help answer important questions, such as (1) What is the morphology of the adsorbate on mineral surface? (2) What is the impact of the collector's dosage, the contacting time and the solution pH on the adsorption of the collector on the mineral surface? This information will help to clarify the reaction and adsorption mechanisms of xanthate on bornite changing with aqueous solutions, and therefore its impact on bornite flotation.

2. Materials and Methods

2.1. Materials

Research-grade bornite (Cu_5FeS_4) and chalcopyrite ($CuFeS_2$) were obtained from Wards Natural Science Establishment Inc. Mineral samples were finely polished by consecutively using #800, #1200 and #2400 sandpaper, and then diamond-polishing paste of 10, 5, 2.5 and 1 microns. (MTI Inc., Richmond, CA, USA) Mineral samples were further cleaned by rinsing thoroughly with ethanol and water. A 1.2 cm × 1.2 cm sample was used for the surface characterization, i.e., AFM and ATR-FTIR analysis. The DI (deionized) water used in the present work had a conductivity of 18.2 MΩ·cm^{-1} at 22 °C and a surface tension of 72.8 mN/m at 22 °C. Potassium amyl xanthate (PAX, >98%), potassium ethyl xanthate (KEX, >98%) and NaOH (>99%) were obtained from Alfa Aesar and used without further purification. Xanthate solutions were freshly prepared at various concentrations and pH levels as needed each time right before an experiment was carried out.

2.2. AFM Surface Image and Force Measurements

AFM surface image measurements were carried out with a Digital Instrument Nanoscope IIID (Veeco, San Jose, CA, USA) AFM using the contact mode at room temperature (22 ± 1 °C). SNL cantilevers were obtained from Veeco (San Jose, CA, USA). Triangular Si_3N_4 cantilevers with a nominal spring constant of 0.12~0.58 N/m were used for both AFM imaging and force measurements. For the force measurements, the separation distance (H) between the probe and the substrate (bornite plate) was measured by monitoring the deflection of the cantilever.

To study the mineral surface in water, surface image measurements were carried out after 5 mL DI water was gently injected into an AFM fluid cell. Extreme care was taken to avoid the entrapment of air in the cell. After force data and surface images were collected in water, a 10 mL solution of a specific chemical's concentration was flushed through the liquid cell, and the cell was left undisturbed for the adsorption of chemicals on mineral surface. AFM image analysis and force measurement were commenced after the exposure of the mineral plate to the chemical's solution for a specific time. The AFM images as reported in this study, which were processed by no image modification other than being flattened, include both height and deflection images obtained in the contact mode. The same silicon nitride probe used for the force measurement was also applied to obtain the AFM image of the mineral plate in the solutions at different conditions.

2.3. ATR-FTIR Measurement

A Nicolet 6700 Fourier-transform infrared spectrometer (Thermo Electron North America LLC, West Palm Beach, FL, USA) equipped with the Smart iTR accessory was used to collect the mid-infrared spectra. The system was equipped with a liquid-nitrogen-cooled DTGS KBr detector and a diamond ATR crystal with an angle of incidence of $45°$ to ensure the signals were detected. First, the mineral was put on the stage with the freshly polished surface, which was pressed and fastened toward the ATR crystal to collect the background spectra. Second, the same sample was removed from the stage, and a 5 mL xanthate solution of different concentrations was carefully dipped onto the fresh surface and left untouched for a specific time. Third, the mineral sample was tilted to remove most of the solution and gently blown with ultrapure N_2 gas for the removal of residual water. Finally, the mineral sample was again pressed against the ATR crystal, and the ATR-FTIR spectra were collected. The intensities in the spectra were shown in a relative value under the same scale. All spectra were collected at room temperature with no further treatment made toward the spectra except the baseline correction.

3. Results

3.1. AFM Image of Minerlal Surface in Various Xanthate Solutions

Figure 1 shows the surface images of a bare bornite surface obtained in air. Figure 1A is the 5 μm × 5 μm height image with a data scale of 20 nm, which shows that the solid surface was quite smooth despite some scratch lines on the sample surface due to surface polishing. A smooth bare mineral surface is beneficial for the identification and analysis of the adsorbate when the surface contacts the solutions of various collectors. Figure 1B is the 3D image of Figure 1A. Figure 1C is the section analysis of Figure 1A, which confirms that the polished bornite surface was quite smooth. Figure 1D is the deflection image of Figure 1A with a 10 nm data scale.

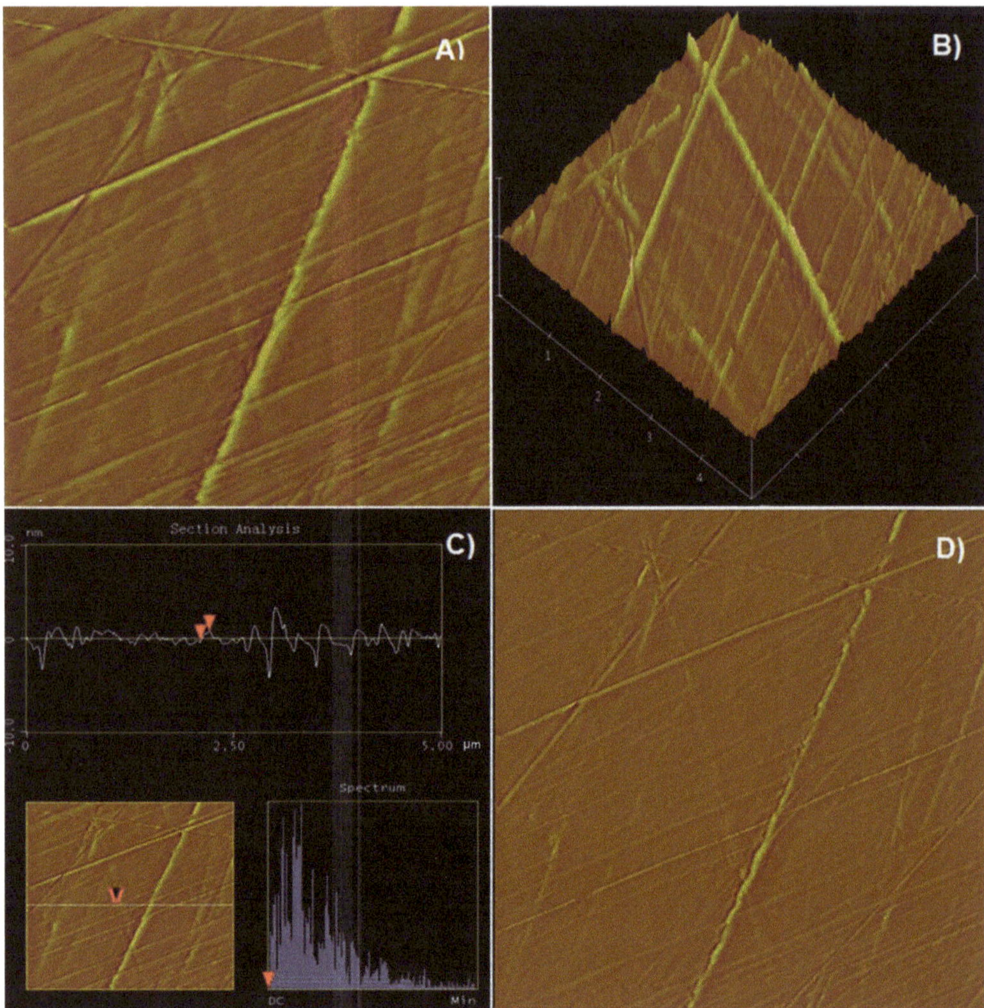

Figure 1. AFM images of a bornite surface in air. (**A**) The 5 μm × 5 μm height image with a data scale of 20 nm, (**B**) the 3D image, (**C**) the section analysis (the red arrows indicate the top and the bottom of an average asperity) and (**D**) the deflection image with a data scale of 10 nm.

Figure 2 shows the surface images of a bare bornite surface which contacted nanopure water in an AFM liquid cell for 10 min. Figure 2A is the 10 μm × 10 μm height image, which shows that the solid surface was still largely smooth, with little adsorbate on the sample surface detected by the AFM probe. Figure 2B is the 3D image of Figure 2A. Figure 2C is the section analysis of Figure 2A. Figure 2D is the deflection image of Figure 2A with a 10 nm data scale.

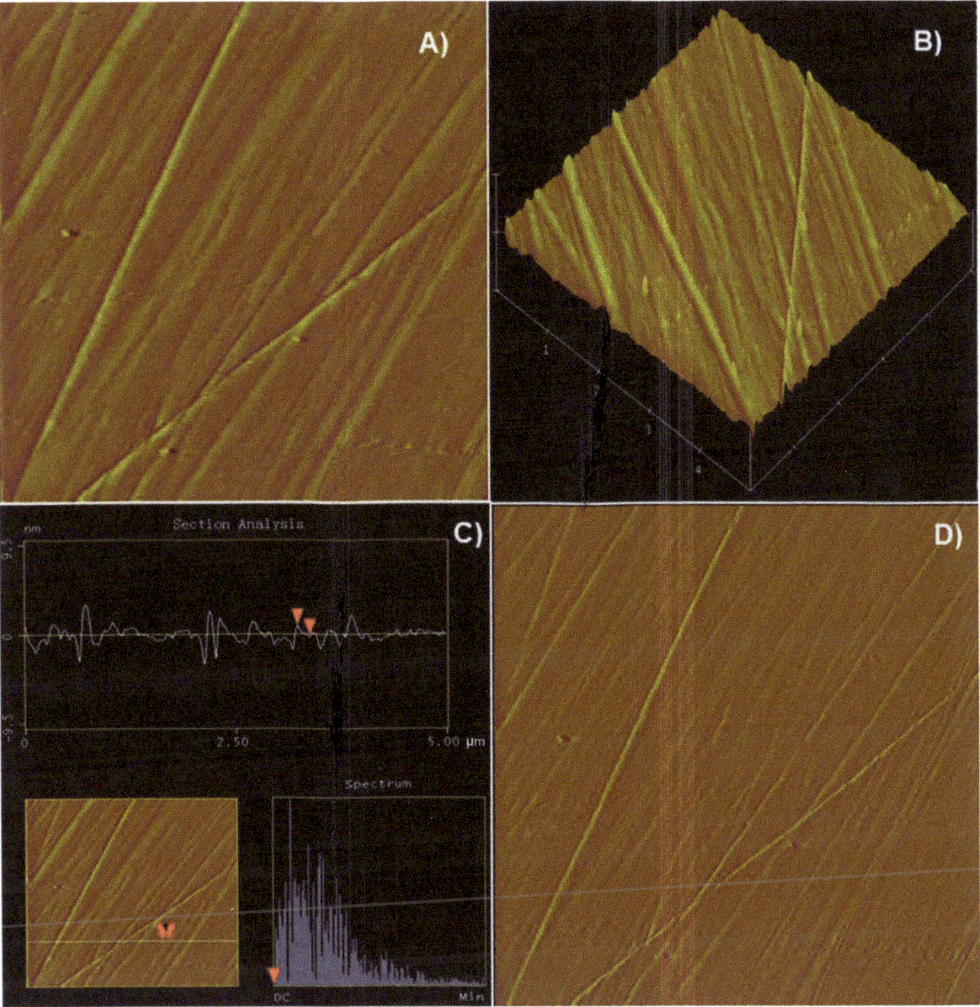

Figure 2. AFM images of a bornite surface soaked in water for 10 min. (**A**) The 5 μm × 5 μm height image with a data scale of 20 nm, (**B**) the 3D image, (**C**) the section analysis (the red arrows indicate the top and the bottom of an average asperity) and (**D**) the deflection image with a data scale of 10 nm.

Figure 3A is the height image of a bornite surface which contacted the 5×10^{-5} M KEX solution at pH 6 for 10 min. Compared to Figure 2A, a significant amount of adsorbate can be observed on the bornite when the mineral surface contacted the xanthate solution for 10 min. Figure 3B is the 3D image of Figure 3A. Figure 3C is the section analysis of Figure 3A, which clearly shows that the surface roughness increased due to the adsorption. Figure 3D is the deflection image of Figure 3A with a 10 nm data scale.

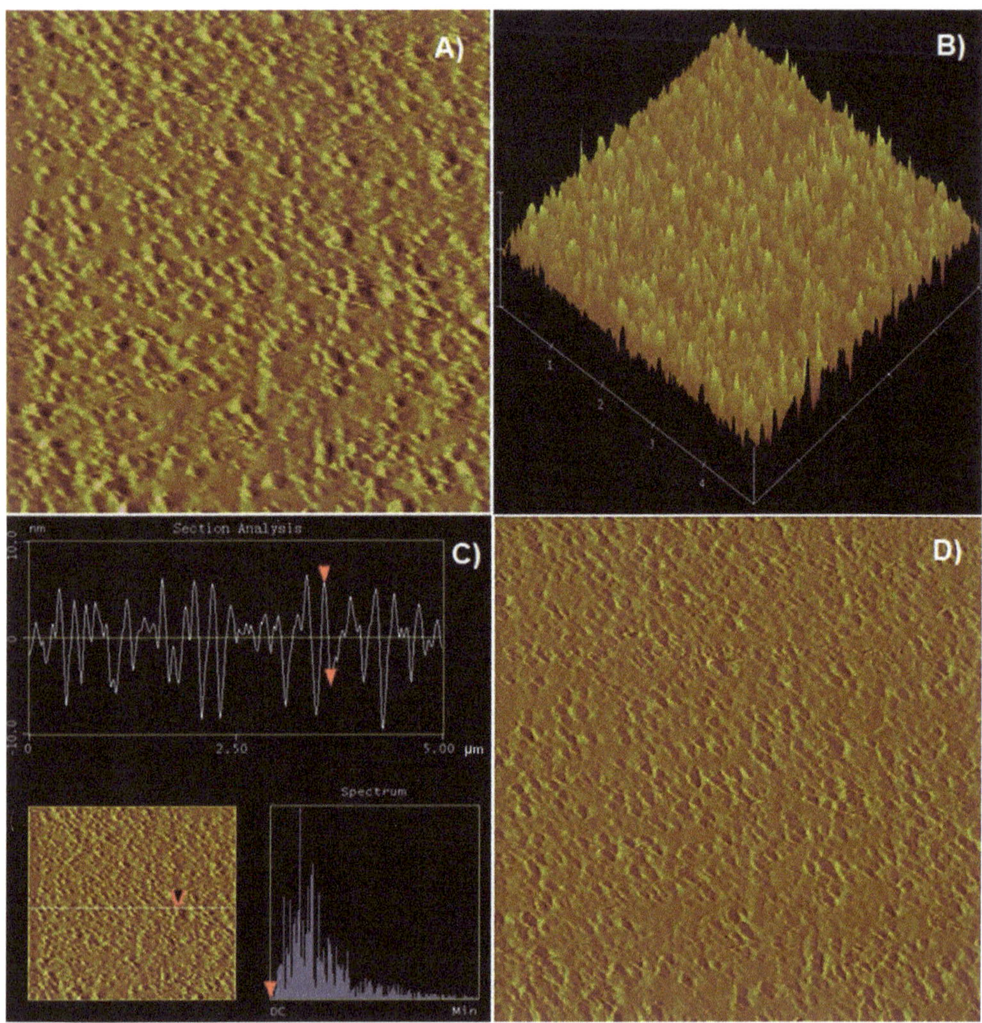

Figure 3. AFM images of a bornite surface soaked in the 5×10^{-5} M KEX solution at pH 6 for 10 min. (**A**) The 5 μm × 5 μm height image with a data scale of 20 nm, (**B**) the 3D image, (**C**) the section analysis (the red arrows indicate the top and the bottom of an average asperity) and (**D**) the deflection image with a data scale of 10 nm.

Figure 4A is the height image of a bornite surface, which was in contact with the 1×10^{-4} M KEX solution at pH 6 for 10 min. Compared to Figure 2A, a significant amount of adsorbate can be observed on the bornite when the mineral surface contacted the 1×10^{-4} M KEX solution for 10 min. In addition, compared to Figure 3A, the mineral surface became rougher, suggesting that more precipitates were formed at the solid/liquid interface when the KEX's concentration increased from 5×10^{-5} M to 1×10^{-4} M. Figure 4B is the 3D image of Figure 4A. Figure 4C is the section analysis of Figure 4A, which confirms that the surface roughness increased due to the adsorption of the collector. Figure 4D is the deflection image of Figure 4A with a 10 nm data scale.

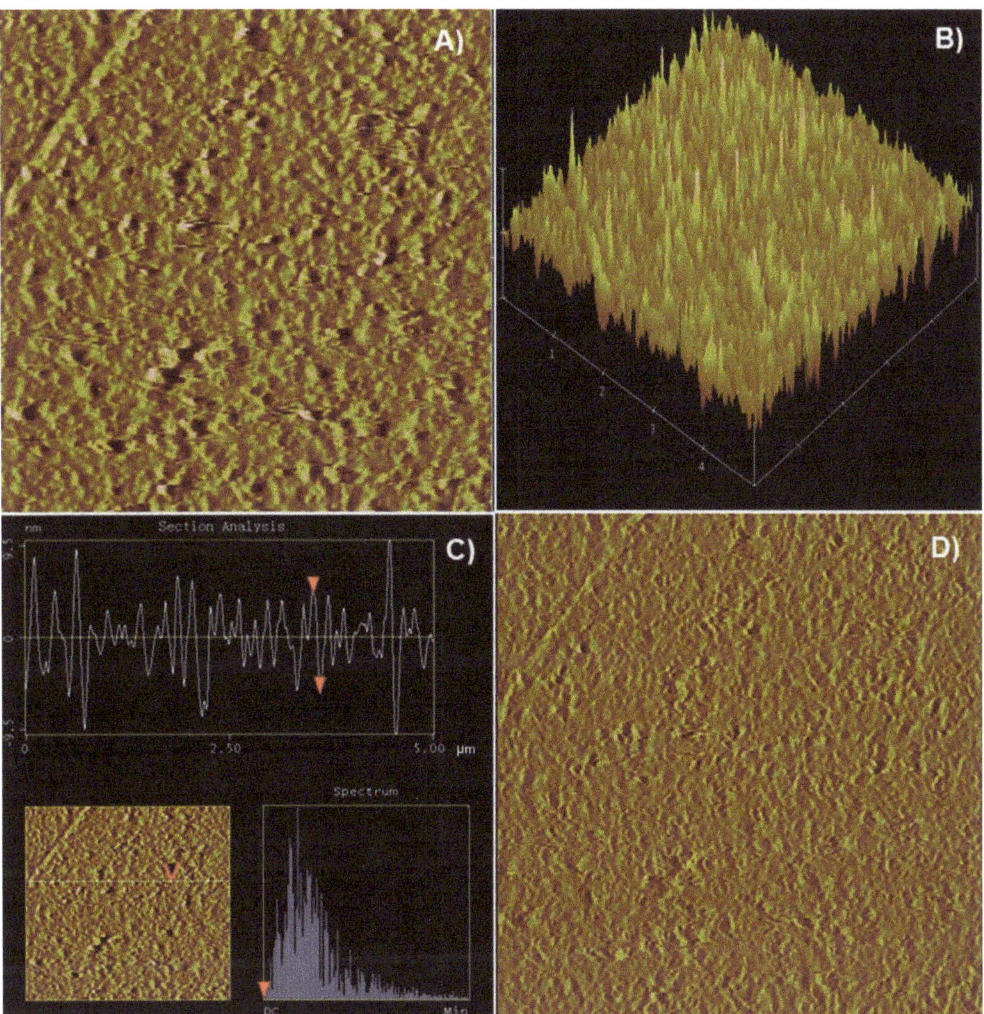

Figure 4. AFM images of a bornite surface soaked in the 1×10^{-4} M KEX solution at pH 6 for 10 min. (**A**) The 5 μm × 5 μm height image with a data scale of 20 nm, (**B**) the 3D image, (**C**) the section analysis (the red arrows indicate the top and the bottom of an average asperity) and (**D**) the deflection image with a data scale of 10 nm.

Figure 5A is the height image of a bornite surface soaked in the 1×10^{-4} M KEX solution at pH 6 for 20 min and further rinsed with 10 mL ethanol and 10 mL water consecutively. The images were finally obtained when the mineral sample contacted the water. Compared to Figure 4A, it can be observed that the precipitates, as observed from Figure 4A, still existed on the bornite surface, and were not dissolved or rinsed off the mineral surface. Figure 5B is the 3D image of Figure 5A. Figure 5C is the section analysis of Figure 5A. Figure 5D is the deflection image of Figure 5A with a 10 nm data scale.

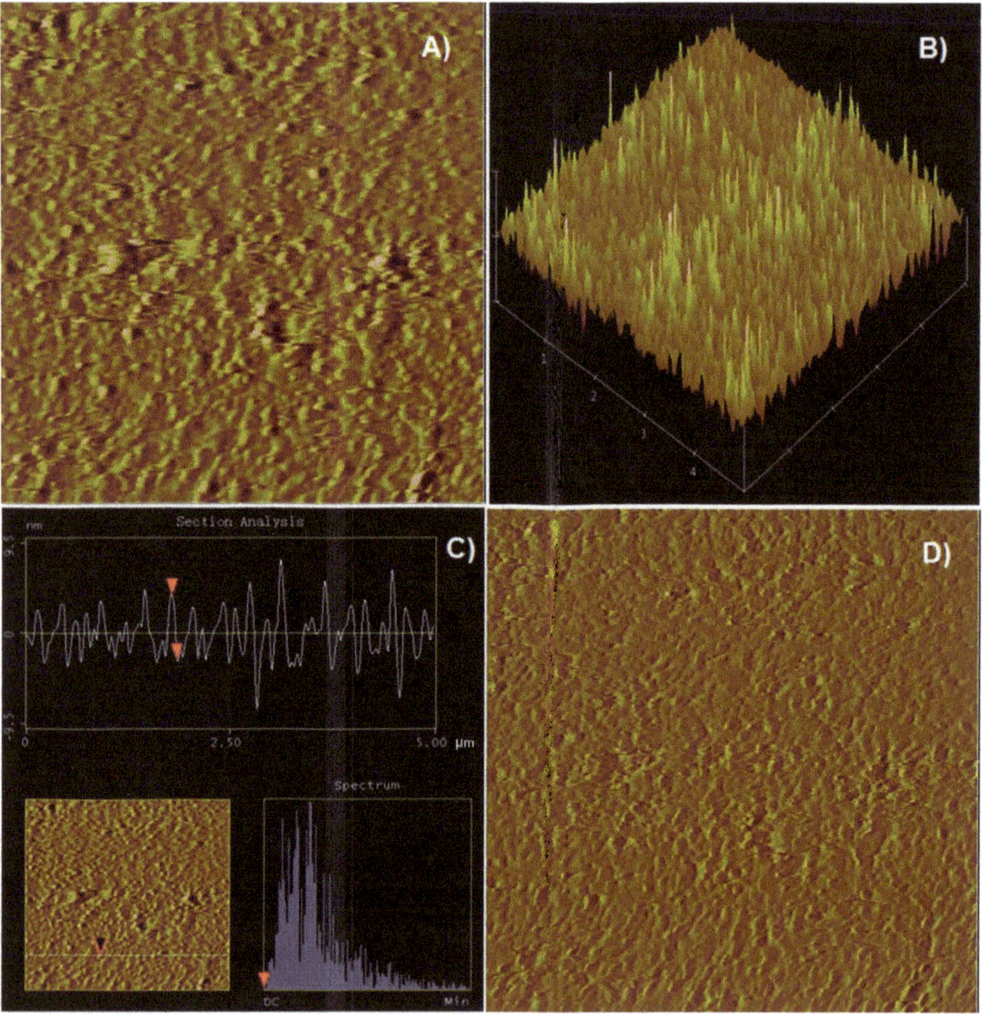

Figure 5. AFM images of a bornite surface soaked in the 1×10^{-4} M KEX solution at pH 6 for 20 min and further rinsed with ethanol and water. (**A**) The 5 μm × 5 μm height image with a data scale of 20 nm, (**B**) the 3D image, (**C**) the section analysis (the red arrows indicate the top and the bottom of an average asperity) and (**D**) the deflection image with a data scale of 10 nm.

Figure 6 shows the AFM images of a bornite surface soaked in the 5×10^{-4} M KEX solution at pH 6 for 10 min. By comparing Figure 6A, the height image, to Figure 2A, a significant amount of bornite can be observed after the mineral surface contacted the xanthate solution. In addition, comparing Figure 6A to Figures 3A and 4A, when the KEX's concentration increased, more precipitates formed at the solid/liquid interface, and the mineral surface became much rougher. Figure 6B is the 3D image of Figure 6A. Figure 6C, the section analysis of Figure 6A, confirms that the surface roughness increased greatly when the bornite surface contacted a high concentration of KEX. Figure 6D is the deflection image of Figure 6A with a 10 nm data scale.

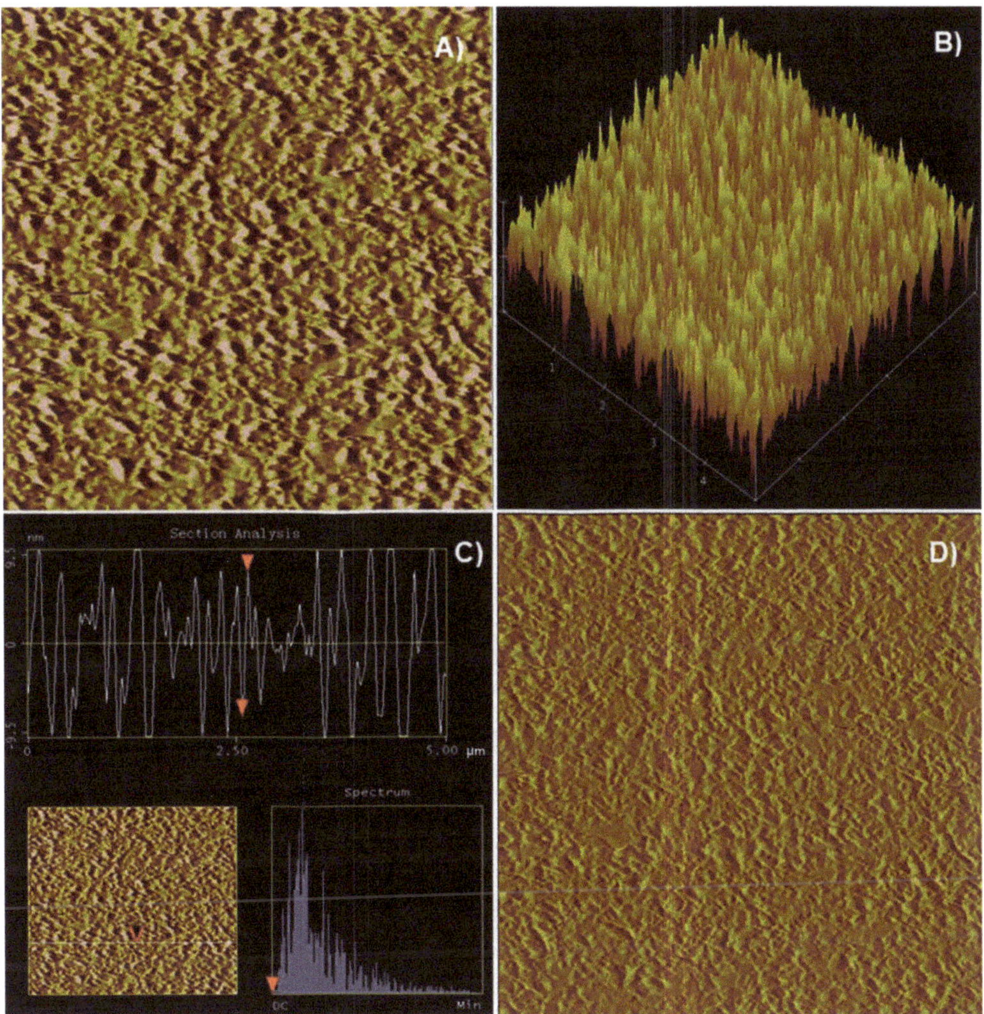

Figure 6. AFM images of a bornite surface soaked in the 5×10^{-4} M KEX solution at pH 6 for 10 min. (**A**) The 5 μm × 5 μm height image with a data scale of 20 nm, (**B**) the 3D image, (**C**) the section analysis (the red arrows indicate the top and the bottom of an average asperity) and (**D**) the deflection image with a data scale of 10 nm.

To verify the adsorption of the KEX on bornite in the solution, after Figure 6 was obtained, a 3 μm × 3 μm area was scanned once, applying a much larger scan force to intentionally remove the adsorbate. Further, the same position was scanned again in a 5 μm × 5 μm area, applying a normal scan force. The result is shown in Figure 7. From Figure 7A, the height image, a 3 μm × 3 μm 'window' can be observed in the center of the image due to the removal of some adsorbate from mineral surface under the previously applied large scan force. That is, the 'window' in the center with a low profile is the bornite surface covered by the adsorbate, which was partially removed by the applied large scan force. The surrounding area with a high profile is the mineral surface covered by the adsorbate, which was not disturbed by the large scan force. By comparing Figure 7B, the 3D image of Figure 7A, to Figure 6B, a pit on the mineral surface can be easily observed, with adsorbate covering the surrounding area. Figure 7C, the section analysis of Figure 7A, shows the height difference between the 'window' (as shown by the green markers) and the surrounding area being covered with adsorbate (as shown by the red markers). Figure 7D is the deflection image of Figure 7A with a 10 nm data scale.

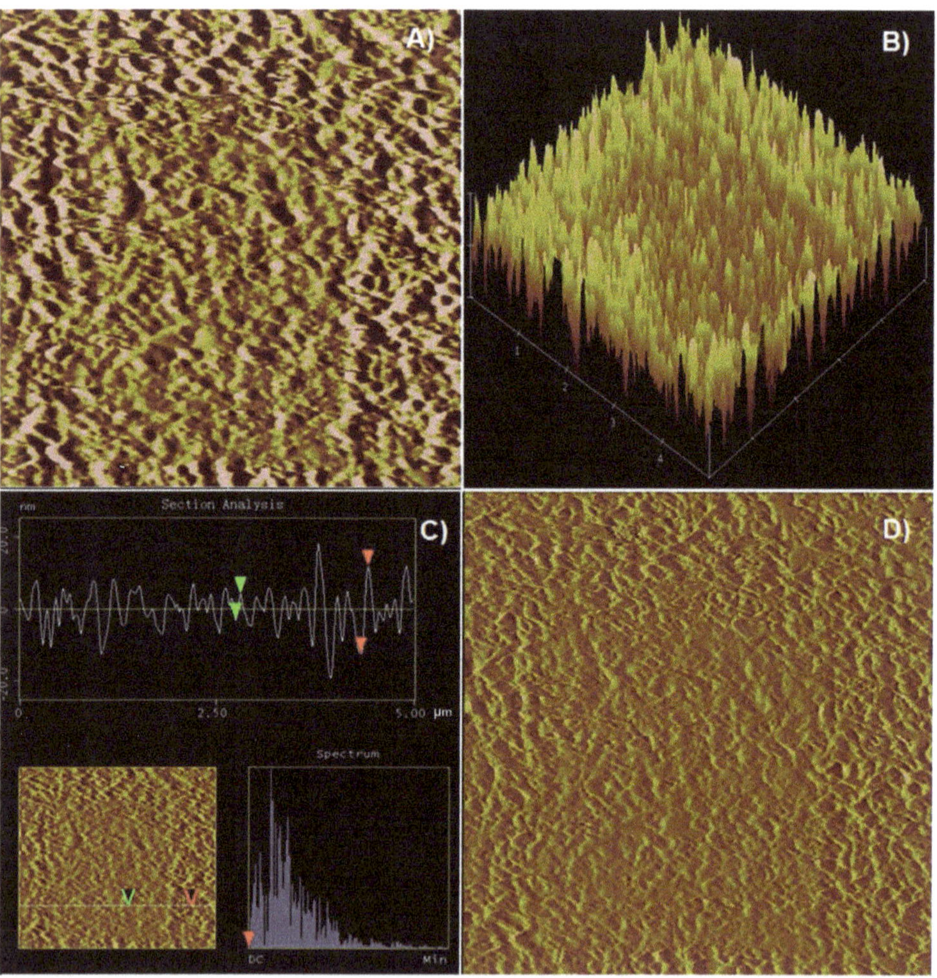

Figure 7. AFM images of a bornite surface soaked in the 5×10^{-4} M KEX solution at pH 6 for 20 min. (**A**) The 5 μm × 5 μm height image with a data scale of 20 nm, (**B**) the 3D image, (**C**) the section analysis and (**D**) the deflection image with a data scale of 10 nm.

Figure 8 shows the AFM images of a bornite surface soaked in the 1×10^{-5} M PAX solution at pH 6 for 10 min. By comparing Figure 8A, the height image, to Figure 2A, some precipitates can be observed on the bornite when the mineral surface contacted the PAX solution. Figure 8B is the 3D image of Figure 8A. Figure 8C, the section analysis of Figure 8A, shows that the surface roughness increased due to the adsorption of PAX. Figure 8D is the deflection image of Figure 8A with a 10 nm data scale.

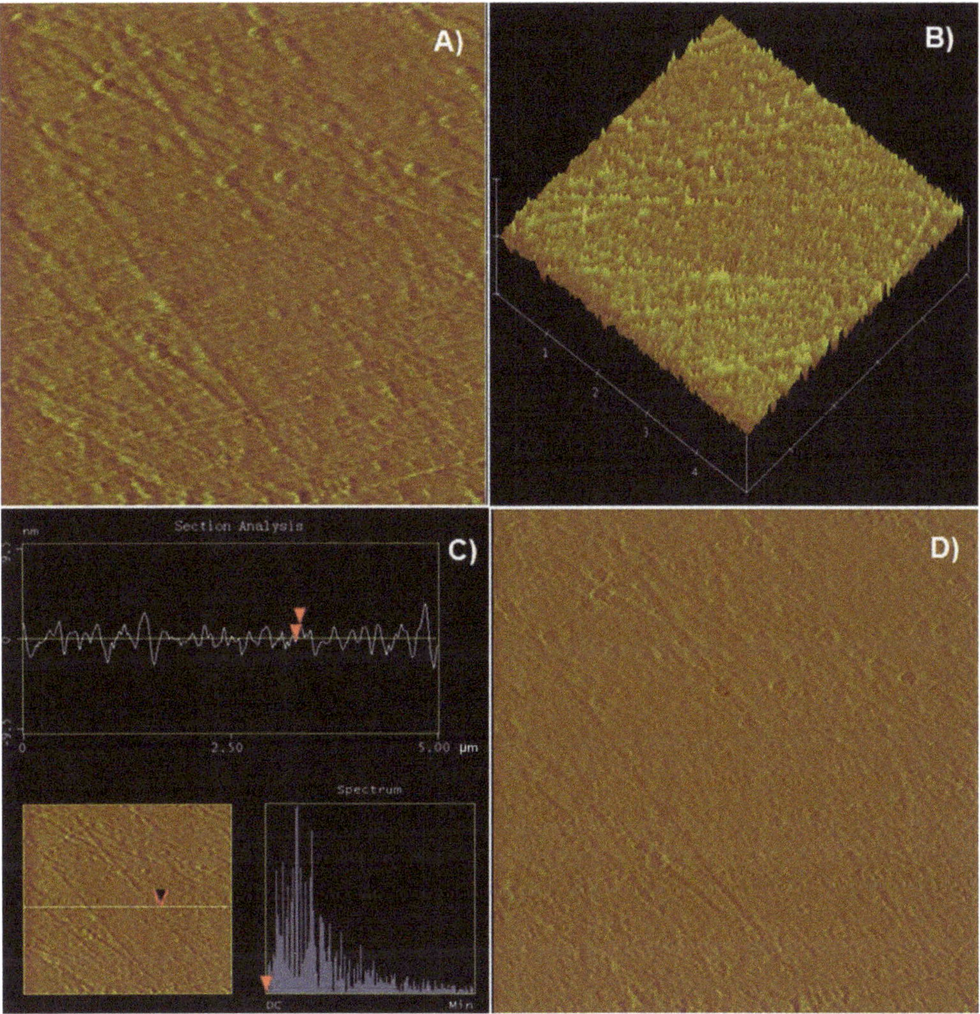

Figure 8. AFM images of a bornite surface soaked in the 1×10^{-5} M PAX solution at pH 6 for 10 min. (**A**) The 5 μm × 5 μm height image with a data scale of 20 nm, (**B**) the 3D image, (**C**) the section analysis (the red arrows indicate the top and the bottom of an average asperity) and (**D**) the deflection image with a data scale of 10 nm.

Figure 9 shows the AFM images of a bornite surface soaked in 5×10^{-5} M PAX solution at pH 6 for 10 min. By comparing Figure 9A, the height image, to Figure 2A, a significant amount of adsorbate can be observed on the bornite when the mineral surface contacted the 5×10^{-5} M PAX solution for 10 min. By comparing Figure 9A to Figure 8A, it can be seen that, when the PAX concentration increased from 1×10^{-5} M to 5×10^{-5} M, the mineral surface became much rougher, with more precipitates forming at the solid/liquid interface. Figure 9B is the 3D image of Figure 9A. Figure 9C is the section analysis of Figure 9A, which confirms that the surface roughness increased with the increasing PAX concentration. Figure 9D is the deflection image of Figure 9A with a 10 nm data scale.

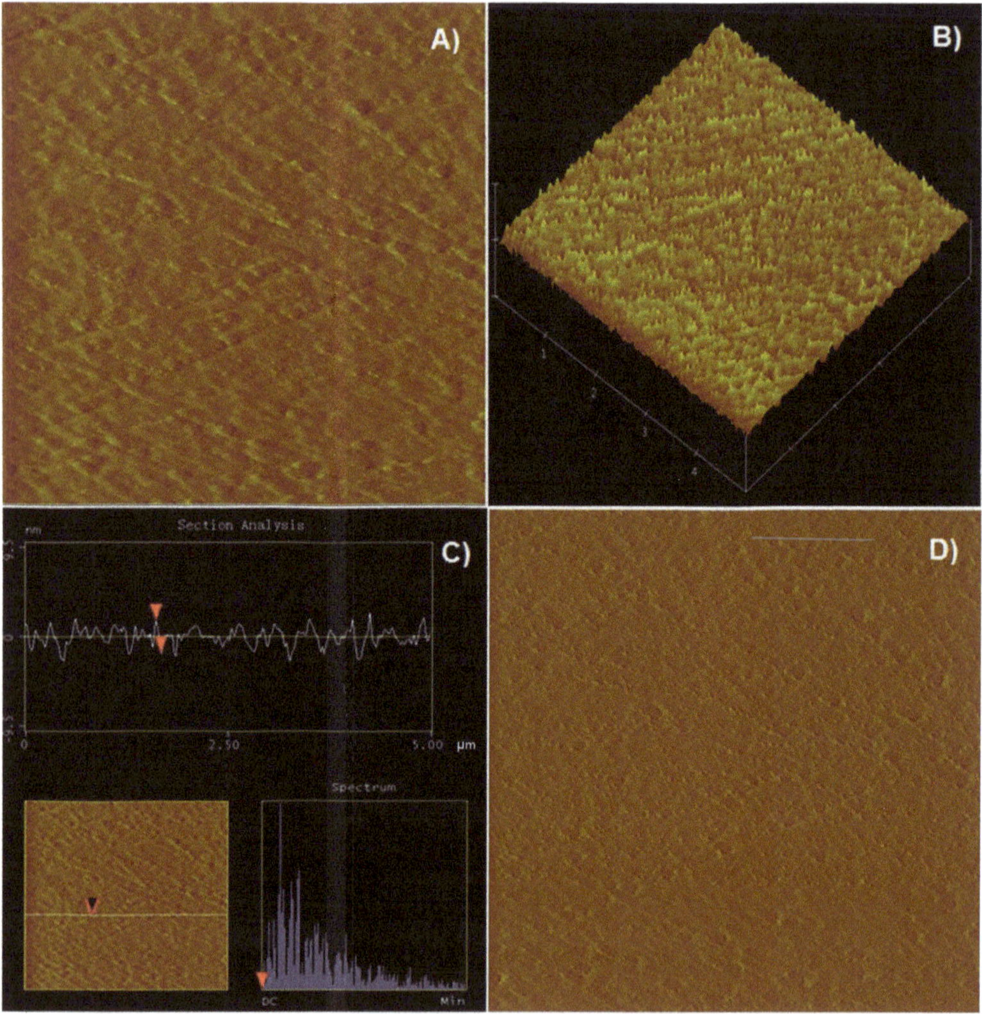

Figure 9. AFM images of a bornite surface soaked in the 5×10^{-5} M PAX solution at pH 6 for 10 min. (**A**) The 5 μm × 5 μm height image with a data scale of 20 nm, (**B**) the 3D image, (**C**) the section analysis (the red arrows indicate the top and the bottom of an average asperity) and (**D**) the deflection image with a data scale of 10 nm.

Figure 10 shows the AFM images of a bornite surface soaked in the 1×10^{-4} M PAX solution at pH 6 for 10 min. By comparing Figure 10A, the height image, to Figure 2A, a significant amount of adsorbate can be observed on the bornite when the mineral surface contacted the 1×10^{-4} M PAX solution for 10 min. By comparing Figure 10A to Figures 8A and 9A, when the PAX concentration increases, the mineral surface became much rougher, with more precipitates forming at the solid/liquid interface. Figure 10B is the 3D image of Figure 10A. Figure 10C, the section analysis of Figure 10A, shows that the surface roughness increased with the increasing PAX concentration. Figure 10D is the deflection image of Figure 10A with a 10 nm data scale.

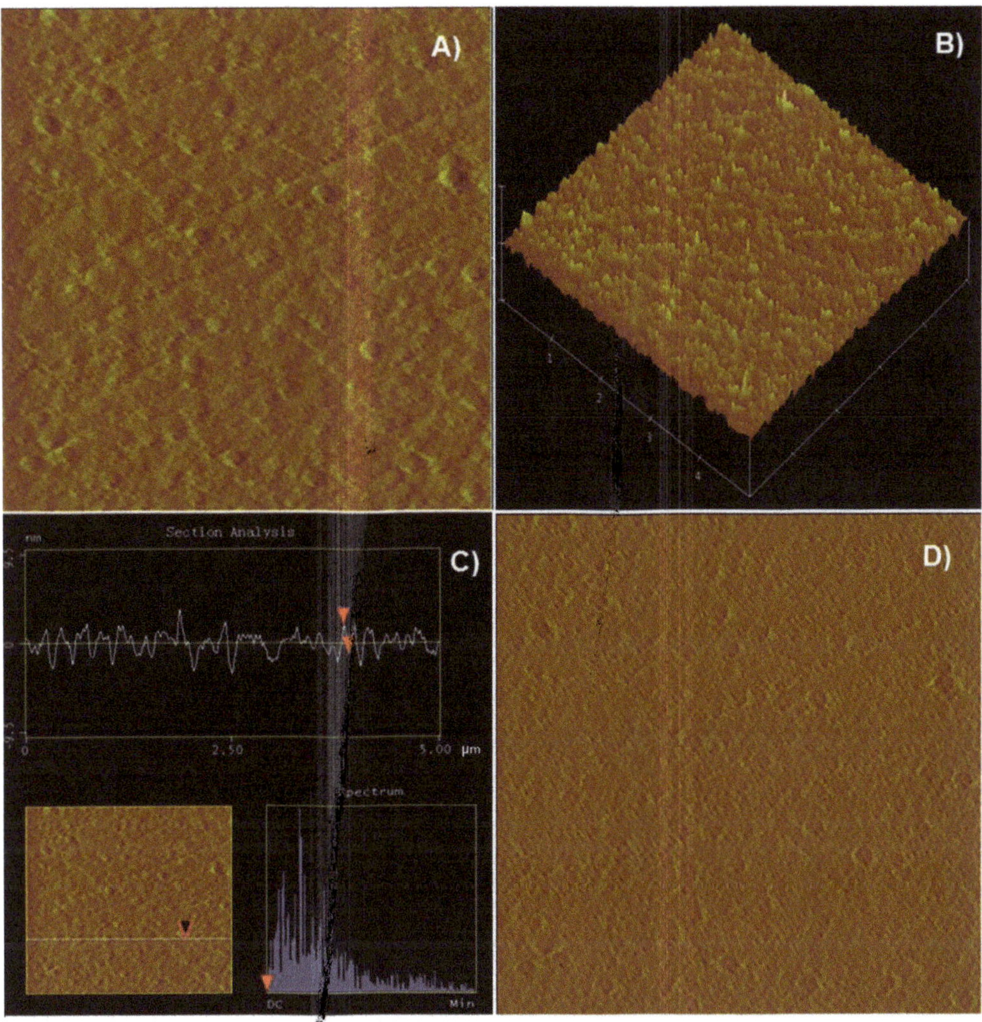

Figure 10. AFM images of a bornite surface soaked in the 1×10^{-4} M PAX solution at pH 6 for 10 min. (**A**) The 5 μm × 5 μm height image with a data scale of 20 nm, (**B**) the 3D image, (**C**) the section analysis (the red arrows indicate the top and the bottom of an average asperity) and (**D**) the deflection image with a data scale of 10 nm.

To verify the adsorption of the PAX on bornite in the solution, after Figure 10 was obtained, a 2 μm × 2 μm area was scanned once, applying a much larger scan force to intentionally remove the adsorbate. Further, the same position was scanned again in a 5 μm × 5 μm area, applying a normal scan force. The result is shown as Figure 11. As shown in Figure 11A, the height image, a 2 μm × 2 μm 'window' is shown in the center of the image due to the partial removal of the adsorbate from the mineral surface under the previously applied large scan force. That is, the 'window' in the center with a low profile is the bornite surface covered by the adsorbate, which was partially removed by the applied large scan force. The surrounding area with a high profile is the mineral surface covered by the adsorbate, which was not disturbed by the large scan force. This finding is as that shown in Figure 7. By comparing Figure 11B, the 3D image of Figure 11A, to Figure 10B, a pit on mineral surface can be easily observed, with adsorbate covering the surrounding area. Figure 11C is the section analysis of Figure 11A, which shows the height difference between the 'window' and the surrounding area covered with adsorbate (indicated by the red markers). Figure 11D is the deflection image of Figure 11A with a 10 nm data scale. The adsorbate strongly combined with the mineral surface strongly, and a quite large scan force had to be applied during the experiment. Therefore, the obtained 'window' was slightly deformed, as shown in Figure 11A.

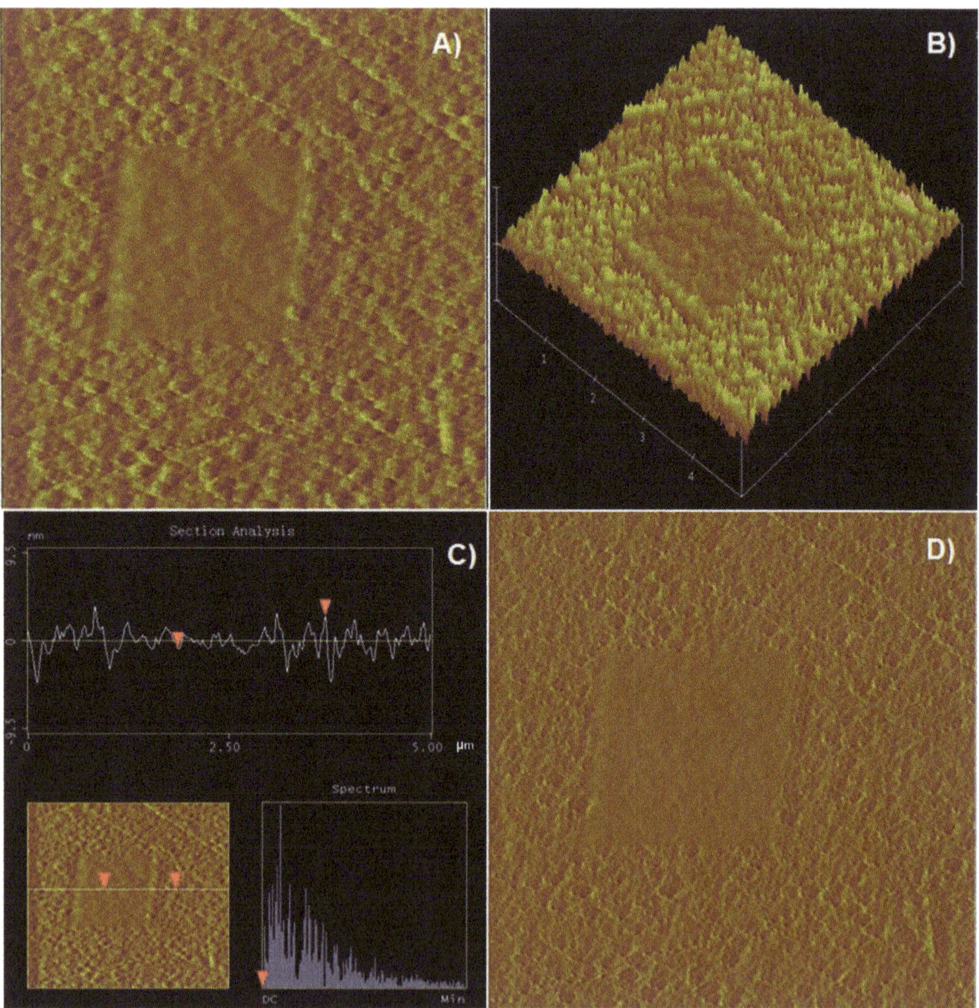

Figure 11. AFM images of a bornite surface soaked in 1×10^{-4} M PAX solution at pH 6 for 20 min. (**A**) The 5 µm × 5 µm height image with a data scale of 20 nm, (**B**) the 3D image, (**C**) the section analysis and (**D**) the deflection image with a data scale of 10 nm. The 2 µm × 2 µm blank 'window' in the center of the image occurred due to the removal of the adsorbate from the mineral surface under the intentionally applied large scan force.

Figure 12 shows the AFM images of a bornite surface soaked in the 5×10^{-4} M KEX solution at pH 10. Figure 12A–C are the height images with a 20 nm data scale obtained after the mineral surface contacted the KEX solution, respectively, for 5, 10 and 20 min. Similar to Figure 6, in all the images, a significant amount of adsorbate can be observed on the bornite. Figure 12D is the 1 μm × 1 μm (large magnification) height image collected right after Figure 12C was obtained.

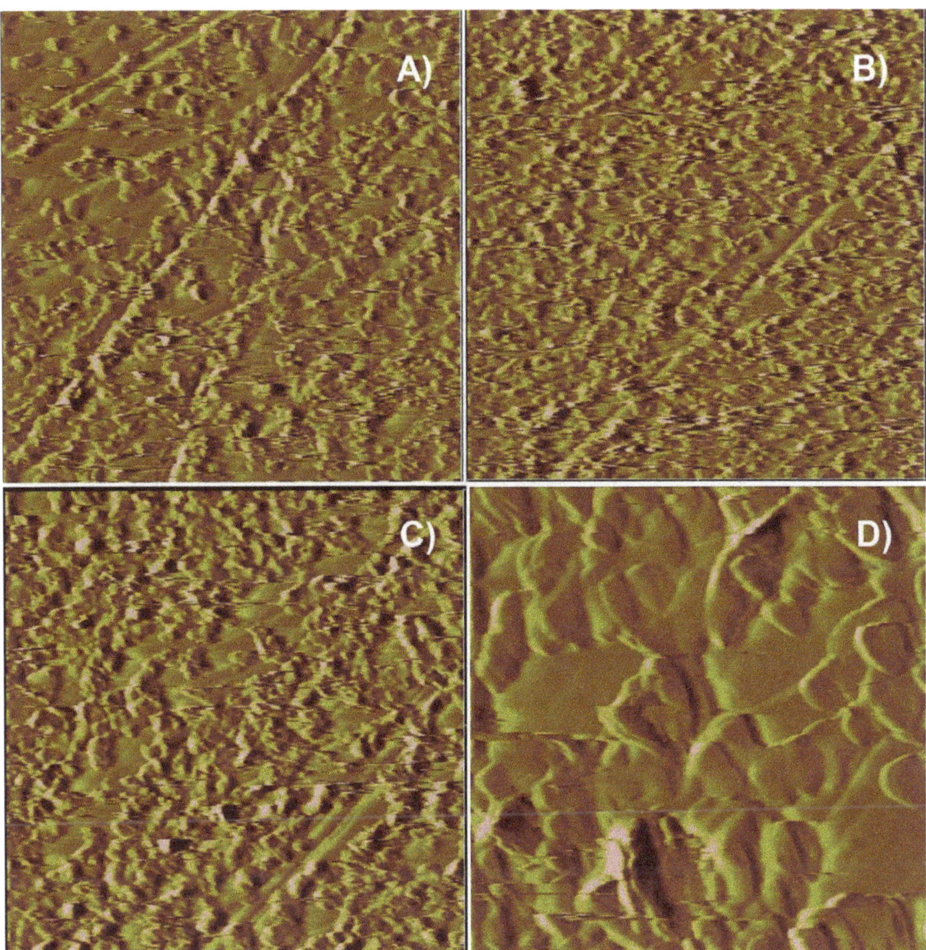

Figure 12. AFM height images (5 μm × 5 μm) with a data scale of 20 nm of a bornite surface soaked in the 5×10^{-4} M KEX solution at pH 10 (**A**) for 5 min, (**B**) for 10 min, (**C**) for 20 min and (**D**) for 20 min (1 μm × 1 μm).

Figure 13 shows the AFM images of a bornite surface soaked in the 1×10^{-4} M PAX solution at pH 10. Figure 13A–C are the height images with a 20 nm data scale obtained after the mineral surface contacted the PAX solution, respectively, for 5, 10 and 20 min. Similar to Figure 10, in all the images, a significant amount of adsorbate can be observed on the bornite. Figure 13D is the 1 μm × 1 μm (large magnification) height image collected right after Figure 13C was obtained.

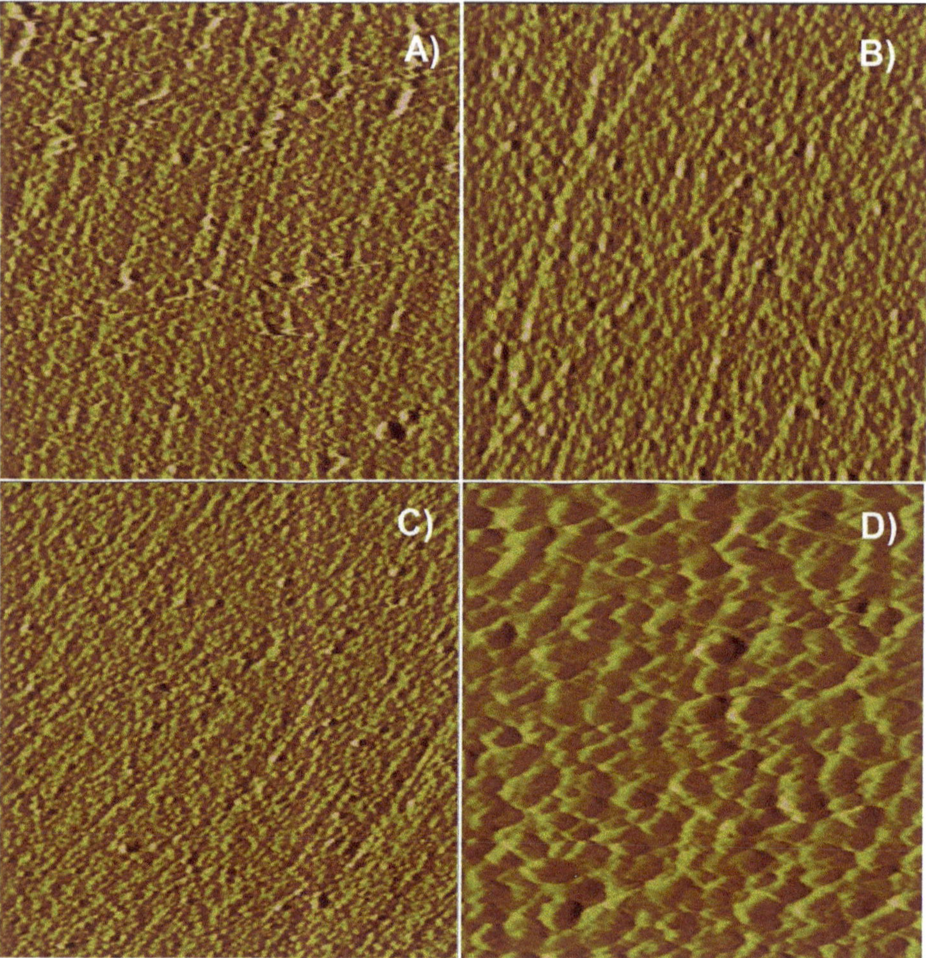

Figure 13. AFM height images (5 μm × 5 μm) with a data scale of 20 nm of a bornite surface soaked in the 1×10^{-4} M PAX solution at pH 10 (**A**) for 5 min, (**B**) for 10 min, (**C**) for 20 min and (**D**) for 20 min (1 μm × 1 μm).

To clearly show the difference of the adsorbate in the morphologies of bornite and chalcopyrite, the adsorption of xanthate on chalcopyrite was also studied, and the AFM image is shown in Figure 14. Figure 14A,B were obtained with 5×10^{-4} M KEX. Figure 14C,D were obtained with 5×10^{-5} M PAX. The AFM images clearly show that there was patch-like adsorbate on the chalcopyrite surface, which was flat with smooth and round edges. This morphology fits well with the fact that oily dialkyl dixanthogen is generally insoluble in water, and the circular boundary is the direct result of the high interfacial tension between the hydrophobic dixanthogen and water [17,18]. In addition, this adsorbate had a completely different morphology as the one shown in Figures 3–13.

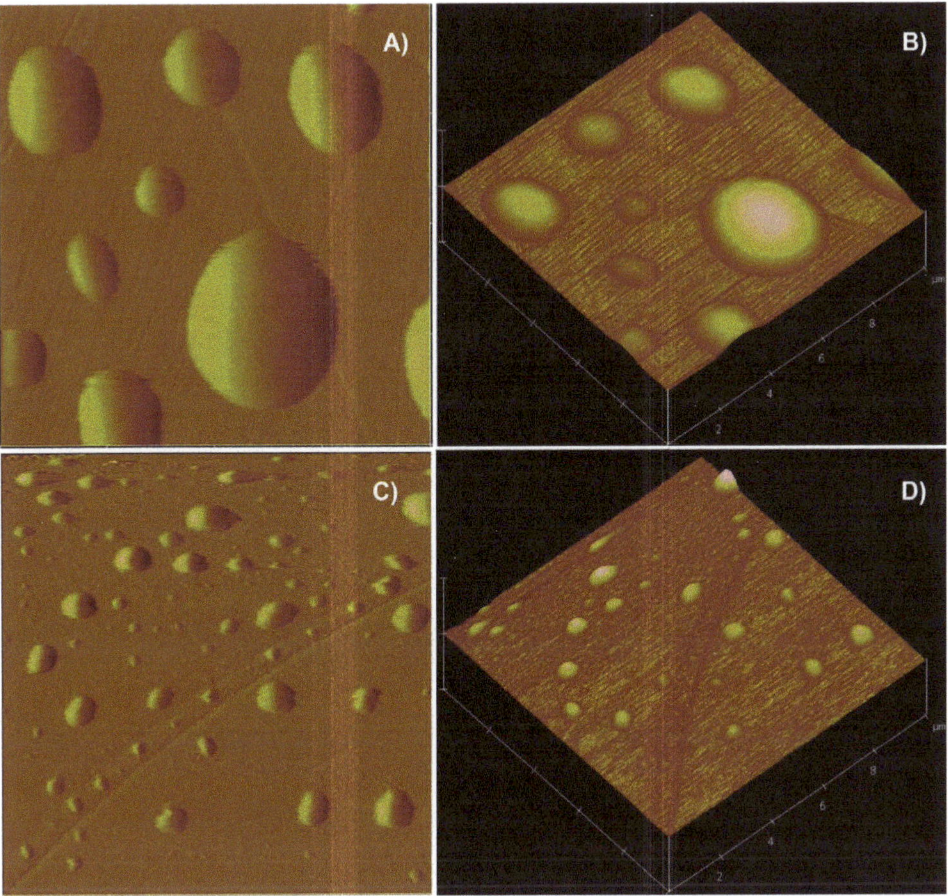

Figure 14. AFM images of a chalcopyrite surface in the xanthate solution at pH 6 for 30 min. (**A**) The 10 μm × 10 μm deflection image with a data scale of 200 nm at 5×10^{-4} M KEX, (**B**) the 3D image of (**A**), (**C**) the 10 μm × 10 μm deflection image with a data scale of 200 nm at 5×10^{-5} M PAX and (**D**) the 3D image of (**C**).

3.2. AFM Surface Force Measurement

The interaction force (F) between an AFM probe and a polished bornite plate is measured when the plate contacts various PAX solutions at pH 6 for 10 min. Using an AFM force measurement, one can obtain both the approach force curve and the retract force curve, which are shown as Figures 15 and 16, respectively.

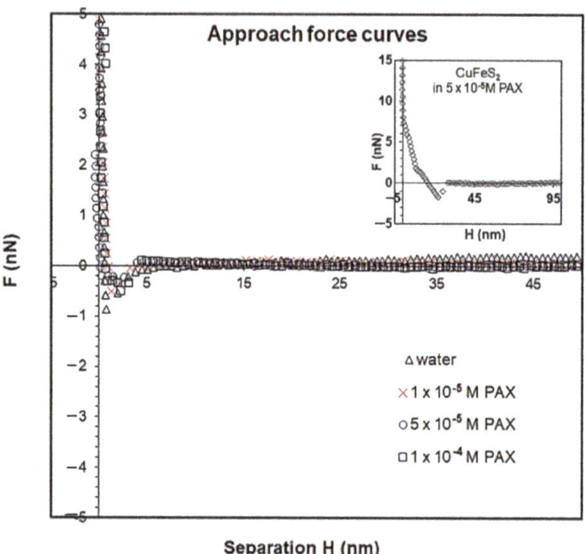

Figure 15. The approach force (F) measured between an AFM probe and a bornite plate soaked in solutions at pH 6 as a function of the separation (H) between the probe and the plate by applying an AFM force measurement. (Δ) in water; ($\times$) in the 1×10^{-5} M PAX solution, (o) in the 5×10^{-5} M PAX solution and ($\square$) in the 1×10^{-4} M PAX solution. The inlet ($\Diamond$) shows the approach force curve obtained with $CuFeS_2$ in the 5×10^{-5} M PAX solution.

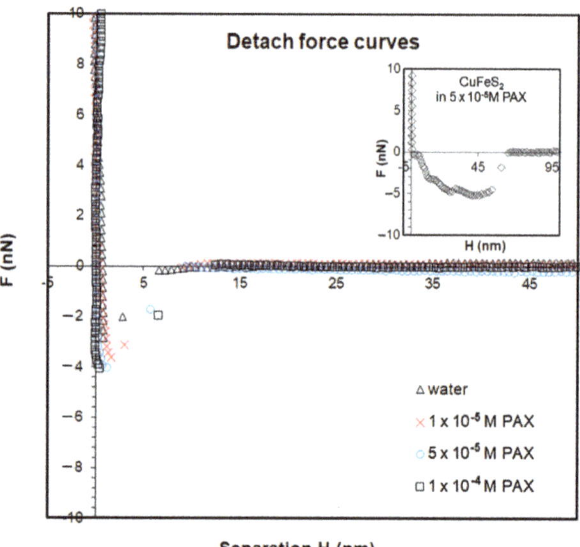

Figure 16. The detach force (F) measured between an AFM probe and a bornite plate soaked in solutions at pH 6 as a function of the separation (H) between the probe and the plate by applying an AFM force measurement. (Δ) in water, ($\times$) in the 1×10^{-5} M PAX solution, (o) in the 5×10^{-5} M PAX solution and ($\square$) in the 1×10^{-4} M PAX solution. The inlet ($\Diamond$) shows the detach force curve obtained with $CuFeS_2$ in the 5×10^{-5} M PAX solution.

Figure 15 shows that the approach force curves measured in water and various PAX solutions were similar to each other. The "jump-in" occurred where the separation was less than 5 nm, which was within the range of the van der Waals force. Figure 16 shows that the detach force measured between an AFM probe and bornite surface in water was about 3 nN. The value increased slightly to 4 nN when in the 1×10^{-5} M PAX solution. Further increasing the concentration of the PAX solution did not significantly change the detach force. The fact that the "jump-off" point was sharp and that the "jump-off" point occurred where the separation was close to 0 nm suggest that the adsorbate is physically rigid in nature. The inlet shows the detach force curve obtained with $CuFeS_2$ in 5×10^{-5} M PAX solution, and the detach force was about 5.2 nN. The fact that the "jump-off" point was not sharp and that the "jump-off" point occurred at above 50 nm confirms that the oily dixanthogen is deformable.

3.3. AFT-FTIR Results

Figure 17 shows the ATR-FTIR spectra of the adsorbate on bornite after the mineral surface 5×10^{-4} M KEX solution at pH 6 for 1 h. On the spectra, the main peaks shown at 1195 cm^{-1} and 1126 cm^{-1} were due to the bonds of C-O-C, and the peaks at 1049 cm^{-1} and 1032 cm^{-1} were due to the bonds of S-C-S. The results are in line with the FTIR spectra of cuprous xanthate as reported by Poling [23] and Leppinen et al. [24]. That is, the obtained FTIR spectra, as shown in Figure 17, confirms that the adsorbate on bornite in xanthate solutions is essentially CuX, with no dixanthogen detected.

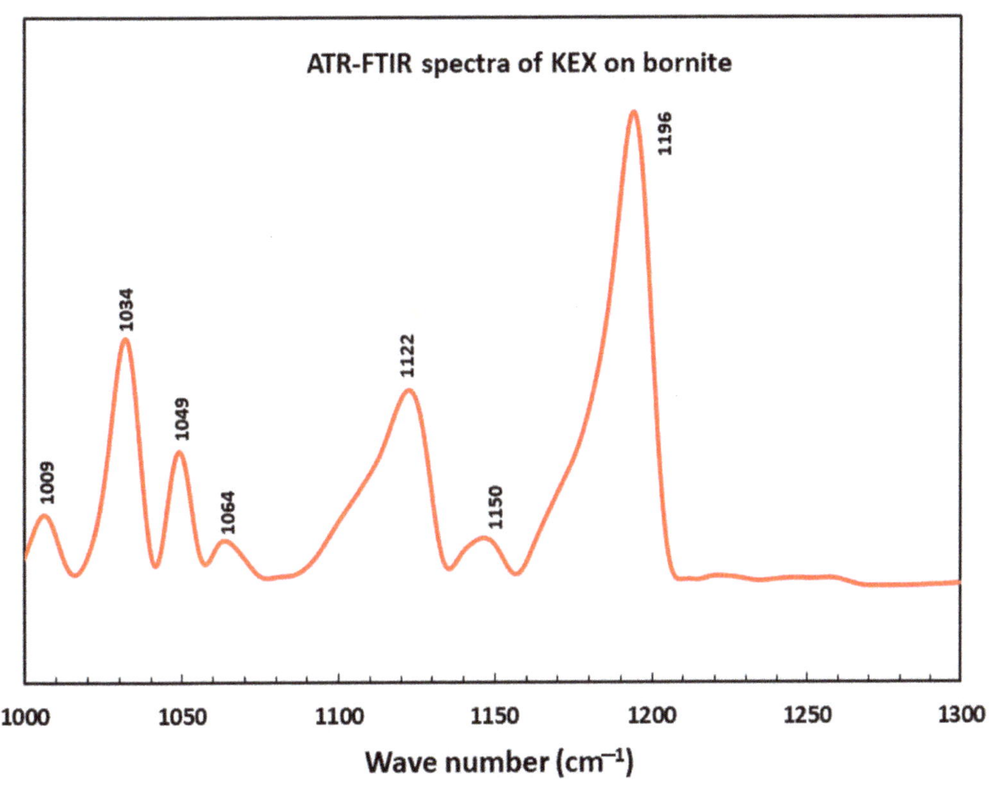

Figure 17. ATR-FTIR spectra of a bornite surface after it contacted the 5×10^{-4} M KEX solution at pH 6 for 1 h.

4. Discussion

4.1. Adsorption of Xanthate on Bornite Surface

As shown in Figures 3–13, a significant amount of adsorbate can be observed on the bornite surface when it contacted xanthate solutions for a specific time. The roughness analysis of the AFM images of a bornite surface in xanthate solutions is summarized and listed as Table 1. In general, during the same timeframe, i.e., 10 min, surface roughness increased when the concentration of xanthate increased. For example, at pH 6, when the concentration of KEX increased from 5×10^{-5} M to 1×10^{-4} M, the roughness Ra value increased from 1.919 nm to 2.372 nm, and the value increased to 2.412 nm when the concentration was further increased to 5×10^{-4} M. A similar trend was also observed for the case of PAX, although the change in values was not as significant as the one as obtained with KEX. The change of the morphology of the bornite surface cannot be attributed to the reaction of the mineral surface with water, because the AFM images obtained with the addition of xanthate solutions are completely different from those shown in Figure 2, which were captured within same timeframe. Therefore, the adsorbate shown in Figures 3–13 must be due to the adsorption of xanthate at the bornite/liquid interface.

Table 1. Roughness analysis of the AFM images of a bornite surface in xanthate solutions.

Xanthate	pH	Concentration (M)	Image Source	Ra(Sa) (nm) *	Rms(Sq) (nm) **
KEX	6	5×10^{-5} M	Figure 3	1.919	2.596
KEX	6	1×10^{-4} M	Figure 4	2.372	3.072
KEX	6	1×10^{-4} M	Figure 5	2.412	3.225
KEX	6	5×10^{-4} M	Figure 6	4.529	5.866
PAX	6	1×10^{-5} M	Figure 8	0.665	0.861
PAX	6	5×10^{-5} M	Figure 9	0.684	0.897
PAX	6	1×10^{-4} M	Figure 10	0.777	1.016
KEX	10	5×10^{-4} M	Figure 12A	2.664	3.907
KEX	10	5×10^{-4} M	Figure 12B	3.166	4.441
KEX	10	5×10^{-4} M	Figure 12C	3.464	4.734
PAX	10	1×10^{-4} M	Figure 13A	2.737	3.570
PAX	10	1×10^{-4} M	Figure 13B	2.846	3.723
PAX	10	1×10^{-4} M	Figure 13C	2.792	3.655

Note: * Ra(Sa): arithmetic average of the absolute values of the surface height deviations. ** Rms(Sq): root mean square average of height deviations taken from the mean image data plane.

The adsorption of metal xanthate with a low solubility and the oxidation of xanthate into dixanthogen on a sulfide mineral surface in an aqueous solution, as summarized by Leja [5] and Woods [6], are generally considered the main mechanisms for the increase in hydrophobicity of sulfide minerals in flotation. Previous AFM studies with chalcopyrite and pyrite [17–19] have shown that the adsorbed dixanthogen on sulfides in an aqueous solution demonstrates patches with smooth and round edges, which fits well with the fact that oily dixanthogen is generally insoluble in water and that the circular boundary is the direct result of the high interfacial tension between hydrophobic dixanthogen and water. In addition, under ambient conditions, i.e., room temperature and normal pressure, dialkyl dixanthogen is usually in a liquid form with a low melting point. [25] During an AFM scanning process, a minimal force must be applied because of the softness of dixanthogen [17–19]. In this investigation, as shown in Figures 3–13, the adsorbate had no smooth and round edges, and the morphology was completely different compared to that observed for dixanthogen, which is shown in Figure 14. In addition, the adsorbate on the bornite surface was not 'soft,' and its morphology was not disturbed, even under an elevated scan force. For example, as shown in Figures 7–11, a large scan force must be applied to remove the adsorbate and open a 'window' in the center of the AFM image. In addition, Figures 15 and 16 show that the adsorbate on the bornite in the xanthate solution was rigid, with sharp "jump-in" and "jump-off" points on the force curves. Therefore, we ruled out the possibility that observed adsorbate on the bornite was dixanthogen.

This finding, as obtained from the AFM imaging analysis results, is in line with what has been previously reported. For example, Allison et al. [3] reported that xanthate adsorbed on bornite in the form of metal xanthate with low solubility, i.e., CuX. It has been suggested that the final adsorption products on sulfides are highly associated with the semiconductor type of sulfide minerals. That is, dixanthogen is usually formed on n-type minerals, while metal xanthate is observed on p-type minerals. Bornite is classified as a p-type mineral [26], which favors the formation of metal xanthate.

According to Buckley et al. [9], the adsorption of xanthate on bornite is an electrochemical process depending on the potential. When the potential is above −0.35 v, bornite is oxidized in water and yields an iron–free copper sulfide, the reaction of which is shown as follows:

$$Cu_5FeS_4 + 3H_2O = Cu_5S_4 + Fe(OH)_3 + 3H^+ + 3e \tag{1}$$

Comparing Figures 1 and 2, the bornite surface that contacted the water was rougher than the surface that contacted air. In the present investigation, the solution potential was not controlled, and the potential value of DI water was higher than −0.35 V. These results suggest that the bornite surface will undertake oxidation reaction to some extent when it contacts water following the reaction, as shown by Equation (1).

Zachwieja et al. [10] also proposed that the adsorption of xanthate on bornite is due to the following simplified anodic reaction simplified, introducing insoluble cuprous xanthate on the bornite surface:

$$Cu_5S_4 + nX^- = nCuX + Cu_{5-n}S_4 + ne^- \tag{2}$$

That is, xanthate adsorbs on bornite mainly in the form of insoluble cuprous xanthate at a low solution potential. The production of dixanthogen on bornite occurs only when the potential is above the rest potential of X/X_2 couple. In the present study, the solution potential was not controlled, and the value was generally below −0.1 V. In addition, as mentioned before, no noticeable dixanthogen was observed from the obtained AFM images. Therefore, the adsorbate, as shown in Figures 3–11, is mainly cuprous alkyl xanthate with a low solubility.

In addition, the ATR-FTIR results of the adsorption of KEX on bornite show that the main peaks on the obtained spectra were at 1195 cm^{-1} (C–O–C), 1126 cm^{-1} (C–O–C), 1049 cm^{-1} (S–C–S) and 1032 cm^{-1} (S–C–S) (Figure 16). The results are almost identical to those that have been reported for the FTIR spectra of CuX by Poling [23] and Leppinen et al. [24]. In addition, the fact that the characteristic peaks of ethyl xanthate dixanthogen (X_2), namely those at 1020 cm^{-1} and 1260 cm^{-1}, were not observed on the spectra as obtained rules out the existence of X_2 on the bornite surface. Therefore, the irregular adsorbate, as shown in Figures 3–11, is basically insoluble cuprous xanthate (CuX). In this sense, the adsorption behavior of xanthate on the bornite is very similar to the adsorption behavior applicable for the case of chalcocite/xanthate systems [20].

4.2. Effect of the Hydrocarbon Chain of Xanthate

In froth flotation, the rank and concentration of a collector are two important parameters in determining the collectivity and selectivity of the collector. In general, a xanthate collector with a high rank has a high collectivity and, therefore, a low selectivity, and vice versa. In the present work, the effect of the collector's rank on the adsorption of xanthate on bornite was studied using both KEX and PAX as collectors. Figures 3–11 show that, when the collector's concentration is constant, PAX adsorbs on the bornite surface with a significantly higher surface coverage and a more uniform layer structure. This can be attributed to the fact that the formation of cuprous amyl xanthate occurs at a lower surfactant concentration than that for cuprous ethyl xanthate because of the lower-solubility product of cuprous amyl xanthate. For example, it has been reported that the solubility product of cuprous amyl xanthate and cuprous ethyl xanthate is 8.0×10^{-22} and 5.2×10^{-20}, respectively. [27,28] Allison et al. [3] also reported that the percentage of reacted xanthate increased by one-fold, with the carbon number of xanthate increasing from 2 to 5 at the same surfactant concentration. Therefore, the fact that cuprous amyl xanthate forms a more uniform layer than cuprous ethyl xanthate is due to the longer hydrocarbon chain of the former and, therefore, an increased lateral hydrophobic attraction between hydrocarbon chains. Therefore, the higher-rank xanthate, i.e., PAX, is more powerful than the lower rank-xanthate, i.e., KEX, for adsorption on the bornite surface. Therefore, the higher-rank xanthate provides a highly improved flotation collectivity.

Allison et al. [3] observed that "no product of reaction with the methyl and ethyl homologues could be detected, although both reacted very extensively with the surface." They further explained this by stating that "the reaction products are not detected because the lower homologues of cuprous xanthate are extremely insoluble in CS_2 and most other solvents and consequently are not extracted from the surface." As shown in Figures 3–7 obtained with the present work, the KEX did adsorb on the bornite intensively, with the mineral surface being fully covered. The binding of the adsorbate, i.e., cuprous ethyl xanthate, and bornite is very strong, and the adsorbate was not removed even after applying a large scan force. In addition, Figure 5 shows that the adsorbate was not extracted from the bornite surface by rinsing with ethanol alcohol.

4.3. Effect of the Concentration of Xanthate

According to Equation (2), increasing the concentration of xanthate, i.e., the reactant, makes the reaction moves in the normal direction, resulting in more reaction product, i.e., CuX, produced on mineral surface. This concentration effect of xanthate is clearly shown in the AFM images obtained with the present work. For example, Figures 3, 4 and 6 clearly show that the amount of adsorbate increases greatly in surface coverage and the height of the adsorbate, when the concentration of KEX increases from 5×10^{-5} M to 5×10^{-4} M. The same conclusion can also be drawn for the case of PAX. By comparing Figures 8–10, one can see that when the concentration of PAX increases from 1×10^{-5} M to 1×10^{-4} M, the amount of adsorbate increases at a same contacting time.

4.4. Effect of Adsorption Time

Figures 6, 7, 10 and 11 show that the height of the adsorbate and the surface coverage increased when the adsorption time increased from 10 min to 20 min in the xanthate solution at pH 6. In addition, as shown in Figure 12A–C and Figure 13A–C, the adsorption increased when the adsorption time increased from 5 min to 10 min and further to 20 min in the xanthate solution at pH 10. The trend was much more evident in the case of KEX. All these images clearly show that the adsorption of the xanthate on bornite increased with the adsorption time. The finding is in line with a common industrial practice of copper ore beneficiation, which involves adding collectors in a mill to increase the adsorption time, as well as the benefit from the adsorption of the collector on the freshly exposed mineral surface.

4.5. Impact of Xanthate on Bornite Flotation

In froth flotation, specific chemicals, i.e., collectors or promoters, are added to the pulp to increase the surface hydrophobicity of a target mineral. This results in the increase of both the attractive hydrophobic force and the adhesion force between the mineral particles and bubbles. The former can facilitate a particle/bubble attachment, and the latter can retard a particle/bubble detachment, which are both beneficial for froth flotation.

According to Cassie's equation [29]:

$$cos\theta = f_1 cos\theta_1 + f_2 cos\theta_2 \tag{3}$$

where θ_1 is the contact angle for component 1 with a surface area fraction f_1, θ_2 is the contact angle for component 2 with surface area fraction f_2 and θ is the contact angle of the composite material. In addition, $f_1 + f_2 = 1$ for the case of the adsorption of the collector on the bornite surface, assuming that component 1 is the bare bornite and component 2 is the adsorbate, i.e., CuX, θ_2 should be larger than θ_1. Therefore, increasing f_2 and/or θ_2 will increase θ, i.e., the hydrophobicity of bornite with adsorbate, resulting in a better flotation.

In the present study, as shown in the AFM images, it is clear that xanthate can effectively adsorb on a bornite surface, as xanthate showed an almost full coverage at a concentration above 1×10^{-5} M PAX. This suggests that the adsorption of cuprous xanthate resulted in a large f_2, which is beneficial for a large θ. Increasing the dosage of xanthate and adsorption time will increase the cuprous xanthate's surface coverage, i.e., f_2, and it is also beneficial for to increase the surface hydrophobicity.

The force measurement results show that the detach forces measured in the xanthate solutions were larger than the force measured with water. In general, a large detach force suggests a large adhesion between a probe and substrate through the media. Following the Derjaguin approximation [30], it is predicted that a larger adhesion force will be achieved when the probe/liquid/bornite interfacial tension increases. In the present investigation, the Si_3N_4 AFM probe was inert in water, and it did not directly react with xanthate. Therefore, the surface energy of the Si_3N_4 probe did not change. In addition, in the present study, the surface tension of water media remained the same because the short-chain xanthate surfactant was used at a very low concentration. Therefore, the increase in the detachment force, as shown by the AFM force measurement, was mainly due to the increase of the interfacial tension between bornite and water, suggesting an increase in the hydrophobicity of bornite because of the adsorption of the cuprous xanthate on the mineral surface. We also suggest carrying out an AFM "colloid force" measurement by directly measuring the interaction force between a hydrophobic "colloid probe" and the adsorbate on bornite surface. The results will help to better understand the interaction between a bubble and a bornite particle in xanthate solutions in froth flotation. Such an investigation of force measurement, which is beyond the scope of the present study, is recommended for a future work.

5. Conclusions

AFM surface image measurements were applied to study the adsorption of xanthate on bornite in an aqueous solution in situ. The AFM images showed that the xanthate adsorbed on the mineral surface strongly when bornite contacted the KEX and PAX solution. The ATR-FTIR result confirms that that the adsorbate was essentially cuprous xanthate. Increasing the hydrocarbon chain length of xanthate increased the collectivity of the collector by increasing the surface coverage of the cuprous xanthate on the mineral surface at a lower concentration. Both increasing the chemical dosage and increasing the adsorption time will increase the surface coverage of CuX on mineral surface, which contributes to a better flotation by increasing the surface hydrophobicity of bornite.

Funding: This research received no external funding.

Data Availability Statement: The data presented in this study are available on request from the corresponding author. The data are not publicly available due to ethical.

Acknowledgments: J. Zhang is grateful to Freeport-McMoRan Copper & Gold, Inc. for sponsoring the Freeport McMoRan Copper and Gold Chair in Mineral Processing in the Department of Mining and Geological Engineering in the University of Arizona. Reviewers' insightful comments and valuable suggestions are greatly appreciated.

Conflicts of Interest: The authors declare no conflict of interest.

References

1. Gaudin, A.M.; Schuhmann, R., Jr. The action of potassium n-amyl xanthate on chalcocite. *J. Phys. Chem.* **1936**, *40*, 257–275. [CrossRef]

2. Poling, G.W.; Leja, J.J. Infrared study of xanthate adsorption on vacuum deposited films of lead sulfide and metallic copper under conditions of controlled oxidation. *Phys. Chem.* **1963**, *67*, 2121–2126. [CrossRef]
3. Allison, S.A.; Goold, L.A.; Nicol, M.J.; Granville, A.D. A determination of the products of reaction between various sulfide minerals and aqueous xanthate solution, and a correlation of the products with electrode rest potentials. *Metall. Trans.* **1972**, *3*, 2613–2618. [CrossRef]
4. Fuerstenau, M.C. Sulphide mineral flotation. In *Principles of Flotation*; King, R.P., Ed.; South African Institute of Mining and Metallurgy Monograph Series: Johannesburg, South Africa, 1982; pp. 159–182.
5. Leja, J. *Surface Chemistry of Flotation*; Plenum Press: New York, NY, USA, 1982.
6. Woods, R. Electrochemistry of sulphide flotation. In *Principles of Mineral Flotation, The Wark Symposium*; Jones, M.H., Woodcock, J.T., Eds.; AIMM: Melbourne, Victoria, Australia, 1984; pp. 91–115.
7. Mielczarski, J.; Suoninen, E. XPS study of ethyl xanthate adsorbed onto cuprous sulphide. *Surf. Interface Anal.* **1984**, *6*, 34–39. [CrossRef]
8. Mielczarski, J. XPS study of ethyl xanthate adsorption on oxidized surface of cuprous sulfide. *J. Colloid Interface Sci.* **1987**, *120*, 201–209. [CrossRef]
9. Buckley, A.N.; Hamilton, I.C.; Woods, R. Lnvestigation of the Surface Oxidation of Bornite by Linear Potential Sweep Voltammetry and X-ray Photoelectron Spectroscopy. *J. Appl. Electrochem.* **1984**, *14*, 63–74. [CrossRef]
10. Zachwieja, J.B.; Walker, G.W.; Richardson, P.E. Electrochemical flotation of sulfides: The bornite-ethylxanthate system. *Miner. Metall. Process.* **1987**, *4*, 146–151. [CrossRef]
11. Hangone, G.; Bradshaw, D.; Ekmekci, Z. Flotation of a copper sulphide ore from Okiep using thiol collectors and their mixtures. *J. SAIMM* **2005**, *105*, 199–206.
12. Dhar, P.; Thornhill, M.; Rao Kota, H. Investigation of Copper Recovery from a New Copper Ore Deposit (Nussir) in Northern Norway: Dithiophosphates and Xanthate-Dithiophosphate Blend as Collectors. *Minerals* **2019**, *9*, 146. [CrossRef]
13. Polkin, S.I.; Kuzkin, S.F.; Golov, V.M. Application of radiography to studies on the mechanism of interaction between flotation reagents and mineral surfaces. Non-Ferr. Metal: Moscow, Russia, 1955.
14. Plaksin, I.N.; Shafeyev, R.S.; Zaiteseva, S.P. Applications of autoradiography to studies of flotation reagents disposition on mineral surfaces. *Proc. Acad. Sci. S.S.S.R.* **1956**, *108*.
15. Kim, B.S.; Hayes, R.A.; Prestidge, C.A.; Ralston, J.; Smart, R.S.C. In-situ scanning tunneling microscopy studies of galena surfaces under flotation-related conditions. *Colloids Surf. A* **1996**, *117*, 117–129. [CrossRef]
16. Smart, R.S.C.; Amarantidis, J.; Skinner, W.; Prestidge, C.A.; Vanier, L.L.; Grano, S.R. Surface analytical studies of oxidation and collector adsorption in sulfide mineral flotation, in Wandelt, K. and Thurgate, S. (ed), Solid–Liquid Interfaces. *Top. Appl. Phys.* **2003**, *85*, 3–62.
17. Zhang, J.; Zhang, W. An AFM study of chalcopyrite surface in aqueous solution. In Proceedings of the SME Annual Meeting preprint, Phoenix, AZ, USA, 21–24 February 2010.
18. Zhang, J.; Zhang, W. *The Adsorption of Collectors on Chalcopyrite Surface Studied by an AFM in 'Separation Technologies'*; Young, C., Luttrell, G., Eds.; SME: Englewood, CO, USA, 2012; pp. 65–73. ISBN 978-0-87335-339-7.
19. Zhang, J.; Zhang, W. An atomic force microscopy study of the adsorption of collectors on chalcopyrite. In *Microscopy: Advances in Scientific Research and Education*; Méndez-Vilas, A., Ed.; Formatex Research Center: Badajoz, Spain, 2014; Volume 2, pp. 967–973. ISBN 978-84-942134-4-1.
20. Zhang, J.; Zhang, W. An AFM study of the adsorption of collector on chalcocite. In Proceedings of the SME Annual Meeting Preprint 15-138, Denver, CO, USA, 15–18 February 2015.
21. Zhang, J.; An, D.; Withers, J. A Micro-Scale Investigation of the Adsorption of Collectors on Bastnaesite. *Min. Metall. Explor.* **2019**, *36*, 957.
22. An, D.; Zhang, J. A Study of Temperature Effect on the Xanthate's Performance during Chalcopyrite Flotation. *Minerals* **2020**, *10*, 426. [CrossRef]
23. Poling, G.W. Infrared Studies of Adsorbed Xanthates. Ph.D. Thesis, University of Alberta, Edmonton, AB, Canada, 1963.
24. Leppinen, J.O.; Basilio, C.I.; Yoon, R.H. In-situ FTIR study of ethyl xanthate adsorption on sulfide minerals under conditions of controlled potential. *Int. J. Miner. Process.* **1989**, *26*, 259–274. [CrossRef]
25. Rao, S.R. *Xanthate and Related Compounds*; Dekker: New York, NY, USA, 1971.
26. Shuey, R.T. *Semiconducting Ore Minerals*; Elsevier: Amsterdam, The Netherlands, 1975.
27. Kakovsky, I.A. Physiochemical properties of some flotation reagents and their salts with ions of heavy non-ferrous metals. In Proceedings of the Second International Congress of Surface Activity, London, UK, 1 January 1957; pp. 225–237.
28. Ackerman, P.K.; Harris, G.H.; Klimpel, R.R.; Aplan, F.F. Evaluation of flotation collectors for copper sulfides and pyrite, III. Effect of xanthate chain length and branching. *Int. J. Miner. Process.* **1987**, *21*, 141–156. [CrossRef]
29. Cassie, A.; Baxter, S. Wettability of porous surfaces. *Trans. Faraday Soc.* **1944**, *40*, 546–551. [CrossRef]
30. Derjaguin, B.V. Friction and adhesion. IV. The theory of adhesion of small particles. *Kolloid-Zeitschrift* **1934**, *69*, 155–164. [CrossRef]

Article

Assessment of Operational Effectiveness of Innovative Circuit for Production of Crushed Regular Aggregates in Particle Size Fraction 8–16 mm

Tomasz Gawenda , Agata Stempkowska * , Daniel Saramak , Dariusz Foszcz , Aldona Krawczykowska and Agnieszka Surowiak

Department of Environmental Engineering, Faculty of Civil Engineering and Resource Management, AGH University of Science and Technology, Mickiewicza 30 Av., 30-059 Cracow, Poland; gawenda@agh.edu.pl (T.G.); dsaramak@agh.edu.pl (D.S.); foszcz@agh.edu.pl (D.F.); aldona.krawczykowska@agh.edu.pl (A.K.); asur@agh.edu.pl (A.S.)
* Correspondence: stemp@agh.edu.pl

Abstract: The purpose of this paper is to analyze a modern and unique technological system producing common aggregates at the Imielin Dolomite Mine. The installation was built on the basis of inventions of AGH UST and consists of an impact crusher, innovative screens WSR and WSL, light fraction separator SEL and hard fraction separator SET, low-pressure hydrocyclone NHC and infrastructure. The study was carried out on the crusher and screen on the example of production of aggregates with grain size 8–16 mm from dolomite, granite, limestone, sandstone, and gravel. The results showed that cubic aggregates with a low content of irregular grains of less than 1% can be produced in this technological system.

Keywords: regular aggregate; innovative installation; separation

Citation: Gawenda, T.; Stempkowska, A.; Saramak, D.; Foszcz, D.; Krawczykowska, A.; Surowiak, A. Assessment of Operational Effectiveness of Innovative Circuit for Production of Crushed Regular Aggregates in Particle Size Fraction 8–16 mm. *Minerals* **2022**, *12*, 634. https://doi.org/10.3390/min12050634

Academic Editors: Shuai Wang, Xingjie Wang and Jia Yang

Received: 7 April 2022
Accepted: 13 May 2022
Published: 17 May 2022

Publisher's Note: MDPI stays neutral with regard to jurisdictional claims in published maps and institutional affiliations.

1. Introduction

Requirements for aggregate quality and minimization of energy consumption pose increasing challenges for raw material processing plants. A demanding market may in future require large quantities of products with narrow grain size distribution and specific grain shapes. The physical and chemical properties of the material, such as density, hardness, strength, and structure, depend on the place of exploitation (the origin of the raw material), and therefore, are generally invariable in the processing operations in mineral processing. On the other hand, specific particle size and shape or surface structure are achievable depending on the processing methods used during their manufacture (especially crushing) and can directly affect other properties [1–3]. The most important rock raw materials for civil engineering are fractured aggregates produced from magmatic, sedimentary, and metamorphic rocks [4–6]. The protection of the earth's natural resources and the increased use of industrial wastes have resulted in parallel research on the use of recycled and artificial aggregates [7]. Aggregates of the highest quality (e.g., basalt, melaphyre) are used in various branches of construction. The created precast elements that carry enormous dynamic loads, are subject to direct abrasion and adverse weather conditions. These structures should be made of aggregates with low abrasiveness, high strength, and resistance to water and frost. In addition, these aggregates should be characterized by particle shape close to spherical or cubic, sharp edges and rough fracture surfaces. All types of aggregates are commonly used for road pavement layers and structural elements of all types of concrete [8–12].

In the ongoing scientific research on the issues of mineral aggregate production [13,14], it has been noted that the hardness of the raw rock affects its shape. In general, the more compact the raw material, the more difficult it is to obtain cubic particles from it [15–18].

Moreover, in the finer fractions of the crushing products, the largest number of elongated and flat grains is obtained. Therefore, it is advisable to use impact crushers (e.g., with a vertical shaft) at the final stages of crushing. As the crushing rate increases, the percentage of irregular (especially in fine fractions) aggregates also increases, thus crushers should be operated at a crushing rate that is not too high. However, all of these important factors influence the necessity to expand and use multi-stage systems, which usually lead to increasing both the investment and operating costs. There are many studies on the influence of grain geometry on the parameters (such as grinding ability) and properties of aggregates and materials made from them [19–22]. However, there is no data on how to produce cubic aggregates with a proportion of up to 100% apart from publications and patents by the author Gawenda et al. [14,23–26], and a few publications on how to increase the proportion of regular grains [13,27,28], especially in the industrial utilization of aggregate products in building industry [29–31].

Full-Scale Prototype of an Innovative Technological Circuit

In order to increase the quality of aggregates and reduce the number of crushing stages, research was undertaken as part of the implementation of the NCBiR application project: Action 4.1 of the Operational Programme Intelligent Development 2014–2020. HTS Gliwice in cooperation with scientific and research units (AGH in Krakow, ICIMB Łukasiewicz Research Network) built a full-scale prototype of an innovative technological system for refining mineral aggregates. The tests were carried out in real conditions. The installation of the technological system was placed in the Imielin Dolomite Mine and consisted of crushing, enriching, and classifying machines, as shown in Figure 1.

Figure 1. Technological line in the Imielin Dolomite Mine, image from a drone.

The process line includes:

I. *Formator*, which is an installation for the production of regular and irregular fractions consisting of: Impact crusher, three-deck rotary specialized vibrating screen (WSR), three-deck linear specialized vibrating screen (WSL), infrastructure (feeding and buffer cages, control and power supply, belt conveyors, chutes, pumps);
II. Light contaminant separator (SEL);
III. Separator of difficult-to-separate fractions (SET);
IV. Low pressure hydrocyclone (NHC).

A schematic of the pilot plant used for process testing on an industrial scale is shown in Figure 2. This is the unique plant in the world that includes the production of innovative refined aggregates with regular grains in the range of 2–8 mm and 8–16 mm and a sand fraction of 0.1–2 mm.

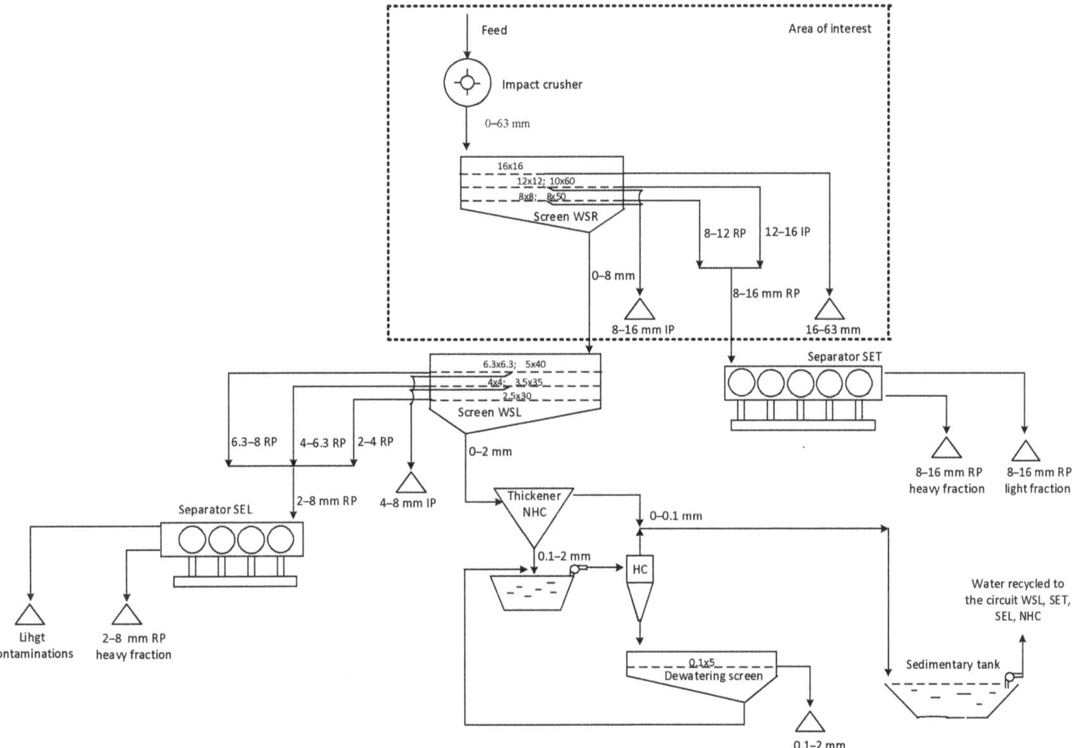

Figure 2. A simplified flow diagram of the installation for the mineral aggregates refinement in the Imielin Dolomite Mine.

The aim of this paper is to present the operation of the *Formator* technological system that produces 8–16 mm class of shaped aggregates (regular particles = RP) from various rock materials. The analyzed system consisting of an impact crusher and a WSR screen is marked on the technological diagram with a dashed line (Figure 2). It was built in accordance with the concept of the patent PL-231748B1. The assumption for the construction of this system was the possibility of producing aggregates with a content of non-formed grains (irregular particles = IP) below 3%, which is not possible in conventional technological systems.

Table 1 provides a brief comparison of an innovative device with a typical device.

Table 1. Advantages and disadvantages of the innovative and ordinary screens in the technological system for the production of regular aggregates.

Innovative Screen Machine	Typical Screen Machine
Possibility of separation of regular aggregates with a low content of irregular grains <3%.	Lack of possibility of separation of regular aggregates. To achieve this effect, it would be necessary to build two screens sited one behind the other and connected by chutes or conveyor belts, which generates investment costs and the need for land area for development (Gawenda T. 2019: Equipment layout for production of molded aggregates Patent No. PL233689 B1, AGH in Krakow). The first screen separates narrow grain classes and the second screen on slot screens separates aggregates by shape.
The final (over-screen) products have a low content of sub-grain up to 2%.	Sub-grain content in products up to more than 10%.
The use of a screen in the crushing system eliminates one crushing stage in the production of cubic aggregates due to the screening of irregular grains, which can be re-grinded.	Need to divert all of the material to regrind to improve aggregate cubicity.
The need to sow narrow fractions to separate them into regular and irregular grains, which increases the number of beds, but then there is the possibility of separation of the narrow fractions, which stabilizes the uniform composition of the aggregate mix.	The fractions are not separated into regular and irregular grains, but to separate more narrow fractions the number of decks also needs to be increased.
Need for larger screening area (30% larger screen) or reduced capacity.	Higher efficiency.
Need to provide higher toss ratio (min. of 6 g) to eliminate clogging of the screen mesh.	Lower toss ratio.

2. Materials and Methods

Characteristics of the technological unit-*Formator*.

2.1. Impact Crusher

The KU 80/120 impact crusher used in the technological system (Figure 3) was selected due to its advantages related to obtaining products with a lower content of irregular grains and higher strength of aggregates in comparison with products obtained from cone and jaw crushers [27]. Crushing occurs mainly due to the impact of material grains by the rotating slats attached to the rotor and the impact of the material accelerated by the slats on the stationary breaker plates. The adjustable gap between the rotor and plates allows for the adjustment of the size of the obtained fractions to the needs of the user. Properly selected materials, of which the slats and stationary plates are assembled, ensure their high resistance to abrasion, thanks to the low operating costs that the user incurs. Table 2 presents the characteristics of the crusher.

Table 2. Characteristics of the KU 80/120 crusher.

Name	Unit	Value
Inlet Size	[mm]	800 × 1170
Rotor diameter	[mm]	1110
Rotor length	[mm]	1150
Capacity (maximum)	[t/h]	450
Feed grain size range	[mm]	0–700
Hard feed grain size	[mm]	0–400
Crusher weight	[kg]	11,851
Rotor weight	[kg]	4136
Crusher drive power (max.)	[kW]	250
Rotor Speed I	[rpm]	529
Rotor Speed II	[rpm]	670

Figure 3. View of the KU 80/120 crusher and the WSR 3-2.0/6.0 screen.

2.2. Specialized Rotary Vibrating Screen (WSR)

Specialized rotary vibrating screen (WSR 3-2.0/6.0) (Figure 3) is a machine designed for screening the excavated material with transversely tensioned screens. The aggregate delivered to the screen deck is subjected to circular harmonic vibrations, where the material is periodically lifted under the force of gravity and moves along the screen deck. The elements causing the circular motion vibrations are rotating eccentric (shaft and inertial) masses. The vibrating force is directed at an angle of 15° to the base of the screen. The structure of the screen is shown in a schematic diagram in Figure 4. The technical characteristics are provided in Table 3. The screen is built in accordance with the concept of invention patented by AGH. It is an atypical screen (specialized) since it has three decks and produces six fractions. Its characteristic feature is first screening the feed material into fractions of 8–12 and 12–16 mm with the use of two decks equipped with square mesh screens, and then separating the irregular grains from the regular ones with the use of slot screens (rectangular mesh) (Figure 2). In the oversize-screen products, form aggregates are obtained, which, when joined together, form the innovative 8–16 mm RP (regular particles) products. In the undersize-screen products of these deposits, products with an increased proportion of 8–16 mm IP irregular particles are obtained after merging. Moreover, this screen separates the 0–8 mm fraction, which is fed to another specialized WSL screen, also to obtain the shaped aggregates. The fraction above 16 mm is a typical commercial product.

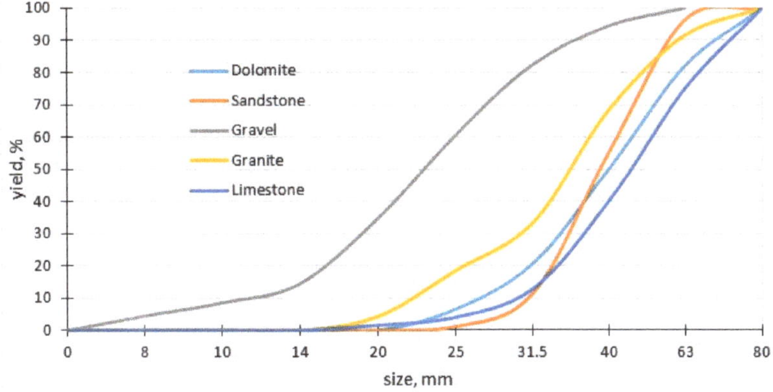

Figure 4. Particle size distribution curves of five feeds sent to the impact crusher.

Table 3. Characteristics of the WSR 3-2.0/6.0 rotary screen.

Name	Unit	Value
Sieve dimensions	[mm]	6000×2000
Working frequency	[Hz]	16.6
Maximum stroke	[mm]	9
Angle of screen inclination	[°]	15
Drive power of screening unit	[kW]	30
Motor rotation	[rpm]	980
Maximum efficiency	[t/h]	350
Screen deck 1	[mm]	half deck-16 × 16 mesh, half a deck-blank screen
Screen deck 2	[mm]	half deck-12 × 12 mesh half deck-10 × (60) 75 gap
Screen deck 3	[mm]	half a deck-8 × 8 mesh half deck-8 × (50) 60 gap

3. Results and Discussion

3.1. Tests on Different Types of Raw Materials

The raw materials selected for testing were various in terms of lithology and grain size distribution. Triassic dolomite, Triassic limestone, sandstone from the Carpathian flysch, gravel of river origin, and granite from the Central Sudety region were used. The graph (Figure 4) presents the particle size distribution curves of the feeds sent for crushing in the impact crusher.

Analyzing the grain size composition of the feed material (Table 4), it should be emphasized that the finest grain size range is the river gravel 0–63 mm and the largest share of coarse grains by limestone is in the range 14–80 mm. The content of out-of-form grains measured with a Schultz caliper in accordance with PN-EN 933-4:2008 [32] (SI shape index) in the feed varied and ranged from about 18% for dolomite to 25% for sandstone.

Table 4. Characteristics of raw materials and operation of the crusher.

Raw Material	Feed Grain Size [mm]	Shape Index SI [%]		Comminution Degree		Crusher Throughput [t/h]
		Feed	Crusher Product	S90	S50	
dolomite	20–80	17.8	14.9	2.8	6.7	80
sandstone	25–80	25.3	13.5	2.8	4.2	80
gravel	0–63	24.3	13.6	1.8	3.3	60
granite	14–80	20.0	15.7	2.2	3.9	100
limestone	14–80	18.6	13.7	3.4	6.0	100

The raw material was crushed using an impact crusher at a rotation speed of 529/min. The outlet gaps between the strips and the impact plates were 20, 40, and 60 mm. The finest grades (S50) obtained were highest for dolomite and limestone (about 6) and lowest for gravel, which is related to the grain size of the feed. The particle size distribution curves of the crusher products are shown in Figure 5.

Table 5 summarizes the content of irregular grains in different fractions of products obtained from the impact crusher. The content of non-formed grains was measured using slotted sieves in accordance with the PN-EN 933-3:2012 [33] standard (flakiness index FI), with grains below 4 mm (outside the scope of the standard) included in the analysis for the purposes of the project. The formed irregular grains in the crushing process depend mainly on the physical and mechanical properties of the raw material (hardness, flakiness, toughness, structure, texture), but also on the type of crusher and its technical and technological parameters [34]. Therefore, it can be seen (Table 5) that the highest proportion of irregular grains (from 15–25%) was observed in the finest fractions of 2–8 mm. This is a known phenomenon that aggregate producers have problems with, as the fine commercial grades are the most difficult to meet the standard requirements. The objective of this project was to reduce the content of unshaped grains to at least 3% in the Formator installation. The proportion of irregular grains also depends on the comminution degree of the raw material. It was observed that the highest content of irregular grains in the 2–8 mm class (FI = 25% sandstone, FI = 24% dolomite) was obtained for high comminution degree S50 values 6.7 and 4.5, and the lowest values FI = 15% (gravel), FI = 17% (granite) for the lowest comminution degree S50 values 3.3 and 3.9. Similar trends are found for the wider grain fraction 2–16 mm where the highest FI ratios are found for dolomite 17.8%, sandstone 16.2%, and limestone 15.2%. Here, the raw materials were crushed at the highest values of comminution degrees (Tables 4 and 5).

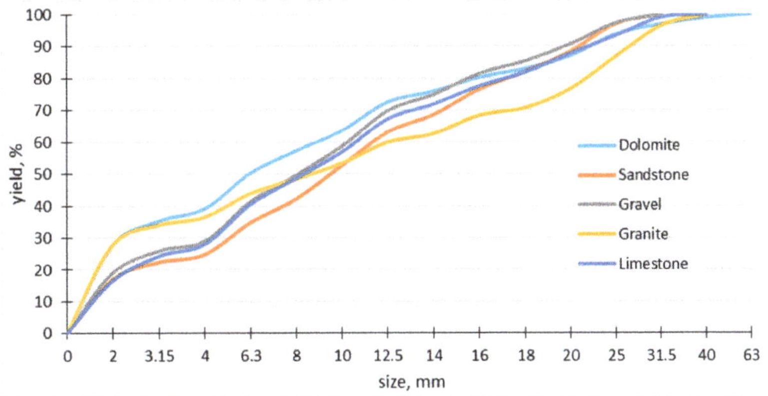

Figure 5. Particle size distribution curves of five comminution products in an impact crusher directed to the WSR screen.

Table 5. Contents of irregular grains in individual fractions obtained in an impact crusher.

Crusher Product [mm]	Flakiness Index FI [%]				
	Dolomite	Sandstone	Gravel	Granite	Limestone
2–8	24.0	25.0	15.3	17.0	18.9
8–16	9.8	9.7	10.8	10.3	11.1
2–16	17.8	16.2	13.0	13.7	15.2
2–31.5	14.9	13.5	13.6	15.7	13.7

The products obtained after crushing in the impact crusher were transported by a belt conveyor to the WSR screen, where they were sieved into fractions, 0–8, 8–12, 12–16, and +16 mm. The 8–12 and 12–16 mm fractions were sifted on slot screens (longitudinal rectangular mesh), from which regular and irregular grains are separated. These fractions are then combined with respect to shape into fractions of 8–16 mm RP (with regular grains) and 8–16 mm IP (with non-formed grains). Figures 6 and 7 present selected photographs of the product received. Figure 8 presents examples of the samples for analysis.

All of the obtained grain composition curves for the 8–16 mm sieving products of the upper-screening (RP) and bottom-screening (IP) are summarized in a graph (Figure 9). From the analyses, it can be concluded that granite with irregular grains had the finest grain size and dolomite with regular grains had the coarsest grain size. The 8–16 mm IP granite had the highest content of sub-grain up to 22%, which was not completely screened (outcrop of fractions below 8 mm). This is related to the higher throughput (about 100 t/h, Table 3). Comparing the remaining diagrams with each other, it should be stated that all of the passing material, i.e., those containing irregular grains, are considerably finer than the retained material on the screen surface, i.e., regular ones, as they contain about 10% of undersize. On the other hand, the retained material with regular grains contains from about 1 to 5% of undersize, which indicates the high quality of the aggregates. The content of 5% of sub-grain was recorded for limestone and granite with the screening capacity of about 100 t/h.

Figure 6. Limestone tests 8–16 mm regular and irregular particles.

Figure 7. Sandstone tests 8–16 mm regular and irregular particles.

Table 5 presents the content of irregular grains for the five final products (FI flakiness index). The final products are the retained material of the WSR 8–16 mm RP screen, on which regular grains retain. All of these products have flakiness indexes of less than 1.5%, the intended goal has been achieved and this means that the content of regular grains in these products exceeds 98%. The lowest contents were obtained for granite (0.2%), limestone (0.3%), and sandstone (0.5%). For gravel, the highest value was 1.3%. The content of irregular grains for the product including under-screen (6.3–16 mm) was also calculated and ranges from 8 to 17%. Moreover, Table 6 summarizes the content of irregular grains that occurs in the screening product of the screen and comes from the process of sifting these grains on a given screening deck (irregular grains screened on the screen fall through its openings and accumulate in the screening product). In accordance with the idea of the invention, these products can be crushed again or used as aggregates. The values of the proportion of irregular grains are not high, as they range from 9% for granite to 17% for gravel. It can be concluded that these values have typical aggregates produced in processing plants [13,14,28]. The partial shares (6.3–16 mm) reduce the content of irregular grains by 1%, with the exception of limestone (increase by 0.4%).

Figure 8. Final product: Limestone 8–16 mm regular (**left**), irregular (**right**).

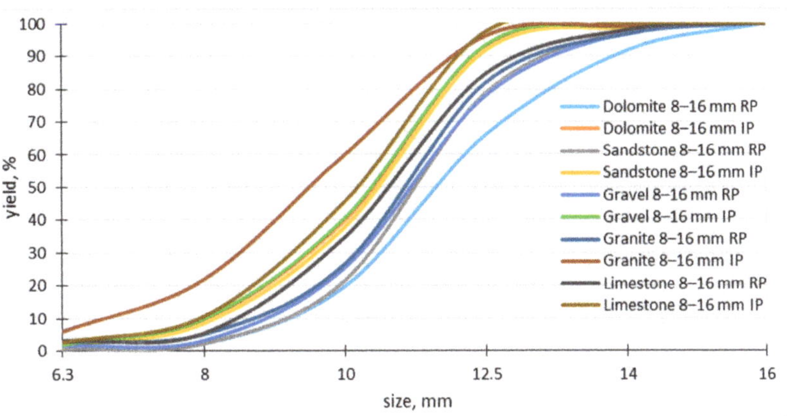

Figure 9. Particle size distribution curves of five screening products in the fraction 8–16 mm (WSR screen).

It is worth noting that the content of irregular grains in typical processing plants reaches about a dozen or more percent in coarse products, and in fine grains they often exceed 20%. For comparison, Table 7 summarizes the content of irregular grains for typical products obtained before refinement in the WSR specialized screen, i.e., these aggregates would be obtained if a typical vibrating screen was used in the technological system. The content of irregular grains would range from 10–11%. In addition, Table 8 summarizes the flow balance of the 8–16 mm product (mass yield) with separation into fractions with regular particles (RP) and irregular particles (IP).

Table 6. The flakiness index FI for the five final products of 8–16 mm of the *Formator* technological system.

Product [mm]	Irregular Particles Content [%]				
	Dolomite	Sandstone	Gravel	Granite	Limestone
8–16 RP upper-screen product	1.0	0.5	1.3	0.2	0.3
8–16 RP upper-screen product with undersize grains (6.3–8)	1.0	0.5	1.3	0.2	0.3
8–16 IP bottom-screen product	10.3	16.1	17.4	9.1	11.8
8–16 IP bottom-screen product with undersize grains (6.3–8)	9.8	15.2	16.5	8.2	12.2

Table 7. The flakiness index (FL) values of products received from conventional crushing and screening of plants compared with those received from the innovative one.

Product [mm]	Irregular Particles Content [%]				
	Dolomite	Sandstone	Gravel	Granite	Limestone
8–16 RP innovative	1.0	0.5	1.3	0.2	0.3
8–16 typical	9.8	9.7	10.8	10.3	11.1

Table 8. Mass balance with separation into fractions with regular particles (RP) and irregular particles (IP).

Product [mm]	Mass Yield of RP and IP in Products [t/h]				
	Dolomite	Sandstone	Gravel	Granite	Limestone
8–16 RP	16.41	22.93	15.53	18.14	25.49
8–16 IP	2.09	4.57	3.57	1.86	3.51
8–16 RP + IP	18.5	27.5	19.1	20.0	29.0

3.2. Optimization of Required Machine Parameters

As a result of the tests, the necessary operating parameters of the machines were optimized for their proper functioning. The most important issues included the solution of the problem of blocking of the mesh screens with difficult grains during the process of screening regular and irregular grains in the WSR screen. Difficult grains are grains of similar size to a given screen opening, which cause clogging of the sieves and a decrease of the screening efficiency. Due to the specificity of the technological process, each narrow fraction that is matched to the appropriate width of the slot contains 100% of grains difficult to sift. Therefore, the first screening tests were characterized by too strong blocking of the mesh with difficult grains, as shown in the photographs (Figure 10).

Factors affecting the efficiency of the screening process include [35–37]:

- Dynamic parameters of the screen deck;
- Layer thickness of the screened material;
- Type of screen deck;
- Properties of the material (such as surface moisture, hardness and compaction, grain shape and range of grain size) to be screened.

Since the last three factors cannot be changed due to the nature of the process, the solution had to be found in the dynamics of the screen. In vibrating screens, the movement of grain on the screen is caused by inertia forces, which are the result of periodic movement of the screen. A value of vibration amplitude that is significantly low influences the lowering of screening process effectiveness by blocking of sieves with grains at a given output or lowering of screening output at assumed screening effectiveness. In screen machines, the transport of the material on the surface of the screen is usually achieved by means of vibrators causing harmonic vibrations or shafts with unbalanced masses, which are directed at an angle to the surface of the screen, while the screen can also be inclined to the horizontal. To parameterize the operation of the screen the dynamic index is determined, which is the ratio of the maximum acceleration of the screen to the acceleration of the earth. The dynamic index of the screen also informs the values of load of the screen construction by inertia forces. Depending on the type of screen motion, we use the feed rate (u_1) (rectilinear screen motion) and toss index (u_2) (circular, elliptical screen motion), which is the ratio of maximum screen acceleration to the acceleration due to gravity.

Figure 10. Difficult grains blocking slot screen decks (toss index 4.4).

For screens with circular vibration trajectory the toss ratio u_2 has the following form [35]:

$$u_2 = \frac{r\omega^2}{g \cos \beta} > 1 \tag{1}$$

where r is the radius of vibration, mm; ω is the angular velocity, rad/s; g is acceleration due to gravity, m/s^2; and β is the angle of inclination of the sieve to the horizontal.

The pitch of the riddle is affected by unbalanced masses, which change the radius of its oscillation. Therefore, this radius can be adjusted by adding or removing masses on the eccentric shaft. Since the radius of vibration measured on the riddle was 4 mm, which provided a toss ratio of more than 4.4, it was recommended to gradually increase the unbalanced masses on the shaft and study the cleaning of the sieves (Figure 11). The data are summarized in Table 9 along with the calculated toss ratio u_2.

Table 9. Operating parameters of WSR screen.

Radius of Vibration r, [mm]	Angular Velocity ω, [rad/s]	Angle of Inclination of the Sieve to the Horizontal, β [°]	Toss Index u_2
4	102.6	15	4.4
5	102.6	15	5.5
6	102.6	15	6.6

The best effect of sieve cleaning of "difficult" grains was obtained for a vibration radius of 6 mm, which raised the toss ratio to a value of 6.6 acceleration due to gravity. A photo of the conducted tests and the sieve cleaning effects are illustrated by the photographs shown in Figure 12.

Figure 11. Change of unbalanced masses on the eccentric shaft of the screen.

Figure 12. Photographs showing the condition of sieve blockage with difficult grains: At a toss index of 5.5 (**left**), at a toss index of 6.6 (**right**).

4. Conclusions

- The installation produced at the Dolomite Mine in Imielin has innovative solutions in terms of products and processes, in accordance with the Oslo handbook [38]. It is a novelty that has not been used to date in the world due to its unique features and functionality as compared with the solutions available on the domestic (Polish) and foreign markets.
- By adjusting the appropriate dimensions of the screen (or screen decks), the capacity of the technological crushing and screening system can be increased by at least 30%. The increase in capacity depends on the content of regular grains, which can be taken out of the system as a final product, since there is no need to further crush them in other crushers.

- Tests conducted on the crushing and production of molded aggregates in the innovative WSR screen have shown that aggregates produced from five different rock materials have very low FI flakiness indexes.
- The final products are the upper-screen products of the 8–16 mm RP deposits, on which the regular grains remain. All of these products have flakiness indexes of less than 1.5% (Table 6), this indicates that the content of regular grains in these products exceeds 98%.
- The assumption for the construction of this system was the possibility of producing aggregates with a content of non-formed grains below 3%. The lowest contents were obtained for granite (0.2%), limestone (0.3%), and sandstone (0.5%). The highest value of 1.3% was obtained for gravel.
- It should be highlighted that the obtained low values of irregular grains practically do not occur in any mineral processing plants, thus they can be considered as innovative high-cubic aggregates.

5. Patents

The following patent granted in Poland is related to this paper: Author: Gawenda T.: Wibracyjny przesiewacz wielopokładowy, AGH w Krakowie. Patent No. PL 231748 B1 granted on 12 June 2018.

Author Contributions: Conceptualization, T.G.; methodology, T.G., A.S. (Agata Stempkowska) and D.S.; validation, D.F. and A.K.; formal analysis, T.G., D.S. and D.F.; investigation, T.G., A.S. (Agata Stempkowska), D.F., A.K. and A.S. (Agnieszka Surowiak); data curation, D.S. and D.F.; writing—original draft preparation, T.G. and A.S. (Agata Stempkowska); writing—review and editing, T.G. and D.S. All authors have read and agreed to the published version of the manuscript.

Funding: The paper is the effect of completing the NCBiR Project, contest no. 1 within the subaction 4.1.4 "Appliaction projects" POIR in 2017, entitled "Elaboration and construction of the set of prototype technological devices to construct an innovative technological system for aggregate beneficiation along with tests conducted in conditions similar to real ones". The Project is co-financed by the European Union from sources of the European Fund of Regional Development within the Action 4.1 of the Operation Program Intelligent Development 2014–2020.

Data Availability Statement: The data presented in this study are available on request from the corresponding author.

Conflicts of Interest: The authors declare no conflict of interest.

References

1. Meddah, M.S.; Zitouni, S.; Belâabes, S. Effect of content and particle size distribution of coarse aggregate on the compressive strength of concrete. *Constr. Build. Mater.* 2010, *24*, 505–512. [CrossRef]
2. Miskovsky, K.; Duarte, M.T.; Kou, S.Q.; Lindqvist, P.A. Influence of the mineralogical composition and textural properties on the quality of coarse aggregates. *J. Mater. Eng. Perform.* 2004, *13*, 144–150. [CrossRef]
3. Gawenda, T.; Saramak, D.; Tumidajski, T. Regression models of rock materials crushing in jaw crushers. 7th National Polish Scientific Conference on Complex and Detailed Problems of Environmental Engineering. *Zesz. Nauk. Wydz. Budownictwai I Inz. Srodowiska* 2005, *22*, 659–670.
4. Adessina, A.; Fraj, A.B.; Barthélémy, J.-F.; Chateau, C.; Garnier, D. Experimental and micromechanical investigation on the mechanical and durability properties of recycled aggregates concrete. *Cem. Concr. Res.* 2019, *126*, 105900. [CrossRef]
5. Piasta, W.; Budzyński, W.; Góra, J. Wpływ rodzaju kruszywa grubego na odkształcalność betonów zwykłych. *Przegląd Bud.* 2012, *7–8*, 35–38.
6. Ostrowski, K.; Sadowski, Ł.; Stefaniuk, D.; Wałach, D.; Gawenda, T.; Oleksik, K.; Usydus, I. The effect of the morphology of coarse aggregate on the properties of self-compacting high-performance fibre-reinforced concreto. *Materials* 2018, *11*, 1372. [CrossRef]
7. Barreto Santos, M.; De Brito, J.; Santos Silva, A. A Review on Alkali-Silica Reaction Evolution in Recycled Aggregate Concrete. *Materials* 2020, *13*, 2625. [CrossRef] [PubMed]

8. Molugaram, K.; Shanker, J.S.; Ramesh, A. A study on influence of shape of aggregate on strength and quality of concrete for buildings and pavements. *Adv. Mater. Res.* **2014**, *941–944*, 776–779. [CrossRef]
9. Ganapati Naidu, P.; Adiseshu, S. Influence of coarse aggregate shape factors on bituminous mixtures. *Int. J. Eng. Res. Appl.* **2011**, *1*, 2013–2024.
10. Akçaoğlu, T. Determining aggregate size & shape effect on concrete microcracking under compression by means of a degree of reversibility method. *Constr. Build. Mater.* **2017**, *143*, 376–386. [CrossRef]
11. Adams, M.P.; Ideker, J.H. Influence of aggregate type on conversion and strength in calcium aluminate cement concrete. *Cem. Concr. Res.* **2017**, *100*, 284–296. [CrossRef]
12. Stempkowska, A.; Gawenda, T.; Naziemiec, Z.; Adam Ostrowski, K.; Saramak, D.; Surowiak, A. Impact of the Geometrical Parameters of Dolomite Coarse Aggregate on the Thermal and Mechanic Properties of Preplaced Aggregate Concrete. *Materials* **2020**, *13*, 4358. [CrossRef] [PubMed]
13. Eloranta, J. Sposoby wpływania na jakość kruszyw. Prezentacja badań firmy Metso Minerals na nośniku CD. In Proceedings of the 2006 VI Konferencja, „Kruszywa Mineralne—Surowce—Rynek—Technologie—Jakość", Szklarska Poręba, Poland, 26–28 April 2006; OWPW: Wrocław, Poland, 2006.
14. Gawenda, T. Production Methods for Regular Aggregates and Innovative Developments in Poland. *Minerals* **2021**, *11*, 1429. [CrossRef]
15. Naziemiec, Z.; Gawenda, T. Ocena efektów rozdrabniania surowców mineralnych w różnych urządzeniach kruszących. In Proceedings of the VI Konferencja *„Kruszywa Mineralne—surowce—rynek—technologie—jakość"*, Szklarska Poręba, Poland, 26–28 April 2006; OWPW: Wrocław, Poland; pp. 83–94.
16. Gawenda, T. *Analiza i Zasady Doboru Kruszarek Oraz Układów Technologicznych w Produkcji Kruszyw Łamanych*; Monografia, Wyd. AGH: Kraków, Poland, 2015.
17. Naziemiec, Z.; Gawenda, T. Badanie procesu kruszenia zamkniętym obiegiem. In Proceedings of the VII Konferencja Kruszywa mineralne: Surowce—rynek—technologie jakość, Szklarska Poręba, Poland, 18–20 April 2007; OWPW: Wrocław, Poland; pp. 107–116.
18. Nowak, A.; Gawenda, T. Analiza porównawcza kruszarek w wielostadialnych układach rozdrabniania skał bazaltowych. *Górnictwo I Geoinżynieria* **2006**, *3*, 267–278.
19. Rocco, C.; Elices, M. Effect of aggregate shape on the mechanical properties of a simple concrete. *Eng. Fract. Mech.* **2009**, *76*, 286–298. [CrossRef]
20. Mora, C.F.; Kwan, A.K.H. Sphericity, shape factor and convexity measurement of coarse aggregate for concerete using Digital image processing. *Cem. Concr. Res.* **2000**, *30*, 351–358. [CrossRef]
21. Strzałkowski, P.; Duchnowska, M.; Kaźmierczak, U.; Bakalarz, A.; Wolny, M.; Karwowski, P.; Stępień, T. Evaluation of the Structure and Geometric Properties of Crushed Igneous Rock Aggregates. *Materials* **2021**, *14*, 7202. [CrossRef]
22. Ostrowski, K. The influence of coarse aggregate shape on the properties of high-performance, self-compacting concrete. *Tech. Trans. Civ. Eng.* **2017**, *5*, 25–33. [CrossRef]
23. Gawenda, T.; Saramak, D.; Stempkowska, A.; Naziemiec, Z. Assessment of Selected Characteristics of Enrichment Products for Regular and Irregular Aggregates Beneficiation in Pulsating Jig. *Minerals* **2021**, *11*, 777. [CrossRef]
24. Gawenda, T. Wibracyjny przesiewacz wielopokładowy, AGH w Krakowie. Patent No. PL 231748 B1, 12 June 2018.
25. Gawenda, T. Układ Urządzeń do Produkcji Kruszyw Foremnych. Patent No. PL233689B1, 7 August 2019.
26. Gawenda, T.; Saramak, D.; Naziemiec, Z. Układ Urządzeń do Produkcji Kruszyw oraz Sposób Produkcji Kruszyw. Patent No. PL233318B1, 7 June 2019.
27. Gawenda, T. Wpływ rozdrabniania surowców skalnych w różnych kruszarkach i stadiach kruszenia na jakość kruszyw mineralnych, Gospodarka Surowcami Mineralnymi Polska Akademia Nauk. *Kom. Gospod. Surowcami Miner.* **2013**, *29*, 53–65. [CrossRef]
28. Naziemiec, Z. *Przeróbka i Badania Kruszyw Mineralnych*; Monografia nr 356, Wyd. AGH: Kraków, Poland, 2019.
29. Janardhana Reddy, K.R.; Das, A. Shape characterisation of aggregates in three dimension. *Eur. J. Environ. Civil Eng.* **2022**, *26*, 345–359. [CrossRef]
30. Kusumawardani, D.M.; Wong, Y.D. The influence of aggregate shape properties on aggregate packing in porous asphalt mixture (PAM). *Constr. Build. Mater.* **2020**, *255*, 119379. [CrossRef]
31. Kusumawardani, D.M.; Wong, Y.D. Effect of Aggregate Shape Properties on Performance of Porous Asphalt Mixture. *J. Mater. Civil Eng.* **2021**, *33*, 8. [CrossRef]
32. *PN-EN 933-4:2008*; Badania Geometrycznych Właściwości Kruszyw—Część 4: Oznaczanie Kształtu Ziaren—Wskaźnik Kształtu. Polski Komitet Normalizacyjny: Warszawa, Poland, 2008.
33. *PN-EN 933-3:2012*; Badania Geometrycznych Właściwości Kruszyw—Część 3: Oznaczanie Kształtu Ziaren za Pomocą Wskaźnika Płaskości. Polski Komitet Normalizacyjny: Warszawa, Poland, 2012.
34. Tumidajski, T.; Naziemiec, Z. Wpływ warunków procesu kruszenia na kształt ziaren kruszyw mineralnych. In Proceedings of the IV Konferencja „Kruszywa Mineralne—Surowce—Rynek—Technologie—jakość", Szklarska Poręba, Wyd, Wrocław, Poland, 14–16 April 2004.
35. Banaszewski, T. *Przesiewacze*; Wydawnictwo Śląsk: Katowice, Poland, 1990.

36. Wodziński, P. *Przesiewanie i Przesiewacze*; Wydawnictwo Politechniki Łódzkiej: Łódź, Poland, 1997.
37. Feliks, J.; Tomach, P.; Foszcz, D.; Gawenda, T.; Olejnik, T. Research on the New Drive of a Laboratory Screen with Rectilinear Vibrations in Transient States. *Energies* **2021**, *14*, 8444. [CrossRef]
38. Ministerstwo Nauki i Szkolnictwa Wyższego, Departament Strategii i Rozwoju Nauki. *Podręcznik Oslo. Zasady Gromadzenia i Interpretacji Danych Dotyczących Innowacji*, 3rd ed.; Organizacja Współpracy Gospodarczej i Rozwoju Urząd Statystyczny Wspólnot Europejskich: Warszawa, Poland, 2006. (In Polish)

Article

Boron Impurity Deposition on a Si(100) Surface in a SiHCl$_3$-BCl$_3$-H$_2$ System for Electronic-Grade Polysilicon Production

Qinghong Yang [1,2,†], Fengyang Chen [1,2,†], Lin Tian [3], Jianguo Wang [4,†], Ni Yang [1,3,*], Yanqing Hou [1,*], Lingyun Huang [1,*] and Gang Xie [2,3]

1   State Key Laboratory of Complex Nonferrous Metal Resources Clean Utilization, Kunming University of Science and Technology, Kunming 650093, China; 20192102032@stu.kust.edu.cn (Q.Y.); fychen@stu.kust.edu.cn (F.C.)
2   Faculty of Metallurgical and Energy Engineering, Kunming University of Science and Technology, Kunming 650093, China; wqs@stu.kust.edu.cn
3   State Key Laboratory of Pressure Hydrometallurgical Technology of Associated Nonferrous Metal Resources, Yuantong North Road 86, Kunming 650031, China; ltian@stu.kust.edu.cn
4   China Copper Co., Ltd., Yuantong North Road 86, Kunming 650031, China; jgwang@stu.kust.edu.cn
*   Correspondence: yqh@stu.kust.edu.cn (N.Y.); hyqsimu@kust.edu.cn (Y.H.); hly@kust.edu.cn (L.H.)
†   These authors contributed equally to this work.

Abstract: A study of boron impurities deposited on a Si(100) surface in a SiHCl$_3$-BCl$_3$-H$_2$ system is reported in this paper, using periodic density functional theory with generalized gradient approximation (GGA). The results show that the discrete distances of BCl$_3$ and SiHCl$_3$ from the surface of the Si(100) unit cell are 1.873 Å and 2.340 Å, respectively, and the separation energies are -35.2549 kcal/mol and -10.64 kcal/mol, respectively. BCl$_3$ and SiHCl$_3$ are mainly adsorbed on the surface of the Si(100) unit cell in particular molecular orientations: the positive position and the hydrogen bottom-two-front position from the analysis of the bond length change and adsorption energy. The adsorption of SiHCl$_3$ and BCl$_3$ is accompanied by a charge transfer from the molecule to the surface of the unit cell of 0.24 and 0.29 eV, respectively. BCl$_3$ reacts more readily than SiHCl$_3$ with the Si(100) surface, resulting in the deposition of boron impurities on the polysilicon surface.

Keywords: electronic-grade polysilicon; boron impurities; chemical vapor deposition; density functional theory; differential charge density

Citation: Yang, Q.; Chen, F.; Tian, L.; Wang, J.; Yang, N.; Hou, Y.; Huang, L.; Xie, G. Boron Impurity Deposition on a Si(100) Surface in a SiHCl$_3$-BCl$_3$-H$_2$ System for Electronic-Grade Polysilicon Production. *Minerals* **2022**, *12*, 651. https://doi.org/10.3390/min12050651

Academic Editors: Kenneth N. Han, Shuai Wang, Xingjie Wang and Jia Yang

Received: 12 April 2022
Accepted: 12 May 2022
Published: 21 May 2022

Publisher's Note: MDPI stays neutral with regard to jurisdictional claims in published maps and institutional affiliations.

1. Introduction

With the rapid development of the electronic information industry, silicon is a very important raw material for semiconductor manufacture, and electronic-grade polysilicon materials play a pivotal role [1,2]. Polysilicon has a high mobility in semiconductor circuits, and consequently it is widely used in such industries as semiconductors, integrated circuits and computer chips [3,4]. The polysilicon materials used in the semiconductor and electronic information industries are electronic-grade polysilicon, and are generally 99.9999% pure Si [5]. For a long time, the production technology for electronic-grade polysilicon has been monopolized by many countries. The demand for electronic-grade polysilicon in China is almost entirely dependent on imports, which are highly dependent on the international environment [6]. Because of my country's insufficient research on semiconductor materials, the purity of polysilicon has never reached the international level, and the lack of sufficiently pure silicon has become a key factor that directly hinders the development of my country's semiconductor industry.

The main processes for the production of polysilicon include chemical vapor methods and metallurgy [7]. The representative of the chemical vapor deposition methods are the modified Siemens method [8], and the fluidized-bed [9], silane [10] and gas-liquid-deposition methods [11]. The main metallurgical methods are the thermite reduction method and the regional hot-melt purification method, but the thermite reduction method is only suitable for use in a laboratory [12,13]. Since the development of the improved Siemens chemical vapor deposition method, no other method can produce the same purity in the product and the process [14], and hence this is currently the method mainly used in industry to produce polysilicon. However, there is an unavoidable problem in production by the improved Siemens method: distilled trichlorosilane ($SiHCl_3$) and a small amount of boron trichloride (BCl_3) enter the Siemens reduction furnace, causing a heterogeneous reaction on the surface of the silicon rod, which results in the doping of boron into the polysilicon [15]. However, research on boron impurities arising in the production of polysilicon by the improved Siemens method is still relatively weak, and research on this process is therefore of great significance.

The impurities in high-purity electronic-grade polysilicon come mainly from the boron impurity in the acceptor, and the content of this needs to be controlled below 0.33 ppb [15]. To control the B doping of polysilicon in chemical vapor deposition, it is important to reduce the content of impurity B in the product and improve its purity [16,17], and this requires in-depth and systematic research on B-doping in the polysilicon deposition process. To study the chemical vapor deposition rate, Jenkinson and Pollard experimented to explore the deposition rate of boron in a BCl_3-H_2 system in a parallel flow reactor [18]. However, under these conditions, the boron deposition rate was limited by the transfer process of BCl_3. To describe the kinetics and reaction mechanism of the BCl_3 hydrogenation process more accurately, Sezgi designed a collision-jet chemical vapor deposition reactor [19]. The role of the reactor was to minimize the impact of the transfer process on the boron deposition rate. On this basis, Sezgi further studied the kinetics of the process of boron chemical vapor deposition in the BCl_3-H_2 system and obtained a deposition rate of B in the range of 75–1350 °C [20]. The Key Laboratory of Ultra-High Temperature Structural Composites of Northwestern Polytechnical University studied the influence of temperature on the kinetics of chemical vapor deposition during the preparation of Si-B-C ceramics. The results showed that the process is controlled by chemical reaction kinetics with an activation energy of 271 kJ/mol, the initial step of the deposition kinetics being BCl_3 (g) + H_2 (g) $\rightarrow$ $HBCl_2$ (g) + HCl (g) [21]. That study also proved that this step is one of the main reactions in the thermodynamic analysis of ZrB_2 synthesized using the $ZrCl_4$-BCl_3-H_2 system [22].

The doping mechanism of B in polysilicon chemical vapor deposition has been studied both domestically and abroad. To accurately and economically reflect the actual industrial situation, this study uses the method of simulation by quantum chemistry calculation to study the B-doping control mechanism in the process of polysilicon chemical gas deposition in a vacuum environment. The properties of different atomic positions of BCl_3 and $SiHCl_3$ on the Si(100) surface are calculated and analyzed using the first-principles method, and the most favorable state is determined by comparing the different adsorption energies. At the same time, the optimal adsorption density of states is calculated and the adsorption reaction mechanism is obtained.

2. Calculational Methods

On the basis of functional theory, the plane-wave method employing a supersoft pseudopotential was adopted, and a GGA-BLYP functional (a Becke–Lee–Yang–Parr functional incorporating the generalized gradient approximation) was used for the approximate calculation [23]. The DMol^3 module [24] in the Material Studio software package (Version 8.0, Accelrys Company, San Diego, CA, America) was used for the calculation of the gas-phase reaction, which includes the optimization of the $SiHCl_3$, BCl_3 and H_2 molecules and the calculation of the energy and frequency. The CASTEP (Cambridge Serial Total Energy Package) module [25] was used to calculate the gas–solid reaction, involving mainly the calculation of the relevant parameters of the adsorption reaction of the adsorbed molecules on the surface of the basic model and a search for the transition state of the entire adsorption reaction. For the Si(100) surface, a 7-layer structure was adopted, with the cutoff energy selected as 450 eV and the k-point grid set to $3 \times 3 \times 1$ [26]. In the supersoft pseudopotential method, the convergence criteria of the pseudopotential have to be selected [27]. The convergence criteria for energy, force and maximum displacement are selected at the same time, with values chosen as energy (2×10^{-5} eV/atom), force (0.05 eV/Å) and maximum displacement (2×10^{-3} Å). The internal stress between atoms must be less than 0.1 GPa [28]. The path of the Brillouin zone was set to GFQZG, the complete LST/QST method was adopted for the transition state search, and a vacuum layer with a thickness of 12 Å was selected to eliminate interaction between the layers. The error in the surface energy was limited to 0.05 J/m^2. The formula used to calculate the surface energy is shown in Equation (1) [29]:

$$\sigma = \frac{E_{slab} - nE_{bulk}}{2A} \tag{1}$$

where E_{slab} is the total energy of the unit-cell of the model, E_{bulk} is the energy of a single atom in the unit cell, n is the number of atoms in the unit cell, and A is the total surface area of the unit cell.

The mechanism of SiHCl$_3$ decomposition on the low index plane of the polycrystal silicon is performed the same as silicon [30]. Therefore, the adsorption energy (the change in the total energy before and after adsorption) of the SiHCl$_3$ and BCl$_3$ molecules on the surface of the Si(100) unit cell is calculated using Equation (2) [31]:

$$E_{adsorption} = E_{surface*} - E_{surface} - E_{molecules} \tag{2}$$

where $E_{adsorption}$ represents the adsorption energy of BCl$_3$ or SiHCl$_3$ onto the surface of the Si(100) unit cell, $E_{surface*}$ represents the energy of the adsorption base surface, $E_{surface}$ is the energy of the surface without any interactions, and the total energy of the molecules before adsorption is represented by $E_{molecules}$.

Figure 1a shows the geometric structure of the model Si(100) unit cell. Its surface contains three types of Si atoms: top, bridge and acupoint Si atoms. To study the adsorption of related molecules on the polysilicon (100) surface, a complete unit cell surface must first be established. The thickness of the vacuum layer was set to 12 Å to avoid layer-to-layer interactions. In the process of establishing the model, in order to satisfy the stability of the model and the accuracy of the experiment at the same time, we did not impose any restrictions on the upper three layers of atoms but fixed the lower four layers of atoms. Simultaneously, to saturate the covalent bonds of each silicon atom in the bottom layer, two H atoms were added to each Si atom in the bottom layer, adding hydrogen atom at the bottom can increase the stability of the model, and the silica–hydrogen bond at the bottom is conducive to the stable convergence of the model, and the surface and near-surface layer atoms were fully relaxed. The supercell chemical formula for the entire unit cell is Si$_{83}$H$_9$. After surface optimization, the optimization curve finally reached a stable convergence state, as shown in Figure 1b, which proved that the surface structure met the experimental requirements.

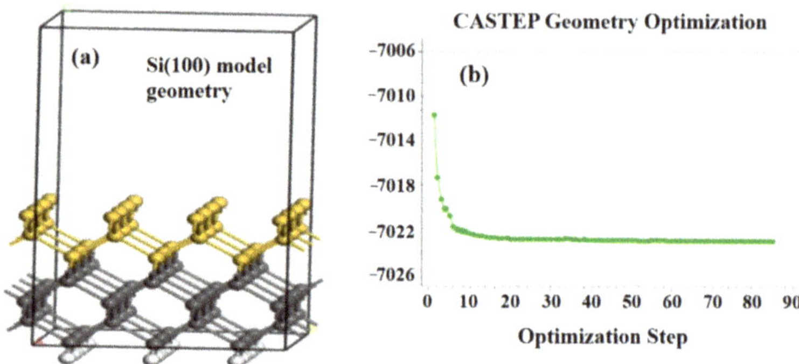

Figure 1. Model geometry and optimization curve of Si(100) surface. (**a**) is the Si(100) model geometry; (**b**) is the model optimization convergence curve.

3. Results and Discussion

3.1. Adsorption of BCl₃ on the Surface of Polysilicon Si(100)

The molecular structure of BCl_3 is a plane regular triangle. The structure of BCl_3 adsorbed on the Si(100) surface was obtained by placing BCl_3 molecules on a super cell (3 × 3) surface and fully relaxing it. The surface optimization of BCl_3 with different structures was performed as follows. A BCl_3 molecule was used to establish stable adsorption models for three orientations: the side position, lateral position and positive position of the molecule. The changes in the distance before and after adsorption were compared and analyzed. The results are shown in Figure 2.

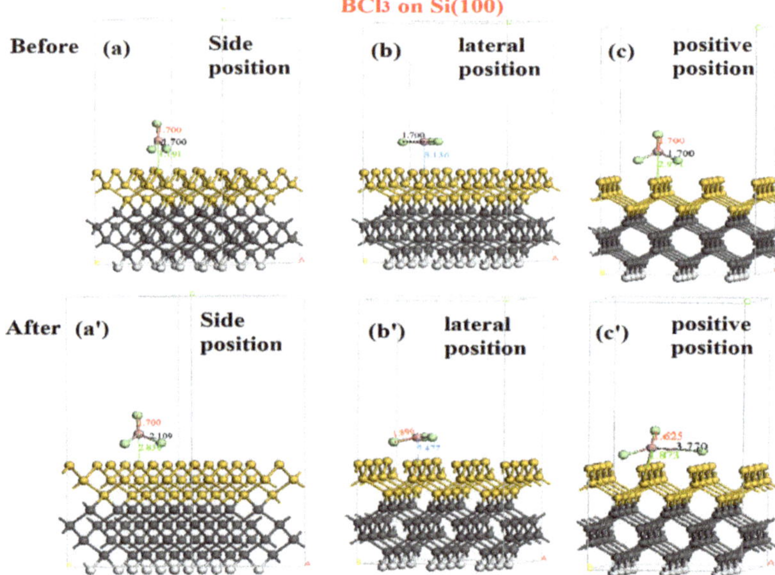

Figure 2. Diagrams of the molecular structure and distance of BCl_3 before and after adsorption on the Si(100) surface. Diagrams (**a–c**) show the adsorption structure before reaction, (**a'–c'**) the adsorption structure after the reaction, ((**a,a'**) show side position of the molecule, (**b,b'**), lateral position of the molecule, and (**c,c'**) positive position of the molecule).

Changes in the distance between the molecules and the surface before and after the reaction for the three adsorption structures were observed. It was found that when BCl_3 molecules are adsorbed on the Si(100) surface of the unit cell in the three possible structures, the distance before and after the adsorption has changed correspondingly, and the distance between the molecules and the surface of the unit cell in the three structural configurations is correspondingly shortened. This shows that an adsorption reaction between the BCl_3 molecule and the Si(100) surface occurs readily. The distances of the ortho-adsorbed BCl_3 molecule before and after optimization were markedly different, at 2.971 Å and 1.873 Å, respectively. The B atom in BCl_3 forms a covalent bond with the Si(100) surface and one of the B–Cl bonds is elongated, indicating that the Si atom has an attractive effect on Cl. Table 1 shows the change in distance before and after adsorption in each of the three positions.

Table 1. Distances before and after the adsorption of BCl_3 on the Si(100) surface.

Adsorption Mode	Distance before Optimization (Å)	Optimized Distance (Å)
Side position of the molecule	3.191	2.836
Lateral position of the molecule	3.139	2.447
Positive position of the molecule	2.917	1.873

The adsorption energies of the three adsorption structures were then calculated. The smaller the adsorption energy, the easier it is to adsorb, and vice versa. The calculated results for the adsorption energy of the side, lateral and positive positions of the molecule are shown in Table 2.

Table 2. Adsorption energy of BCl_3 after adsorption on Si(100) surface.

Adsorption Mode	$E_{adsorption}$ (kcal/mol)
Side position of the molecule	−29.2613
Lateral position of the molecule	−23.5846
Positive position of the molecule	−35.3549

According to the data in Table 2, the adsorption energies of the three adsorption modes on the Si(100) surface are all negative, indicating that BCl_3 molecules are readily adsorbed by the Si(100) surface and that the adsorption reaction occurs easily on the surface. The relative adsorption energies of the three molecular orientations are as follows: E (lateral position) > E (side position) > E (positive position). It can be seen that the orientation in which BCl_3 molecules are most easily adsorbed on the Si(100) surface is the positive position. Thus, the change in the molecular adsorption distance after adsorption was verified, and the most favorable orientation for adsorption was the positive position.

Differential electron density analysis (Figure 3a) and calculation of the charge distribution (Figure 3b) of BCl_3 molecules adsorbed on the Si(100) surface were also carried out. The color of the Si atoms depicted at the bridge sites on the Si(100) surface is denser, indicating that a large amount of charge is transferred in this local area and that the charge has migrated after the adsorption reaction has occurred. The B–Cl bond on the right side is clearly elongated, and an electron cloud is formed between the Cl atom and the silicon atom at the adjacent bridge site on the right, indicating that the silicon atom at the adjacent bridge site on the right has an attractive effect on the Cl atom.

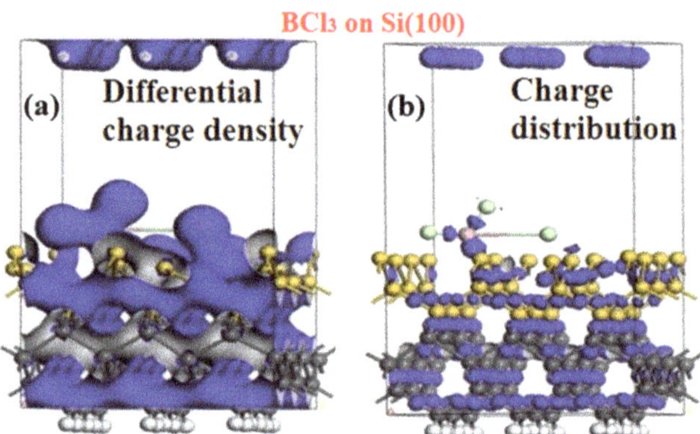

Figure 3. Differential charge density (**a**) and charge distribution (**b**) of BCl_3 adsorbed on the Si(100) surface of the unit cell.

3.2. Adsorption of SiHCl₃ Molecules on the Surface of Polysilicon Si(100)

The molecular structure of $SiHCl_3$ is tetrahedral, and we discuss here the adsorption process of optimized $SiHCl_3$ on the Si(100) surface. Figure 4 shows six possible adsorption positions on the surface, which can be described as the hydrogen top-one-positive position, hydrogen bottom-one-front position, hydrogen top-two-front position, hydrogen side-one-front position, hydrogen side-two-front position, and hydrogen bottom-two-front position. After the optimization calculation, the distance between the molecule and the surface was found to have been shortened, indicating that $SiHCl_3$ is prone to react with the surface of the Si(100) unit cell. This conclusion can also be drawn from the changes in the distance before and after adsorption at the six adsorption sites listed in Table 3. When $SiHCl_3$ is adsorbed at the second bottom-front position of the hydrogen, its distance changes from 3.000 Å to 2.340 Å. The Si atom of $SiHCl_3$ forms a covalent bond with the Si atom at the bridging position, indicating that the $SiHCl_3$ molecule has been adsorbed on the Si(100) unit cell. Simultaneously, it is observed that the Si–Cl bond of the $SiHCl_3$ molecule is distinctly elongated, which indicates that the Cl atom is attracted by the Si atom adjacent to the bridge site.

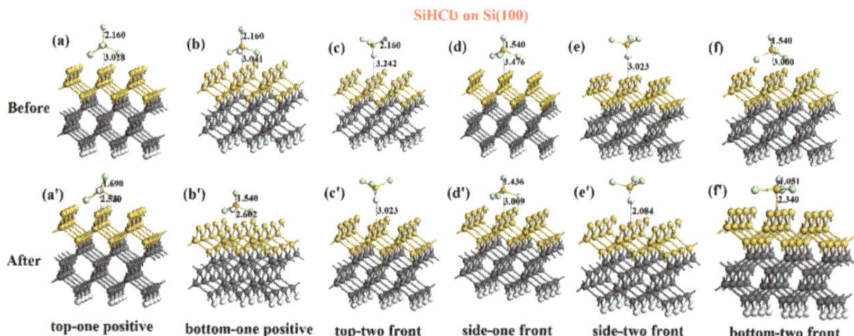

Figure 4. Diagram of molecular structures and distances of $SiHCl_3$ before and after adsorption on the Si(100) surface. Diagrams (**a–f**) show the adsorption structures before reaction, (**a′–f′**) those after reaction ((**a–f,a′–f′**) show hydrogen in the top-one-positive, bottom-one-front, top-two-front, side-one-front, side-two-front, and bottom-two-front positions, respectively).

Table 3. Distances before and after the adsorption of $SiHCl_3$ on the Si(100) surface.

Adsorption Mode	Distance before Optimization (Å)	Optimized Distance (Å)
Hydrogen top-one-positive position	3.018	2.530
Hydrogen bottom-one-front position	3.041	2.602
Hydrogen top-two-front position	3.242	3.023
Hydrogen side-one-front position	3.476	3.009
Hydrogen side-two-front position	3.023	2.804
Hydrogen bottom-two-front position	3.000	2.340

Table 4 shows the calculated adsorption energy of $SiHCl_3$ on the Si(100) surface. The adsorption energies obtained for the six adsorption structures are all negative, indicating that SiHCl3 reacts readily with the Si(100) surface. The order of the adsorption energies of the several positions is as follows: E(hydrogen side-two-front position) > E(hydrogen top-two-front position) > E(hydrogen top-one-positive position) > E(hydrogen bottom-one-front position) > E(hydrogen side-one-front position) > E(hydrogen bottom-two-front position). The adsorption energy (E) of $SiHCl_3$ is greatest when the adsorption structure of the second hydrogen base is in the front position, indicating that adsorption is more likely to occur in this orientation, at the front position of the hydrogen base 2. From the calculated adsorption energies, it can be seen that the adsorption of $SiHCl_3$ molecules in the hydrogen bottom-two-front position corresponds to the most favorable orientation, and the calculated adsorption distances confirm this conclusion.

Table 4. Adsorption energy of $SiHCl_3$ after adsorption on the Si(100) surface.

Adsorption Mode	$E_{adsorption}$ (kcal/mol)
Hydrogen top-one-positive position	−4.02
Hydrogen bottom-one-front position	−6.07
Hydrogen top-two-front position	−3.41
Hydrogen side-one-front position	−8.92
Hydrogen side-two-front position	−2.47
Hydrogen bottom-two-front position	−10.64

To gain a deeper understanding of the charge distribution in the adsorption process of $SiHCl_3$, we can use the electronic levels to characterize the reaction. The differential electron density and charge distribution of the $SiHCl_3$ molecules adsorbed on the surface of the Si(100) unit cell from the face center downward were calculated. The calculated results are shown in Figure 5a,b. From the results shown in the figure, it can be seen that the color representing the charge concentrated in the bond between the Si atoms on the Si(100) surface and the Si atom in $SiHCl_3$ is darker than that for other bonds. This not only shows that electrons have been transferred after adsorption, but also shows that a large amount of charge is transferred in this local area. Looking at the image of the adsorbed $SiHCl_3$ molecule, it is seen that the three Cl atoms and the bonding angles between the H and Si atoms in the molecule have changed, and both the Cl and H atoms show an upward posture. In addition, it can be seen that the silicon atoms on the surface and the two Si atoms at the right and back of the bridge silicon atoms adjacent to the bonded Si atoms and the corresponding Cl atoms form an electron cloud. This proves that the two have an attractive effect.

Figure 5. Differential charge density (**a**) and charge distribution (**b**) after $SiHCl_3$ is adsorbed on the Si(100) surface.

3.3. Co-Adsorption of BCl₃ and SiHCl₃ on the Surface of Polysilicon Si(100)

A BCl_3 molecule and a $SiHCl_3$ molecule were placed together on the surface of the Si(100) supercell to investigate the adsorption reaction. Figure 6 shows the structural changes before and after the co-adsorption of BCl_3 and $SiHCl_3$ on the Si(100) surface. Compared to the structure before adsorption in Figure 6a, it can be concluded from the structure after adsorption (Figure 6b) that the Si atom in the $SiHCl_3$ molecule forms a covalent bond with the B atom in BCl_3. A Cl atom in each of the BCl_3 and $SiHCl_3$ molecules undergoes a dissociation reaction and is adsorbed on the Si atom at the nearest bridge site on the Si(100) surface. The bond lengths of the other Cl–B bonds in the BCl_3 molecule are significantly elongated, and the B atom is adsorbed on the Si(100) surface at the same time. These results indicate that BCl_3 and $SiHCl_3$ molecules are placed on the Si(100) surface simultaneously. After the adsorption reaction, it is evident that BCl_3 is more easily adsorbed on the Si(100) surface than is $SiHCl_3$. When BCl_3 and $SiHCl_3$ coexist on the Si(100) surface, they are more likely to have an effect than $SiHCl_3$ alone on the surface. Moreover, their simultaneous presence there will be accompanied by a dissociation process; in which each dissociates a Cl atom, and the Cl atom in the BCl_3 molecule is attracted by the Si(100) surface.

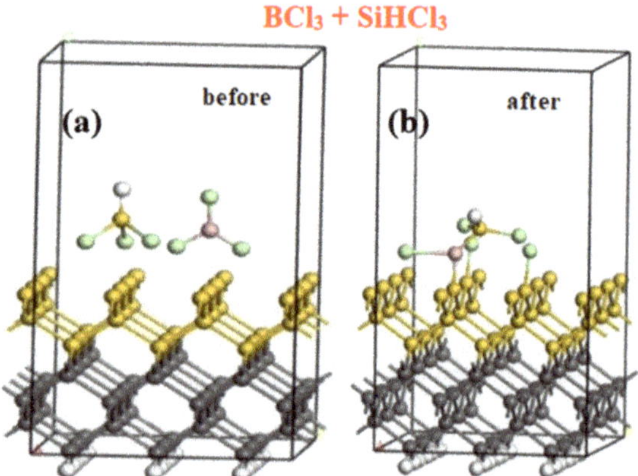

Figure 6. Structure changes before (**a**) and after (**b**) the co-adsorption of BCl_3 and $SiHCl_3$.

A microscopic electronic analysis was also performed of the charge distribution and differential electron density when BCl_3 and $SiHCl_3$ are present together on the surface of polysilicon. The darker blue color between BCl_3 and $SiHCl_3$ in Figure 7b indicates that there is a large amount of electron transfer, and it can be seen that the B–Cl bond in the BCl_3 molecule is distinctly elongated, and a dissociated Cl atom is adsorbed on the Si(100) surface, indicating that it charge transfer has occurred during the adsorption process. From the charge density results in Figure 7a, indicating charge transfer between the surface and the molecules, we can also see where the electrons accumulate, which proves the occurrence of the adsorption reaction.

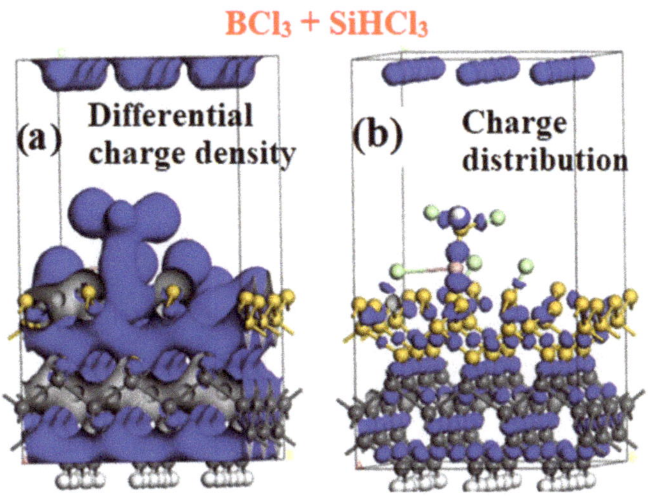

Figure 7. Differential charge density (**a**) and charge distribution (**b**) before and after the co-adsorption of BCl_3 and $SiHCl_3$ on the surface.

3.4. Electronic Structure Analysis

To further understand the bonding mechanism of the adsorption of BCl_3 and $SiHCl_3$ on the Si(100) surface, the adsorption density of states, partial density of states, and Mulliken charge distribution were calculated, and the results are shown in Figure 8 and Table 5. It can be observed that the Si3s orbital lies far away in energy from the Fermi level on the surface of the Si(100) unit cell. After the adsorption reaction, the electrons of the Si3s orbital are not easily transferred. The valence band where electron transfer occurs is Si3p, which constitutes the valence band on the left side of the Fermi level on the surface of the Si(100) unit cell. Although it is theoretically impossible for the Si3s orbital to transfer electrons, a very small amount of electron transfer does occur from the Si3s orbital after the reaction. The valence state on the right side of the Fermi level is composed of Si3d orbitals, and the bandgap between the top of the valence band and the bottom of the conduction band is 1.502 eV. After the reaction, the bandgap is reduced to 0.087 eV. The narrowing of the bandgap indicates that the Si is more likely to react, and its chemical properties become more active, which in turn indicates that it is more likely to interact with BCl_3 and $SiHCl_3$ molecules.

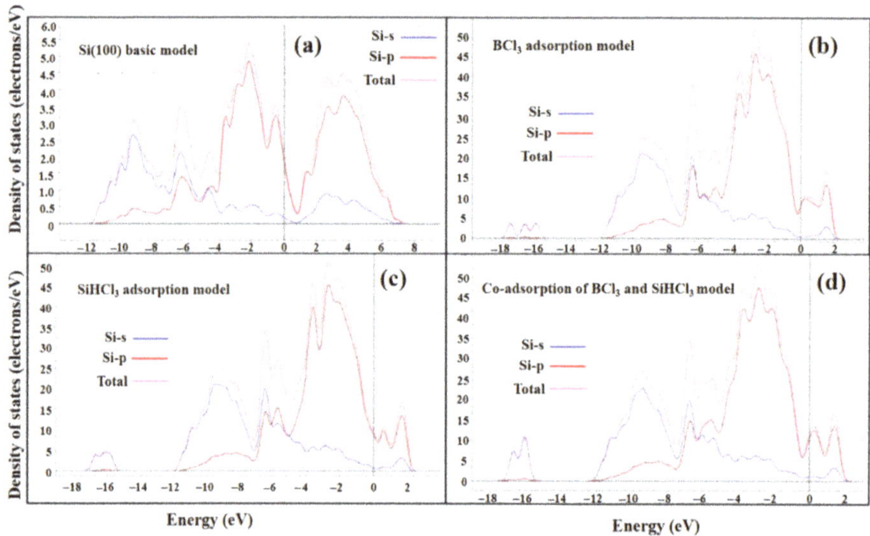

Figure 8. State density changes of BCl_3 and $SiHCl_3$ molecules after the adsorption on the Si(100) cell surface. (**a**) is the density of states of Si(100); (**b**) is the density of states for the BCl_3 adsorption model; (**c**) is the density of states for the $SiHCl_3$ adsorption model; (**d**) is the density of states for the co-adsorption of BCl_3 and $SiHCl_3$.

Table 5. Mulliken charge co-adsorption of BCl_3 and $SiHCl_3$ molecules on the surface of Si(100) unit cell.

Project	Species	S	p	d	Total	Charge (eV)	Change (eV)
Ideal BCl_3	B	0	6.26	2.44	10.87	1.13	
	Cl1	1.96	5.33	0	7.28	−0.28	
	Cl2	1.96	5.33	0	7.28	−0.28	0
	Cl3	1.96	5.32	0	7.28	−0.28	
Ideal $SiHCl_3$	Si	3.18	7.42	3.36	13.96	1.13	
	Cl1	2.08	6.12	0	8.20	−0.32	
	Cl2	2.08	6.12	0	8.20	−0.33	0
	Cl3	2.08	6.12	0	8.20	−0.35	
	H	1.76	4.28	0	6.04	−0.30	
Si(100) surface adsorption BCl_3	B	0	6.23	2.17	10.68	1.32	
	Cl1	1.95	5.36	0	7.34	−0.31	
	Cl2	1.95	5.32	0	7.27	−0.27	0.29
	Cl3	1.95	5.33	0	7.28	−0.28	
Si(100) surface adsorption $SiHCl_3$	Si	2.21	6.14	2.13	10.48	1.02	
	Cl1	1.95	5.39	0	7.34	−0.34	
	Cl2	1.95	5.39	0	7.24	−0.34	0.24
	Cl3	1.95	5.33	0	7.29	−0.29	
	H	1.95	5.36	0	7.34	−0.31	

By analyzing the electronic density of states of BCl_3 and $SiHCl_3$ molecules adsorbed on the surface of the Si(100) unit cell in Figure 8, it is found that after the two molecules are adsorbed on the surface of the unit cell, their band gaps are narrowed and the electrons inside the molecules are transferred to the outermost electron layer and to the conduction band, which is more reactive. Based on molecular chemical structure theory, we know that the B atom in the BCl_3 molecule has three coordination sites, and the electron configuration of B is $1s^2 2s^2 2p^1$, showing that B is an electron deficient atom. Six electrons are provided by three Cl atoms, and each of the three Cl atoms also provides a filled 2p orbital that is side-by-side with an empty p orbital of the B atom and is thus able to form a strong π bond. The formation of bonds by sp^2 hybridization can explain the three B–Cl bonds in BCl_3. Similarly, for the $SiHCl_3$ molecule, electrons are transferred to the outermost electron layer. The Si atom in the $SiHCl_3$ molecule has four coordination sites and the outermost electron configuration is $3s^2 3p^2$, which is eventually completed by the four covalent bonds formed with Si. This can explain the origin of the four covalent bonds, and it can be concluded that the spatial configuration of $SiHCl_3$ is a that of a regular tetrahedron.

From the Mulliken charge distribution of the two molecules shown in Table 5, it can be observed that when $SiHCl_3$ is adsorbed on the surface of the Si(100) unit cell, charge transfer occurs, and there are two charge transfer paths. The first is the transfer of 1.30 eV of charge from the Si atom of the $SiHCl_3$ molecule to the surface of the unit cell. The second is charge transfer from the surface of the unit cell to the three Cl atoms and the H atom of the $SiHCl_3$ molecule, a charge transfer of 1.06 eV. After fitting and offsetting, it is found that during the entire adsorption reaction of $SiHCl_3$ on the surface of the Si(100) unit cell, the Si atom in the $SiHCl_3$ has transferred a net 0.24 eV of charge to the surface of the Si(100) unit cell. When BCl_3 is adsorbed on the surface of the Si(100) unit cell, significant charge transfer also occurs, and there are two charge transfer paths. The first is a transfer of 1.52 eV from the B atom in BCl_3 to the surface of the unit cell. The second is a transfer of 1.23 eV from the surface of the unit cell to the three Cl atoms and one H atom in the BCl_3 molecule. After fitting and offsetting, during the entire adsorption reaction of BCl_3 on the surface of the Si(100) unit cell, the B atom in BCl_3 is seen to have transferred 0.29 eV of charge to the surface of the Si(100) unit cell.

According to the results for transfer of electrons between $SiHCl_3$, BCl_3, and the surface of polysilicon during the adsorption process, both molecules transfer charge to the surface of the unit cell during adsorption, with BCl_3 transferring 0.05 eV more charge to the surface of the Si(100) unit cell than $SiHCl_3$ does. It can be shown that the surface of the Si(100) unit cell has a definite adsorption effect on BCl_3 and $SiHCl_3$ molecules, and that the B atoms in BCl_3 are more easily adsorbed on the Si(100) surface, which means that B and polysilicon will be deposited on the silicon rod together during the production process. The above results also explain the phenomenon that when BCl_3 and $SiHCl_3$ molecules are co-adsorbed on the surface of the Si(100) unit cell, the Si atoms in the $SiHCl_3$ molecules first form a covalent bond with the B atoms in BCl_3.

The B impurities may deposit on the surface of the silicon rod competing with polysilicon and the trace B-compound impurities may be enriched by the influence of transport phenomena. The above two phenomena are the main reasons causing the high-impurity B in the silicon production. It is obvious that the adsorption of BCl_3 on silicon to cause the deposition of B on silicon surface. Therefore, transport phenomenon investigation may be effective method to reduce boron impurity inside poly-silicon, such as increasing mole fraction of H_2 and temperature.

4. Conclusions

(1) BCl_3 and $SiHCl_3$ are mainly adsorbed on the surface of the Si(100) unit cell in the positive position and the hydrogen bottom-two-front position of the molecule, respectively. The discrete distances of BCl_3 and $SiHCl_3$ from the surface of the Si(100) unit cell are 1.873 Å and 2.340 Å, respectively, and the separation energies are −35.2549 kcal/mol and −10.64 kcal/mol, respectively.

(2) Compared with $SiHCl_3$, BCl_3 reacts more easily with the Si(100) surface, and when BCl_3 and $SiHCl_3$ coexist, BCl_3 reacts more readily than $SiHCl_3$ with the Si(100) surface. When BCl_3 and $SiHCl_3$ are present simultaneously, the gas phase reaction is accompanied by a dissociation process, in which each molecule dissociates a Cl atom that is adsorbed on the Si(100) surface. At the same time, a distinctly elongated B–Cl bond shows that the Si(100) surface also has an attractive effect on Cl atoms.

(3) After the adsorption of $SiHCl_3$ and BCl_3, 0.24 and 0.29 eV of charge, respectively, are found to have been transferred from the molecule to the surface of the unit cell. Both BCl_3 and $SiHCl_3$ are readily adsorbed on the surface of the Si(100) unit cell, but BCl_3 is more easily adsorbed. These results confirm that the B atom in BCl_3 in the adsorption model forms a covalent bond with the Si atom on the Si(100) unit cell surface, and the Si atom in the $SiHCl_3$ molecule forms a covalent bond with the B atom in BCl_3.

Author Contributions: Conceptualization, L.H. and Y.H.; Data curation, F.C., Y.H. and N.Y.; Investigation, J.W., Y.H. and N.Y.; Methodology, N.Y. and G.X.; Software, L.T. and L.H.; Validation, Q.Y.; Writing—Original draft, Q.Y.; Writing—Review and editing, Q.Y., J.W. and Y.H. All authors have read and agreed to the published version of the manuscript.

Funding: This research was funded by the National Natural Science Foundation of China Project (No. 21566015 and No. 52074141).

Institutional Review Board Statement: Not applicable.

Informed Consent Statement: Informed consent was obtained from all subjects involved in the study.

Data Availability Statement: Data sharing not applicable. No new data were created or analyzed in this study. Data sharing is not applicable to this article.

Conflicts of Interest: The authors declare no conflict of interest.

References

1. Chen, J.; Zhang, X. Prediction of thermal conductivity and phonon spectral of silicon material with pores for semiconductor device. *Phys. B Condens. Matter* **2021**, *614*, 413034. [CrossRef]
2. Chen, P.; Zeng, Q.; Guo, W.; Chen, J. The source, enrichment and precipitation of ore-forming elements for porphyry Mo deposit: Evidences from melt inclusions, biotite and fluorite in Dasuji deposit, China. *Ore Geol. Rev.* **2021**, *135*, 104205. [CrossRef]
3. Cheng, D.; Gao, Y.; Liu, R. Finite element analysis on processing stress of polysilicon cut by diamond multi-wire saw. *Mater. Sci. Semicond. Process.* **2021**, *131*, 105860. [CrossRef]
4. Xiong, Q.M.; Chen, Z.; Huang, J.-T.; Zhang, M.; Song, H.; Hou, X.F.; Li, X.B.; Feng, Z.J. Preparation, structure and mechanical properties of Sialon ceramics by transition metal-catalyzed nitriding reaction. *Rare Met.* **2020**, *39*, 589–596. [CrossRef]
5. Zhou, L.; Li, S.; Tang, Q.; Deng, X.; Wei, K.; Ma, W. Effects of solidification rate on the leaching behavior of metallic impurities in metallurgical grade silicon. *J. Alloys Compd.* **2021**, *882*, 160570. [CrossRef]
6. Gao, M.; Liu, B.; Zhang, X.; Zhang, Y.; Li, X.; Han, G. Ultrathin MoS_2 nanosheets anchored on carbon nanofibers as free-standing flexible anode with stable lithium storage performance. *J. Alloys Compd.* **2022**, *894*, 162550. [CrossRef]
7. O'Mara, W.; Herring, R.B.; Hunt, L.P. *Handbook of Semiconductor Silicon Technology*; Crest Publishing House: Kettering, UK, 2007.
8. Davis, K.O.; Brooker, R.P.; Seigneur, H.P.; Rodgers, M.; Rudack, A.C.; Schoenfeld, W.V. Pareto analysis of critical challenges for emerging manufacturing technologies in silicon photovoltaics. *Sol. Energy* **2014**, *107*, 681–691. [CrossRef]
9. Chanlaor, P.; Limtrakul, S.; Vatanatham, T.; Ramachandran, P.A. Modeling of Chemical Vapor Deposition of Silane for Silicon Production in a Spouted Bed via Discrete Element Method Coupled with Eulerian Model. *Ind. Eng. Chem. Res.* **2018**, *57*, 12096–12112. [CrossRef]
10. Filtvedt, W.O.; Holt, A.; Ramachandran, P.A.; Melaaen, M.C. Chemical vapor deposition of silicon from silane: Review of growth mechanisms and modeling/scaleup of fluidized bed reactors. *Sol. Energy Mater. Sol. Cells* **2012**, *107*, 188–200. [CrossRef]
11. Shi, S.; Guo, X.; An, G.; Jiang, D.; Qin, S.; Meng, J.; Li, P.; Tan, Y.; Noor ul Huda Khan Asghar, H.M. Separation of boron from silicon by steam-added electron beam melting. *Sep. Purif. Technol.* **2019**, *215*, 242–248. [CrossRef]
12. Wang, C.; Wang, T.; Li, P.; Wang, Z. Recycling of $SiCl_4$ in the manufacture of granular polysilicon in a fluidized bed reactor. *Chem. Eng. J.* **2013**, *220*, 81–88. [CrossRef]
13. Safarian, J. Thermochemical Aspects of Boron and Phosphorus Distribution Between Silicon and BaO-SiO_2 and CaO-BaO-SiO_2 lags. *Silicon* **2019**, *11*, 437–451. [CrossRef]
14. Bye, G.; Ceccaroli, B. Solar grade silicon: Technology status and industrial trends. *Sol. Energy Mater. Sol. Cells* **2014**, *130*, 634–646. [CrossRef]
15. Bullón, J.; Margaria, T.; Miranda, A.; Miguez, J. Physical and chemical analysis of the upgraded metallurgical silicon for the solar industry. In *Silicon Form the Chemical and Solar Industry IX*; Norwegian University of Science and Technology: Oslo, Norway, 2008.
16. Haupfear, E.A.; Schmidt, L.D. Kinetics of boron deposition from BBR_3 + H_2. *Chem. Eng. Sci.* **1994**, *49*, 2467–2481. [CrossRef]
17. Liu, W.; Huang, F.; Liao, Y.; Zhang, J.; Ren, G.; Zhuang, Z.; Zhen, J.; Lin, Z.; Wang, C. Treatment of CrVI-Containing $Mg(OH)_2$ Nanowaste. *Angew. Chem.* **2008**, *120*, 5701–5704. [CrossRef]

18. Jenkinson, J.P.; Pollard, R. Thermal diffusion effects in chemical vapor deposition reactors. *J. Electrochem. Soc.* **1984**, *131*, 2911. [CrossRef]

19. Sezgi, N.A.; Ersoy, A.; Dogu, T.; Özbelge, H.Ö. CVD of boron and dichloroborane formation in a hot-wire fiber growth reactor. *Chem. Eng. Process. Process Intensif.* **2001**, *40*, 525–530. [CrossRef]

20. Asli Sezgi, N.; Dogu, T.; Onder Ozbelge, H. Mechanism of CVD of boron by hydrogen reduction of BCl_3 in a dual impinging-jet reactor. *Chem. Eng. Sci.* **1999**, *54*, 3297–3304. [CrossRef]

21. Liu, Y.; Zhang, L.; Cheng, L.; Zeng, Q.; Zhang, W.; Yang, W.; Feng, Z.; Li, S.; Zeng, B. Uniform design and regression analysis of LPCVD boron carbide from BCl_3–CH_4–H_2 system. *Appl. Surf. Sci.* **2009**, *255*, 5729–5735. [CrossRef]

22. Deng, J.; Cheng, L.; Zheng, G.; Su, K.; Zhang, L. Thermodynamic study on co-deposition of ZrB_2–SiC from $ZrCl_4$–BCl_3–CH_3SiCl_3–H_2–Ar system. *Thin Solid Film.* **2012**, *520*, 7030–7034. [CrossRef]

23. Mahammedi, N.A.; Ferhat, M.; Belkada, R. Prediction of indirect to direct band gap transition under tensile biaxial strain in type-I guest-free silicon clathrate Si46: A first-principles approach. *Superlattices Microstruct.* **2016**, *100*, 296–305. [CrossRef]

24. Delley, B. An all-electron numerical method for solving the local density functional for polyatomic molecules. *J. Chem. Phys.* **1990**, *92*, 508–517. [CrossRef]

25. Clark, S.J.; Segall, M.D.; Pickard, C.J.; Hasnip, P.J.; Probert, M.I.; Refson, K.; Payne, M.C. First principles methods using CASTEP. *Z. Für Krist. Cryst. Mater.* **2005**, *220*, 567–570. [CrossRef]

26. Zhang, Y.F.; Viñes, F.; Xu, Y.J.; Li, Y.; Li, J.Q.; Illas, F. Role of Kinetics in the Selective Surface Oxidations of Transition Metal Carbides. *J. Phys. Chem. B* **2006**, *110*, 15454–15458. [CrossRef] [PubMed]

27. Vanderbilt, D. Soft self-consistent pseudopotentials in a generalized eigenvalue formalism. *Phys. Rev. B* **1990**, *41*, 7892. [CrossRef] [PubMed]

28. Wang, Z.; Chen, X.; Cheng, Y.; Niu, C. Adsorption and Deposition of Li_2O_2 on the Pristine and Oxidized TiC Surface by First-principles Calculation. *J. Phys. Chem. C* **2015**, *119*, 25684–25695. [CrossRef]

29. Sun, W.; Ceder, G. Efficient creation and convergence of surface slabs. *Surf. Sci.* **2013**, *617*, 53–59. [CrossRef]

30. Peng, M.; Shi, B.; Han, Y.; Li, W.; Zhang, J. Crystal facet dependence of SiHCl3 reduction to Si mechanism on silicon rod. *Appl. Surf. Sci.* **2022**, *580*, 152366. [CrossRef]

31. Galán, O.A.L.; Carbajal-Franco, G. Energy profiles by DFT methods for CO and NO catalytic adsorption over ZnO surfaces. *Catal. Today* **2021**, *360*, 38–45. [CrossRef]

Article

Detoxification of Arsenic-Containing Copper Smelting Dust by Electrochemical Advanced Oxidation Technology

Meng Li [1,2,3], Junfan Yuan [1], Bingbing Liu [1], Hao Du [2], David Dreisinger [3], Yijun Cao [1,4,*] and Guihong Han [1,*]

[1] School of Chemical Engineering, Zhengzhou University, Zhengzhou 450001, China; li.meng@zzu.edu (M.L.); moguizihuan@163.com (J.Y.); liubingbing@zzu.edu.cn (B.L.)

[2] National Engineering Laboratory for Hydrometallurgical Cleaner Production Technology, Key Laboratory of Green Process and Engineering, Institute of Process Engineering, Chinese Academy of Sciences, Beijing 100190, China; hdu@ipe.ac.cn

[3] Department of Materials Engineering, The University of British Columbia, Vancouver, BC V6T1Z4, Canada; david.dreisinger@ubc.ca

[4] Henan Province Industrial Technology Research Institute of Resources and Materials, Zhengzhou University, Zhengzhou 450001, China

* Correspondence: yijuncao@126.com (Y.C.); hanguihong@zzu.edu.cn (G.H.)

Abstract: A large amount of arsenic-containing solid waste is produced in the metallurgical process of heavy nonferrous metals (copper, lead, and zinc). The landfill disposal of these arsenic-containing solid waste will cause serious environmental problems and endanger people's health. An electrochemical advanced oxidation experiment was carried out with the cathode modified by adding carbon black and polytetrafluoroethylene (PTFE) emulsion. The removal rate of arsenic using advanced electrochemical oxidation with the modified cathode in 75 g/L NaOH at 25 °C for 90 min reached 98.4%, which was significantly higher than 80.69% of the alkaline leaching arsenic removal process. The use of electrochemical advanced oxidation technology can efficiently deal with the problem of arsenic-containing toxic solid waste, considered as a cleaner and efficient method.

Keywords: arsenic; copper smelting dust; electrochemical advanced oxidation technology; iron-free Fenton-like reaction

Citation: Li, M.; Yuan, J.; Liu, B.; Du, H.; Dreisinger, D.; Cao, Y.; Han, G. Detoxification of Arsenic-Containing Copper Smelting Dust by Electrochemical Advanced Oxidation Technology. *Minerals* **2021**, *11*, 1311. https://doi.org/10.3390/min11121311

Academic Editors: Shuai Wang, Xingjie Wang, Jia Yang and Carlito Tabelin

Received: 15 October 2021
Accepted: 22 November 2021
Published: 24 November 2021

Publisher's Note: MDPI stays neutral with regard to jurisdictional claims in published maps and institutional affiliations.

1. Introduction

The output of nonferrous metals is increasing by the year, and most of the heavy nonferrous metal concentrates contain arsenic. The arsenic in these heavy nonferrous metal minerals is mainly combined with copper, lead, zinc, and sulfur [1,2]. In the nonferrous metal metallurgical process, arsenic usually enters smoke, smelting residues, and acid in the form of waste slag, wastewater, and exhaust gas, respectively [3]. Compared with elemental arsenic, arsenic compounds are more toxic. Arsenic can enter the human body through air, water, and other ways, causing great harm to the human skin, digestive, and respiratory system [4]. Long-term exposure to arsenic substances can cause skin cancer, lung cancer, bladder cancer, and kidney cancer [5]. Tumor tissue analyses confirmed that it is one of the human carcinogens. Arsenic-containing copper smelting dust (ARCD) is a solid hazardous waste formed through processes such as copper smelting, converting, and anode slime smelting [6]. Arsenic mainly forms compounds with copper and sulfur in copper sulfide concentrates. During the copper smelting process, part of the arsenic will enter the copper smelting dust. If copper smelting dust is recycled directly to copper smelting without arsenic removal, arsenic will accumulate, reducing the main metallurgy production efficiency and affecting the entire smelting process [7]. Further, if the ARCD is not appropriately treated, it will have a more significant impact on the environment and human health.

The main methods for removing arsenic from arsenic-containing solid waste include alkaline leaching, acid leaching, and roasting. The alkaline leaching process mainly converts arsenic from solid waste residue to the liquid phase by alkali to realize the separation of arsenic from other metals [8–10]. The acid leaching process usually uses sulfuric acid and hydrochloric acid to leach the arsenic-containing material and then separates the arsenic from the arsenic-containing material through solid–liquid separation [11]. The roasting method mainly includes oxidation roasting, reduction roasting, and solidification calcination. The roasting arsenic removal process is used mainly to volatilize the arsenic in the form of arsenic trioxide at high temperatures to separate it from other valuable metals [12,13]. The As(III) in the solid waste is converted to As(V) into the liquid phase and then removed by precipitation, adsorption, or other methods [14–17]. Most of the arsenic in ARCD exists as As(III). Converting As(III) to As(V) can reduce its toxicity. The standard methods for oxidizing As(III) to As(V) include H_2O_2 oxidation, O_3 oxidation, and others [18–21]. However, the oxidation of these oxidants may lead to the formation of toxic by-products, and the cost is relatively high. The electrochemical advanced oxidation technology is characterized by the generation of H_2O_2, hydroxyl radicals, active chlorine species with strong oxidizing ability, etc. [22]. The use of electrochemical advanced oxidation technology can be even more efficient and convenient for the oxidation of As(III) to As(V) [23–25].

The electrochemical advanced oxidation method mainly generates H_2O_2 in situ; namely, when the oxygen passes into the cathode, the H_2O_2 is in-situ generated through the two-electron oxygen reduction reaction (ORR) [26]. The H_2O_2 generated in situ at the cathode can reduce the transportation and storage costs of H_2O_2 during the reaction process, and the oxidation efficiency may increase [26]. Advanced electrochemical oxidation has the advantages of high efficiency, green environmental protection, and controllable process of oxidation method. This approach uses different electrode materials to generate oxidants. Compared with oxidation methods such as ozone oxidation, ultraviolet oxidation, and microwave oxidation catalysis, the electrochemical advanced oxidation method has the advantages of rapid reaction and simple operation and can generate active oxygen groups in situ for material oxidation. At present, the most widely used cathode materials are carbonaceous due to their high stability, corrosion resistance, and high conductivity. Carbonaceous materials mainly use graphite, carbon nanotubes, and carbon felt (CF). The carbon felt electrode has a porous structure conducive to the transmission of oxygen, causing higher in situ generation of H_2O_2 [27,28].

The modification of carbon felt can increase the electrochemical activity of the electrode and promote the generation of H_2O_2 by introducing active oxygen-containing groups on the surface. Treated with KOH firstly, then calcined at a high temperature, the oxygen-containing functional groups, the specific surface area, and the microporous structure of the carbon felt will be increased [29]. The surface of carbon felt is doped with nonmetallic elements, and the electrochemical performance of the electrode can be enhanced by doping with nonmetallic elements such as N, P, and S. After doping these nonmetal elements, the number of H_2O_2 produced can be increased, the secondary pollution in the electrochemical reaction can be reduced, and the pH range of electrochemical experiments can be expanded [30,31]. Metal materials have a strong electrical conductivity and catalytic activity, so the electrical conductivity and electrochemical performance of the electrode are enhanced by introducing metal ions. By introducing Fe, Cu, and other elements, the electrode gains higher electrochemical performance [32].

The carbon felt is modified by carbon black and PTFE [2]. Under the action of carbon black, the conductivity of the carbon felt electrode, the active sites of the reaction, and the electrochemical performance of the carbon felt cathode would be improved. The carbon felt electrode was characterized and analyzed using a scanning electron microscope, nitrogen adsorption–desorption test, and water contact angle. The modified carbon felt electrode was used to remove arsenic from the ARCD by electrochemical advanced oxidation method.

2. Experimental Procedures

2.1. Materials

The ARCD collected from Western Mining Group Co., Ltd. (Xining, China) was dried at 60 °C for 24 h, then it was crushed, ground, and sieved. The particle size of $-74{\sim}+48$ μm was selected as the experimental material. All chemicals used in this experiment were of analytical grade. The experimental water had high purity. The experimental carbon felt of Inner Mongolia Wanxing Carbon Co., Ltd., the carbon black of the US CABOT Cabot Black VXC-72, and the PTFE of Daikin Fluorochemicals (China) Co., Ltd. (Changshu, China) were used in the experiment.

2.2. Methods

The carbon felt was ultrasonically cleaned in a mixed solution of acetone, ethanol, and ultrapure water for 30 min and then dried in a vacuum drying oven at 60 °C for 24 h. The carbon felt sample was labeled as the original carbon felt. A certain proportion of PTFE emulsion (60%) and carbon black was mixed with 30 mL of ultrapure water and 1 mL of n-butanol and then ultrasonically mixed for 10 min. The carbon felt was immersed in the mixture and ultrasonicated for 30 min. The ultrasonic carbon felt was dried in a vacuum for 24 h. The dried carbon felt was calcined at 360 °C for 1 h. The calcined carbon felt was labeled as modified carbon felt.

The experiment was carried out in a 250 mL undivided three-electrode cell at 25 °C, and the temperature was controlled by an electrochemical workstation (PARSTAT 4000A). The original carbon felt and the modified carbon felt were used as the working electrode, the platinum sheet was used as the counter electrode, and the mercury-oxide mercury electrode was used as the reference electrode. The electrolyte was the NaOH solution with a concentration of 75 g/L. Before electrochemical experiments, oxygen was pumped into the solution for 30 min to saturate the oxygen in the electrolyte. The ARCD was added to the electrolyte at a concentration of 500 mg/L. The arsenic removal experiment was conducted at the condition of original carbon felt and modified carbon felt.

The arsenic in the leaching solution was removed by calcium salt, and $Ca(OH)_2$ was added to an exact amount of 100 mL arsenic leaching solution. The mixed solution was placed in a constant temperature water bath at 65 °C for 2 h with the rotation speed of 300 r/min and then filtered by a water circulating vacuum pump. The filtrate was added 0.075 mol/L aluminum chloride as a flocculant in a constant temperature water bath for further arsenic removal [33,34]. The filtrate was stirred at a speed of 120 r/min for 5 min and then stirred at a speed of 80 r/min for 15 min. After the mixture was complete, the filtering operation was performed. Finally, the obtained final filtrate can be recycled for the leaching process after adjusting the concentration, and the final residue can be recycled for the copper smelting process.

2.3. Characterization and Analysis

The phases of ARCD were determined by X-ray diffraction (XRD, PANalytical Empyrean, Almelo, The Netherlands), and the chemical composition was determined by inductively coupled optical emission spectrometry (ICP-OES, PerkinElmer Optima 7300 V, Richmond, CA, USA). The carbon felt and modified carbon felt were characterized by a scanning electron microscope (SEM, FEI Quanta 250, Hillsboro, OH, USA) equipped with energy dispersive X-ray spectrometry (EDS, EDAX Genesis, Richmond, CA, USA). Further, its microscopic morphology, the specific surface area, and pore size distribution were measured by the BET with N_2 adsorption and desorption analyzer. The hydrophilicity and hydrophobicity of the electrode surface were measured by a contact angle meter (Theta, Biolin, Espoo, Finland), and the pH value of the electrolyte was obtained using a pH meter.

The concentration of H_2O_2 generated in the electrolyte was measured with a double-beam UV-visible spectrophotometer using a potassium titanium oxalate developer. Electron spin resonance spectroscopy (ESR, Bruker EMX12, Karlsruhe, Germany) was used to measure the ·OH content in the electrolyte, and DMPO was added as a quencher for OH for comparison experiments.

The leaching efficiencies of the target metal were obtained as follows:

$$\eta = \frac{C_{As(\text{solution})} \cdot V}{C_{As(\text{solid})} \cdot M} \times 100\% \tag{1}$$

where $C_{As(\text{solution})}$ represents the As content in the solution, V represents the electrolyte volume, $C_{As(\text{solid})}$ corresponds to the arsenic content in the ARCD, and M is the content of the ARCD in the electrolyte.

3. Results and Discussion

3.1. Characterization of the Spent Catalyst of the ARCD

The ARCD contains a variety of metal elements. As shown in Table 1, the content of As_2O_3 reaches 8.78%, and the content of CuO, ZnO, PbO, and Bi_2O_3 is 15.31%, 1.95%, 17.7%, and 6.62%, respectively. The XRD shows that the main phases in the smelting dust are As_2O_5, As_2O_3, and CuAsS. The XPS analysis shows that the primary forms of arsenic in the material are As(III) and As(V). The contents of As(V) and As(III) in the ARCD are 59% and 41%, respectively. As shown in Figure 1d, the copper in the ARCD mainly exists in Cu(I) and Cu(II). The ARCD was analyzed by SEM-EDS, which shows that As coexisted with Cu, Pb, and Zn.

Table 1. The main compositions of the ARCD (* Data from XRF and other data from ICP, wt.%).

Element	As_2O_3	CuO	ZnO	PbO	Bi_2O_3	Fe_2O_3	* SO_3	Others
Content	8.78	15.31	1.95	17.7	6.62	1.63	20.73	27.38

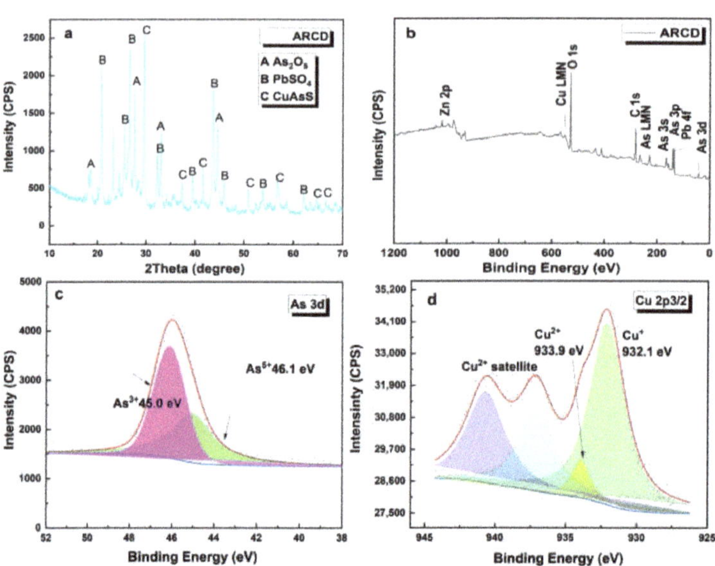

Figure 1. *Cont.*

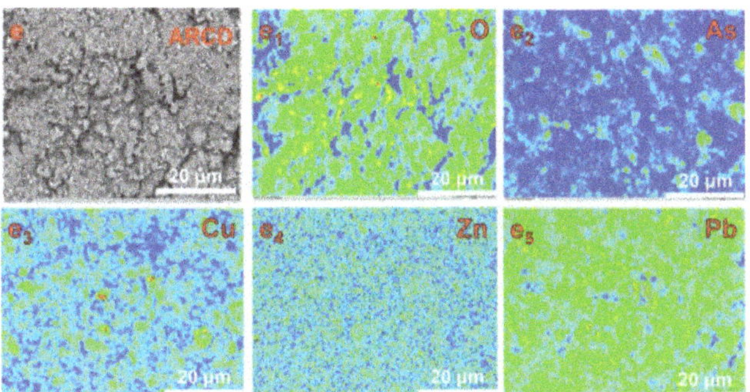

Figure 1. (**a**) XRD pattern of ARCD, (**b**) XPS spectra of ARCD, full spectrum, (**c**) As 3d, (**d**) Cu 2p3/2, (**e**) SEM-EDS of ARCD.

3.2. Alkaline Leaching Mechanisms

The arsenic in the ARCD was converted to arsenate by the NaOH leaching process and entered the liquid phase. The leaching process was mainly conducted according to the following (2)–(4) reactions. The thermodynamic analysis of the reaction shows that the reaction can be carried out at 10–100 °C. As can be seen from Figure 2, reactions (2)–(4) can be carried out under the experimental conditions.

$$As_2O_3 + 2NaOH = 2NaAsO_2 + H_2O \tag{2}$$

$$As_2O_3 + 6NaOH + O_2 = 2Na_3AsO_4 + 3H_2O \tag{3}$$

$$As_2O_5 + 6NaOH = 2Na_3AsO_4 + 3H_2O \tag{4}$$

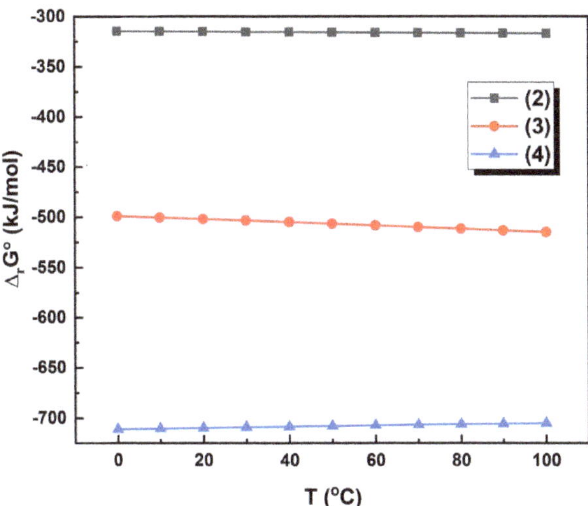

Figure 2. The plots of $\Delta rG°{\sim}T$ for the Equations (2)–(4) (Drawn by HSC 5.0 chemistry software, Outokumpu Research, Finland).

The Eh-pH diagram was used to analyze the thermodynamics of the elements in the leaching process of ARCD. Figure 3 shows the Eh—pH diagrams of As, Pb, Zn, and Cu, respectively. Arsenic exists in the form of As(s), $HAsO_2$(aq), $H_2AsO_3^-$, $HasO_4^{2-}$, and AsO_4^{3-} under alkaline conditions. At higher pH and higher potential, $H_2AsO_3^-$ oxidized to AsO_4^{3-} which shows that the As in the ARCD will enter the solution. Pb exists in the form of Pb^{2+}, $PbOH^+$, $HPbO_2^-$, and $PbO(s)$ under alkaline conditions. Under higher pH conditions, Pb mainly exists in the form of $HPbO_2^-$, which means Pb can dissolve in the alkaline solution. With the change of pH, Zn varies in the form of $Zn(OH)_2$, $HZnO_2^-$, and ZnO_2^{2-}, and Zn could dissolve in the solution when the alkaline strength increases.

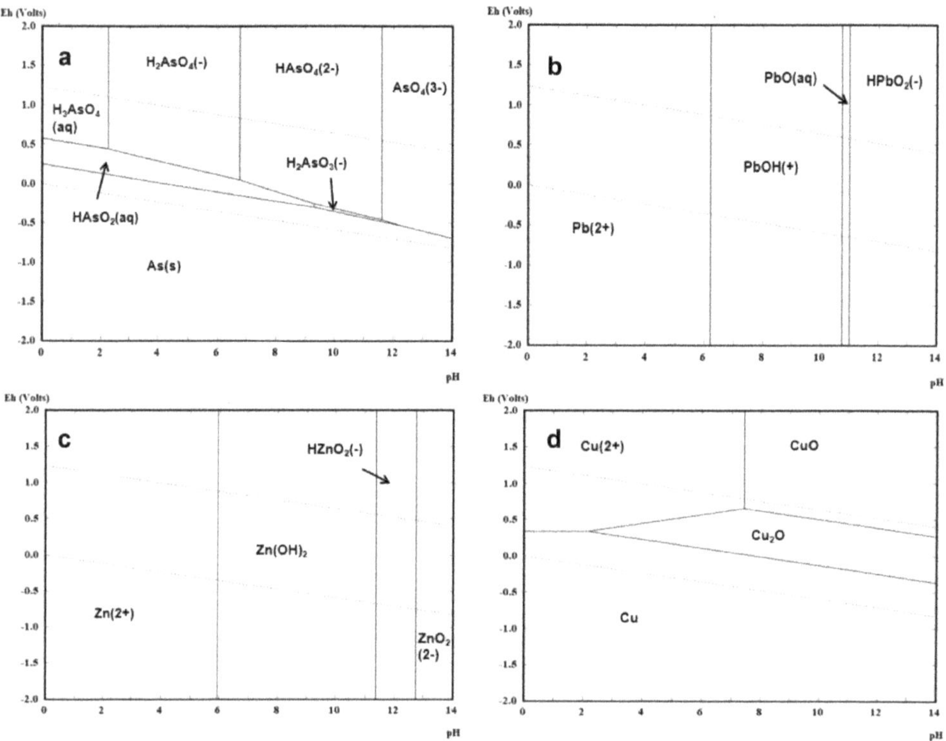

Figure 3. (a) Eh—pH diagrams of As, (b) Pb, (c) Zn, and (d) Cu, respectively.

3.3. The Alkaline Leaching Process of the ARCD

The alkaline leaching process was optimized to increase the leaching rate of arsenic. The leaching rate of arsenic was optimized via varying the temperature, NaOH concentration, and time to finally obtain the optimum conditions, as can be seen from Figure 4; NaOH of 75 g/L, the temperature of 70 °C, and leaching time of 90 min. The arsenic removal rate from ARCD of 94% was obtained.

After alkaline leaching, the leaching residue was analyzed by XRD and XPS. Figure 5a,b shows that as the NaOH concentration increases, the peak intensity of arsenic decreases until it finally disappears. At the same time, $Cu(OH)_2$ phase was found in the slag, which indicates that $Cu(OH)_2$ is formed after NaOH leaching and remains in the slag. The XPS shows that the arsenic in the ARCD exists in the form of trivalent and pentavalent. Figure 5d indicates that, after alkaline leaching, arsenic exists in the trivalent form in the leaching residue. This means that this part of trivalent arsenic needs to be removed by the oxidation leaching process.

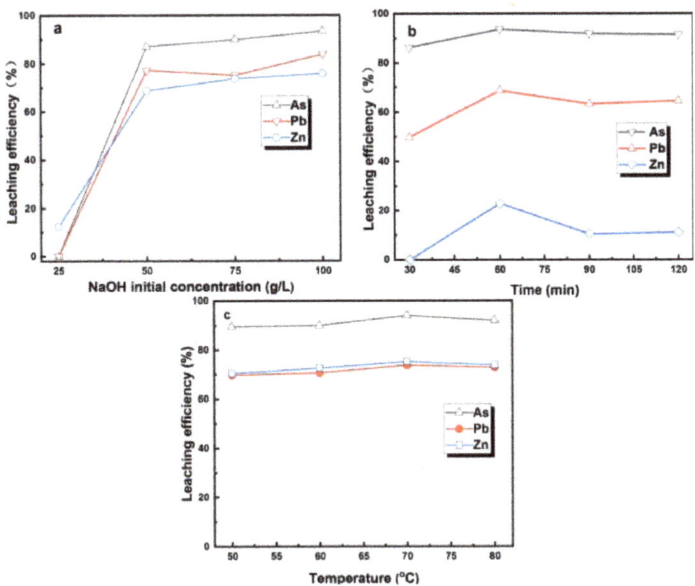

Figure 4. (**a**) Effect of NaOH concentration on the leaching efficiencies (70 °C, 10:1 mL/g and 90 min), (**b**) experimental reaction duration (70 °C, 10:1 mL/g and 75 g/L) and (**c**) reaction temperature (75 g/L, 10:1 mL/g and 90 min).

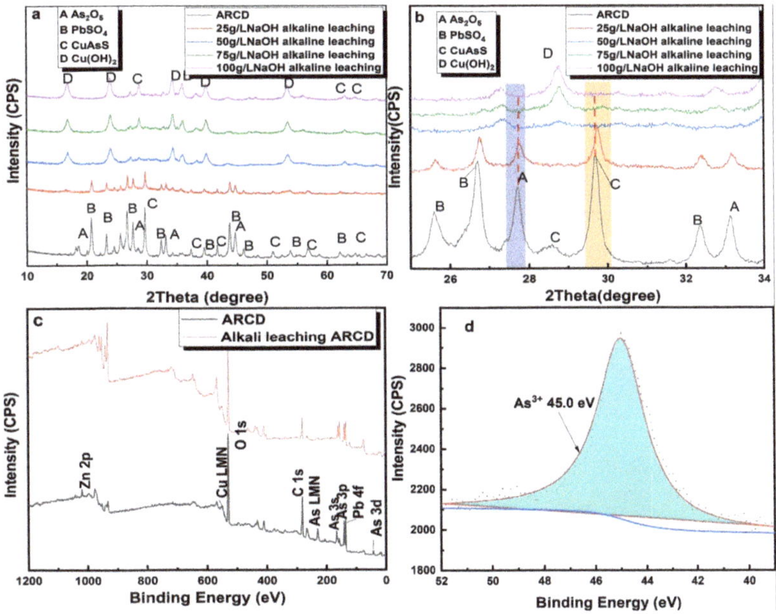

Figure 5. (**a**,**b**) XRD pattern, (**c**) XPS spectra, full spectrum, (**d**) As 3d of the alkaline leaching residue.

3.4. Electrochemical Advanced Oxidation Treatment of ARCD

3.4.1. Characteristics of the Cf and Modified Carbon Felt

The morphologies of carbon felt and modified carbon felt were obtained by SEM analysis. Figure 6a,b shows that the fibers on the surface of the original carbon felt (RCF) are smooth, and the surface of the pretreated carbon felt has no adhesion and impurities. As shown in Figure 6c,d, there are many particles on the surface of the carbon felt modified by carbon black and PTFE. The interconnection of these particles dramatically changes the surface structure of the carbon felt electrode, increasing the surface area of the electrode and the gas—liquid contact interface of the carbon felt electrode [2]. The modified carbon felt may enhance the generation of H_2O_2 when used as a cathode. As shown in Figure 6c,d, after modification, the contact angle increased from 109.64° to 122.33°, and the cathode overflow reduced [2].

Figure 6g,h shows the N_2 adsorption/desorption isotherm and pore size distribution of the original carbon felt and the modified carbon felt, respectively. Compared with the original carbon felt, the modified carbon felt shows a larger specific surface area. The specific surface areas of the modified carbon felt CF—1:1, CF—1:3, and CF—1:5 are 2.3978 m^2/g, 3.7009 m^2/g, and 6.9583 m^2/g, respectively. As shown in Figure 6g,h, the total pore volume is 0.015595 cm^3/g, 0.066591 cm^3/g, and 0.135 cm^3/g, respectively. According to the pore size distribution, the modified carbon felt has a smaller pore size, and the nanopore structure on the electrode surface is increased.

3.4.2. Performance of the Modified CF Electrode

The cyclic voltammetry (CV) curves of the original carbon felt and modified carbon felt is shown in Figure 7. After the CV test, it can be seen that an oxygen reduction peak appears at −0.5 V, indicating an oxygen reduction reaction in Figure 7a. Compared with the original carbon felt, the modified carbon felt has a greater peak current density. With the increase of the scanning speed, the oxidation peak moves in the direction of positive voltage, and the reduction peak moves in the direction of negative voltage (Figure 7b). The oxygen reduction peak current of the modified carbon felt electrode was significantly improved, meaning the improved electrochemical performance.

Figure 6. *Cont.*

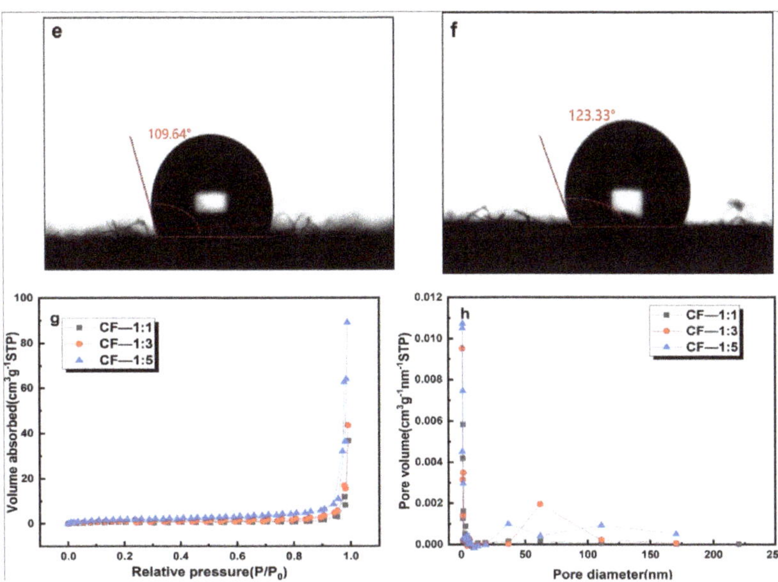

Figure 6. Morphological changes of modified CF are as follows: (**a**) low magnification and (**b**) high magnification SEM images of RCF, (**c**) low-power and (**d**) high—power SEM images of modified CF, (**e**) Contact angle of the original CF electrode, (**f**) Contact angle of the modified CF electrode, (**g**) pore size distribution, and (**h**) N_2 adsorption/desorption isotherms.

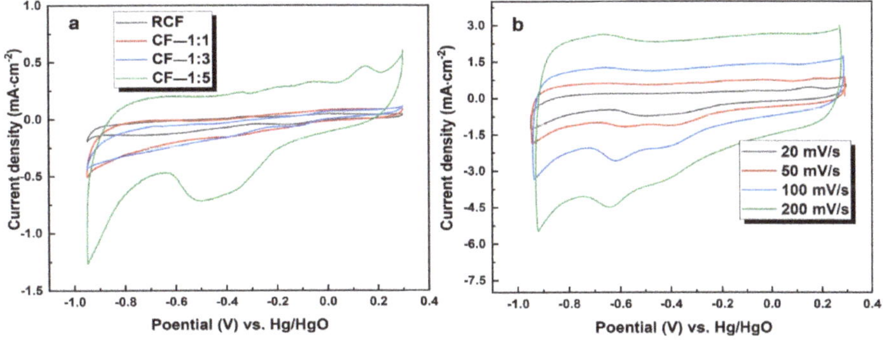

Figure 7. (**a**) CV diagram of RCF and modified CF, (**b**) changes of CF—1:5 electrode with scavenging speed electrode CV.

3.4.3. Influencing Factors of Hydrogen Peroxide Generation

The carbon felt cathode was modified by adjusting CB:PTFE ratio. The influence of the electrodes on the generation of H_2O_2 was carried out through different modified electrodes. As the addition amount continued to increase, the oxygen reduction current and the current density both gradually increased, resulting in the increase in generated H_2O_2 content. The content of H_2O_2 generated at CB:PTFE of 1:5 reached 148.048 mg/L. As shown in Figure 8a, the amount of H_2O_2 produced by CF—1:5 is almost 2.6 times that of unmodified carbon felt.

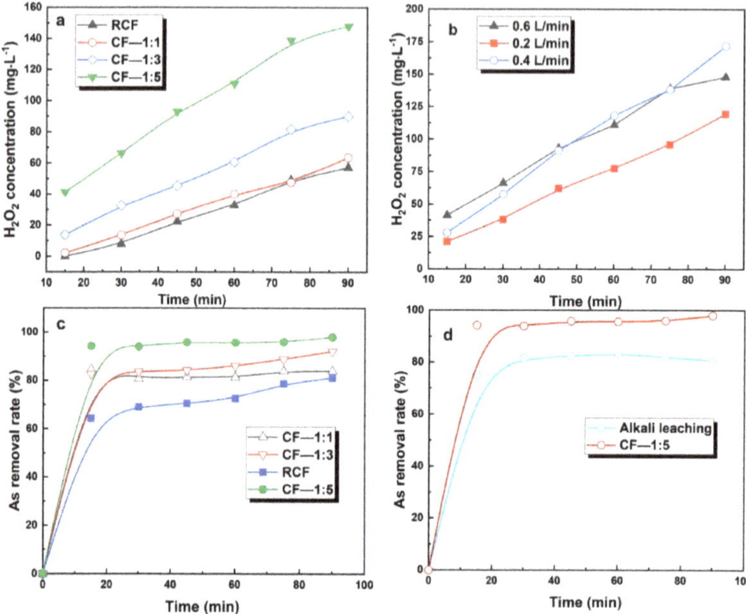

Figure 8. (**a**) Content of H_2O_2 in situ generated by different electrodes, (**b**) the change of H_2O_2 generated by CF—1:5 with different oxygen flow rate, (**c**) the change of arsenic removal rate from the ACRD by different electrodes, (**d**) and the difference between alkaline leaching and electrochemical advanced oxidation leaching.

The increasing trend of H_2O_2 production was carried out by adjusting the oxygen flow rate, and the results were given in Figure 8b. As the oxygen flow rate increases from 0.2 L/min to 0.6 L/min, the generated H_2O_2 also increases, and 0.4 L/min was selected as the optimized oxygen flow rate.

3.4.4. Performance of the Modified Electrode in Detoxification of ARCD

A two—electrode system is adopted in the detoxification experiments; a carbon felt electrode, a modified electrode (working electrode), and a graphite rod electrode (counter electrode). The electrochemical oxidation experiment was carried out at the voltage of 1 V. The experimental conditions were as follows: 150 mL of 75 g/L NaOH, solid to liquid ratio of 500 mg/L, experiment temperature of 25 °C, experiment time of 90 min, and the oxygen flow rate of 0.4 L/min. After 90 min of the electrolysis experiment, the removal rate of arsenic reached 98.04%. By comparing pretreated carbon felt and modified carbon felt, the removal rate of arsenic increased from 84.5% to 98.04% at CF of 1:5 (shown in Figure 8c).

3.5. Electrochemical Oxidation Leaching Mechanisms

The ESR tests were applied in the electrochemical oxidation leaching process to detect the reactive oxygen species, and DMPO was used as a trapping reagent. As shown in Figure 9, when 0.02 mol/L DMPO was added to the electrolyte, significant active ·OH signals were detected under the condition of the oxygen flow rate of 0.4 L/min. Due to the presence of OH, the leaching rate of arsenic can be further improved with respect to alkaline leaching. Furthermore, the arsenic leaching process can be considered as an advanced oxidation process (AOPs). When oxygen is not supplied, the generation of ·OH cannot be detected, further showing that only the transition metals can catalyze the conversion of HO_2^- into ·OH.

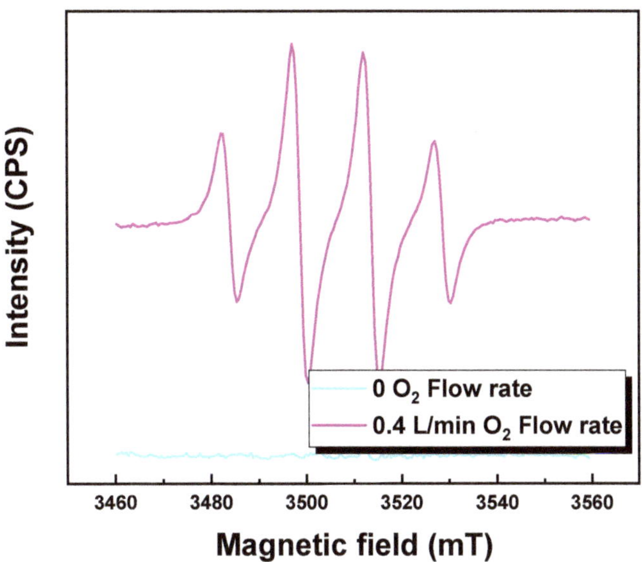

Figure 9. ESR tests of ·OH during the electrochemical oxidation leaching process.

Figure 10 shows the schematic diagram of the experimental setup and arsenic oxidation by reactive oxygen species. A large amount of HO_2^- with oxidizing properties was in situ generated in the alkaline solution system through the oxidation reaction (5) at the cathode [35–37]. In addition, Cu(I) was oxidized by HO_2^- to Cu(II), triggering the production of ·OH, followed by Cu(II) conversion into copper hydroxide by OH^- into the final residue, also detected by XRD (Figure 5), and no Cu(I) was detected in the final residue. Therefore, the electrochemical AOPs can be considered an iron-free Fenton-like reaction and can be described by reaction (6) [38,39]. The OH species have very high activity and can oxidize As(III) through reaction (7). The arsenic in the ACRD is oxidized into the liquid phase, and the removal rate of arsenic is further improved compared with alkaline leaching;

$$O_2 + H_2O + 2e^- \rightarrow HO_2^- + OH^- \tag{5}$$

$$Cu^+ + HO_2^- \rightarrow Cu^{2+} + \cdot OH \tag{6}$$

$$AsO_3^{3-} + \cdot OH \rightarrow AsO_4^{3-} + H_2O \tag{7}$$

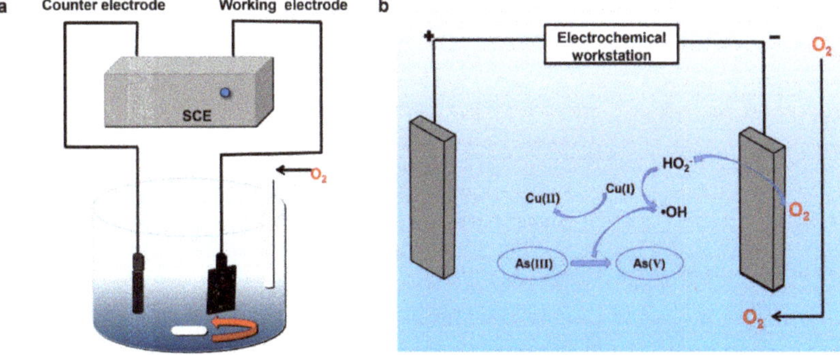

Figure 10. (**a**) Schematic diagram of the experimental setup and (**b**) Schematic diagram of the oxidation process of low arsenic by reactive oxygen species.

3.6. The Difference between Alkaline Leaching Process and Electrochemical Advanced Oxidation Leaching Process

The purpose of alkaline leaching of arsenic-containing waste is to transfer arsenic from solid waste to the leachate. However, the alkaline leaching process has disadvantages such as a large amount of waste liquid, easy generation of H_2S gas, and low arsenic leaching efficiency. The oxidation treatment of the waste residue containing low valent arsenic is beneficial to transforming the arsenic from a stable form to an unstable form (e.g., a water-soluble state, an exchange state, and a carbonate bond state), thereby increasing the leaching rate of arsenic. The electrochemical advanced oxidation method uses reactive oxygen groups with strong oxidation properties generated in situ to oxidize low arsenic in the ARCD to the leaching solution, which improves the removal rate of arsenic. The alkaline leaching process has the problem of incomplete removal of arsenic. The use of an electrochemical advanced oxidation process can increase the removal rate of arsenic without secondary pollution. The alkaline leaching process cannot oxidize the As(III). The advanced oxidation technology is used to oxidize the low valence arsenic in the solid waste and transfer it into the leaching solution, increasing the arsenic removal rate. As shown in Figure 8d, the removal rate of arsenic is only 80.69%, but it reaches 98.04% through advanced electrochemical oxidation.

4. Conclusions

The use of an electrochemical advanced oxidation leaching process can effectively remove arsenic from arsenic-containing solid waste. By adding carbon black and PTFE to improve the electrochemical performance of the carbon felt, the specific surface area of the carbon felt, the electrochemical performance, and the content of in suit generated H_2O_2 are all increased. The arsenic in the ARCD can be transferred to the solution by the electrochemical AOPs; therefore, the high-efficiency detoxification of the ARCD is realized. Under the condition of the voltage of 1.0 V, the electrolysis duration of 90 min, the oxygen flow rate of 0.4 L/min, and the CF of 1:5, the removal rate of arsenic reaches 98.4%. In addition, the electrochemical oxidation leaching method avoids secondary pollution and can be considered a clean and efficient arsenic removal method.

Author Contributions: Data curation, B.L.; writing—original draft preparation, M.L.; writing—review and editing, J.Y.; supervision, D.D. and H.D.; project administration, Y.C.; funding acquisition, G.H. All authors have read and agreed to the published version of the manuscript.

Funding: This work was funded by [the National Natural Science Foundation of China] (Grant No. 52004252), [the National Key Research and Development Program of China] (Grant No. 2018YFC1901601), [the Postdoctoral Research Grant in Henan Province] (Grant No. 201902016), [Henan Province Key R&D and Promotion Special (Scientific and Technical) Project] (Grant No. 212102310600) and [the Key Research Project of Henan Province Colleges and Universities] (Grant No. 20A440011).

Institutional Review Board Statement: Not applicable.

Informed Consent Statement: Not applicable.

Data Availability Statement: Not applicable.

Acknowledgments: This work was financially supported by the National Natural Science Foundation of China (Grant No. 52004252), the National Key Research and Development Program of China (Grant No. 2018YFC1901601), the Postdoctoral Research Grant in Henan Province (Grant No. 201902016), Henan Province Key R&D and Promotion Special (Scientific and Technical) Project (Grant No. 212102310600) and the Key Research Project of Henan Province Colleges and Universities (Grant No. 20A440011).

Conflicts of Interest: The authors declare no conflict of interest.

References

1. Yang, B.; Zhang, G.L.; Deng, W.; Ma, J. Review of arsenic pollution and treatment progress in nonferrous metallurgy industry. *Adv. Mater. Res.* **2013**, *634–638*, 3239–3243. [CrossRef]
2. Yu, F.; Zhou, M.; Yu, X. Cost-effective electro-Fenton using modified graphite felt that dramatically enhanced on H_2O_2 electro-generation without external aeration. *Electrochim. Acta* **2015**, *163*, 182–189. [CrossRef]
3. Liu, D.G.; Min, X.B.; Ke, Y.; Chai, L.Y.; Liang, Y.J.; Li, Y.C.; Yao, L.W.; Wang, Z.B. Co-treatment of flotation waste, neutralization sludge, and arsenic-containing gypsum sludge from copper smelting: Solidification/stabilization of arsenic and heavy metals with minimal cement clinker. *Environ. Sci. Pollut. Res.* **2018**, *25*, 7600–7607. [CrossRef]
4. Chung, J.Y.; Yu, S.D.; Hong, Y.S. Environmental source of arsenic exposure. *J. Prev. Med. Public Health* **2014**, *47*, 253. [CrossRef] [PubMed]
5. Choong, T.; Chuah, T.G.; Robiah, Y.; Koay, F.; Azni, I. Arsenic toxicity, health hazards and removal techniques from water: An overview. *Desalination* **2007**, *217*, 139–166. [CrossRef]
6. Xu, B.; Ma, Y.; Gao, W.; Yang, J.; Jiang, T. A review of the comprehensive recovery of valuable elements from copper smelting open-circuit dust and arsenic treatment. *JOM* **2020**, *72*, 3860–3875. [CrossRef]
7. Jaroíková, A.; Ettler, V.; Mihaljevi, M.; Drahota, P.; Culka, A.; Racek, M. Characterization and pH-dependent environmental stability of arsenic trioxide-containing copper smelter flue dust. *J. Environ. Manag.* **2018**, *209*, 71–80. [CrossRef] [PubMed]
8. Guo, X.Y.; Yi, Y.; Shi, J.; Tian, Q.H. Leaching behavior of metals from high-arsenic dust by $NaOH$–Na_2S alkaline leaching. *Trans. Nonferrous Met. Soc. China* **2016**, *26*, 575–580. [CrossRef]
9. Liu, W.; Han, J.; Ou, Z.; Wu, D.; Qin, W. Arsenic and antimony extraction from high arsenic smelter ash with alkaline pressure oxidative leaching followed by Na_2S leaching. *Sep. Purif. Technol.* **2019**, *222*, 53–59.
10. Chen, Y.; Liao, T.; Li, G.; Chen, B.; Shi, X. Recovery of bismuth and arsenic from copper smelter flue dusts after copper and zinc extraction. *Miner. Eng.* **2012**, *39*, 23–28. [CrossRef]
11. Wang, Y.; Fang, X.; Deng, P.; Rong, Z.; Cao, S. Study on thermodynamic model of arsenic removal from oxidative acid leaching. *J. Mater. Res. Technol.* **2020**, *9*, 3208–3218. [CrossRef]
12. Zhong, D.P.; Li, L.; Tan, C. Separation of arsenic from the antimony-bearing dust through selective oxidation using CuO. *Metall. Mater. Trans. B* **2017**, *48*, 1308–1314. [CrossRef]
13. Zhong, D.P.; Lei, L.I. Separation of arsenic from arsenic—Antimony-bearing dust through selective oxidation—Sulfidation roasting with CuS. *Trans. Nonferrous Met. Soc. China* **2020**, *30*, 223–235. [CrossRef]
14. Shan, C.; Tong, M. Efficient removal of trace arsenite through oxidation and adsorption by magnetic nanoparticles modified with Fe-Mn binary oxide. *Water Res.* **2013**, *47*, 3411–3421. [CrossRef]
15. Hu, C.Y.; Lo, S.L.; Kuan, W.H. High concentration of arsenate removal by electrocoagulation with calcium. *Sep. Purif. Technol.* **2014**, *126*, 7–14. [CrossRef]
16. Jin, Y.; Fei, L.; Tong, M.; Hou, Y. Removal of arsenate by cetyltrimethylammonium bromide modified magnetic nanoparticles. *J. Hazard. Mater.* **2012**, *227–228*, 461–468. [CrossRef]
17. Bui, T.H.; Kim, C.; Hong, S.P.; Yoon, J. Effective adsorbent for arsenic removal: Core/shell structural nano zero-valent iron/manganese oxide. *Environ. Sci. Pollut. Res.* **2017**, *24*, 24235–24242. [CrossRef] [PubMed]
18. Li, W.; Han, J.; Liu, W.; Qin, W.; Cai, L. Arsenic removal from lead-zinc smelter ash by $NaOH$-H_2O_2 leaching. *Sep. Purif. Technol.* **2019**, *209*, 128–135. [CrossRef]
19. Zhao, Y.; Qiu, W. Arsenic oxidation and removal from flue gas using H_2O_2/$Na_2S_2O_8$ solution. *Fuel Process. Technol.* **2017**, *167*, 355–362. [CrossRef]
20. Tubić, A.; Agbaba, J.; Dalmacija, B.; Rončević, S.; Ivančev-Tumbas, I. Effects of O_3, O_3/H_2O_2 and coagulation on natural organic matter and arsenic removal from typical northern Serbia source water. *Sep. Sci. Technol.* **2010**, *45*, 2453–2464. [CrossRef]
21. Kim, D.H.; Bokare, A.D.; Min, S.K.; Choi, W. Heterogeneous catalytic oxidation of As(III) on nonferrous metal oxides in the presence of H_2O_2. *Environ. Sci. Technol.* **2015**, *49*, 3506. [CrossRef]
22. Oturan, M.A. Electrochemical advanced oxidation technologies for removal of organic pollutants from water. *Environ. Sci. Pollut. Res.* **2014**, *21*, 8333–8335. [CrossRef] [PubMed]
23. Luo, Y.; Wu, Y.; Huang, C.; Xiao, D.; Chu, P.K. Graphite felt incorporated with MoS_2/rGO for electrochemical detoxification of high-arsenic fly ash. *Chem. Eng. J.* **2019**, *382*, 122763. [CrossRef]
24. Zheng, S.; Zhang, Y.; Wei, J. Reinforced As(III) oxidation by the in-situ electro-generated hydrogen peroxide on MoS_2 ultrathin nanosheets modified carbon felt in alkaline media. *Electrochim. Acta* **2017**, *252*, 245–253.
25. Wang, X.Q.; Liu, C.P.; Yuan, Y.; Li, F.B. Arsenite oxidation and removal driven by a bio-electro-Fenton process under neutral pH conditions. *J. Hazard. Mater.* **2014**, *275*, 200–209. [CrossRef]
26. Xia, G.; Lu, Y.; Xu, H. Electrogeneration of hydrogen peroxide for electro-Fenton via oxygen reduction using polyacrylonitrile-based carbon fiber brush cathode. *Electrochim. Acta* **2015**, *158*, 390–396. [CrossRef]
27. Sopaj, F.; Rodrigo, M.A.; Oturan, N.; Podvori Ca, F.I.; Pinson, J.; Oturan, M.A. Influence of the anode materials on the electrochemical oxidation efficiency. Application to oxidative degradation of the pharmaceutical amoxicillin. *Chem. Eng. J.* **2015**, *262*, 286–294. [CrossRef]
28. Tian, J.; Olajuyin, A.M.; Mu, T.; Yang, M.; Xing, J. Efficient degradation of rhodamine B using modified graphite felt gas diffusion electrode by electro-Fenton process. *Environ. Sci. Pollut. Res.* **2016**, *23*, 11574–11583. [CrossRef]

29. Chen, S.; Tang, L.; Feng, H.; Zhou, Y.; Zeng, G.; Lu, Y.; Yu, J.; Ren, X.; Peng, B.; Liu, X. Carbon felt cathodes for electro-Fenton process to remove tetracycline via synergistic adsorption and degradation. *Sci. Total. Environ.* **2019**, *670*, 921–931. [CrossRef]

30. Liu, X.; Yang, D.; Zhou, Y.; Zhang, J.; Lin, T. Electrocatalytic properties of N-doped graphite felt in electro-Fenton process and degradation mechanism of levofloxacin. *Chemosphere* **2017**, *182*, 306–315. [CrossRef]

31. Li, K.; Liu, J.; Li, J.; Wan, Z. Effects of N mono- and N/P dual-doping on H_2O_2, OH generation, and MB electrochemical degradation efficiency of activated carbon fiber electrodes. *Chemosphere* **2018**, *193*, 800–810. [CrossRef] [PubMed]

32. Ganiyu, S.O.; Le, T.H.; Bechelany, M.; Esposito, G.; Hullebusch, E.V.; Oturan, M.A.; Cretin, M. A hierarchical cofe-layered double hydroxide modified carbon-felt cathode for heterogeneous electro-Fenton process. *J. Mater. Chem. A* **2017**, *5*, 3655–3666. [CrossRef]

33. Camacho, J.; Wee, H.Y.; Kramer, T.A.; Autenrieth, R. Arsenic stabilization on water treatment residuals by calcium addition. *J. Hazard. Mater.* **2009**, *165*, 599–603. [CrossRef] [PubMed]

34. Bindi, L.; Catelani, T.; Chelazzi, L.; Bonazzi, P. Reinvestigation of the crystal structure of lautite, CuAsS. *Acta Crystallogr. Sect. E Struct. Rep. Online* **2008**, *64*, i22. [CrossRef]

35. Zhang, X.; Tian, J.; Han, H.; Sun, W.; Hu, Y.; Wang, T.; Yang, Y.; Cao, X.; Tang, H. Arsenic removal from arsenic-containing copper and cobalt slag using alkaline leaching technology and $MgNH_4AsO_4$ precipitation. *Sep. Purif. Technol.* **2020**, *238*, 116422. [CrossRef]

36. Zhan, M.; Xi, Y.; Xian, Q.; Kong, L. Photosensitized degradation of bisphenol a involving reactive oxygen species in the presence of humic substances. *Chemosphere* **2006**, *63*, 378–386. [CrossRef]

37. Burns, J.M.; Cooper, W.J.; Ferry, J.L.; King, D.W.; Dimento, B.P.; Mcneill, K.; Miller, C.J.; Miller, W.L.; Peake, B.M.; Rusak, S.A. Methods for reactive oxygen species (ROS) detection in aqueous environments. *Aquat. Sci.* **2012**, *74*, 683–734. [CrossRef]

38. Liu, B.; Wang, X.; Yu, X.; Zheng, S.; Du, H.; Dreisinger, D.; Yi, Z. A promising approach to recover a spent scr catalyst: Deactivation by arsenic and alkaline metals and catalyst regeneration. *Chem. Eng. J.* **2018**, *342*, 1–8. [CrossRef]

39. Bokare, A.D.; Choi, W. Review of iron-free Fenton-like systems for activating H_2O_2 in advanced oxidation processes. *J. Hazard. Mater.* **2014**, *275*, 121–135. [CrossRef]

minerals

Article

Discussion on Criterion of Determination of the Kinetic Parameters of the Linear Heating Reactions

Kui Li [1], Wei Zhang [1,*], Menglong Fu [2], Chengzhi Li [2] and Zhengliang Xue [2]

[1] The State Key Laboratory of Refractories and Metallurgy, Wuhan University of Science and Technology, Wuhan 430081, China; likui@wust.edu.cn

[2] Key Laboratory for Ferrous Metallurgy and Resources Utilization of Ministry of Education, Wuhan University of Science and Technology, Wuhan 430081, China; fml@wust.edu.cn (M.F.); chengzhi.li@wust.edu.cn (C.L.); xuezhengliang@wust.edu.cn (Z.X.)

* Correspondence: wei_zhang@wust.edu.cn

Abstract: Generally, the linear correlation coefficient is one of the most significant criteria to appraise the kinetic parameters computed from different reaction models. Actually, the optimal kinetic triplet should meet the following two requirements: first, it can be used to reproduce the original kinetic process; second, it can be applied to predict the other kinetic process. The aim of this paper is to attempt to prove that the common criteria are insufficient for meeting the above two purposes simultaneously. In this paper, the explicit Euler method and Taylor expansion are presented to numerically predict the kinetic process of linear heating reactions. The mean square error is introduced to assess the prediction results. The kinetic processes of hematite reduced to iron at different heating rates (8, 10 and 18 K/min) are utilized for validation and evaluation. The predicted results of the reduction of $Fe_2O_3 \rightarrow Fe_3O_4$ indicated that the inferior linear correlation coefficient did provide better kinetic predicted curves. In conclusion, to satisfy the above two requirements of reproduction and prediction, the correlation coefficient is an insufficient criterion. In order to overcome this drawback, two kinds of numerical prediction methods are introduced, and the mean square error of the prediction is suggested as a superior criterion for evaluation.

Keywords: correlation coefficient; kinetic parameters; criterion; hematite; reduction

Citation: Li, K.; Zhang, W.; Fu, M.; Li, C.; Xue, Z. Discussion on Criterion of Determination of the Kinetic Parameters of the Linear Heating Reactions. *Minerals* **2022**, *12*, 81. https://doi.org/10.3390/min12010081

Academic Editors: Shuai Wang, Xingjie Wang, Jia Yang and M Akbar Rhamdhani

Received: 6 December 2021
Accepted: 6 January 2022
Published: 10 January 2022

1. Introduction

Overall, to elucidate the reaction mechanisms and describe the kinetic process of the experimental data, the kinetic triplet (i.e., the activation energy, E; the pre-exponential factor, A; the reaction model, $f(\alpha)$) needs to be evaluated using kinetic analysis methods. In the practical application, the kinetic triplet is a momentous parameter for the resource recovery process of waste ore in mineral engineering and metallurgical engineering [1–6]. The kinetic analysis methods can be classified into the following two categories: one is isothermal kinetics and the other is nonisothermal kinetics. Prior to isothermal kinetic studies, thermogravimetric analyses are widely applied to characterize various materials, due to some advantages presented by many researchers [7,8]. The model-fitting method was widely used to investigate the reaction mechanism in quantities of literature from thermal-stimulated experimental data, based on a constant heating temperature program (TGA, DSC, DTA, etc.) [9–11]. Model-fitting approaches allow for the use of various reaction models ($f(\alpha)$) to fit experimental data, and then each reaction model produces a single pair of E and A [12]. Because of the physical interpretation of activation energy, its values derived from different models are usually proved to be in a rational range, with the same order of magnitudes. Identifying the optimal kinetic triplet from the above-produced kinetic parameters is thorny and strenuous work. Commonly, the kinetic parameters chosen by the linear correlation coefficient can accurately reproduce the original kinetic curves, whereas, for the practical application, the kinetic parameters should successfully predict the other kinetic curves with the same mechanism. Therefore, is it reasonable to select the kinetic parameters according to the correlation coefficient only? This question is seldom presented, but is a real problem for the application of nonisothermal analysis. Actually, the most persuasive criterion to choose kinetic parameters is satisfaction to the experimental and predicted data. However, the kinetic parameters obtained using the model-fitting method are rarely applied to predict the nonisothermal kinetic process.

The reduction reaction of hematite is one of the most widely investigated reactions in history [13–17]. In general, there are two kinds of reaction mechanisms for the hematite reduction reaction, which depends on the reaction temperature. The first mechanism is that hematite converts to magnetite and then directly to iron below 576 °C. Nevertheless, the reaction mechanism will experience $Fe_2O_3 \rightarrow Fe_3O_4 \rightarrow Fe_xO \rightarrow Fe$ above 576 °C [18]. In this paper, we take the low-temperature (below 576 °C) reduction of hematite as an example to clarify that the correlation coefficient is an insufficient criterion to assess the kinetic triplet.

2. Theoretical Models

2.1. Nonisothermal Kinetics

Before the instruments for nonisothermal measurements (TGA, DSC, DTA, etc.) were invented, the concepts of solid-state kinetics had already been established [19–21]. Nearly all kinetics analysis methods are based on the following equation [22,23]:

$$\frac{d\alpha}{dt} = k(T)f(\alpha) \tag{1}$$

where α denotes the extent of the conversion, t [min] is the reaction time, $k(T)$ [min^{-1}] is the reaction constant, expressed by the Arrhenius equation, and $f(\alpha)$ denotes the reaction model. Some of the most common reaction models are presented in Table 1 [22].

Table 1. Reaction models of solid-state kinetics.

	Reaction Model	Code	$f(\alpha)$	$g(\alpha)$
1	Power law	p4	$4\alpha^{3/4}$	$\alpha^{1/4}$
2	Power law	P3	$3\alpha^{2/3}$	$\alpha^{1/3}$
3	Power law	P2	$2\alpha^{1/2}$	$\alpha^{1/2}$
4	Power law	P2/3	$2/3\alpha^{-1/2}$	$\alpha^{3/2}$
5	One-dimensional diffusion	D1	$1/2\alpha^{-1}$	α^2
6	Mampel (first order)	F1	$1 - \alpha$	$-\ln(1 - \alpha)$
7	Awrami-Erofeev	A4	$4(1 - \alpha)[-\ln(1 - \alpha)]^{3/4}$	$[-\ln(1 - \alpha)]^{1/4}$
8	Avrami-Erofeev	A3	$3(1 - \alpha)[-\ln(1 - \alpha)]^{2/3}$	$[-\ln(1 - \alpha)]^{1/3}$
9	Avrami-Erofeev	A2	$2(1 - \alpha)[-\ln(1 - \alpha)]^{1/2}$	$[-\ln(1 - \alpha)]^{1/2}$
10	Three-dimensional diffusion	D3	$3/2(1 - \alpha)^{2/3}[1 - (1 - \alpha)^{1/3}]^{-1}$	$[1 - (1 - \alpha)^{1/3}]^2$
11	Contracting sphere	R3	$3(1 - \alpha)^{2/3}$	$1 - (1 - \alpha)^{1/3}$
12	Contracting cylinder	R2	$2(1 - \alpha)^{1/2}$	$1 - (1 - \alpha)^{1/2}$
13	Two-dimensional diffusion	D2	$[-\ln(1 - \alpha)]^{-1}$	$(1 - \alpha)\ln(1 - \alpha) + \alpha$

The temperature dependence can be presented through the Arrhenius equation, as follows:

$$k(T) = A\exp\left(\frac{-E}{RT}\right) \tag{2}$$

where A [min^{-1}] is the pre-exponential factor, E [J/mol] is the activation energy of the reaction, R is the gas constant, and T [K] is the reaction temperature.

Combining Equations (1) and (2) yields Equation (3), as follows:

$$\frac{d\alpha}{dt} = A\exp\left(\frac{-E}{RT}\right)f(\alpha) \tag{3}$$

For nonisothermal conditions, a constant heating rate β is usually adopted, thus the following equation is applied:

$$\beta = \frac{dT}{dt} \tag{4}$$

where β [K/min] is the linear heating rate. Combining Equations (3) and (4) means that the explicit temporal dependence in Equation (3) is eliminated through the following trivial transformation:

$$\frac{d\alpha}{dT} = \frac{A}{\beta}\exp\left(-E/RT\right)f(\alpha) \tag{5}$$

Taking the natural logarithm of Equation (3) leads to the following:

$$\ln\left(\frac{d\alpha/dt}{f(\alpha)}\right) = \ln A - \frac{E}{RT} \tag{6}$$

Using the Sharp–Wentworth method [24], a plot of $\ln\left(\frac{d\alpha/dt}{f(\alpha)}\right)$ against $\frac{1}{T}$ should result in a straight line, with the slope of $-E/R$ and the intercept of $\ln A$, then the activation energy and pre-exponential factor can be calculated. The reaction is regularly regarded as the first order or other apparent reaction orders gained from the Freeman–Carroll [8] method.

The integral of Equation (5) yields the following:

$$g(\alpha) = \int_0^\alpha \frac{d\alpha}{f(\alpha)} = \frac{A}{\beta}\int_{T_0}^{T} \exp(-E/RT)dT \tag{7}$$

where $g(\alpha)$ is the integral form of the $1/f(\alpha)$ (shown in Table 1), and T_0 is the initial reaction temperature. The right-hand expression of Equation (5) has no analytical solution, and many researchers [7,25,26] have approached the analytical solutions with some mathematical simplifications. One of the most popular mathematical simplifications is the Coats–Redfern equation [7], which is as follows:

$$\ln\frac{g(\alpha)}{T^2} = \ln\left(\frac{AR}{\beta E}\left[1 - \frac{2RT_m}{E}\right]\right) - \frac{E}{RT} \tag{8}$$

where T_m is the mean experimental temperature, and the Arrhenius parameters (E and A) can be derived from the plot of $\ln\frac{g(\alpha)}{T^2}$ against $\frac{1}{RT}$.

2.2. The Criterion of Determination

Regular kinetic analysis has the following two major purposes: one is theoretical and the other is practical. Theoretically, each member of the kinetic triplet (A, E, and $f(\alpha)$) represents a different physical concept [27]. A is associated with the frequency of the vibrations of the activated complex and E represents the energy barrier [28]. $f(\alpha)$ or $g(\alpha)$ is regarded as the reaction mechanism [29]. Practically, the kinetic triplet needed to provide a mathematical description of the kinetic process. Based on the mathematical description, not only the original kinetics data can be reproduced, but the other kinetic processes could also be numerically predicted [27]. When adopting the Sharp–Wentworth or Coats–Redfern methods to deal with experimental data, a set of kinetic triplets can be derived in different reaction models. Because of the existence of the "compensation effect", there is a strong linear correlation between the values E and $\ln A$ computed from the corresponding reaction model $f(\alpha)$ [30]; most of the kinetic triplets can satisfactorily reproduce the original experimental data [12]. However, not all kinetic triplets can make a satisfying prediction. Therefore, how to choose optimal kinetic parameters is a significant problem.

For a same reaction, various fitting models will produce different coefficients of linear correlation, r. In many papers [31–33], the linear correlation, r, is usually estimated as a criterion for selection. A single pair of A and E is then commonly chosen, corresponding to a reaction model that gives the maximum absolute value of the correlation coefficient, $|r_{max}|$ [12]. Actually, the maximum value of $|r|$ does not necessarily represent the most probable model in statistics [12,34–36]. Vyazovkin suggested combining the linear correlation, r, and the result of the isothermal kinetics prediction, based on nonisothermal kinetics parameters, to pick the kinetic triplet [12]. The prediction formula is as follows:

$$t_\alpha = \frac{g(\alpha)}{A \exp(-E/RT_0)} \tag{9}$$

where T_0 is the constant temperature (isothermal condition) and t_α is the time to reach the extent of conversion at the temperature T_0. Nevertheless, sometimes the method will be invalid. Whether the nonisothermal kinetic triplet can be applied to the isothermal condition is controversial [12,37]. Hence, a new idea that uses nonisothermal kinetics parameters to predict the nonisothermal kinetics process may be more reliable. Two numerical methods are introduced to evaluate the criterion, explained below.

2.3. Explicit Euler Method

When the activation energy, reaction order, and the pre-exponential factor are determined, the results of the numerical prediction can be obtained by using the explicit Euler method.

Combining Equations (3) and (4), we can obtain the following equation:

$$\frac{d\alpha}{dt} = f_k = A \exp\left(\frac{-E}{R(T_0 + t_k\beta)}\right) f(\alpha) \tag{10}$$

Based on the explicit Euler formula, the numerical prediction formula can be obtained as follows:

$$\begin{cases} \alpha_{k+1} = \alpha_k + hf_k \\ \alpha_0 = \alpha(t_0) \qquad (k = 0, 1, 2\cdots, m-1) \\ h = b/m \end{cases} \tag{11}$$

where k denotes the node of the Euler formula, f_k [min^{-1}] denotes the slope of function $\alpha(t)$ at time t_k [min], h [min] is the step length of the Euler formula, m is the number of nodes, and b [min] denotes the total reaction time. Regularly, α_0 equals zero.

2.4. Taylor Expansion Method

Supposing the activation energy, reaction order, and the pre-exponential factor are given, the numerical solution of Equation (7) can be calculated using the Taylor expansion method, which is as follows:

$$g(T) = \frac{A}{\beta} \int_{T_0}^{T} \exp(-E/RT)dT \tag{12}$$

where $g(T)$ is the original function of the above temperature integral. Equation (13) is deduced by second-order Taylor expansion of the original function, $g(T)$, as follows:

$$g(T_0 + \Delta T) \approx g(T_0) + g'(T_0)\Delta T + \frac{1}{2!}g''(T_0)\Delta T^2 \tag{13}$$

where the following applies:

$$g'(T) = \frac{A}{\beta} \exp(-E/RT) \tag{14}$$

$$g''(T) = \frac{AE \exp(-E/RT)}{\beta RT^2} \tag{15}$$

2.5. Mean Square Error

In order to appraise the deviation between the experimental value and the predicted value, the mean square error is defined as Equation (16), which is as follows:

$$MSE = \frac{1}{N} \sum_{t=1}^{N} \left(y_{\exp} - y_{\mathrm{pre}}\right)^2 \tag{16}$$

where MSE denotes the mean square error, N denotes the number of nodes, $y_{\exp}$ is the conversion rate of the experiment, and y_{pre} is the predicted conversion rate by the explicit Euler or Taylor expansion methods.

It is worth emphasizing that the predicted data are simulated from the kinetic parameters derived from the other experimental data. For the same reaction, the smaller the mean square error is, the more accurate the prediction is.

3. Experimental Procedure

In order to validate the numerical prediction model, nonisothermal experiments are conducted and a differential thermal analyzer of STA 409 from NETZSCH (manufactured by NETZSCH-Gerätebau GmbH, Germany) is used as apparatus. The arrangement of the gas tube, Al_2O_3 crucible (sample carrier), gas monitor, thermal analyzer, and other auxiliary equipment is shown in Figure 1.

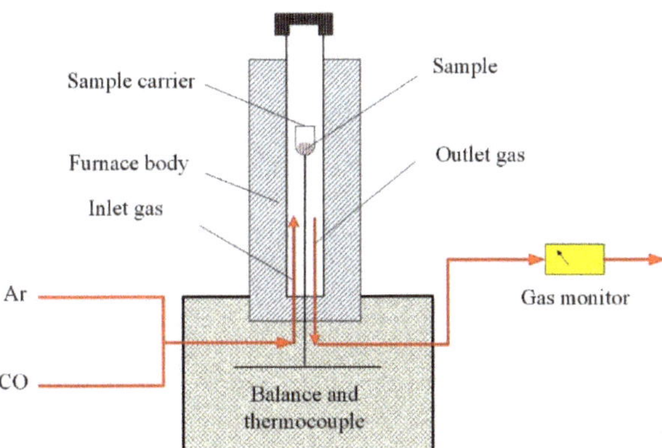

Figure 1. Experimental apparatus.

Chemical pure hematite powder with a purity of 99.0 wt.% (from Sinopharm Chemical Reagent Co., Ltd., Shanghai, China) was used as raw material. The powder was placed in the Al_2O_3 crucible, then it was appropriately placed on the weight sensor of the analyzer, and all the gas tubes were well sealed. Before being linearly heated, furnace atmosphere was cleansed from air with high-pure Ar gas, which was then substituted with pure CO at a flowrate of 30 mL/min. Consequently, the raw materials with masses of 370, 823, 610 and 335 mg were heated at heating rates of 3, 8, 10 and 18 K/min, respectively. When the reaction temperature was heated to 576 °C, the material was cooled to room temperature in an inert Ar atmosphere.

4. Results and Discussion

4.1. Fitting Results

Based on the principle of thermodynamics, when the reaction temperature is below 576 °C [38], the reduction of hematite can be divided into two steps. Firstly, hematite is reduced to magnetite, then magnetite is reduced to iron.

$$3Fe_2O_3(s) + CO(g) = 2Fe_3O_4(s) + CO_2(g) \quad \Delta G^\theta = -42.227 - 0.048T(K) \qquad (17)$$

$$\frac{1}{4}Fe_3O_4(s) + CO(g) = \frac{3}{4}Fe(s) + CO_2(g) \quad \Delta G^\theta = -7.960 + 0.010T(K) \qquad (18)$$

The Gibbs energies of the reduction reactions of hematite to magnetite and magnetite to iron are negative under experimental conditions. Therefore, the above two reactions can occur in the experimental temperature range. The predominance area diagram of iron oxide reduction reactions is shown in Figure 2 [38].

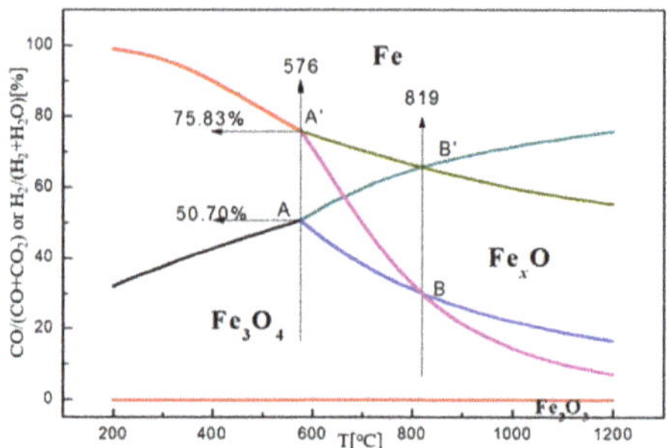

Figure 2. The predominance area diagram of iron oxide reduction reactions.

As shown in Figure 2, the reduction products of hematite are determined by reaction temperature and partial pressure of reduction gas. Point A and point A' in Figure 2 are three-line intersections of the reactions reduced by CO and H$_2$, respectively. Figure 2 can be divided into three parts by point A. On the upper left part of point A, the expected reduction product is Fe; on the right part of point A, Fe$_x$O is thermodynamically stable. Below the point A, Fe$_3$O$_4$ is thermodynamically predominant. As a result, the low-temperature reduction of hematite includes two processes, i.e., Fe$_2$O$_3 \to$ Fe$_3$O$_4$ and Fe$_3$O$_4 \to$ Fe.

The TG curves and DTG curves of the experimental data are shown in Figure 3.

According to the appearance of the DTG curve for the 3 K/min heating rate, we can observe that there are two troughs (329.1 °C and 438.0 °C), a peak (349.0 °C), and a significant inflection point (530.0 °C). The above phenomenon indicates that there are three reactions occurring. The first reaction began at 253.0 °C, and then the reaction rate reached the maximum at 329.1 °C. From 329.1 °C to 438.0 °C, the reaction rate gradually decreases until it reaches zero. The mass loss of the first reaction is 4.78%, which is larger than the theoretical mass loss of Fe$_2$O$_3$ reduced to Fe$_3$O$_4$ (3.33%). Practically, because of the non-stoichiometry of magnetite, the actual mass loss of Fe$_2$O$_3 \to$ Fe$_3$O$_4$ is 4.7% [18]. Therefore, the first reaction is hematite reduced to magnetite. The change from 329.1 °C to 438.0 °C denotes that the reaction of Fe$_3$O$_4$ reduced to Fe is proceeding. When the reaction temperature reaches about 530 °C, the shape of the DTG curve changes, which alludes to a new reaction. The new reaction may be the carbon deposition reaction (2CO $\to$ C + CO$_2$) or the carburization reaction (Fe$_3$O$_4$ + 6C $\to$ Fe$_3$C + 5CO). Therefore, we chose the temperature range of 329.1 °C to 520.0 °C to investigate the reaction of Fe$_3$O$_4$ reduced to Fe. The shapes of the DTG curves for 8, 10, 18 K/min are similar to the shape of the DTG curve for 3 K/min, which indicates there are two reactions occurring (Fe$_2$O$_3 \to$ Fe$_3$O$_4$ and then Fe$_3$O$_4 \to$ Fe) during heating at 8, 10, 18 K/min. The selected temperature ranges of the above two reactions in each heating rate are shown in Table 2.

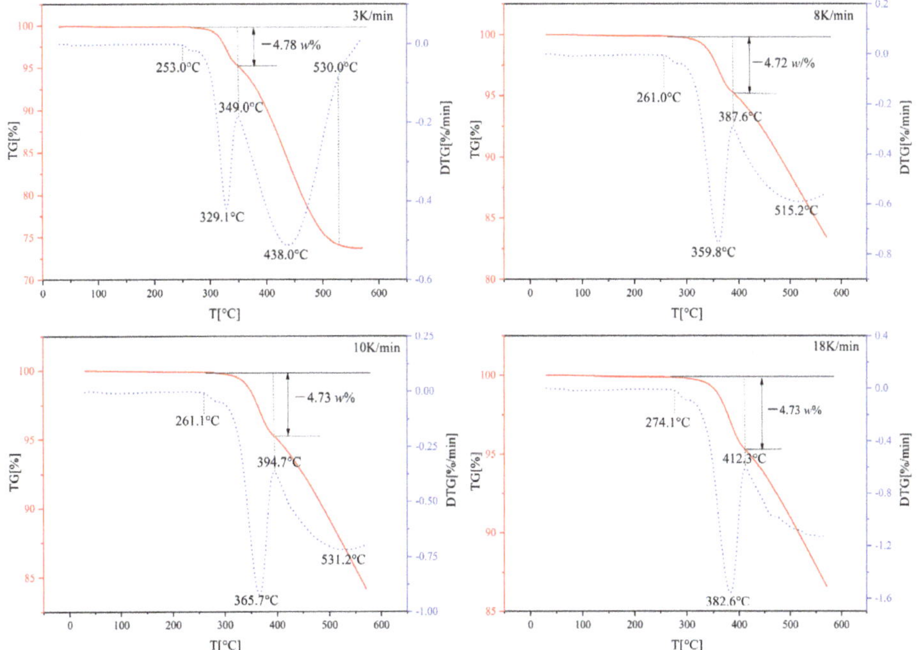

Figure 3. TG curves and DTG curves of experimental data. Red: TG curve. Blue: DTG curve.

Table 2. The selected temperature ranges of the two reactions.

Heating Rate/(K/min)	$Fe_2O_3 \rightarrow Fe_3O_4$ (°C)	$Fe_3O_4 \rightarrow Fe$ (°C)
3	253.0–329.1	329.1–520.0
8	261.0–387.6	387.6–520.0
10	261.1–394.7	394.7–520.0
18	274.1–412.3	412.3–520.0

The kinetic parameters of the hematite reduction reaction and the corresponding correlation coefficients at 3, 8, 10 and 18 K/min heating rates are displayed in Tables 3 and 4.

Table 3. The kinetic parameters of the reaction of $Fe_2O_3 \rightarrow Fe_3O_4$ based on the optimal correlation coefficient.

	Sharp–Wentworth				Coats–Redfern			
β/(K/min)	E (J/mol)	Reaction Order	ln (A/min)	r^2	E (J/mol)	Reaction Order	ln (A/min)	r^2
3	246,385	First	47.68	0.9992	176,498	First	33.28	0.9857
8	179,333	First	35.25	0.9994	144,763	First	26.37	0.9875
10	179,316	First	35.35	0.9989	147,216	First	26.78	0.9906
18	171,360	First	34.15	0.9981	132,533	First	23.81	0.9796

Table 4. The kinetic parameters of the reaction of $Fe_3O_4 \rightarrow Fe$ based on the optimal correlation coefficient.

$\beta/(K/min)$	Sharp–Wentworth				Coats–Redfern			
	E (J/mol)	Reaction Order	ln (A/min)	r^2	E (J/mol)	Reaction Order	ln (A/min)	r^2
3	72,576	First	9.28	0.9994	81,382	First	10.57	0.9984
8	56,040	First	6.48	0.9969	142,386	Third	21.39	0.9908
10	58,440	First	7.32	0.9919	147,765	Third	22.38	0.9917
18	63,649	First	8.44	0.9941	171,202	Third	26.53	0.9933

4.2. Results from Explicit Euler Method

When the activation energy, reaction mechanism, and pre-exponential factor of a group of thermogravimetric data have been obtained in advance, the dependence of α on t can be calculated using Equations (10) and (11) for an arbitrary heating rate experiment.

Figure 4 shows comparisons between the experimental data and the predicted results, using different kinetics parameters of Fe_2O_3 reduced to Fe_3O_4 by the explicit Euler method. In the sub-image of Figure 4, the reaction mechanisms are both first orders, and the kinetics parameters are achieved from the experimental data with the heating rate of 3 K/min; the pre-exponential factor and activation energy for the Sharp–Wentworth models are 5.12×10^{20} min^{-1} and 246.39 kJ/mol, and for the Coats–Redfern models they are 2.85×10^{14} min^{-1} and 176.50 kJ/mol, respectively.

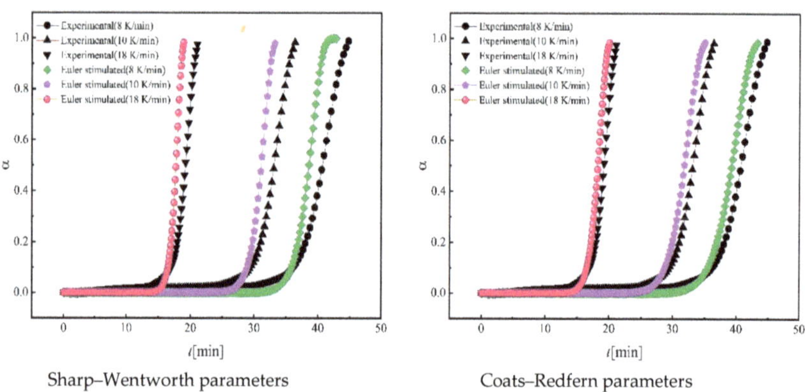

Sharp–Wentworth parameters   Coats–Redfern parameters

Figure 4. Experimental and predicted data using Sharp–Wentworth and Coats–Redfern parameters of $Fe_2O_3 \rightarrow Fe_3O_4$ by explicit Euler method.

Figure 5 shows comparisons between the experimental data and the predicted results for the reaction of magnetite reduced to iron. The predicted results are based on kinetic parameters of the heating rate of 3 K/min. The reaction mechanisms of magnetite reduced to iron for the Sharp–Wentworth and Coats–Redfern methods are both first orders. The kinetic parameters used for the prediction are as follows: Sharp–Wentworth's pre-exponential factor and activation energy are 10,708 min^{-1} and 72.58 kJ/mol, respectively, and the pre-exponential factor and activation energy of Coats–Redfern are 38,901 min^{-1} and 81.38 kJ/mol, respectively.

For the reaction of hematite reduced to magnetite, according to Figure 4, the results of the Coats–Redfern prediction are closer to the experimental values. The mean square errors of the Sharp–Wentworth and Coats–Redfern methods are calculated by Equation (18). Based on the formula of the mean square error, the number of nodes with heating rates of 8, 10 and 18 K/min are 70, 65 and 53, respectively.

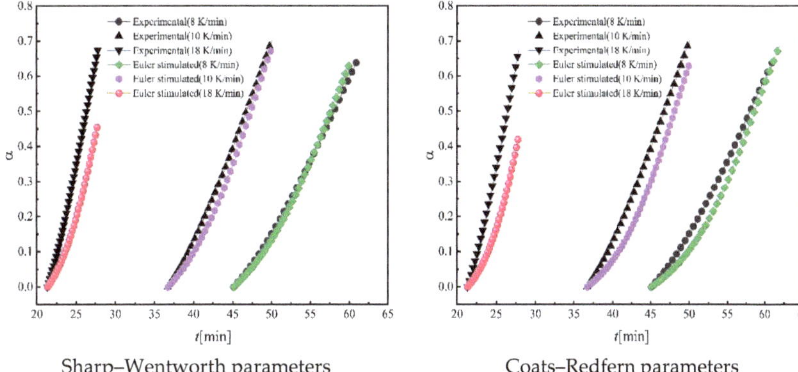

Sharp–Wentworth parameters   Coats–Redfern parameters

Figure 5. Experimental and predicted data using Sharp–Wentworth and Coats–Redfern parameters of $Fe_3O_4 \rightarrow Fe$ by explicit Euler method.

According to the results of the mean square error in Table 5, for the same heating rate, the mean square value of the Coats–Redfern method is lower than that of the Sharp–Wentworth method. Therefore, when the heating rate is 3 K/min, the kinetic parameters of hematite reduced to magnetite obtained by Coats–Redfern are superior to those obtained by the Sharp–Wentworth method.

Table 5. The mean square errors of the reaction of $Fe_2O_3 \rightarrow Fe_3O_4$ for the explicit Euler method.

	Sharp–Wentworth			Coats–Redfern		
β	8 K/min	10 K/min	18 K/min	8 K/min	10 K/min	18 K/min
N	70	65	53	70	65	53
MSE	0.0463	0.0563	0.0720	0.0171	0.0174	0.0191

As for the reaction of magnetite reduced to iron, according to the mean square error formula, the number of nodes with heating rates of 3, 8 and 18 K/min are 83, 88 and 66, respectively.

The mean square errors of the Sharp–Wentworth method are smaller than those of the Coats–Redfern method in Table 6. Consequently, the kinetic parameters of magnetite reduced to iron, with a 3 K/min heating rate, calculated by the Sharp–Wentworth method, are more preferable than those by the Coats–Redfern method. Significantly, whether the data of the Sharp–Wentworth method or the Coats–Redfern method are used, the predicted results of 18 K/min are much different from the experimental data. This may be a result of the high heating rate or the speedy reduction process, which leads to a significant influence of inner diffusion on the dominant chemical reduction process.

Table 6. The mean square errors of the reaction of $Fe_3O_4 \rightarrow Fe$ for the explicit Euler method.

	Sharp–Wentworth			Coats–Redfern		
β	8 K/min	10 K/min	18 K/min	8 K/min	10 K/min	18 K/min
N	83	88	66	83	88	66
MSE	0.0003	0.0013	0.0185	0.0019	0.0054	0.0262

4.3. Results from Taylor Expansion Method

For a group of experimental data, the activation energy, reaction order, and the pre-exponential factor are gained in advance, since the left-hand side of Equation (7) can be integrated and the right-hand can be solved by using the recursive method, as Equations (13)–(15) demonstrated. With the help of a computer program, the calculated values of the reacted ratio α against time t can be found. Finally, the kinetic processes of those experiments with other heating rates could also be numerically predicted.

Figure 6 depicts the predicted results for the reaction of hematite reduced to magnetite by using different kinetic triplets. The Taylor expansion method adopts the same parameters as the explicit Euler method.

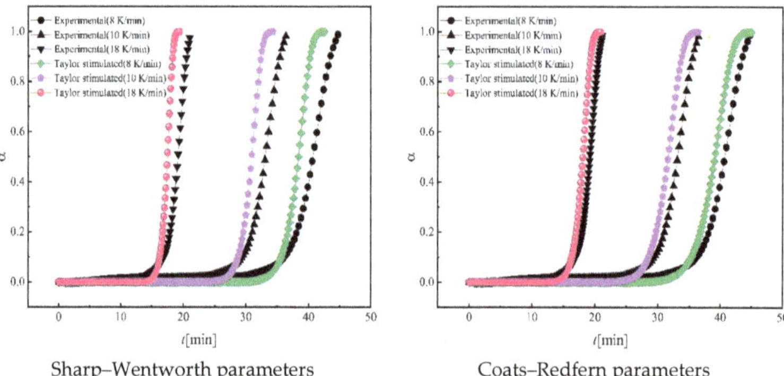

Sharp–Wentworth parameters   Coats–Redfern parameters

Figure 6. Experimental and predicted data using Sharp–Wentworth and Coats–Redfern parameters of $Fe_2O_3 \rightarrow Fe_3O_4$ using the explicit Euler method.

According to Figure 6 and Table 7, the same conclusion can be drawn, which is that the kinetic parameters of hematite reduced to magnetite, with a heating rate of 3 K/min, obtained by the Coats–Redfern method, are superior to those obtained by the Sharp–Wentworth method.

Table 7. The mean square errors of the reaction of $Fe_2O_3 \rightarrow Fe_3O_4$ for Taylor expansion.

	Sharp–Wentworth			**Coats–Redfern**		
β	8 K/min	10 K/min	18 K/min	8 K/min	10 K/min	18 K/min
N	70	65	53	70	65	53
MSE	0.0531	0.0618	0.0873	0.0200	0.0213	0.0264

Figure 7 shows comparisons between the experimental and predicted data for the reaction of magnetite reduced to iron with the Taylor expansion method. In order to compare with the explicit Euler method, the same nodes are selected to calculate the mean square errors.

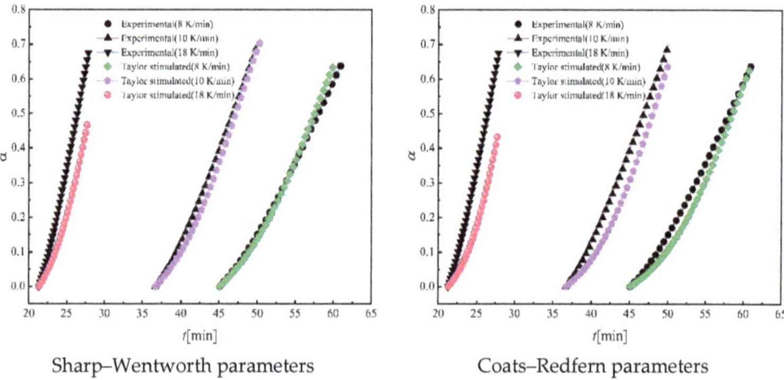

Sharp–Wentworth parameters   Coats–Redfern parameters

Figure 7. Experimental and predicted data using Sharp–Wentworth and Coats–Redfern parameters of $Fe_3O_4 \rightarrow Fe$ by Taylor expansion.

Based on the data of the mean square errors in Table 8, the result predicted by using the Sharp–Wentworth method is better than that predicted by the Coats–Redfern method. Consequently, the kinetic parameters of magnetite to iron with a 3 K/min heating rate, obtained by Sharp–Wentworth, are superior. When the heating rate is 18 K/min, the predicted results are still negative.

Table 8. The mean square errors of the reaction of $Fe_3O_4 \rightarrow Fe$ for Taylor expansion.

	Sharp–Wentworth			Coats–Redfern		
β	8 K/min	10 K/min	18 K/min	8 K/min	10 K/min	18 K/min
N	70	65	53	70	65	53
MSE	0.0003	0.0010	0.0166	0.0016	0.0047	0.0238

5. Conclusions

Model-fitting approaches are extensively used to obtain kinetics parameters in various fields. Normally, the correlation coefficient becomes the primary criterion to evaluate the kinetic parameters of different reaction models. One takes for granted that the kinetic parameters selected by the criterion are optimal. However, when the kinetic predicted results are taken into account, the actual situation may be unexpected. In this paper, the low-temperature reduction reaction of hematite was conducted in four different heating rate conditions (3, 8, 10 and 18 K/min). In order to evaluate the prediction ability of the kinetics parameters, the kinetics parameter of 3 K/min heating rate is adopted to predict the kinetic curves of 8, 10 and 18 K/min heating rates. For the reduction step of $Fe_2O_3 \rightarrow Fe_3O_4$, the prediction using the kinetics parameters of the Coats–Redfern ($r = 0.9857$) method should have been inferior to that of the Sharp–Wentworth ($r = 0.9992$) method, according to the correlation coefficient criterion. However, exactly the opposite is true for the sake of practical prediction. The kinetic parameters of Coats–Redfern ($r = 0.9857$) provide a better prediction result. In conclusion, although the correlation coefficient is widely used as the decisive criterion to assess kinetics parameters, it is shown to be insufficient when the capability of the practical prediction of kinetics parameters is taken into account.

Author Contributions: Conceptualization, W.Z.; methodology, M.F.; investigation, C.L.; resources, Z.X.; writing—original draft preparation, K.L.; writing—review and editing, W.Z. and K.L. All authors have read and agreed to the published version of the manuscript.

Funding: This research was funded by the National Natural Science Foundation of China (NSFC), grant numbers 51804228 and 52104340, and the China Scholarship Council (CSC), grant number 201908420169.

Acknowledgments: This work is financially supported by the National Natural Science Foundation of China (NSFC), grant numbers 51804228 and 52104340, and the China Scholarship Council (CSC), grant number 201908420169.

Conflicts of Interest: The authors declare no conflict of interest.

Abbreviations

α	Conversion rate	-
β	Linear heating rate	K/min
b	Total reaction time	min
f	Slope	min^{-1}
m	Number of nodes	-
t	Reaction time	min
h	Step length	min
$f(\alpha)$	Reaction model	-
$g(\alpha)$	The integral form of $f(\alpha)$	-
$k(T)$	Reaction constant	min^{-1}
A	Pre-exponential factor	min^{-1}

E	Activation energy	J/mol
N	Mean squared number of nodesnodes	-
T	Reaction temperature	K
MSE	Mean square error	-
TGA	Thermal gravimetric analyzer	-
DSC	Differential scanning calorimetry	-
DTA	Differential thermal analysis	-

References

1. Yuan, S.; Zhang, Q.; Yin, H.; Li, Y. Efficient iron recovery from iron tailings using advanced suspension reduction technology: A study of reaction kinetics, phase transformation, and structure evolution. *J. Hazard. Mater.* **2021**, *404*, 124067. [CrossRef]
2. Wan, X.; Shi, J.; Klemettinen, L.; Chen, M.; Taskinen, P.; Jokilaakso, A. Equilibrium phase relations of CaO–SiO$_2$–TiO$_2$ system at 1400 °C and oxygen partial pressure of 10−10 atm. *J. Alloys Compd.* **2020**, *847*, 156472. [CrossRef]
3. Zhang, J.; Zhang, W.; Xue, Z. Oxidation Kinetics of Vanadium Slag Roasting in the Presence of Calcium Oxide. *Miner. Process. Extr. Metall. Rev.* **2017**, *38*, 265–273. [CrossRef]
4. Liu, P.; Liu, C.; Hu, T.; Shi, J.; Zhang, L.; Liu, B.; Peng, J. Kinetic study of microwave enhanced mercury desorption for the regeneration of spent activated carbon supported mercuric chloride catalysts. *Chem. Eng. J.* **2021**, *408*, 127355. [CrossRef]
5. Peng, X.; Liu, W.; Liu, W.; Zhao, L.; Zhang, N.; Gu, X.; Zhou, S. Fluorite enhanced magnesium recovery from serpentine tailings: Kinetics and reaction mechanisms. *Hydrometallurgy* **2021**, *201*, 105571. [CrossRef]
6. Zhang, W.; Dai, J.; Li, C.; Yu, X.; Xue, Z.; Saxén, H. A Review on Explorations of the Oxygen Blast Furnace Process. *Steel Res. Int.* **2021**, *92*, 1–23. [CrossRef]
7. Coats, A.W.; Redfern, J.P. Kinetic parameters from thermogravimetric data. *Nature* **1964**, *201*, 68–69. [CrossRef]
8. Freeman, E.; Carroll, B. The application of thermoanalytical techniques to kinetics: The thermogravimetric evaluation of the kinetics of the decomposition of calcium oxalate monohydrate. *J. Phys. Chem.* **1975**, *62*, 394–397. [CrossRef]
9. Wang, T.X.; Ding, C.Y.; Lv, X.W.; Xuan, S.W.; Li, G. Reduction kinetics of MgO-doped calcium ferrites under CO–N$_2$ atmosphere. *J. Iron Steel Res. Int.* **2019**, *26*, 1265–1272. [CrossRef]
10. Huang, Z.C.; Wu, K.; Hu, B.; Peng, H.; Jiang, T. Non-Isothermal Kinetics of Reduction Reaction of Oxidized Pellet Under Microwave Irradiation. *J. Iron Steel Res. Int.* **2012**, *19*, 1–4. [CrossRef]
11. Han, G.H.; Jiang, T.; Zhang, Y.B.; Huang, Y.F.; Li, G.H. High-temperature oxidation behavior of vanadium, titanium-bearing magnetite pellet. *J. Iron Steel Res. Int.* **2011**, *18*, 14–19. [CrossRef]
12. Vyazovkin, S.; Wight, C.A. Model-free and model-fitting approaches to kinetic analysis of isothermal and nonisothermal data. *Thermochim. Acta* **1999**, *340–341*, 53–68. [CrossRef]
13. Darken, L.S.; Gurry, R.W. The System Iron-Oxygen. I. The Wüstite Field and Related Equilibria. *J. Am. Chem. Soc.* **1945**, *67*, 1398–1412. [CrossRef]
14. Xing, L.Y.; Zou, Z.S.; Qu, Y.X.; Shao, L.; Zou, J.Q. Gas–Solid Reduction Behavior of In-flight Fine Hematite Ore Particles by Hydrogen. *Steel Res. Int.* **2019**, *90*, 1–10. [CrossRef]
15. Zheng, H.; Schenk, J.; Spreitzer, D.; Wolfinger, T.; Daghagheleh, O. Review on the Oxidation Behaviors and Kinetics of Magnetite in Particle Scale. *Steel Res. Int.* **2021**, *92*, 1–13. [CrossRef]
16. Pei, Z.; Peimin, G.; Dianwei, Z. study on reduction sequenece of hematite at low-temperature non-equilibrium state. *Iorn Steel* **2006**, *41*, 12–15.
17. Wimmers, O.J.; Arnoldy, P.; Moulijn, J.A. Determination of the reduction mechanism by temperature-programmed reduction: Application to small Fe$_2$O$_3$ particles. *J. Phys. Chem.* **1986**, *90*, 1331–1337. [CrossRef]
18. Zhang, W.; Zhang, J.H.; Zou, Z.S.; Li, Q.; Qi, Y.H. Influences of non-stoichiometry on thermodynamics and kinetics of iron oxides reduction processes. *Ironmak. Steelmak.* **2014**, *41*, 715–720. [CrossRef]
19. Hedvall, J.A. Reaktionsfähigkeit Fester Stoffe. *Angew. Chem.* **1938**, *51*, 150. [CrossRef]
20. Garner, W.E. *Chemistry of the Solid State*; Butterworths Scientific Publications: London, UK, 1955. [CrossRef]
21. Jost, W. Diffusion und Chemische Reaktionen in Festen Stoffen. *Nature* **1938**, *142*, 776. [CrossRef]
22. Vyazovkin, S.; Burnham, A.K.; Criado, J.M.; Pérez-Maqueda, L.A.; Popescu, C.; Sbirrazzuoli, N. ICTAC Kinetics Committee recommendations for performing kinetic computations on thermal analysis data. *Thermochim. Acta* **2011**, *520*, 1–19. [CrossRef]
23. Han, Y.; Li, T.; Saito, K. Comprehensive method based on model free method and IKP method for evaluating kinetic parameters of solid state reactions. *J. Comput. Chem.* **2012**, *33*, 2516–2525. [CrossRef]
24. Sharp, J.H.; Wentworth, S.A. Kinetic analysis of thermogravimetric data. *Anal. Chem.* **1969**, *41*, 2060–2062. [CrossRef]
25. Ozawa, T. Kinetic analysis of derivative curves in thermal analysis. *J. Therm. Anal.* **1970**, *2*, 301–324. [CrossRef]
26. Sestak, J.; Satava, V.; Wendlandt, W.W. The study of heterogeneous processes by thermal analysis. *Thermochim. Acta* **1973**, *7*, 333–556. [CrossRef]
27. Vyazovkin, S. Model-free kinetics: Staying free of multiplying entities without necessity. *J. Therm. Anal. Calorim.* **2006**, *83*, 45–51. [CrossRef]
28. Glasstone, K.S.; Laidler, J.; Eyring, H. The Theory of Rate Processes. *Nature* **1941**, *149*, 509–510.
29. Galwey, A.K.; Brown, M. *Thermal Decomposition of Ionic Solids*; Elsevier Science: Amsterdam, The Netherlands, 1999.

30. Brown, M.E.; Dollimore, D.; Galwey, A.K. *Comprehensive Chemical Kinetics.Reactions in the Solid State*; Elsevier: Amsterdam, The Netherlands, 1980; Volume 22.
31. Sardari, A.; Alamdari, E.K.; Noaparast, M.; Shafaei, S.Z. Kinetics of magnetite oxidation under non-isothermal conditions. *Int. J. Miner. Metall. Mater.* **2017**, *24*, 486–492. [CrossRef]
32. Barile, C.; Casavola, C.; Vimalathithan, P.K.; Pugliese, M.; Maiorano, V. Thermomechanical and morphological studies of CFRP tested in different environmental conditions. *Materials* **2018**, *12*, 63. [CrossRef]
33. Wang, F.; Qian, D.S.; Xiao, P.; Deng, S. Accelerating cementite precipitation during the non-isothermal process by applying tensile stress in GCr15 bearing steel. *Materials* **2018**, *11*, 2403. [CrossRef]
34. Johnson, N.L.; Leone, F.C. *Statistics and Experimental Design in Engineering and the Physical Sciences*; Wiley: New York, NY, USA, 1977.
35. Harnett, D.L. *Statistical Methods*; Addison-Wesley: Reading, UK, 1982.
36. Draper, N.R.; Smith, H. *Applied Regression Analysis*; Wiley: New York, NY, USA, 1981.
37. Vyazovkin, S.; Wight, C.A. Isothermal and nonisothermal reaction kinetics in solids: In search of ways toward consensus. *J. Phys. Chem. A* **1997**, *101*, 8279–8284. [CrossRef]
38. Zhang, W.; Zhang, J.; Li, Q.; He, Y.; Tang, B.; Li, M.; Zhang, Z.; Zou, Z. Thermodynamic analyses of iron oxides redox reactions. In Proceedings of the 8th Pacific Rim International Congress on Advanced Materials and Processing, PRICM 8, Waikoloa, HI, USA, 4–9 August 2013; Marquis, F., Ed.; Wiley: Waikoloa, HI, USA, 2013; pp. 777–789.

Article

Effect of Basicity on the Sulfur Precipitation and Occurrence State in Kambara Reactor Desulfurization Slag

Renlin Zhu [1], Jianli Li [1,2,3,*], Jiajun Jiang [2], Yue Yu [2] and Hangyu Zhu [2]

1 The State Key Laboratory of Refractories and Metallurgy, Wuhan University of Science and Technology, Wuhan 430081, China; zhurenlin522425@wust.edu.cn
2 Hubei Provincial Key Laboratory for New Processes of Ironmaking and Steelmaking, Wuhan University of Science and Technology, Wuhan 430081, China; 0204039@wust.edu.cn (J.J.); yuyue@wust.edu.cn (Y.Y.); zhuhy@wust.edu.cn (H.Z.)
3 Key Laboratory for Ferrous Metallurgy and Resources Utilization of Ministry of Education, Wuhan University of Science and Technology, Wuhan 430081, China
* Correspondence: jli@wust.edu.cn

Abstract: Kambara Reactor (KR) desulfurization slag used as slag-making material for converter smelting can promote early slag melting in the initial stage and improve the efficiency of dephosphorization. However, its direct utilization as a slagging material can increase the sulfur content in molten steel since KR desulfurization slag contains 1~2.5% sulfur. Therefore, this research focuses on the effect of basicity on the precipitation behavior and occurrence state of sulfur in KR desulfurization slag in order to provide an academic reference for the subsequent removal of sulfur from slag through an oxidizing atmosphere. The solidification process of slag was simulated by the Factsage8.0. The slag samples were analyzed by X-ray diffraction (XRD) and scanning electron microscope (SEM), and the amount of CaS grains was analyzed using Image-ProPlus6.0 software. The thermodynamic calculation showed that the crystallization temperature of CaS in the molten slag gradually decreased with the increase in basicity, and the CaS crystals in the molten slag mainly existed in the matrix phase and at the silicate grain boundaries. A large number of CaS grains were precipitated along the silicate grain boundary in low-basicity (R = 2.5 and 3.0) slags and fewer CaS grains were precipitated along the silicate grain boundary, while the CaS grain density in the matrix phase was higher in the high-basicity (R = 3.5, 4.0, 4.5) slag. With the increase in basicity, the number of CaS grains gradually decreased, and the CaS grain sizes in slag sample increased gradually. The sulfur in the synthetic slag was in the form of CaS crystals and the amorphous phase, and the content of amorphous sulfur gradually increased with increasing basicity.

Keywords: KR desulfurization slag; basicity; CaS; precipitation; occurrence

Citation: Zhu, R.; Li, J.; Jiang, J.; Yu, Y.; Zhu, H. Effect of Basicity on the Sulfur Precipitation and Occurrence State in Kambara Reactor Desulfurization Slag. *Minerals* **2021**, *11*, 977. https://doi.org/10.3390/min11090977

Academic Editors: Chiharu Tokoro, Shuai Wang, Xingjie Wang and Jia Yang

Received: 12 August 2021
Accepted: 4 September 2021
Published: 8 September 2021

Publisher's Note: MDPI stays neutral with regard to jurisdictional claims in published maps and institutional affiliations.

1. Introduction

Sulfur is a detrimental impurity in molten steel that can easily form long strips or flake-shaped sulfide inclusions, causing a decrease in the transverse compressive strength and plasticity of steel [1]. In addition, the FeS formed in molten steel easily shapes a low-temperature eutectic at the grain boundary of ferrite and austenite, which reduces the hot workability of steel [2]. Lv et al. [3] conducted a study about the effect of sulfur on the properties of high-manganese austenitic steel and discovered that an increase in sulfur content leads to a decrease in the ductility and plasticity temperature range of high-manganese austenitic steel. To reduce the effect of sulfur on the performance of steel, metallurgical enterprises often pretreat blast furnace hot metal for desulfurization outside the furnace. At present, the commonly used domestic hot metal desulfurization pretreatment processes outside the furnace mainly include the KR desulfurization process and the injection desulfurization process [4,5]. The KR desulfurization process involves inserting the poured cross-shaped refractory stirring head into the molten iron to a certain depth, and then driving it at a certain speed to rotate in the molten iron to produce a vortex. Then, the desulfurizer and the molten iron are fully contacted to achieve a desulfurization effect [6]. In the injection desulfurization process, inert gas and desulfurizer are injected into the molten iron at the same time by a spray gun, and the inert gas plays the main role of carrying the desulfurizer and stirring the molten iron [7,8]. Due to the superior dynamic conditions and high desulfurization efficiency of the KR mechanical stirring desulfurization process, it has gradually replaced the injection desulfurization process and has become the most widely used hot metal pretreatment desulfurization process in China [9]. However, the widespread use of the KR desulfurization process in China is leading to an increase in the output of KR desulfurization slag, causing the accumulation of KR desulfurization slag, occupying a large amount of land resources, and polluting the surrounding waters and atmospheric environment of enterprises. It can be concluded that improving the comprehensive utilization rate of KR desulfurization slag is extremely important for environmentally friendly production within enterprises.

The composition of KR desulfurization slag comprises CaO, SiO_2, MgO, Al_2O_3, CaF_2, CaS and MFe, and the sulfur content is 1.0~2.5% [10]. At present, the treatments of KR desulfurization slag in domestic and foreign enterprises mainly include recovering MFe from the slag and reusing the tailing as sintering ingredients, building materials etc. [11–14]. Han et al. [11] used the combined process of grinding classification–gravity, separation–magnetic and separation to extract iron from desulfurization slag of Angang, and the TFe grade of the iron concentrate reached 86.32%, the S grade decreased to 0.21%, and an iron recovery rate of 78.48% in the slag was achieved. Sheng et al. [12] studied the comprehensive utilization of KR desulfurization slag in neutralizing acid mine wastewater and found that KR desulfurization slag can effectively neutralize acid mine wastewater, although there is little f-CaO in the slag. Fujino et al. [14] reported that the application of KR desulfurization slag in sintering production can reduce coking coal consumption, increase the melt, and reduce CO_2 emission in the sintering process. However, if the use of KR desulfurization slag exceeds 15%, the permeability of the sintering bed will be decreased. In addition, Nakai et al. [15] showed that adding 70% KR desulfurization slag for hot metal pretreatment desulfurization has the same desulfurization capacity as using pure desulfurizer (CaO-5%CaF_2) and can save 40% desulfurizer consumption. In summary, the above research has promoted the technical development of the comprehensive utilization of KR desulfurization slag to varying degrees, but the above process does not consider the removal of sulfur in slag to improve the value of the comprehensive utilization of KR desulfurization slag. Furthermore, the economic benefit of the above processes for the utilization of KR desulfurization slag is low. CaF_2 and CaS in the slag easily produce the toxic gases HF and H_2S due to reacting with the moisture in air, resulting in secondary pollution. Therefore, developing new techniques for the comprehensive utilization of KR desulfurization slag is of great significance to the long-term development of enterprises.

The main component of KR desulfurization slag is CaO, which is mostly contained in the form of silicate ($2CaO \cdot SiO_2$, $3CaO \cdot SiO_2$) and f-CaO in KR slag [16]. If KR desulfurization slag replaces the reactive lime for converter slag making, it can promote early slag melting in the initial stage of converter smelting and can improve the efficiency of converter dephosphorization. Relevant researchers [17–19] have reported that C2S can form a stable solid solution C2S-C3P ($2CaO \cdot SiO_2$-$3CaO \cdot P_2O_5$) with C3P ($3CaO \cdot P_2O_5$) to improve the stability of phosphorus in the molten slag. To this end, the reuse of KR desulfurization slag in converter smelting process has been studied. Jiang et al. [20] concluded that the average sulfur content is 0.0136% higher than in the normal converter smelting process. It can be seen that reusing KR desulfurization slag for converter smelting directly increases the sulfur content and reduces the quality of molten steel. Therefore, in order to compensate for the shortcomings of the current smelting process, the thermodynamic database Factsage8.0 was used herein to simulate the solidification process of slag, as well as XRD and SEM-EDS were used to analyze and detect the mineral phase and microstructure of slag to investigate the effect of basicity on the sulfur precipitation behavior and occurrence state of KR desulfurization slag in the form of synthetic slag to provide a theoretical basis for the subsequent removal of sulfur in KR desulfurization slag through an oxidizing atmosphere and realize the comprehensive utilization of KR desulfurization slag in the converter smelting process.

2. Experimental Procedure

2.1. Synthetic Slag Composition

According to the existing conclusions and the desulfurization principle of the KR desulfurization process, it can be determined that sulfur in slag mainly exists in the form of a CaS phase [21]. Therefore, the sulfur in synthetic slag was added in the form of CaS, and the mass fraction was 3.38% (1.5% S). CaO, SiO_2, Al_2O_3, MgO and CaF_2 were all analytical pure chemical reagents. According to industrial KR desulfurization slag, the basicity of synthetic slag was set to 2.5, 3.0, 3.5, 4.0, and 4.5. The mass fractions of Al_2O_3 and MgO in the synthetic slag were controlled to be 6.00% and 3.00%, respectively. The CaO and SiO_2 contents in the synthetic slag were added according to binary basicity. The mass ratio of CaF_2 and CaO in the slag was 1:9. The chemical compositions of synthetic slags with different basicities in the CaO-SiO_2-CaF_2-CaS-based system are shown in Table 1.

Table 1. Chemical composition of molten slag, wt.%.

Sample	CaO	SiO_2	Al_2O_3	MgO	CaS	CaF_2	$CaO\%/SiO_2\%$
S1	59.23	23.69	6.00	3.00	3.38	6.58	2.50
S2	61.96	20.65	6.00	3.00	3.38	6.88	3.00
S3	64.07	18.31	6.00	3.00	3.38	7.12	3.50
S4	65.76	16.44	6.00	3.00	3.38	7.31	4.00
S5	67.13	14.92	6.00	3.00	3.38	7.46	4.50

2.2. Preparation Process of Synthetic Slag

The laboratory preparation of synthetic slag mainly includes the laboratory preparation of CaS and synthetic slag samples with different basicities in Table 1.

In the beginning, CaS added into synthetic slag was prepared through high-temperature carbothermal reduction in the laboratory because CaS is apt for oxidization. The raw materials for preparing CaS are carbon powder (purity > 99.85%) and $CaSO_4 \cdot 2H_2O$ (purity > 99.95%). $CaSO_4 \cdot 2H_2O$ was first heated at 200 °C for 2 h to dehydrate and form $CaSO_4$, and then the dehydrated $CaSO_4$ was mixed with carbon powder in a molar ratio of 1:2 and loaded into the corundum crucible. Then, the corundum crucible was placed in the constant temperature zone of the carbon-tube furnace with nitrogen atmosphere at a flow rate of 1L/min. The heating rate was set to 10 °C/min, and the reaction temperature was set to 1100 °C. Finally, after holding at 1100 °C for 2 h, the sample was cooled to room temperature in a nitrogen atmosphere and the CaS purity was calculated. The principle of CaS preparation in the laboratory are shown in Equations (1) and (2).

$$CaSO_4 \cdot 2H_2O = CaSO_4 + 2H_2O \tag{1}$$

$$CaSO_4 + 2C = CaS + 2CO_2 \tag{2}$$

Second, the quality of the synthetic slag sample was determined to be 100 g, and the analytical pure chemical reagents were precisely weighed according to the different slag compositions shown in Table 1. Furthermore, the chemical reagents were fully stirred and evenly screened twice with 35 mesh and 200 mesh standard screens. The mixed synthetic slag was placed into the corundum crucible, and the corundum crucible was placed in the constant temperature zone of the carbon-tube furnace. The nitrogen flow rate of the carbon-tube furnace was set to 1 L/min, the reaction temperature was set to 1600 °C, and the heating rate was 15 °C/min. The slag sample was extracted by a quartz glass tube when the holding times reached 15 min.

2.3. Analysis Method

In order to clarify the evolution process of mineral composition and slag microstructure with basicity in different slag samples. The samples S1–S5 were detected by X-ray diffraction analyzer (XRD, Rigaku/SmartLab SE, Tokushima, Tokyo, Japan). The detection data of the X-ray diffractometer were analyzed by Search-Match software, and the data were plotted by Origin2019. Moreover, the microstructure of different slag samples was analyzed by SEM-X spectrometry (SEM-EDS; ThermoFisher/Apreo S HiVac, Wyman Street, Waltham, MA, USA). Finally, the amount of CaS grains in different slag samples were analyzed by Image-ProPlus 6.0 (IPP6.0, Media Cybernetics, Bethesda, MD, USA).

3. Thermodynamic Calculation of Synthetic Slag Solidification

The Equlib module of the thermodynamic database FactSage (FactSage 8.0, GTT, Germany) was used to simulate the equilibrium solidification process of the S3 sample with FToxid and FTsalt. In the software, the pure liquids and pure solids of the compound species of the production were chosen. The simulation temperature range was 1100~1600 °C, and the calculation step was set as 10 °C/min. The results are shown in Figure 1. When the temperature was 1600 °C, two phases coexisted in the molten slag, namely, the liquid phase and the MeO#1 phase (CaO solid solution). When the temperature decreased to 1420 °C, the C2S phase began to precipitate from the liquid phase; when the temperature was lower than 1340 °C, the sulfur in the residual liquid began to precipitate in the form of CaS; when the temperature was lower than 1320 °C, the MeO#2 phase (MgO solid solution) and the $12CaO \cdot 7Al_2O_3 \cdot CaF_2$ phase began to precipitate from liquid phase; when the temperature decreased to 1275 °C, the precipitation content of MeO#1 and C2S increased simultaneously. When the temperature decreased to 1150 °C, all phases in the synthetic slag were completely precipitated. It can be seen from Figure 1 that the MeO#2 phase and CaS were mainly precipitated in the form of pure substance during solidification, and no chemical reaction occurred with the other components in the molten slag, while the sulfur in the slag completely precipitated into the CaS phase. Moreover, Al_2O_3 reacted with CaF_2 and CaO in the slag to form the low-melting point compound $12CaO \cdot 7Al_2O_3 \cdot CaF_2$.

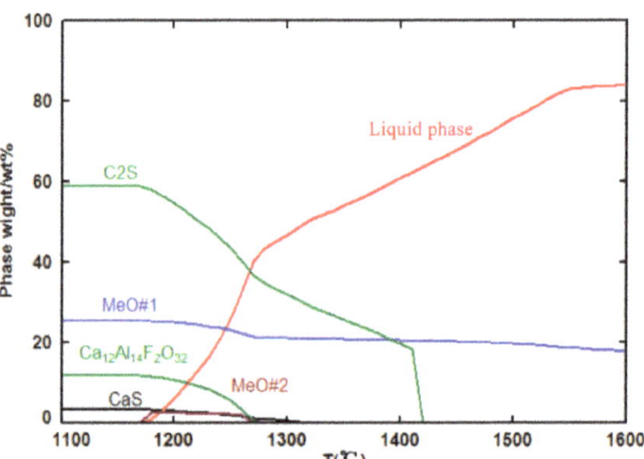

Figure 1. Solidification simulation analysis of synthetic slag S3.

Second, the amount of precipitation of each phase vs. basicity is plotted in Figure 2, indicating that the amount of precipitation amount of the CaS, MeO#2 and $12CaO \cdot 7Al_2O_3 \cdot CaF_2$ phases in the molten slag remained the same quality with an increase in basicity. This is because the contents of MgO and CaS in the different synthetic slags were 3.00 and 3.38 g, respectively, and the sulfur and magnesium in the slag were able to completely precipitate. According to the thermodynamic calculation in Figure 1, MgO and CaS did not react with the other components in the slag during solidification, and they could completely precipitate into MeO#2 phase (MgO solid solution) and CaS during solidification. Additionally, the complete precipitation of $12CaO \cdot 7Al_2O_3 \cdot CaF_2$ in slag was unaffected by the change in basicity, mainly due to CaO being the main composition in the synthetic slag, in the molten state, Al_2O_3 in slags with different basicities completely reacted with CaO and CaF_2 to form $12CaO \cdot 7Al_2O_3 \cdot CaF_2$. Moreover, when the basicity increased from 2.5 to 4.5, the C2S phase decreased from 72.62% to 48.05%, and the amount of precipitation of MeO#1 phase increased from 10.25% to 34.82% in the solid slag. The amount of precipitation of C2S phase in the slag was negatively correlated with the increase in basicity, while the amount of precipitation of the MeO#1 phase was positively correlated.

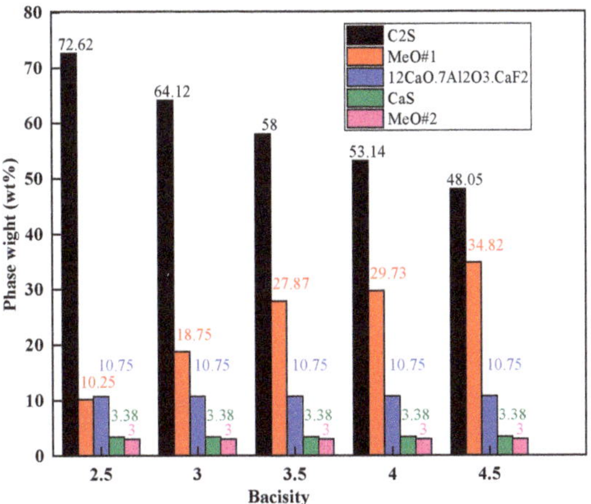

Figure 2. Relationship between the total amount of precipitation of each phase in the solid slag and the basicity.

Since the slag samples were prepared in the laboratory at 1600 °C, the thermodynamic analysis of the molten slag phase at 1600 °C was calculated to provide a reference for the mineral structure in molten slags with different basicities; the calculation result is shown in Figure 3. It was observed that all slag samples were composed of MeO#1 and liquid phases at 1600 °C. With the increase in basicity from 2.5 to 4.5, the liquid phase content in the molten slag decreased from 96.26% to 73.63%, while the MeO#1 phase content increased from 3.74% to 26.37%. It can be concluded that the supersaturation of MeO#1 phase in molten slag increased with the increase in basicity at 1600 °C.

Moreover, the process of sulfur precipitation in molten slag is related to the crystallization temperature of its sulfur; thus, the thermodynamic calculation of the crystallization temperature of sulfur in slag can provide thermodynamic a theory for the analysis of sulfur precipitation behavior in molten slag. The crystallization temperature of CaS in different basicities slag was calculated as shown in Figure 4. When the basicity was 2.5, the highest crystallization temperature of CaS was 1340 °C. When the basicity was 4.5, the crystallization temperature of CaS was 1305 °C. Thus, it can be seen that the crystallization temperature of CaS was a strong function of basicity. The increase in basicity tended to promote a lower crystallization temperature, and the temperature decreased from 1340 to 1305 °C.

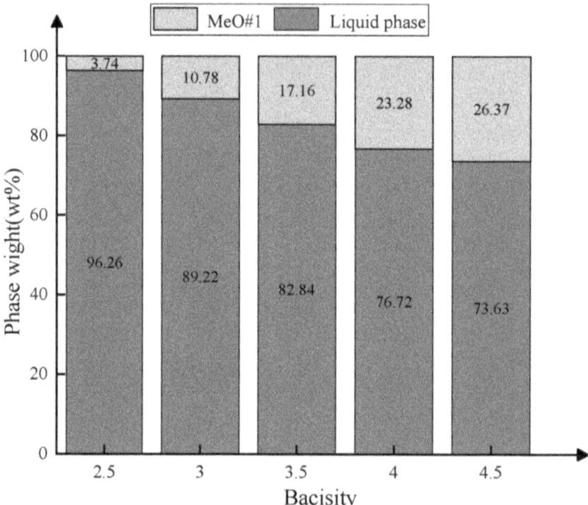

Figure 3. Relationship between the molten slag phase and the basicity at 1600 °C.

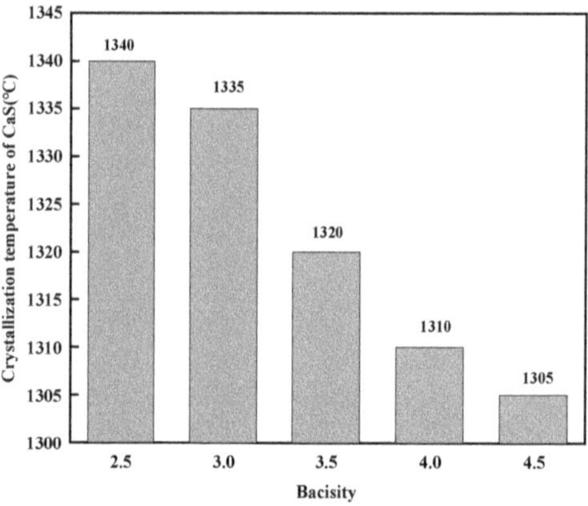

Figure 4. Diagram of the relationship between the crystallization temperature and the basicity of CaS in molten slag.

Based on the above solidification simulation analysis of each phase in the molten slag, the sulfur in the slag was precipitated in the form of CaS minerals, and the basic phases in the molten slag were C_2S, MeO#1, MeO#2, $12CaO \cdot 7Al_2O_3 \cdot CaF_2$ and CaS. Two phases coexisted in the molten slag at 1600 °C, and the particle concentration of the MeO#1 phase increased with the increase of basicity. During the equilibrium solidification process, the sulfur in the slag completely precipitated in the form of a CaS phase, and the crystallization temperature of CaS in the slag gradually decreased with the increase in basicity.

4. Results

4.1. Analysis of Synthetic Slag

Figure 5 presents the XRD analysis results of the synthetic slag samples with different basicities. The corresponding phase was C3S at diffraction angles of 32.3°, 32.7°, 34.4°, 41.5°, 51.9° and 62.4°. The corresponding phase was C2S at diffraction angles of 32.3, 32.7and 41.5°. The corresponding phase was CaS at diffraction angles of 31.5°, 45.2° and 60.2°, and MeO#2 at diffraction angles of 36.3°, 43.4° and 62.5°. When a slag basicity of R ≤ 3.5, the characteristic peak of the MeO#1 phase didn't appear in the solid slag. When the basicity increased to 4.0, the characteristic peak of the MeO#1 phase began to appear at

2θ of 32.5°, 37.2°, and 54.5 °, and the intensity of the characteristic peak of the MeO#1 phase gradually enhanced with the increase in basicity. This was mainly because the basicity of the slag was less than or equal to 3.5, the CaO in the slag was in an unsaturated state in the actual melting process, and CaO completely reacted with the other components in the slag to form silicate in the heat preservation process. Furthermore, when the basicity was greater than or equal to 4.0, the MeO#1 phase in the slag gradually increased due to the basicity being greater than or equal to 4.0, the supersaturation of the CaO content in the slag gradually increased, which caused the content of the MeO#1 phase that precipitated in the slag during solidification to gradually increase. Based on the above analysis, it can be concluded that the basic mineral phases of the solid slag were the silicate phase (C_2S and C_3S), the MeO#2 phase and the CaS phase, when the basicity was 2.5, 3.0 or 3.5; the basic mineral phases of the solid slag included the silicate phase (C_2S and C_3S), the MeO#1 phase (CaO solid solution), the MeO#2 phase (MgO solid solution) and the CaS phase, when the basicity was 4.0 or 4.5.

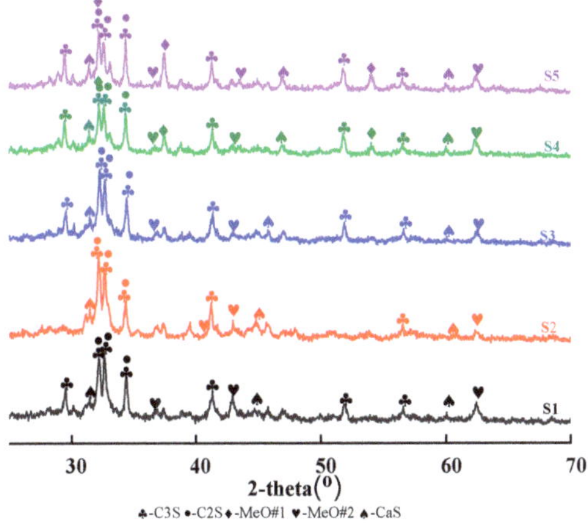

Figure 5. XRD diagram of synthetic slags with different basicities.

In addition, the thermodynamic calculations in Figures 1 and 2 show that the main phase in solid slag was the C2S phase, but the experimental results conclude that the silicate phase were C_2S and C_3S in solid slag. The main reason for the phenomenon is that the Equilib module of the thermodynamic database Factsage8.0 lacks C3S data. Therefore, the thermodynamic calculation results showed that the slag contained only C2S phase.

4.2. Microstructure of the Slag Samples

Figure 6 shows the synthetic slag samples with different basicities prepared at 1600 °C and 1 L/min nitrogen for 15 min. According to the results of EDS, C2S, C3S, matrix, MeO#1, MeO#2 phase and CaS phases were determined in solid slag samples, a-C3S, b-C2S, c-matrix, d-MeO#2, e-CaS and f-MeO#1 phase. When the basicity was 2.5, the solid slag mainly contained long-strip C3S phase and small particle C2S-phase, while the CaS and black MeO#2 phase dispersed in the silicate grain boundary and matrix phase. When the basicity was 3.0, the C2S phase in the molten slag was round structure and the content was more than that in S1, while the content of the C3S phase was less than that in S1. The MeO#2 phase in slag sample S2 was mainly distributed at the grain boundary of the C2S phase or C3S phase. In addition, the CaS grains in S2 were mainly distributed at the grain boundaries of the round cake-like C2S phase, and the CaS grains were larger than those in S1. When the basicity of the molten slag was 3.5, the C3S phase was mainly distributed in the slag with a large area of block structures, and a small amount of dispersion was distributed in the matrix phase, while there was no obvious single C2S phase in the solid slag. Furthermore, the MeO#2 phase began to grow up in the black block structure, and the larger dimensional CaS grains began to increase. When the basicity of the slag was 4.0 and 4.5, the content of the C3S phase in the slag was more than that of S3, the grain area of the MeO#2 phase was larger than that in S3, while the white CaS phase was distributed at the grain boundary of the C3S phase in strip and particle shapes. Moreover, the porous oval MeO#1 phase began to appear in S4 slag samples, the number of MeO#1 phase grains in S5 was higher than that in the S4. XRD analysis results showed that the slag with different basicity contained C2S phase, but EDS analysis of slag only S1, S2 contained a large number of C2S phase. S3, S4, S5 did not contain C2S phase due to the main silicate product of S3, S4 and S5 slag samples with high basicity being C3S in the heat preservation process.

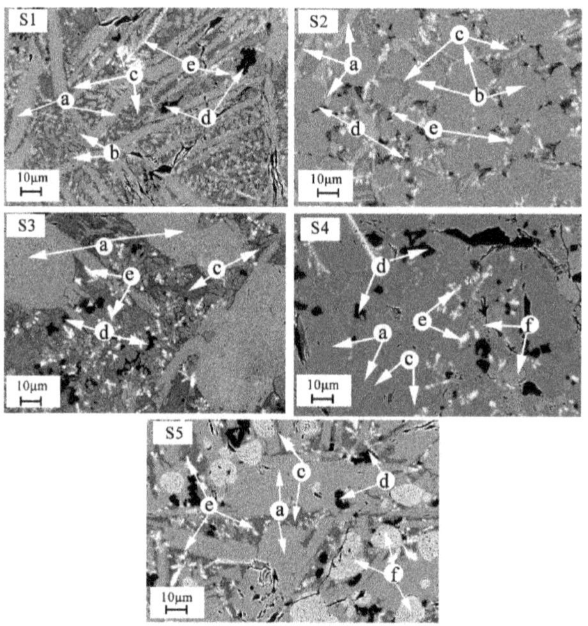

Figure 6. Micro analysis chart of slag samples with different basicity. a, C3S; b, C2S; c, matrix phase; d, MeO#2 phase; e, CaS phase; f, MeO#1 phase. **S1**—with bacisity of 2.5, **S2**—with bacisity of 3.0, **S3**—with bacisity of 3.5, **S4**—with bacisity of 4.0, **S5**—with bacisity of 4.5.

Moreover, by comparing and analyzing the mineral structure of S1–S5 slag samples, it was found that the number of CaS grains that precipitated along the silicate grain boundary in the S1 and S2 slag samples gradually decreased, while the CaS grains that precipitated in the matrix phase of S3–S5 slag samples were higher than S1 and S2 slag samples. The grain size of CaS increased gradually in the slag with the increase in basicity. The matrix phase in the solid slag was mainly distributed along the precipitated phase, and the distribution of the matrix phase in molten slag was mainly controlled by the precipitated phase. In addition, the number of CaS grains in the slag was statistically analyzed by image-ProPlus6.0, and the results are shown in Figure 7. When the slag basicity was 2.5, the number of CaS grains in the slag was 246, and when the slag basicity was 4.5, the number of CaS grains in the slag was 99. Thus, it can be seen that the number of CaS grains in molten slag gradually increased with the increase in basicity.

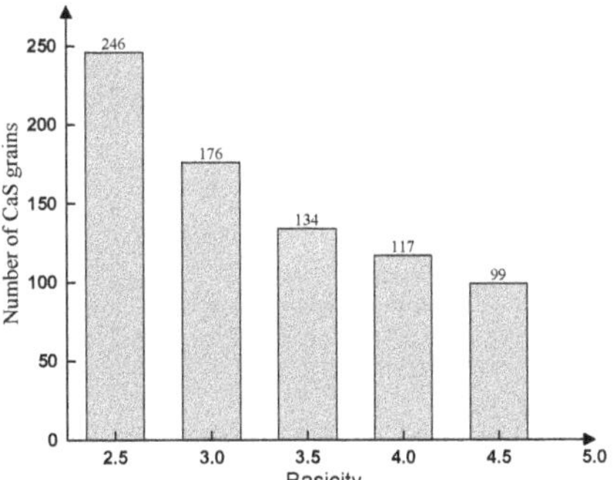

Figure 7. Number of CaS grains in molten slag.

5. Discussion

5.1. Effect of Basicity on the Sulfur Precipitation Behavior

When the basicity of the slag was 2.5–3.0, the slag contained separate C2S and C3S phases. This is mainly because the silicate formed by the slag reaction was in the C2S and C3S zones of CaO-SiO$_2$ phase diagram. When the slag basicity was equal to 3.5, 4.0 or 4.5, the silicate phase in the slag comprised a large area of massive the C3S phase. This was mainly because the silicate product in the slag was completely located in the C3S area in the CaO-SiO$_2$ phase diagram; the silicate product in the slag was mainly the C3S phase at 1600 °C. Moreover, the CaS grain size increased gradually with the increase in basicity due to the complex silicate network structure of slag decreased with the increase in basicity, which decreased the viscosity of slag, and promoted the growth of CaS grain in the slag. Certain scholars [22,23] have reported that an increase of CaO content in the slag led to the decrease of bridge oxygen content, which makes silicate network structure become simpler and decreases the viscosity of slag. Herein, the number of CaS grains in the slag samples decreased gradually with the increase in basicity due to the initial crystallization temperature of CaS decreased with the increase in basicity in the molten slag (Figure 4), which led to a decrease of the supercooling zone of CaS crystallization in the molten slag, leading to an increase of critical nucleation of CaS, and resulting in a gradual decrease in the nucleation rate of CaS crystals.

In addition, the precipitation behavior of the sulfur in the molten slags with different basicities was related to the nucleation mode of sulfur during solidification. The precipitation process of the sulfur in the molten slag included homogeneous and heterogeneous nucleation. Based on the above analysis of the microstructure of slags with different basicities, the sulfur in the solid slag was mainly distributed in the form of CaS in the matrix phase and grew along the grain boundary of the silicate. The CaS grains in the slags with low basicities (R = 2.5 and 3.0) mainly grew along the silicate grain boundary due to a large number of silicate solid particles existed in the matrix phase, which provided favorable conditions for the heterogeneous nucleation of sulfur in the matrix phase during precipitation. Jiang et al. [24] concluded that the existence of solid-phase ZrO$_2$ particles in CaO-Al$_2$O$_3$ slag causes heterogeneous nucleation of slag during solidification. Furthermore, the matrix phase of high-basicity slags (R = 3.5, 4.0 and 4.5) contained an amount of CaS grains, and several CaS grains grew along the silicate grain boundaries. This was mainly because the area of single silicate grains in the high-basicity slags was larger than that in the low-basicity slags, which reduced the contact area between the matrix phase and silicate solid particles, resulting in the precipitation process of CaS in high-basicity slags being mainly homogeneous nucleation. This implied that the precipitation of sulfur in the matrix phase of the low-basicity slags was mainly heterogeneous nucleation, while the CaS in high-basicity slags was mainly homogeneous nucleation.

5.2. Effect of Basicity on the Occurrence State of Sulfur in Slag

The micro-area distribution of the sulfur in the S1 and S3 slag samples in Figure 8 were analyzed by scanning electron microscopy (SEM-EDS), as shown. Through the surface scanning analysis data, it was found that the matrix phase around the CaS grains contained dispersed sulfur elements; thus, it can be determined that the slag matrix phase in the solidification process contained unprecipitated amorphous sulfur. The sulfur of the synthetic slag existed in two different states, namely, CaS crystal and amorphous sulfur. The sulfur elements distributed in the aggregated state of the molten slag were CaS crystals, while the sulfur elements distributed in the dispersed state occurred in the matrix phase as an amorphous structure. Comparing the microzone distribution of the sulfur elements in Figure 8a,b, it can be discovered that the amorphous structure density of the sulfur elements in the S3 slag sample was higher than that in the S1 slag sample, and the content of amorphous sulfur in the slag increased with an increase in basicity. The key reason of such a phenomenon was that the crystallization temperature of the CaS in the molten slag decreased with the increase in basicity and the supercooling zone of the CaS crystallization decreased during solidification in the same cooling system (air cooling), which reduced the amount of precipitation of the CaS phase and increased the amorphous sulfur content in matrix phase. Moreover, the thermodynamic calculation of sulfur precipitation during solidification in Figure 2 indicated that the sulfur in the slag was completely precipitated in the form of CaS, but the surface scanning distribution analytical results showed that the amorphous sulfur content of matrix phase increased with the increase in basicity. This was mainly because the kinetic condition of S^{2-} diffusion in the matrix phase deteriorated with the decrease in slag temperature during solidification, while certain S^{2-} in the matrix phase failed to diffuse to the surface of the CaS crystal nucleus to form CaS crystals and occurred in the form of an amorphous structure in matrix phase.

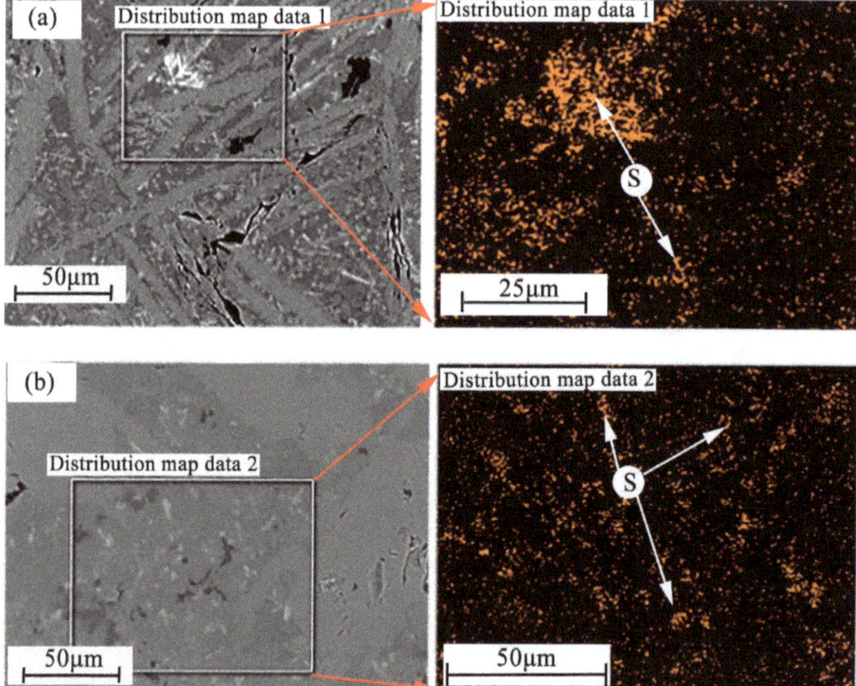

Figure 8. Surface scanning analysis of the sulfur elements in molten slag. (**a**) S1 with a basicity of 2.5, (**b**) S3 with a basicity of 3.5.

Based on the above, research status of KR desulfurization slag and the experimental results are discussed. At present, the comprehensive utilization process of KR desulfurization slag in China is not considered to remove sulfur from KR desulfurization slag and to reuse it in converter smelting processes. Therefore, the results are beneficial to provide a theoretical basis for the subsequent removal of sulfur from slag through oxidizing atmosphere.

6. Conclusions

(1)   According to the calculated results of molten slag solidification process based on thermodynamic database FactSage8.1, it was concluded that sulfur in KR desulfurization slag was mainly precipitated in the form of CaS, and the crystallization temperature of CaS in slag decreased with the increase of basicity;

(2)   CaS grains mainly precipitated along silicate grain boundaries in low-basicity ($R = 2.5$ and 3.0) slags, and the precipitation behavior of CaS was mainly heterogeneous nucleation. There were fewer CaS grains precipitated along silicate grain boundaries in molten slags with high basicity ($R = 3.5, 4.0$ and 4.5), and the precipitation behavior of sulfur in matrix phase was mainly homogeneous nucleation;

(3)   The number and the size of CaS grains decreased and increased respectively with the increase of the slag basicity.

Author Contributions: R.Z. drafted the manuscript and conducted the experiments, J.J. helped to conduct the experimental work, and J.L. modified and polished the draft. Y.Y. and H.Z. helped to read the manuscript. All authors have read and agreed to the published version of the manuscript.

Funding: The work was supported by the National Natural Science Foundation of China (No. 51974210, 52074197), the Hubei Provincial Natural Science Foundation (No. 2019CFB697), and the State Key Laboratory of Refractories and Metallurgy.

Acknowledgments: The authors would like to thank the National Natural Science Foundation of China and the Hubei Provincial Natural Science Foundation for their financial supports.

Conflicts of Interest: The authors declare no conflict of interest.

References

1.   He, M.; Wang, N.; Chen, M.; Chen, M.; Li, C.F. Distribution and motion behavior of desulfurizer particles in hot metal with mechanical stirring. *J. Powder Technol.* **2019**, *361*, 455–461. [CrossRef]
2.   Zhao, D.G.; Li, J.S.; Wang, S.H.; Liu, H.W.; Tong, Z.X.; Wu, F.P. Practical research on desulfurization in converter with the analysis of sulfur in steel for civil engineering. *J. Adv. Mater. Res.* **2012**, *56*, 17–20. [CrossRef]
3.   Lv, B.; Zhang, F.C.; Li, M.; Hou, R.J.; Qian, L.H.; Wang, T.S. Effects of phosphorus and sulfur on the thermoplasticity of high manganese austenitic steel. *J. Mater. Sci. Eng. A* **2010**, *527*, 5648–5653. [CrossRef]
4.   Iwamasa, P.K.; Fruehan, R.J. Effect of FeO in the slag and silicon in the metal on the desulfurization of hot metal. *J. Metall. Mater. Trans. B* **1997**, *28*, 47–57. [CrossRef]
5.   Wang, J.; Ge, W.S.; Chen, L.; Huang, S.Q.; Fan, J.R.; Zhong, Z.H. Study on technology of KR desulfurization for hot metal with vanadium and titanium. *J. Adv. Mater. Res.* **2012**, *581*, 1077–1082. [CrossRef]
6.   Nakai, Y.; Sumi, I.; Uno, H.; Kikuchi, N.; Kishimoto, Y. Effect of flux dispersion behavior on desulfurization of hot metal. *J. ISIJ Int.* **2010**, *50*, 403–410. [CrossRef]
7.   Marissa, V.R.; Antonio, R.S.; Rodolfo, M.; Miguel, A.H.; Federico, C.A.; Javier, C.A. Hot metal pretreatment by powder injection of lime-based reagents. *J. Steel Res. Int.* **2001**, *72*, 173–182.
8.   Barron, M.A.; Hilerio, I.; Medina, D.Y. Modeling and simulation of hot metal desulfurization by powder injection. *Open J. Appl. Sci.* **2015**, *5*, 295–303. [CrossRef]
9.   Zhang, M.L.; Xu, A.J. Comparison of application of KR method with that of injection method in hot metal desulphurzation. *J. Steelmaking* **2009**, *25*, 73–77. (In Chinese)
10.   Tong, Z.B.; Ma, G.J.; Cai, X.; Xue, Z.L.; Wang, W. Characterization and valorization of kanbara reactor desulfurization waste slag of hot metal pretreatment. *Waste Biomass. Valoriz.* **2016**, *7*, 1–8. [CrossRef]
11.   Han, Y.X.; Wang, L.Y.; Li, L.X.; Ren, F.; Zhao, C.Y.; Liu, X. Comprehensive utilization of slag from desurfurization and salg skimming processes of AnSteel. *J. Min. Metall. Eng.* **2009**, *29*, 29–32. (In Chinese)
12.   Sheng, G.H.; Huang, P.; Wang, S.S.; Chen, G.G. Potential reuse of slag from the kambara reactor desulfurization process of iron in an acidic mine drainage treatment. *J. Environ. Eng.* **2014**, *140*, 232–236. [CrossRef]

13. Maekelae, M.; Watkins, G.; Poeykioe, R.; Nurmesniemi, H.; Dahl, O. Utilization of steel, pulp and paper industry solid residues in forest soil amendment: Relevant physicochemical properties and heavy metal availability. *J. Hazard. Mater.* **2012**, *207*, 21–27. [CrossRef]
14. Kazuya, F.; Koichiro, O.; Taichi, M.; Eiki, K. Effective utilization of KR slag in iron ore sintering process. *J. Iron Steel Inst. Jpn.* **2017**, *103*, 357–364.
15. Nakai, Y.; Kikuchi, N.; Iwasa, M.; Nabeshima, S.; Kishimoto, Y. Development of slag recycling process in hot metal desulfurization with mechanical stirring. *J. Steel Res. Int.* **2009**, *80*, 727–732.
16. Wu, Q.F.; Bao, Y.P.; Lin, L.; XU, G.P.; Cheng, H.G.; Xin, C.P. Ineralogy characteristics and sulfur behavior of KR desulfurization slag. *J. China Metall.* **2015**, *25*, 44–47. (In Chinese)
17. Xue, H.M.; Li, J.; Xia, Y.J.; Wang, Y. Mechanism of phosphorus enrichment in dephosphorization slag produced using the technology of integrating dephosphorization and decarburization. *J. Met.* **2021**, *11*, 216–230.
18. Pahlevani, F.; Kitamura, S.Y.; Shibata, H.; Shibata, H.; Maruoka, N. Distribution of P_2O_5 between solid solution of $2CaO \cdot SiO_2$-$3CaO \cdot P_2O_5$ and liquid phase. *J. ISIJ Int.* **2010**, *50*, 822–829. [CrossRef]
19. Kitamura, S.Y.; Saito, S.; Utagawa, K.; Shibata, H.; Robertson, D.G.C. Mass transfer of P_2O_5 between liquid slag and solid solution of $2CaO \cdot SiO_2$ and $3CaO \cdot P_2O_5$. *J. Trans. Iron Steel Inst. Jpn.* **2009**, *49*, 1838–1844. [CrossRef]
20. Jiang, T.F.; Zhu, L.; Liu, F.G.; Zhao, X.D.; Luo, Y.Z. Recycling of desulfurization slag in pretreatment of hot metal. *J. Iron Steel* **2017**, *52*, 24–27. (In Chinese)
21. Xu, J.F.; Wang, X.H.; Huang, F.X.; Liu, C.Y. Analysis of phase and distribution of sulfur in KR desulfurization slag. *J. Iron Steel* **2015**, *50*, 15–18. (In Chinese)
22. Zhou, L.J.; Wang, W.L.; Ma, F.J.; Jin, L.; Wei, J.; Hiroyuki, M.; Fumitaka, T. A Kinetic study of the effect of basicity on the mold fluxes crystallization. *J. Metall. Mater. Trans. B* **2012**, *43*, 354–362. [CrossRef]
23. Dong, J.M.; Tsukihashi, F. Recent advances in understanding physical properties of metallurgical slags. *J. Met. Mater. Int.* **2017**, *23*, 1–19.
24. Jiang, B.B.; Wang, W.L.; Sohn, I.; Wei, J. A kinetic study of the effect of ZrO_2 and CaO/Al_2O_3 ratios on the crystallization behavior of a CaO-Al_2O_3-based slag system. *J. Metall. Mater. Trans. B* **2014**, *45*, 1057–1067. [CrossRef]

 *minerals*

Article

Effect of Magnesium on the Hydrophobicity of Sphalerite

Gloria I. Dávila-Pulido *, Adrián A. González-Ibarra *, Mitzué Garza-García and Danay A. Charles

Escuela Superior de Ingeniería, Universidad Autónoma de Coahuila, Blvd. Adolfo López Mateos s/n, Nueva Rosita C.P. 26800, Coahuila, Mexico; mitzue.garza@uadec.edu.mx (M.G.-G.); d_charles@uadec.edu.mx (D.A.C.)
* Correspondence: gloriadavila@uadec.edu.mx (G.I.D.-P.); gonzalez-adrian@uadec.edu.mx (A.A.G.-I.)

Abstract: The use of untreated recycled water has negative effects in the flotation of zinc sulfide ores due to the presence of dissolved species, such as magnesium and calcium. Although it has been found that magnesium is a more potent depressant than calcium, it has not been investigated in this role or for the effect of adding sodium carbonate. The results of an investigation to evaluate the effect of magnesium on the hydrophobicity of Cu-activated sphalerite conditioned with Sodium Isopropyl Xanthate (SIPX) are presented. Zeta potential of natural and Cu-activated sphalerite as a function of the conditioning pH and Cu(II) concentration, respectively, was first evaluated. Later, the effect of pH and presence of magnesium on the contact angle of Cu-activated sphalerite conditioned with SIPX was studied; it was also evaluated the effect of sodium carbonate to counteract the effect of magnesium. Cu-activation enhances the zeta potential of sphalerite up to a concentration of 5 mg/L. Contact angle tests, thermodynamic simulation, and surface analysis showed that magnesium hydroxide precipitates on the sphalerite surface at pH 9.6, decreasing its hydrophobicity. Addition of sodium carbonate as alkalinizing agent precipitates the magnesium in the form of a species that remained dispersed in the bulk solution, favoring the contact angle of Cu-activated sphalerite and, consequently, its hydrophobicity. It is concluded that the use of sodium carbonate as alkalinizing agent favors the precipitation of magnesium as hydromagnesite ($Mg_5(OH)_2(CO_3)_4 \cdot 4H_2O$) instead of hydroxide allowing the recovery of sphalerite.

Keywords: sphalerite; hydrophobicity; water chemistry; contact angle; magnesium

Citation: Dávila-Pulido, G.I.; González-Ibarra, A.A.; Garza-García, M.; Charles, D.A. Effect of Magnesium on the Hydrophobicity of Sphalerite. *Minerals* **2021**, *11*, 1359. https://doi.org/10.3390/min11121359

Academic Editors: Shuai Wang, Xingjie Wang, Jia Yang and Hyunjung Kim

Received: 2 November 2021
Accepted: 29 November 2021
Published: 1 December 2021

Publisher's Note: MDPI stays neutral with regard to jurisdictional claims in published maps and institutional affiliations.

1. Introduction

The management of water resources has become an important issue in sulfide concentrators due to the increasing requirements to use high proportions of recycle water from tailings dams, thickener overflows, dewatering, and filter products. Further, the extraction and transportation of fresh water are usually expensive. As a result, the mineral processing industries have adopted water recycling as a common practice in most of the countries, even in those where water scarcity does not represent a current problem. Accordingly, recent studies have focused on the treatment and effects of using recycled water to concentrate sulfide ores [1–5].

Recycled water may contain chemical species that contribute to the already complex chemistry of the flotation process. The ions found in the recycled water interact with the mineral surface causing inadvertent activation of undesirable minerals or depression of valuable ones, affecting the recovery and selectivity of mineral concentration [2,6]. The species commonly present in such water are colloidal material (e.g., silicates, clays, metal hydroxides, gypsum, etc.), base metal ions, hydroxo-complexes, sulfate, sulfite, thiosalts, chloride, magnesium, calcium, sodium, and potassium; as well as traces of organic-based substances as frothers, collectors, and depressants [7].

The role of calcium and magnesium in the flotation of sphalerite have been already studied since both are nearly always present. Calcium has received more attention and although important advances have been made, there are still phenomena to elucidate.

Levay et al. [2] evaluated the role of several water chemistry parameters (e.g., chemical composition, pH, dissolved oxygen, redox potential, and microbiological factors) on the flotation performance of an industrial operation. The authors associate the presence of calcium, magnesium, and iron salts with the precipitation of compounds that adversely affects the recovery of valuable sulfides. The researchers delved into the nature of the calcium precipitates, but did not investigate the magnesium ones.

El-Ammouri et al. [8] conducted a study making use of settling and turbidity techniques to determine the effect of magnesium (50 mg/L) and pH on the aggregation behavior of sphalerite and chalcopyrite samples as well as industrial slurries. According to the authors, adsorption of magnesium hydroxo-complex (i.e., $MgOH^+$) and magnesium hydroxide (i.e., $Mg(OH)_2$) at pH above 9 causes aggregation of the particles by an electrostatic bridging mechanism and when the particles become fully covered by abstracted magnesium hydroxide (between pH 10 and 11) repulsive forces appear between them and aggregation diminishes.

Lascelles et al. [9] studied the effect of magnesium (50 mg/L) on the flotation behavior of Cu-activated sphalerite in the range of pH from 8 to 13 and compared it with the effect of calcium (500 mg/L). The authors found that the flotation of Cu-activated sphalerite with ethyl xanthate was depressed by magnesium at pH above 10. Since copper adsorption and xanthate uptake were not substantially affected by the presence of magnesium, depression was attributed to the adsorption of magnesium species, particularly the precipitated hydroxide; this phenomenon was confirmed by XPS. The authors concluded that magnesium is a more potent depressant than calcium since the later did not show a significant contribution to depression at pH below 12.

This work was aimed to study the effect of magnesium on the hydrophobicity of Cu-activated sphalerite conditioned with Sodium Isopropyl Xanthate (SIPX). Although some studies have been carried out to determine the pH range in which magnesium affect the floatability of Cu-activated sphalerite, it is necessary to delve into the contact angle behavior and thermodynamics of the responsible magnesium species. The effect of adding sodium carbonate as alkalinizing agent has not been evaluated and neither has the mechanism by which it counteracts the negative effect of magnesium. For this purpose, the zeta potential of natural sphalerite as a function of the conditioning pH and Cu-activated sphalerite at pH 9 as a function of Cu(II) concentration was first evaluated; at this stage, sphalerite surface analysis by Scanning Electron Microscopy and Energy-Dispersive Spectrometry (SEM-EDS) to evaluate the adsorption of $Cu(OH)_2$ was carried out. Later, the effect of pH and presence of magnesium on the contact angle of Cu-activated sphalerite further conditioned with SIPX was studied; it was also evaluated the effect of adding sodium carbonate (Na_2CO_3) to counteract the effect of magnesium. Contact angle results were complemented with thermodynamic simulation of the $Mg-H_2O$ and $Mg-CO_3-H_2O$ systems and characterization by Schottky Field Emission ultra-high-resolution Scanning Electron Microscopy Energy Dispersive X-ray Spectrometry (Schottky FE-SEM-EDXS) to identify the nature of the formed precipitates and evaluate their texture.

2. Materials and Methods

2.1. Materials and Reagents

Sphalerite specimens used are high-pure samples collected from the mining district of Bismark mine (Ascensión, Chihuahua, Mexico). Elemental analysis (wt.%) of the specimens obtained by Atomic Absorption Spectroscopy (AAS, PerkinElmer AAnalyst 200) and Inductively Coupled Plasma (ICP, Perkin Elmer Optima 8300) are as follows: 59.07% Zn, 29.09% S, 0.03% Cu, 5.71% Fe, 0.012% Pb, and 6% insoluble species. X-ray diffraction (Philips X-Pert) analysis revealed that the only major mineral species present in the sample is sphalerite.

The presence of magnesium and its concentration range was noted after a sampling campaign to determine the process water quality of four lead-zinc sulfide ores flotation plants. Chemical species were determined by AAS and ICP and the results were useful to determine the magnesium concentration to evaluate its effect on sphalerite hydrophobicity.

Magnesium was added as magnesium nitrate ($Mg(NO_3)_2$) while copper used to activate sphalerite was copper sulphate ($CuSO_4$), both reagent grade. Purified sodium isopropyl xanthate (SIPX) was used as collector. In all the experiments deionized and deoxygenated water of a constant ionic strength of 10^{-2} mol/L $NaNO_3$ (reagent grade) were used. The pH was regulated by adding 0.01 mol/L solutions of sodium hydroxide (NaOH) and hydrochloric acid (HCl), as required. Sodium carbonate (Na_2CO_3) added as alternative alkalinizing agent was also reagent grade.

To properly study the effect of magnesium on the hydrophobicity of sphalerite the contribution of factors such as particle size, mineral composition, and water chemistry were kept constant.

2.2. Zeta Potential

Zeta potential measurements were carried out using a Pen Kem zeta potential analyzer (Lazer Zee Meter 501), which allows the direct reading of the zeta potential calculated from the electrophoretic mobility of the particles in suspension and the Smoluchowski equation. The zeta potential of natural and Cu-activated sphalerite was measured from a suspension of 1 g/L of ground mineral (ionic strength of 10^{-3} mol/L $NaNO_3$) as a function of pH and Cu(II) concentration, respectively. The suspensions were conditioned for 15 min under nitrogen bubbling in a stoppered Erlenmeyer flask under constant magnetic stirring. To minimize mineral oxidation and atmospheric oxygen absorption, sphalerite was ground prior to each measurement and the flask was equipped with a vent which allows nitrogen exhaust. The pH of the suspension was regulated by adding 0.1 mol/L NaOH or HCl, as required.

The measurements were carried out in triplicate so that the average zeta potential obtained and the error bars defining the 95% confidence interval of a student's t-distribution is reposted.

2.3. Contact Angle

The captive bubble technique is the most used method to measure the contact angle of minerals concentrated by froth flotation. It has some advantages, for example, it requires few square millimeters of mineral surface and a small volume of liquid [10]. This technique has been described in detail in previously published research [11,12]. In general, the technique consists in grinding the sample using deionized and deoxygenated water. Followed by conditioning the polished sample under the chemical environment of interest. Subsequently, the sample surface, which is immersed in the aqueous solution, is contacted with an air bubble of about 1 mm diameter. The resulting contact angle is measured with the help of image analysis software. Three chemical environments were evaluated, which are described below.

1. The sphalerite sample was activated with 50 mg/L of Cu(II) at pH 5.4 for 15 min. The sample was then decanted and further conditioned during 5 min with 10^{-4} mol/L SIPX at the pH of interest (i.e., from 5 to 11) and the contact angle was measured in the latter solution.
2. After copper activation as in previous experiment, the sample was conditioned during 5 min in the presence of both SIPX and magnesium (50 mg/L) at the pH of interest (8 to 11.5). Then, the contact angle was measured.
3. After copper activation as in previous cases, the sample was conditioned during 5 min in the presence of SIPX, magnesium, and sodium carbonate (2 g/L). The pH was adjusted at 10.5 and 11 because in previous tests high mineral depression was observed at these values. Then, the contact angle was recorded and measured in a 10^{-3} mol/L $NaNO_3$ solution of the same pH values (i.e., 10.5 and 11) to avoid the optical interference caused by the precipitates.

The generic term Cu(II) is used to refer to both the cation Cu^{2+}, which is the predominant copper species at pH below 6, and to the solid $Cu(OH)_2$, which predominates at pH above 6.

The experiments were carried out in triplicate so that the mean of the measured contact angle error bars defining the 99% confidence interval of a student's t-distribution are reported.

2.4. Thermodynamic Simulation, Surface Analysis, and Characterization

The thermodynamic analysis of the $Mg-H_2O$ and $Mg-CO_3-H_2O$ systems was performed making use of HSC Chemistry v.6 software. In the case of $Mg-CO_3-H_2O$, the computation conducted consisted in adding an accumulation amount of sodium carbonate from 0 to 2 g to 1 L of water of pH 7 containing 50 mg of magnesium (added as $Mg(NO_3)_2$) at 25 °C and 1 atm. The conditions used in the thermodynamic simulation are similar to those encountered in industrial operations, in the sense that not enough time is allowed for the aqueous system to reach equilibrium with the atmospheric CO_2 (0.035% by volume). This fact was simulated by considering a relatively small volume of atmosphere per liter solution (e.g., 22.4 L). The amount of sodium carbonate added in the contact angle measurements and the thermodynamic simulation is similar to that used in the industrial concentrators.

Characterization techniques were employed to complement the experimental results and thermodynamic simulation. In the case of the evaluation of the zeta potential of Cu-activated sphalerite, the mineral surface was analyzed by Scanning Electron Microscopy coupled with Energy-Dispersive Spectrometry (Philips XL30-ESEM) to evaluate the adsorption of $Cu(OH)_2$. The contact angle measurements were also complemented by surface analysis of the Cu-activated sphalerite (SIPX + Mg and SIPX + Mg + Na_2CO_3) by Schottky Field Emission ultra-high-resolution Scanning Electron Microscopy coupled with Energy Dispersive X-ray Spectrometry microanalysis system (JEOL JSM-7800F Prime) to evaluate the texture of the precipitates and chemical composition.

Sample preparation of surface characterization was performed under similar conditions to those used in the experimental tests (i.e., zeta potential and contact angle). First, a sphalerite crystal was activated with Cu(II) and, second, it was conditioned with $SIPX/Mg/Na_2CO_3$ depending the case. The samples were removed from the aqueous solution and left to dry at room temperature in a dissector with the purpose of evaporating the water present on the mineral surface; then a gold coating was applied to the surface by sputtering. The samples were characterized by SEM-EDS and FE ultra-high-resolution SEM-EDXS.

3. Results and Discussion

3.1. Process Water Chemical Charaterization

To carry out this investigation the process water of four lead-zinc sulfide ores concentrators was chemically characterized to determine the proper magnesium concentration, the results are shown in Table 1.

Table 1. Main metals and sulphate ions concentrations present in the process water of four lead-zinc sulfide ores flotation plants.

Plant	Fe (mg/L)	Pb (mg/L)	Zn (mg/L)	Cu (mg/L)	Ca (mg/L)	Mg (mg/L)	SO$_4$ (mg/L)
1	0.54	2.15	112	0.35	587	471	4127
2	0.36	0.44	0.158	–	566	124.7	1809
3	0.003	0.014	0.398	0.573	665	11.2	1910
4	1.59	0.07	1.03	0.011	270	37.2	–

It is generally accepted that the chemical species listed in the Table 1 have an adverse impact on the mineral recovery and selectivity. As can be seen, magnesium concentration varies from 11 to 471 mg/L but, since very high concentrations would make very difficult the experimentation, 50 mg/L was chosen as a proper value. This value is in accordance with previous research [8,9]. It is also observed the presence of chemical species as iron, lead, zinc, copper, calcium, and sulfate; these species are derived from the mineral species contained in the ore like copper sulfides (e.g., chalcopyrite and bornite), pyrite, and non-sulfide gangue.

The process water reported in Table 1 is used in the grinding and flotation circuits, and it is made up of different sources as thickeners overflow, tailings dam, and fresh water. Several potentially detrimental chemical species were detected. However, only magnesium was considered to elucidate its effect on the hydrophobicity of sphalerite.

3.2. Zeta Potential of Sphalerite

The isoelectric point of sphalerite (i.e., IEP) depends on the pH of the preconditioning stage, degree of oxidation of the mineral surface, and iron content [13–17]; IEP values between 2 and 7 are common. Zeta potentials of natural sphalerite as a function of the conditioning pH and Cu-activated sphalerite at pH 9 as a function of Cu(II) concentration were evaluated. It can be observed in Figure 1 that sphalerite develops negative charge along the pH range from 5 to 10. The potential determining ions in absence of oxidation products are those of water, that is hydrogen and hydroxyl ions. It appears that proton adsorption onto sulfur sites is favored in acid solutions, giving rise to a slightly positively charged surface. Under neutral or alkaline conditions, adsorption of hydroxyl ions onto the zinc sites is favored, resulting in a negatively charged surface.

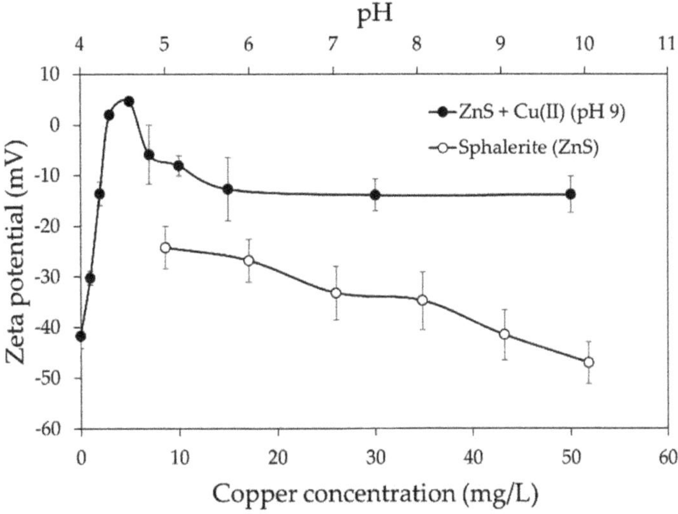

Figure 1. Zeta potential of sphalerite as a function of conditioning pH and Cu-activated sphalerite at pH 9 as a function of Cu(II) concentration (constant ionic strength of 10^{-3} mol/L $NaNO_3$).

It can also be observed in Figure 1 the effect of Cu(II) concentration on the zeta potential of sphalerite at pH 9. Zeta potential increases as the Cu(II) concentration is varied from 0 to 5 mg/L. This behavior is due to the adsorption of copper hydroxide species onto the sphalerite surface and the exchange of Zn^{2+} of the crystal lattice by Cu^{2+}. Then, cupric ions react with the nearby sulfides oxidizing them, leading to the formation of a new surface consisting of cuprous sulfide (Cu_2S) and polysulfides (e.g., Sn^{2-}, or cuprous disulfide (Cu_2S_2)), allowing the formation of copper xanthate and a greater zeta potential. This replacement mechanism also gives rise to the precipitation of Zn^{2+} from the sphalerite lattice as zinc hydroxide, which slowly dissolves and disperses exposing the hydrophobic surface previously formed [16,18].

At concentrations above 5 mg/L Cu(II) the electric charge of sphalerite decreases from positive to negative values (see Figure 1). When Cu(II) concentration is varied from 5 to 15 mg/L, the zeta potential drops from 4.6 to −12.8 mV to later remain practically constant at around −13.8 mV from 15 to 50 mg/L Cu(II). This behavior describes the detrimental effect of adding Cu(II) in excess and it is attributed to the precipitation and accumulation of copper hydroxide onto the sphalerite surface.

The formation of copper xanthate on the surface of sphalerite enhances its hydrophobicity, but this mechanism only occurs until 5 mg/L Cu(II). In Figure 2 can be seen the copper hydroxide adsorption/precipitation on the sphalerite surface when the mineral is conditioned at pH 9 and 50 mg/L Cu(II). The copper hydroxide does not cover the entire surface, but it begins to accumulate. The patches labeled as "1" corresponds to copper hydroxide and the zones labeled as "2" are uncovered sphalerite where no copper content was detected.

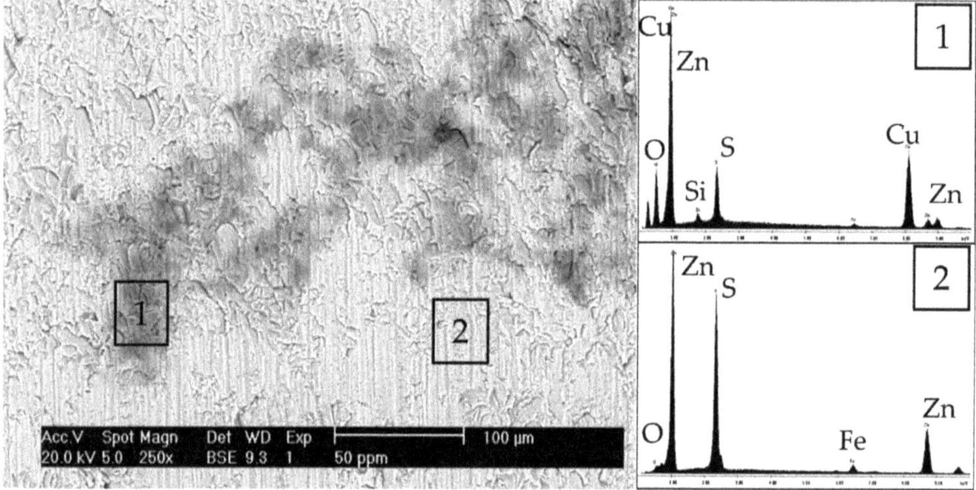

Figure 2. SEM micrograph and microprobe analysis showing Cu(OH)$_2$ adsorption/precipitation onto the sphalerite surface conditioned at pH 9 with 50 mg/L of Cu(II).

It can be seen in Figure 3 that as the concentration of Cu(II) increases from 50 to 200 mg/L, the extent of copper hydroxide patches on the sphalerite surface become bigger and eventually would cover practically the whole surface. This would result in lower recoveries since copper hydroxide is a hydrophilic species. Copper hydroxide adsorption/precipitation on the sphalerite surface can be observed as patches (labeled as "1"); they practically cover the whole sphalerite surface. The zones labeled as "2" are uncovered sphalerite where no copper content was detected. The copper hydroxide patches are dark, and their morphology is flaky.

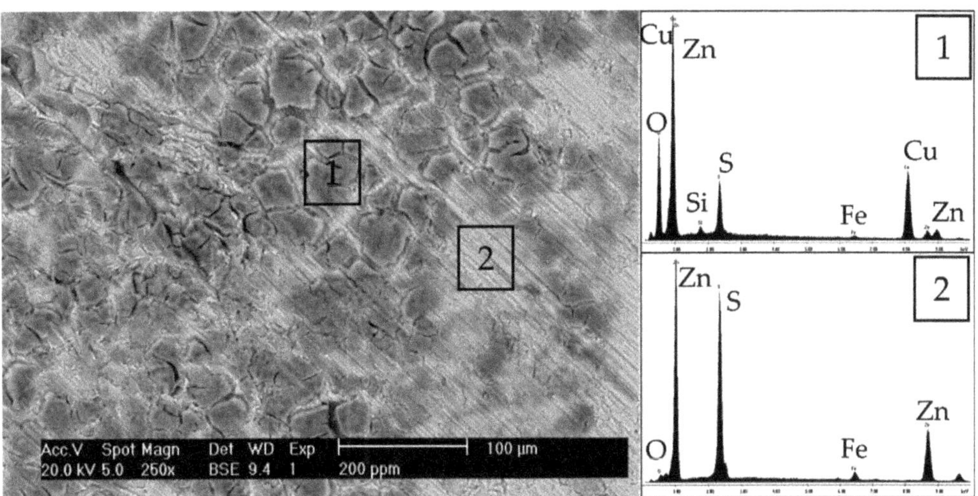

Figure 3. SEM micrograph and microprobe analysis showing Cu(OH)$_2$ adsorption/precipitation onto the sphalerite surface conditioned at pH 9 with 200 mg/L of Cu(II).

3.3. Effect of Magnesium on the Contact Angle of Cu-Activated Sphalerite

It has been suggested that adsorption of magnesium species (e.g., MgOH$^+$ and Mg(OH)$_2$) onto the sphalerite will be favored by a negatively charged mineral surface, which is precisely the case for pH values from 8 to 10. Mirnezami et al. [19] have shown that MgOH$^+$ adsorbs onto sphalerite at pH below 10, while at more alkaline conditions, precipitation and coagulation of Mg(OH)$_2$ also occurs. Thus, decreasing the number of available sites for xanthate adsorption [9]. Magnesium hydroxide (Mg(OH)$_2$) is slightly positively charged at pH from 9.5 to 10, values at which precipitation begins [20]. This will enhance the electrostatic interaction with the negatively charged sphalerite. Copper attraction will result in less adsorption of magnesium hydroxide complexes onto the activated sphalerite as its surface charge would be less negative.

Figure 4 shows the effect of pH on the contact angle of Cu-activated sphalerite conditioned with SIPX (dashed line), Cu-activated sphalerite conditioned with SIPX in presence of magnesium, and Cu-activated sphalerite conditioned with SIPX in presence of magnesium and sodium carbonate.

It can be observed in Figure 4 (dashed line) that the contact angle of Cu-activated sphalerite in absence of magnesium and sodium carbonate increases as the pH of the SIPX conditioning stage is varied from 5 to 9, reaching a maximum of 62°. From to pH 9 to 11 this tendency is reversed to end at approximately 48°. This is because at more alkaline solutions, the Cu(OH)$_2$ patches originated during copper activation (see Figures 2 and 3) slightly hinders the hydrophobicity caused by xanthate adsorption.

Magnesium has a negative effect on the contact angle of Cu-activated sphalerite conditioned with SIPX as a function of pH, as can be seen in Figure 4. At pH from 8 to 9 the effect on the contact angle is not significant, compared with the behavior of Cu-activated sphalerite in absence of magnesium. At these conditions, the interaction between the mineral surface and MgOH$^+$ is negligible since the mineral charge is almost zero and the concentration of MgOH$^+$ is about 10^{-5} mol/L, as revealed by the Mg-H$_2$O distribution diagram of Figure 5a. Otherwise, at pH above 9 the contact angle drop sharply from 55° to 20° as a consequence of the adsorption and precipitation of Mg(OH)$_2$ onto the sphalerite surface. This is supported by the distribution diagram of Figure 5a where is clearly shown that Mg(OH)$_2$ predominates at pH 10 but begins to precipitate at 9.6.

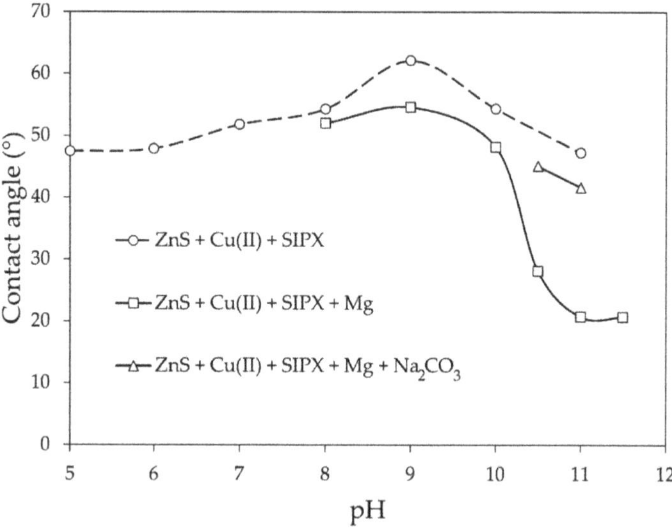

Figure 4. Effect of pH and magnesium (50 mg/L), on the contact angle of Cu-activated (50 mg/L, 15 min) sphalerite further conditioned with SIPX (10^{-4} mol/L, 5 min) at 25 °C. It is also shown the effect of sodium carbonate, added during xanthate conditioning (2 g/L), to counteract the detrimental effect of magnesium.

Sodium carbonate has been used in industrial operations to reduce the concentration of calcium in flotation water since it dissolves gypsum precipitated and heterocoagulated onto the galena and sphalerite particles; it also avoids further gypsum precipitation [21,22]. Calcium and magnesium have chemical similarities, which suggest that magnesium may also precipitate in the form carbonate as calcium does, hence the use of soda ash as alkalinizing agent was tested.

It can be seen in Figure 4 that the use of sodium carbonate has a positive effect on the contact angle of Cu-activated sphalerite conditioned with SIPX in presence of 50 mg/L Mg. For pH values of 10.5 and 11 the contact angles are 45° and 42°, respectively. Thus, the depressing effect of magnesium hydroxide is avoided by employing soda ash as alkalinizing agent since, according with Figure 5b, magnesium precipitates as hydromagnesite ($Mg_5(OH)_2(CO_3)_4 \cdot 4H_2O$) rather than hydroxide. It can be seen in the distribution diagram of the Mg-CO_3-H_2O system of Figure 5b that at pH 9.4 hydromagnesite starts to precipitate and it will keep precipitating while increasing the pH. Apparently, the basic carbonate does not substantially affect sphalerite hydrophobicity since it remains dispersed in the bulk solution as calcium carbonate does in the case of calcium removal from the solution by soda ash addition [22].

Sodium carbonate can be employed in the sphalerite flotation circuit as an operational strategy to avoid the detrimental effect of magnesium on the hydrophobicity of sphalerite. Other option will be to maintain a flotation pH below 10 to avoid precipitation of magnesium hydroxide.

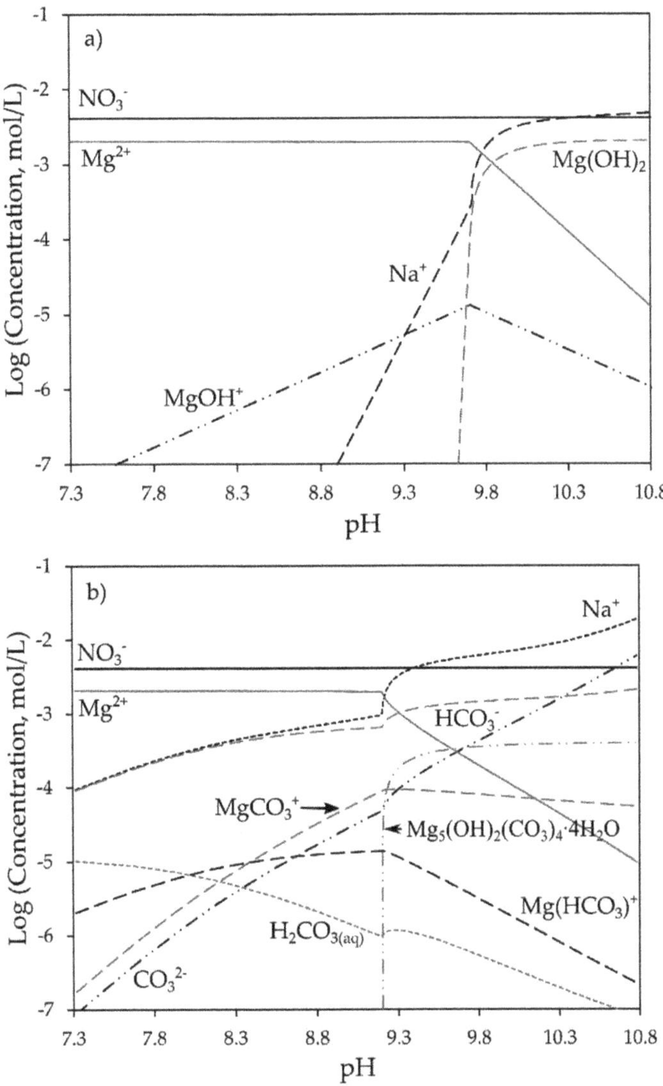

Figure 5. Species distribution diagrams of the (**a**) Mg-H$_2$O and (**b**) Mg-CO$_3$-H$_2$O systems for 50 mg/L of Mg at 25 °C. The latter diagram stands for the cumulative addition of 2 g/L of Na$_2$CO$_3$ to 1 L of solution. Hydromagnesite starts to precipitate when 0.05 g/L of soda ash has been added.

To complement and support the findings made by experimental techniques and thermodynamic simulation, the surface of a Cu-activated sphalerite conditioned with SIPX in presence of: (1) 50 mg/L of Mg and (2) 50 mg/L + 2 g/L Na$_2$CO$_3$ was characterized by Schottky field emission SME-EDSX. The results can be seen in Figures 6 and 7.

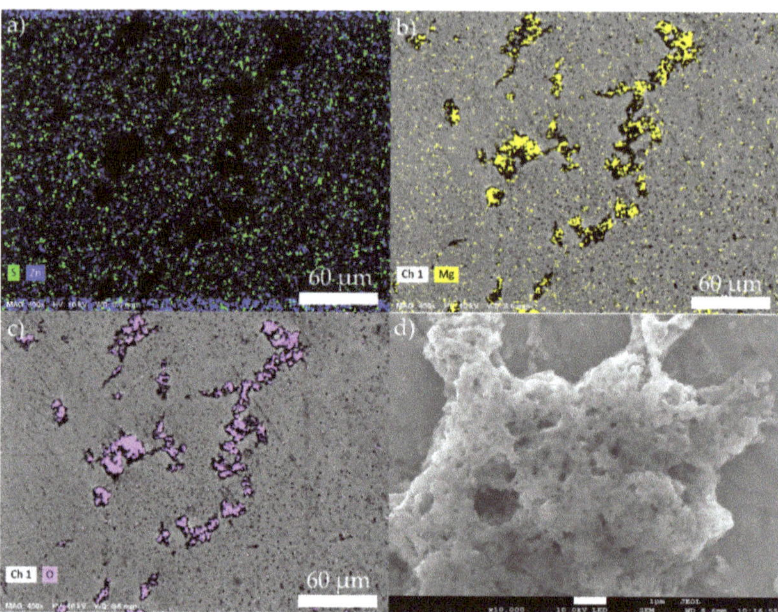

Figure 6. Mapping of location elements for the surface of Cu-activated sphalerite conditioned with SIPX in presence of Mg and morphology of the precipitated $Mg(OH)_2$: (**a**) S and Zn, (**b**) Mg, (**c**) O, (**d**) morphology.

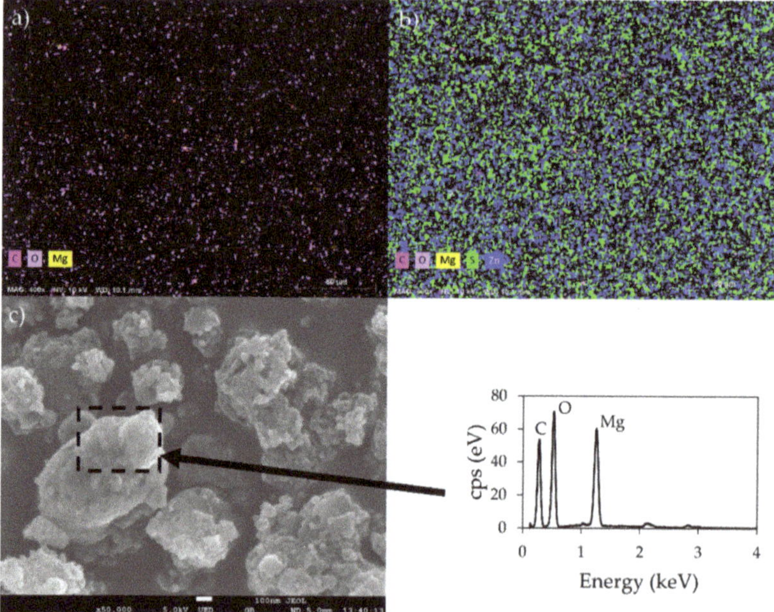

Figure 7. Mapping of location elements for the surface of Cu-activated sphalerite conditioned with SIPX in presence of Mg + Na_2CO_3 and the morphology and composition of the dispersed Mg species: (**a**) C, O, and Mg, (**b**) C, O, Mg, S, and Zn, and (**c**) morphology and EDSX microprobe analysis.

The elemental X-ray maps and high-resolution SEM micrograph in Figure 6 clearly shows the magnesium hydroxide precipitated on the surface of Cu-activated sphalerite conditioned with SIPX. The elemental mapping of Figure 6a shows that the elements that composes the surface are those of sphalerite but, as represented in dark areas, magnesium hydroxide precipitates. The elemental mapping of Figure 5b,c shows that the species precipitated is composed of magnesium and oxygen, supporting the results presented in Figures 4 and 5a. High-resolution SEM micrograph of Figure 6d shows the morphology of the magnesium hydroxide which could cover the entire sphalerite surface affecting the recovery of sphalerite.

The elemental X-ray maps in Figure 7a,b clearly shows that the presence of sodium carbonate avoids the precipitation of magnesium hydroxide on the surface of Cu-activated sphalerite conditioned with SIPX. The elemental composition of the surface corresponds to that of sphalerite although it was conditioned in the presence of 50 mg/L of magnesium. Magnesium hydroxide precipitation/adsorption onto the mineral surface was not detected. These findings support the results of the contact angle experiments and thermodynamic simulation. The addition of sodium carbonate as alkalinizing agent enhances the hydrophobicity of sphalerite in presence of magnesium by preventing or dissolving the precipitated magnesium hydroxide. In the high-resolution SEM micrograph of Figure 7c can be seen the morphology of the dispersed precipitated magnesium compound, which avoids the detrimental effect on the hydrophobicity of sphalerite. According to a EDX microprobe analysis, the magnesium species corresponds to hydromagnesite ($Mg_5(OH)_2(CO_3)_4 \cdot 4H_2O$) as indicated by the thermodynamic simulation presented in Figure 5b. The microprobe analysis was carried out in the area indicated in the dotted box of Figure 7c; magnesium, oxygen, and carbon were detected.

4. Conclusions

The experimental and thermodynamic study, in conjunction with surface characterization, to elucidate the effect of magnesium on the hydrophobicity of sphalerite has shown the following:

1. Cu-activation of sphalerite enhance its zeta potential up to a concentration of 5 mg/L; at concentrations greater than 5 mg/L, copper hydroxide will adsorb/precipitate onto the sphalerite surface decreasing its zeta potential because $Cu(OH)_2$ is hydrophilic in nature.
2. Magnesium has a detrimental effect on the contact angle of Cu-activated sphalerite conditioned with SIPX beginning at pH of approximately 10 due to the precipitation of magnesium hydroxide onto its surface. The precipitation of this species starts at pH 9.6 and rapidly becomes a predominant species as pH increases. If the process pH is not rigorously controlled to keep it below 10, it will cover the whole sphalerite surface diminishing its recovery.
3. The use of sodium carbonate as alkalinizing agent avoids the sodium hydroxide precipitation onto the surface of Cu-activated sphalerite conditioned with SIPX. When sodium carbonate is present, hydromagnesite ($Mg_5(OH)_2(CO_3)_4 \cdot 4H_2O$) begins to precipitate at pH 9.3 and it will keep precipitating at greater pH values, allowing the recovery of sphalerite.

In summary, magnesium has a negative effect on the contact angle of sphalerite, and it could be considered more detrimental than calcium since magnesium hydroxide begins to precipitate at pH 9.6. An operational strategy most be considered since magnesium species are common in the process water, increasingly with a higher proportion of recycled water. A pH lower than 10 must be keep avoiding the precipitation of magnesium hydroxide or sodium carbonate must be added aiming to precipitate the magnesium as hydromagnesite ($Mg_5(OH)_2(CO_3)_4 \cdot 4H_2O$) instead of hydroxide.

Author Contributions: Conceptualization, methodology, writing—original draft preparation, visualization, and formal analysis, supervision, G.I.D.-P.; writing—original draft preparation, investigation, and writing—review and editing, A.A.G.-I.; data curation, visualization, M.G.-G.; methodology, data curation, D.A.C. All authors have read and agreed to the published version of the manuscript.

Funding: This research received no external funding.

Institutional Review Board Statement: Not applicable.

Informed Consent Statement: Not applicable.

Acknowledgments: The authors are grateful to technical and administrative personnel of Cinvestav Saltillo for the SEM-EDS and Schottly field emission SEM-EDSX characterization.

Conflicts of Interest: The authors declare no conflict of interest.

References

1. Rao, S.R.; Finch, J.A. A review of water reuse in flotation. *Miner. Eng.* **1989**, *2*, 65–85. [CrossRef]
2. Levay, G.; Smart, R.S.C.; Skinner, W.M. The impact of water quality on flotation performance. *J. S. Afr. Inst. Min. Metall.* **2001**, *101*, 69–75.
3. Chen, J.; Liu, R.; Sun, W.; Qiu, G. Effect of mineral processing wastewater on flotation of sulfide minerals. *T. Nonferr. Metal. Soc.* **2009**, *19*, 454–457. [CrossRef]
4. October, L.; Corin, K.; Schreithofer, N.; Manono, M.; Wiese, J. Water quality effects on bubble-particle attachment of pyrrhotite. *Miner. Eng.* **2019**, *131*, 230–236. [CrossRef]
5. October, L.L.; Corin, K.C.; Manono, M.S.; Schreithofer, N.; Wiese, J.G. A fundamental study considering specific ion effects on the attachment of sulfide minerals to air bubbles. *Miner. Eng.* **2020**, *151*, 106313. [CrossRef]
6. Coetzer, H.; Du Preez, H.S.; Bredenhann, R. Influence of water resources and metal ions on galena flotation of Rosh Pinah ore. *J. S. Afr. Inst. Min. Metall.* **2003**, *103*, 193–207.
7. Bicak, O.; Ekmekci, Z.; Can, M.; Öztürk, Y. The effect of water chemistry on froth stability and surface chemistry of the flotation of a Cu-Zn sulfide ore. *Int. J. Miner. Process.* **2012**, *102–103*, 32–37. [CrossRef]
8. El-Ammouri, E.; Mirnezami, M.; Lascelles, D.; Finch, J.A. Aggregation index and a methodology to study the role of magnesium in aggregation of sulphide slurries. *CIM Bull.* **2002**, *95*, 67–72.
9. Lascelles, D.; Finch, J.A.; Sui, C. Depressant action of Ca and Mg on flotation of Cu activated sphalerite. *Can. Metall. Q.* **2003**, *42*, 133–140. [CrossRef]
10. Chau, T.T. A review of techniques for measurement of contact angles and their applicability on mineral surfaces. *Miner. Eng.* **2009**, *22*, 213–219. [CrossRef]
11. Dávila-Pulido, G.I.; Uribe-Salas, A. Contact angle study on the activation mechanisms of sphalerite with Cu(II) and Pb(II). *Rev. Metal. Madrid* **2011**, *47*, 329–340.
12. Dávila-Pulido, G.I.; Uribe-Salas, A. Effect of calcium, sulphate and gypsum on copper-activated and non-activated sphalerite surface properties. *Miner. Eng.* **2014**, *55*, 147–153. [CrossRef]
13. Moignard, M.S.; James, R.O.; Healy, T.W. Adsorption of calcium at the zinc sulphide-water interface. *Aust. J. Chem.* **1977**, *30*, 733–740. [CrossRef]
14. Finkelstein, D.P. The activation of sulphide minerals for flotation: A review. *Int. J. Miner. Process.* **1997**, *52*, 81–120. [CrossRef]
15. Sui, C.; Rashchi, F.; Xu, Z.; Kim, J.; Nesset, J.E.; Finch, J.A. Interactions in the sphalerite-Ca-SO$_4$-CO$_3$ systems. *Colloid Surf. A* **1998**, *137*, 69–77. [CrossRef]
16. Fornasiero, D.; Ralston, J. Effect of surface oxide/hydroxide products on the collectorless flotation of copper-activated sphalerite. *Int. J. Miner. Process.* **2006**, *78*, 231–237. [CrossRef]
17. Ikumapayi, F.K. Flotation Chemistry of Complex Sulphide Ores: Recycling of Process Water and Flotation Selectivity. Ph.D. Thesis, Lulea University of Technology, Luleå, Sweden, 2010.
18. Chandra, A.P.; Gerson, A.R. A review of the fundamental studies of the copper activation mechanisms for selective flotation of the sulfide minerals, sphalerite and pyrite. *Adv. Colloid Interfac.* **2009**, *145*, 97–110. [CrossRef] [PubMed]
19. Mirnezami, M.; Hashemi, M.S.; Finch, J.A. Mechanism of alkaline earth metal ion adsorption on sulphide minerals. In Proceedings of the 48th Conference of Metallurgist of CIM, Sudbury, ON, Canada, 23–26 August 2009; pp. 99–106.
20. Liu, H.; Yi, J. Polystyrene/magnesium hydroxide nanocomposite particles prepared by surface-initiated in-situ polymerization. *Appl. Surf. Sci.* **2009**, *255*, 5714–5720. [CrossRef]
21. Dávila-Pulido, G.I. An Experimental Study of the Effect of Calcium Sulfate in the Flotation Water on the Surface Properties of Sphalerite. Ph.D. Thesis, CINVESTAV, Ramos Arizpe, Mexico, 2014.
22. Grano, S.R.; Wong, P.L.M.; Skinner, W.; Johnson, N.W.; Ralston, J. Detection and control of calcium sulfate precipitation in the lead circuit of the Hilton concentrator of Mount Isa Mines Limited. In Proceedings of the XIX International Mineral Processing Congress, San Francisco, CA, USA, 22–27 October 1995; pp. 171–179.

Article

Effect of Sodium Metabisulfite on Selective Flotation of Chalcopyrite and Molybdenite

Yuki Semoto [1], Gde Pandhe Wisnu Suyantara [1,2], Hajime Miki [1,*], Keiko Sasaki [1,2], Tsuyoshi Hirajima [1,3], Yoshiyuki Tanaka [4], Yuji Aoki [3] and Kumika Ura [4]

[1] Department of Earth Resources Engineering, Faculty of Engineering, Kyushu University, Fukuoka 819-0385, Japan; y-semoto19@mine.kyushu-u.ac.jp (Y.S.); pandhe@mine.kyushu-u.ac.jp (G.P.W.S.); keikos@mine.kyushu-u.ac.jp (K.S.); hirajima@mine.kyushu-u.ac.jp (T.H.)

[2] Kyushu University Institute for Asian and Oceanian Studies, Kyushu University, Fukuoka 819-0385, Japan

[3] Sumitomo Metal Mining Co. Ltd., Tokyo 105-8716, Japan; yuji.aoki.m3@smm-g.com

[4] Sumitomo Metal Mining Co. Ltd., Niihama 792-0002, Japan; yoshiyuki.tanaka.u8@smm-g.com (Y.T.); kumika.ura.r3@smm-g.com (K.U.)

* Correspondence: miki@mine.kyushu-u.ac.jp

Abstract: Sodium metabisulfite (MBS) was used in this study for selective flotation of chalcopyrite and molybdenite. Microflotation tests of single and mixed minerals were performed to assess the floatability of chalcopyrite and molybdenite. The results of microflotation of single minerals showed that MBS treatment significantly depressed the floatability of chalcopyrite and slightly reduced the floatability of molybdenite. The results of microflotation of mixed minerals demonstrated that the MBS treatment could be used as a selective chalcopyrite depressant in the selective flotation of chalcopyrite and molybdenite. Furthermore, the addition of diesel oil or kerosene could significantly improve the separation efficiency of selective flotation of chalcopyrite and molybdenite using MBS treatment. A mechanism based on X-ray photoelectron spectroscopy analysis results is proposed in this study to explain the selective depressing effect of MBS on the flotation of chalcopyrite and molybdenite.

Keywords: chalcopyrite; molybdenite; sodium metabisulfite; diesel oil; kerosene; selective flotation

Citation: Semoto, Y.; Suyantara, G.P.W.; Miki, H.; Sasaki, K.; Hirajima, T.; Tanaka, Y.; Aoki, Y.; Ura, K. Effect of Sodium Metabisulfite on Selective Flotation of Chalcopyrite and Molybdenite. *Minerals* **2021**, *11*, 1377. https://doi.org/10.3390/min11121377

Academic Editors: Shuai Wang, Xingjie Wang and Jia Yang

Received: 31 October 2021
Accepted: 3 December 2021
Published: 7 December 2021

Publisher's Note: MDPI stays neutral with regard to jurisdictional claims in published maps and institutional affiliations.

1. Introduction

Molybdenum minerals are often associated with copper sulfide minerals [1]. It is estimated that about 50% of the world's molybdenum production comes from copper and molybdenum (Cu-Mo) ores as a by-product [2,3]. Both copper and molybdenum are important materials in various fields; therefore, the separation of both copper and molybdenum minerals is important. Furthermore, molybdenum minerals play a very important role in making the Cu-Mo processing plant economically viable [4].

Separation of copper and molybdenum sulfide minerals is often carried out in the selective flotation stage by adding a copper depressant (i.e., sodium hydrosulfide (NaHS), sodium sulfide (Na_2S), sodium thiopropionate ($HSCH_2CH_2COONa$), sodium thioglycollate ($HSCH_2COONa$), or Nokes reagent (P_2S_5+NaOH)) [2,5–10]. Other reagents have been developed to replace these toxic and dangerous copper depressants, for instance, by using chitosan [11], lignosulphonate [12], dithiouracil [13], and rhodamine-3-acetic acid [14] as copper depressants in the selective flotation of Cu-Mo sulfide minerals. In addition, various oxidation treatments using plasma pre-treatment, ozone, electrolysis, hydrogen peroxide, and Fenton-like reagent have been applied as selective chalcopyrite depressants in the previous studies [15–19]. However, the effectiveness of these reagents in the Cu-Mo flotation plant has not been reported.

Sulfoxy reagents (i.e., sulfite (SO_3^{2-}), bisulfite (HSO_3^{-}), metabisulfite ($S_2O_5^{2-}$), or sulfur dioxide (SO_2)) have been used as depressants for various minerals [20–27]. These reagents have been commonly used in flotation plants. For instance, sodium sulfite (Na_2SO_3) and sodium metabisulfite ($Na_2S_2O_5$) are the most widely used compounds for the depression of pyrite during the flotation of copper complex ores [2,21,28,29]. However, these studies did not report the depression of copper minerals. The selective depressing effect of Na_2SO_3 on the floatability of chalcopyrite has been investigated by Miki et al. [25] for selective Cu-Mo flotation and by Suyantara et al. [30] for separation of chalcopyrite and enargite using flotation. However, Miki et al. [25] showed that the depression of chalcopyrite in the selective Cu-Mo flotation required a high concentration of Na_2SO_3 (i.e., 0.1 M) at pH 10.8, making this reagent practically and economically unattractive.

On the other hand, Castro et al. [31] showed that sodium metabisulfite has been used to replace lime as a pyrite depressant in the rougher flotation stage of Cu-Mo-Fe ores in Chile. Therefore, it might be more efficient to use sodium metabisulfite for the selective Cu-Mo flotation stage. Chen et al. [32] studied the effect of sodium metabisulfite on the floatability of molybdenite in fresh water and seawater. They reported that sodium metabisulfite exhibited a depressing effect on the recovery of molybdenite in fresh water. However, the effect of sodium metabisulfite on the depression of chalcopyrite in the selective Cu-Mo flotation has been not reported.

The main objective of this study is to investigate the effect of sodium metabisulfite (MBS) on the floatability of chalcopyrite and molybdenite. Furthermore, this work investigated the possibility of using MBS for selective flotation of chalcopyrite and molybdenite. In addition, X-ray photoelectron spectroscopy analysis was performed to assess the effect of MBS on the chemical state on the surface of chalcopyrite and molybdenite.

2. Materials and Methods

2.1. Materials

A high-purity crystal sample of chalcopyrite (Acari Mine, Arequipa, Peru) and molybdenite powder sample supplied by Sumitomo Metal Mining Co., Ltd. (Tokyo, Japan) were used in this study. The chalcopyrite sample was crushed and then hand sorted to minimize the impurity. The chalcopyrite was ground using an agate mortar and pestle, and then dry screened (passing 38 μm screener) prior to all tests. The mean particle sizes of chalcopyrite and molybdenite were 16.1 μm and 10.6 μm, respectively. The chemical and mineralogical compositions of chalcopyrite and molybdenite were measured using X-ray fluorescence (XRF, ZSX Primus 2, Rigaku, Tokyo, Japan) and X-ray diffraction (XRD, Ultima 4, Rigaku, Tokyo, Japan). The chemical and mineralogical analysis results of chalcopyrite and molybdenite are presented in Table 1 and Figure 1, respectively.

Analytical grade sodium metabisulfite (MBS) was employed as a copper depressant in this study. Analytical grade potassium hydroxide (KOH) and hydrochloric acid (HCl) were used as pH modifiers. Industrial grade kerosene and diesel oil were used as collectors. Industrial grade pine oil was used as a frother. All employed reagents were supplied by Wako Chemical Industries, Ltd. (Tokyo, Japan), except for the diesel oil, which was provided by a gas station (ENEOS Co., Ltd., Tokyo, Japan). Millipore (Direct-Q, Merck, Tokyo, Japan) ultra-pure water with a resistivity of 18.2 MΩ.cm was used in all experiments.

Table 1. Chemical compositions of chalcopyrite and molybdenite measured by XRF (%).

Elements	Chalcopyrite	Molybdenite
Al	0.12	0.07
Si	5.74	0.17
S	24.19	41.90
Fe	25.90	0.16
Cu	31.59	-
Zn	-	0.18
Mo	-	51.56

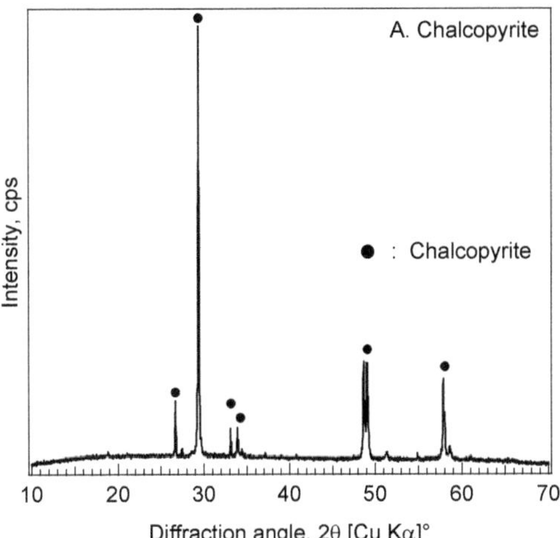

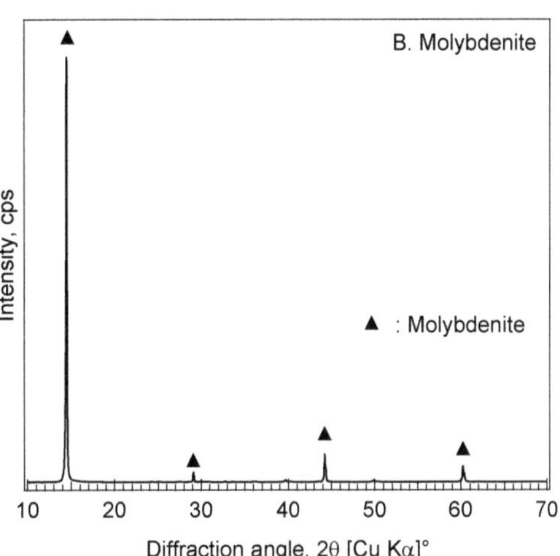

Figure 1. X-ray diffraction patterns of chalcopyrite (**A**) and molybdenite (**B**).

2.2. Microflotation Test

The flotation tests were carried out using single and mixed minerals of chalcopyrite and molybdenite. The mixing ratio was 50% chalcopyrite and 50% molybdenite. The mineral powders (0.6 g) were added to 180 mL of ultra-pure water. The solution pH was controlled at pH 9 for 2 min, and MBS treatment was carried out by adding MBS powder into the suspension. The MBS treatment was performed for 5 min. The MBS concentration varied from 0 mM to 5 mM. Pine oil (0.13 mM) was added to the suspension. The conditioning time for the pine oil was 2 min. To improve the flotation selectivity, a certain amount of emulsified diesel oil or kerosene was employed as collectors. These collectors were added before the addition of MBS powder. The diesel oil or kerosene was emulsified in a homogenizer (HG-200 Homogenizer, AS ONE, Tokyo, Japan) at 20,000 rpm for 30 s. The conditioning time for diesel oil or kerosene was 3 min. The suspension pH was controlled at pH 9 throughout the conditioning process by the addition of KOH or HCl.

The flotation tests were performed by transferring the suspension solution to a Partridge–Smith microflotation device [18]. Flotation was conducted by injecting nitrogen gas through a glass frit at a flow rate of 20 mL/min for 6 min. The float and tailing fractions were collected separately. These fractions were dried in an oven at 110 °C for 12 h and then weighed. The flotation tests were repeated twice, and the average value of recovery of each mineral is reported in this work.

The recovery of the mineral in the flotation of a single mineral system was calculated using Equation (1), where m_{float} and m_{sink} are the mass of the mineral in the float and sink fractions, respectively. Meanwhile, the recovery of chalcopyrite and molybdenite in the flotation of mixed minerals system was calculated based on the assumption that copper and molybdenum belong to chalcopyrite and molybdenite, respectively. The froth and tailing fractions were analyzed by XRF, and the analysis results were used to calculate the recovery of each mineral using Equation (1). Newton efficiency (η) was calculated using Equation (2), where R_c is the recovery of molybdenite in the froth and R_t is the recovery of chalcopyrite in the tailing.

$$\text{Recovery } (\%) = \frac{m_{float}}{m_{float} + m_{sink}} \times 100 \tag{1}$$

$$\eta \ (\%) = R_c - (1 - R_t) \tag{2}$$

2.3. X-ray Photoelectron Spectroscopy (XPS)

X-ray photoelectron spectroscopic analysis was performed to characterize the mineral surface before and after the treatment. The mineral powders (0.1 g) were suspended in 30 mL ultra-pure water. The pH was then adjusted to pH 9. Afterward, the suspension was treated with or without 5 mM MBS for 5 min at pH 9. After the treatment, the mineral sample was filtered, freeze-dried, and stored under a vacuum bag to minimize oxidation. XPS analysis was then carried out using an AXIS-ULTRA (Shimadzu-Kratos Co., Ltd., Manchester, UK) with an Al Kα X-ray source operated at 5 mA and 15 kV. The charge neutralizer was enabled during the analysis. The pressure in the analyzer chamber was 3.1×10^{-7} torr. A survey scan and elemental region scan were performed to extract information on the oxidation states of chemical species on the mineral surface. The collected spectra were analyzed using Casa XPS software (Ver. 2.3.16). Shirley background corrections [33] were used throughout the analyses for carbon (C) 1s, oxygen (O) 1s, iron (Fe) 2p, sulfur (S) 2p, copper (Cu) 2p, and molybdenum (Mo) 3d spectra. The Gaussian–Lorentzian (GL) function was used to deconvolute the spectra. The binding energy was calibrated based on C 1s at a binding energy of 284.6 eV [15,25].

3. Results and Discussion

3.1. Microflotation of Single Mineral

Effects of MBS treatment on the flotation recovery of chalcopyrite and molybdenite in the single mineral system are presented in Figure 2. It should be noted that the microflotation tests were performed in the absence of a collector. The recovery of chalcopyrite significantly decreased from 65% in the absence of MBS to 15% after being treated in 0.5 mM MBS. The recovery of chalcopyrite gradually decreased with increasing concentration of MBS. The recovery of chalcopyrite was 5% after being treated in 5 mM MBS. These microflotation results indicate that the MBS treatment exhibited a strong depressing effect on the floatability of chalcopyrite.

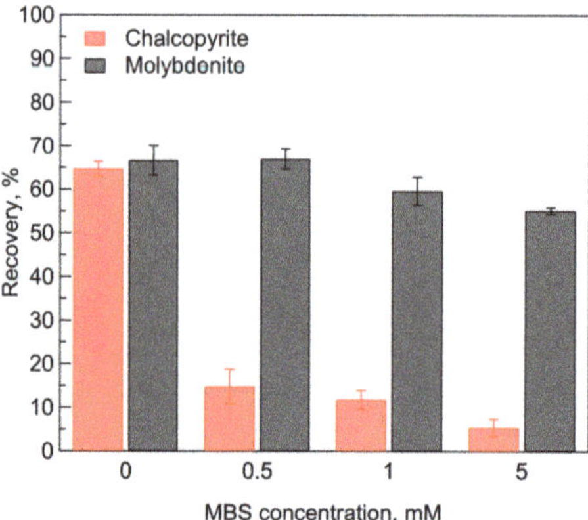

Figure 2. Effects of sodium metabisulfite (MBS) on the recovery of chalcopyrite and molybdenite in the absence of a collector at pH 9.

On the other hand, the addition of 0.5 mM MBS had an insignificant effect on the recovery of molybdenite. However, the addition of a higher concentration of MBS could reduce the recovery of molybdenite. For instance, the recovery of molybdenite slightly decreased from 67% to 55% after being treated in 5 mM MBS. A similar depressing effect of MBS on the floatability of molybdenite in deionized (DI) water has been reported by Chen et al. [32]. Although these microflotation results indicate that MBS treatment slightly depressed the floatability of molybdenite, the depressing effect was less significant compared to that on the floatability of chalcopyrite. In addition, the microflotation results presented in Figure 2 show that MBS treatment has potential as a selective copper depressant in the selective flotation of Cu-Mo.

3.2. XPS Analysis

XPS analysis was performed to explain the flotation results presented in Figure 2 by evaluating the oxidation states of chemical species on the surface of chalcopyrite and molybdenite before and after the MBS treatment. Figure 3 shows the X-ray photoelectron spectra of Cu 2p, S 2p, Fe 2p, and O 1s of chalcopyrite.

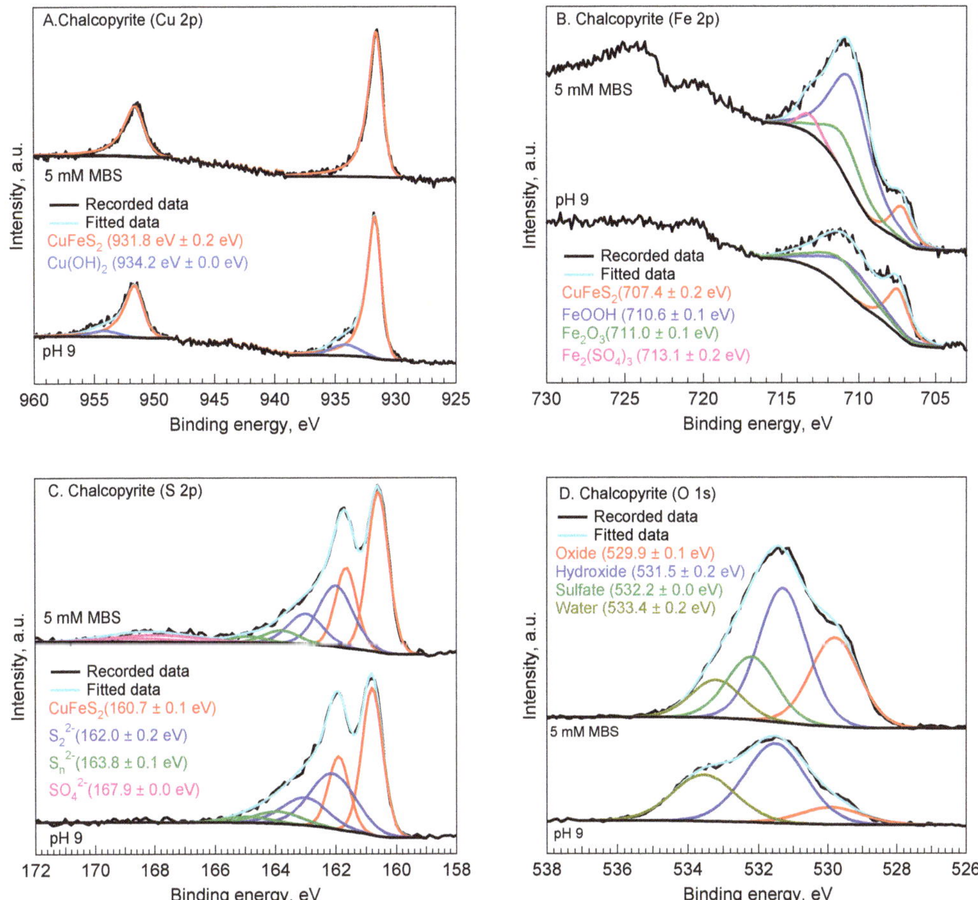

Figure 3. Cu 2p (**A**), Fe 2p (**B**), S 2p (**C**), O 1s (**D**) spectra of chalcopyrite without and with 5 mM MBS treatment at pH 9.

Two Gaussian–Lorentzian (GL) functions could best fit the Cu 2p spectrum of chalcopyrite at pH 9 (Figure 3A). The first GL function shows a peak centered at ca. 931.8 eV, which is in agreement with the Cu(I) of chalcopyrite [34]. The second GL function shows a peak centered at higher binding energy (i.e., 934.2 eV). This peak is attributed to the Cu(II) of Cu(OH)₂ [35]. The presence of Cu(II) is confirmed by the formation of a weak satellite peak located at binding energy 940–945 eV. The presence of Cu(II) on the chalcopyrite surface indicates that the chalcopyrite surface was slightly oxidized at pH 9 in the absence of MBS. In addition, the surface oxidation of chalcopyrite at pH 9 is confirmed by the Fe 2p spectrum. The deconvolution of Fe 2p spectrum of chalcopyrite at pH 9 shown in Figure 3B indicates that in addition to the Fe of chalcopyrite located at 707.4 eV [36], the chalcopyrite surface was covered by FeOOH and Fe_2O_3, as indicated by the peaks located at 710.6 eV and 711.0 eV, respectively [37]. The deconvolution of S 2p spectrum of chalcopyrite at pH 9 (Figure 3C) indicates various peaks located at ca. 160.7 eV, 162.0 eV, and 163.8 eV. These peaks are attributed to chalcopyrite, disulfide (S_2^{2-}), and polysulfide (S_n^{2-}), respectively [36]. The presence of oxide and hydroxide on the chalcopyrite surface is confirmed by the peak located at 529.9 eV [37] and 531.5 eV [38] in the O 1s spectrum of chalcopyrite at pH 9 (Figure 3D). In addition, there is an indication of the presence of adsorbed water on the surface of chalcopyrite, as shown by the peak located at 533.4 eV [39] in Figure 3D.

The MBS treatment reduced the Cu(II) to Cu(I) on the chalcopyrite surface, as indicated by the absence of Cu(II) species in the Cu 2p spectrum of chalcopyrite (Figure 3A). Only the Cu 2p peak of chalcopyrite was observed after the surface was treated in 5 mM MBS. The Fe 2p spectrum in Figure 3B shows that the MBS treatment significantly improved the peak intensity of FeOOH and formed a new peak located at 713.1 eV. This new peak is attributed to $Fe_2(SO_4)_3$ [37]. The formation of sulfate species is confirmed by the appearance of a new peak located at 167.9 eV in the S 2p spectrum in Figure 3C and a new peak located at 532.2 eV in the O 1s spectrum in Figure 3D [36,37]. This sulfate species might be formed as a result of the reaction between MBS and dissolved oxygen in the water, as shown by Equations (3) and (4).

$$Na_2S_2O_5 + H_2O \rightarrow 2Na^+ + 2HSO_3^- \tag{3}$$

$$2HSO_3^- + O_2 \rightarrow 2SO_4^{2-} + 2H^+ \tag{4}$$

Figure 4 presents the X-ray photoelectron spectra of Mo 3d, S 2p, and O 1s of molybdenite. The Mo 3d spectra of the molybdenite surface (Figure 4A) were best fitted with three GL functions. The $3d_{5/2}$ peaks located at ca. 229.5 eV, 229.7 eV, and 232.2 eV are attributed to molybdenum bulk (Mo(IV)) of molybdenite [40], Mo (IV) of molybdenum dioxide (MoO_2) [41], and Mo(VI) of molybdenum trioxide (MoO_3) [42], respectively. The S 2p spectrum of molybdenite shown in Figure 4B indicates a doublet peak with S $2p_{3/2}$ peaks located at 162.3 eV, which is attributed to the monosulfide of molybdenite [19,43]. Figure 4C shows the O 1s spectra of molybdenite. The presence of molybdenum dioxide and molybdenum trioxide is confirmed by the O 1s peaks located at 530.3 eV [44] and 531.4 eV [45], respectively. In addition, there are other peaks located at 532.3 eV and 533.7 eV, which are attributed to the hydroxide of the organic contaminants [30] and adsorbed water (H_2O) [39], respectively.

Similar spectra of Mo 3d, S 2p, and O 1s of molybdenite were obtained after the molybdenite was treated in 5 mM MBS at pH 9. In addition, unlike the formation of sulfate species on the S 2p of chalcopyrite after being treated in 5 mM MBS at pH 9, there was no sulfate formation in the S 2p spectra of molybdenite.

3.3. Proposed Mechanism

Based on the XPS analysis results, a mechanism can be proposed to explain the depressing effect of MBS on the floatability of chalcopyrite. The MBS treatment mainly produces sulfate species as a reaction product when in contact with the dissolved oxygen in water (Equations (3) and (4)). This sulfate species covered the chalcopyrite surface, forming hydrophilic $Fe_2(SO_4)_3$ species on the surface. Furthermore, the MBS treatment improved the intensity of hydrophilic FeOOH and Fe_2O_3 species on the chalcopyrite surface. The presence of these various hydrophilic species on the surface depresses the floatability of chalcopyrite.

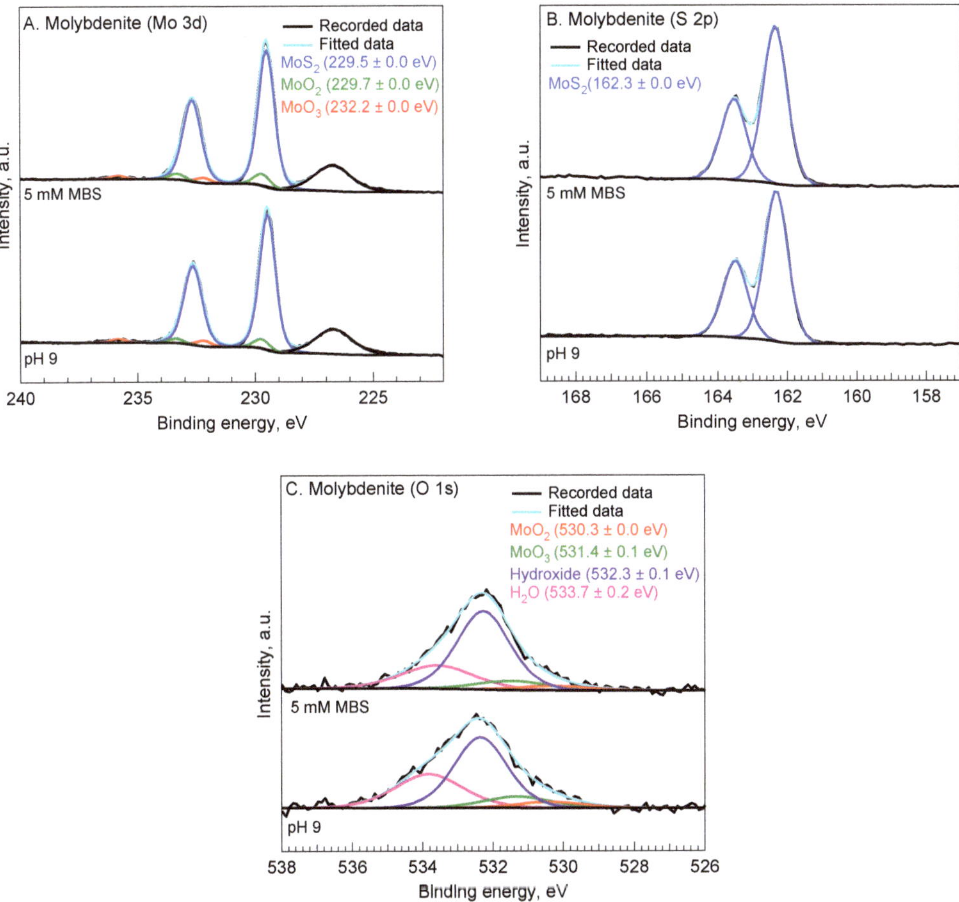

Figure 4. Mo 3d (**A**), S 2p (**B**), O 1s (**C**) spectra of chalcopyrite without and with 5 mM MBS treatment at pH 9.

The XPS results of molybdenite indicate that the MBS had an insignificant effect on the binding energy of chemical species on the surface of molybdenite. However, the flotation results indicate that MBS treatment exhibited a slightly depressing effect on the floatability of molybdenite. This difference in XPS and flotation results could be caused by the crystal structure of molybdenite. The crystal structure of molybdenite shows two types of surfaces: (1) non-polar faces formed by the scission of S–S bonds and (2) polar edges generated by the rupture of strong covalent Mo–S bonds. Based on these bonding types, the faces are hydrophobic and the edges are hydrophilic [46]. These edges might be vulnerable to the MBS treatment than the faces as shown by Chen et al. [32]. They reported that the MBS treatment showed an insignificant effect in the spectra of molybdenite faces, however, it formed sulfate and sulfite species in the S 2p spectra of molybdenite edges after being treated in 1 mM MBS in DI water.

In this study, the molybdenite fine powder was used for flotation tests and XPS analysis, which has a higher edges-to-faces ratio, thus the depressing effect of MBS on the floatability of molybdenite became more apparent. Furthermore, the molybdenite powder used for XPS analysis in this study was not differentiated between the faces and edges of molybdenite, thus the XPS analysis could not identify the effect of MBS treatment on the molybdenite edges. However, the presence of sulfate and sulfite species on the molybdenite edges, as reported by Chen et al. [32], could explain the slightly depressing effect of MBS on the floatability of molybdenite presented in this study.

3.4. Microflotation of Mixed Minerals

The microflotation results of a single mineral, presented in Figure 2, show the possibility of using MBS treatment for selective separation of molybdenite and chalcopyrite. Therefore, the microflotation tests were performed using mixed chalcopyrite and molybdenite. Figure 5 shows the effects of MBS treatment on the recovery of chalcopyrite and molybdenite in the mixed minerals system.

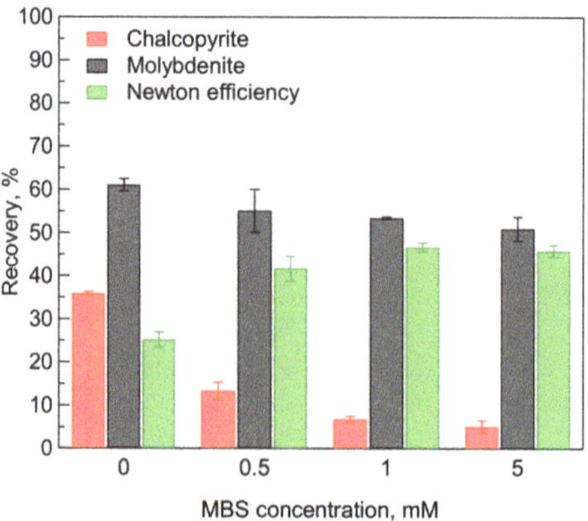

Figure 5. Effect of sodium metabisulfite (MBS) on the recovery of mixed chalcopyrite and molybdenite (1:1 ratio) in the absence of a collector at pH 9.

As expected, the microflotation results of mixed chalcopyrite and molybdenite show a similar trend as shown in the microflotation of a single mineral (Figure 2). The MBS treatment exhibited a strong depressing effect on the recovery of chalcopyrite and slightly reduced the recovery of molybdenite. Furthermore, the depressing effect increased with an increase in the concentration of MBS.

The Newton efficiency is presented in Figure 5 to assess the selectivity of the MBS treatment on the separation of chalcopyrite and molybdenite using flotation. The Newton efficiency increased from 25% before the MBS treatment to 43% after the addition of 0.5 mM MBS. The Newton efficiency reached a maximum value (48%) after the addition of 1 mM MBS and then slightly decreased to 46% after the addition of 5 mM MBS owing to the decreasing recovery of molybdenite. These Newton efficiency results demonstrate that the MBS treatment could be used as a selective depressant of chalcopyrite in the selective flotation of chalcopyrite and molybdenite.

Emulsified kerosene and diesel oil were used as collectors to improve the separation selectivity of chalcopyrite and molybdenite in the flotation of mixed minerals using MBS treatment. These non-polar oil collectors were selected because they are known as selective collectors for molybdenite, owing to a higher affinity on the molybdenite non-polar faces [2,47–50]. In these flotation tests, the concentration of MBS was fixed at 1 mM. This concentration was selected based on the maximum Newton efficiency shown in Figure 5.

Figure 6 shows the effect of emulsified diesel oil and kerosene on the recovery of mixed chalcopyrite and molybdenite after being treated in 1 mM MBS at pH 9. The addition of various concentrations of emulsified diesel oil and kerosene significantly improved the recovery of molybdenite compared to that of chalcopyrite. For instance, the recovery of molybdenite significantly increased from 52% to 82% after the addition of 100 mg/L diesel oil, while the recovery of chalcopyrite slightly increased from 7% to 13% under a similar condition. These results can be attributed to the selective adsorption of diesel oil and kerosene on the molybdenite surface, which was less affected by the MBS treatment than the chalcopyrite surface, as shown by the XPS analysis and flotation results. In addition, Suyantara et al. [48,51] showed that the molybdenite surface became more hydrophobic after the addition of emulsified kerosene, thus molybdenite became more floatable compared to that of chalcopyrite.

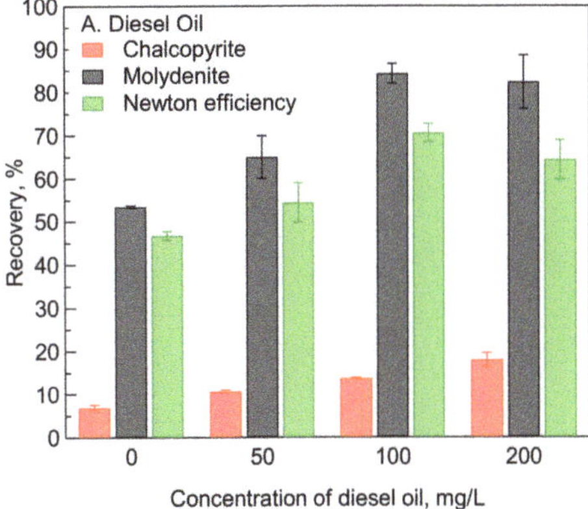

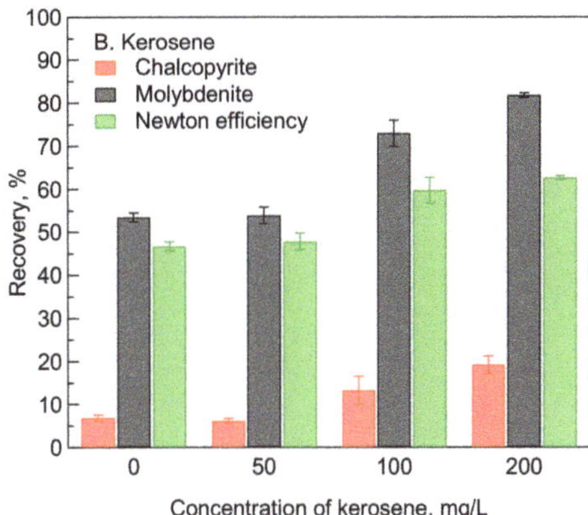

Figure 6. Effect of emulsified diesel oil (**A**) and kerosene (**B**) on the recovery of mixed chalcopyrite and molybdenite after being treated in 1 mM MBS at pH 9.

Figure 6 demonstrates that emulsified diesel oil was more effective to improve the recovery of molybdenite compared to that of emulsified kerosene. For instance, the recovery of molybdenite improved from 52% to 82% after the addition of 100 mg/L emulsified diesel oil, while the recovery of molybdenite increased from 52% to 73% after the addition of 100 mg/L emulsified kerosene. Diesel oil has a better dispersion capability in water (i.e., forming finer and homogeneous emulsion) than kerosene [52]; thus, it is more effectively adsorbed on the molybdenite surface.

Figure 6A indicates that the recovery of molybdenite reached an optimum condition after the addition of 100 mg/L emulsified diesel oil. The addition of a higher concentration (i.e., 200 mg/L) of emulsified diesel oil slightly reduced the recovery of molybdenite. One of the possible reasons for this phenomenon is the addition of excess emulsified diesel oil caused excessive agglomeration of molybdenite particles, causing the molybdenite particle to become too heavy to be carried by the nitrogen bubbles to the froth zone. On the other hand, this phenomenon was not observed if emulsified kerosene was used as a collector (Figure 6B). The maximum recovery of molybdenite was observed after the addition of 200 mg/L emulsified kerosene.

The recovery of chalcopyrite slightly increased with an increase in the concentration of emulsified diesel oil and kerosene (Figure 6). This phenomenon can be caused by the physical adsorption of diesel oil and kerosene on the surface of chalcopyrite, owing to the presence of sulfur species on the slightly oxidized chalcopyrite, as shown from the XPS results. A similar phenomenon has been reported in previous studies [48,53]. In addition, this phenomenon might be caused by the interaction between molybdenite and chalcopyrite (i.e., heterocoagulation and entrapment of chalcopyrite), which requires further investigation. Indeed, Hornn et al. [54] showed that emulsified kerosene can be used to agglomerate the fine particle chalcopyrite, improving its floatability in the agglomeration-flotation.

Table 2 confirms the effectiveness of MBS treatment as a selective depressant of chalcopyrite in the selective flotation of chalcopyrite and molybdenite with the addition of diesel oil and kerosene. In the absence of MBS treatment, the recoveries of chalcopyrite were 30.8% and 43.3% with the addition of 100 mg/L diesel oil and 200 mg/L kerosene, respectively. Meanwhile, the addition of diesel oil and kerosene significantly improved the recoveries of molybdenite, from 61.1% in the absence of collector to 88.0% and 89.5% with the addition of 100 mg/L diesel oil and 200 mg/L kerosene, respectively. With the addition of 1 mM MBS, the recovery of chalcopyrite significantly depressed to 13.7% and 19.1% under similar conditions. On the other hand, the recovery of molybdenite slightly decreased after the MBS treatment.

Table 2. Recovery of chalcopyrite (Cu) and molybdenite (Mo) and Newton efficiency (η) under the optimal condition of diesel oil and kerosene.

No	Collector Condition	MBS Concentration	Recovery, %		η, %
			Cu	Mo	
1	Without collector	0 mM MBS	35.9	61.1	25.2
2	100 mg/L diesel	0 mM MBS	30.8	88.0	57.3
3	100 mg/L diesel	1 mM MBS	13.7	84.2	70.5
4	200 mg/L kerosene	0 mM MBS	43.3	89.5	46.2
5	200 mg/L kerosene	1 mM MBS	19.1	81.7	62.6

Figure 7 shows the Newton efficiency of flotation of mixed chalcopyrite and molybdenite after being treated in various concentrations of emulsified diesel oil and kerosene in 1 mM MBS at pH 9. The Newton efficiency gradually increased with an increase in the concentration of emulsified diesel oil and kerosene. The Newton efficiency reached a maximum value of 70% after the addition of 100 mg/L diesel oil in 1 mM MBS at pH 9. Under a similar concentration, the addition of emulsified kerosene results in lower Newton efficiency (i.e., 60%). Meanwhile, the Newton efficiency reached a maximum value of 62% after the addition of 200 mg/L emulsified kerosene in 1 mM MBS at pH 9. This result indicates that emulsified diesel oil is more effective in improving the selectivity of flotation of chalcopyrite and molybdenite compared to kerosene. As discussed previously, diesel has a higher effectivity owing to it having a better dispersion capability in water than kerosene. Nevertheless, both emulsified diesel oil and kerosene could significantly improve the separation selectivity of selective flotation of chalcopyrite and molybdenite using MBS treatment.

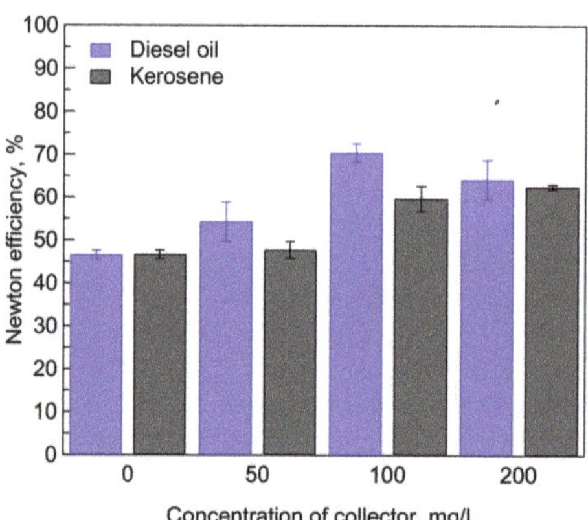

Figure 7. Effects of diesel oil and kerosene addition on the Newton efficiency of selective flotation of chalcopyrite and molybdenite in 1 mM MBS at pH 9.

4. Conclusions

In this study, the effects of MBS treatment on the floatability of chalcopyrite and molybdenite were investigated. The microflotation results of a single mineral show that MBS treatment exhibited a strong depressing effect on the floatability of chalcopyrite and slightly depressed the floatability of molybdenite. The XPS analysis results indicated that the strong depressing effect of MBS on the floatability of chalcopyrite is caused by the formation of hydrophilic species (i.e., sulfate, $Fe_2(SO_4)_3$, FeOOH, and Fe_2O_3) on the surface. On the other hand, MBS treatment showed an insignificant effect on the chemical species on the molybdenite surface.

The microflotation of mixed minerals showed that MBS treatment could be used as a selective depressant of chalcopyrite in the selective flotation of chalcopyrite and molybdenite. Furthermore, the addition of diesel oil or kerosene as a molybdenite collector could significantly improve the separation selectivity (i.e., Newton efficiency) of flotation of chalcopyrite and molybdenite using MBS treatment. The addition of diesel oil could effectively improve the Newton efficiency compared to that of kerosene.

Author Contributions: Conceptualization, Y.S., T.H., H.M. and G.P.W.S.; methodology and experiments, Y.S. and G.P.W.S.; writing—original draft preparation, Y.S.; writing—review and editing, H.M., T.H. and G.P.W.S.; supervision, T.H. and H.M.; project administration, T.H., K.S., H.M., Y.T., Y.A. and K.U.; funding acquisition, Y.T., Y.A., K.S. and H.M. All authors have read and agreed to the published version of the manuscript.

Funding: This research was funded by a Grant-in-Aid for Science Research (JSPS KAKENHI Grant No. JP19H02659 and JP21K18923) from the Japan Society for the Promotion of Science (JSPS)—Japan, and the Sumitomo Metal Mining Co., Ltd., Tokyo, Japan.

Conflicts of Interest: The authors declare no conflict of interest.

References

1. Wills, B.A.; Napier-Munn, T.J. *Wills' Mineral Processing Technology: An Introduction to the Practical Aspects of Ore Treatment and Mineral Recovery*, 7th ed.; Elsevier Science & Technology Books: Burlington, VT, USA, 2006.
2. Bulatovic, S.M. *Handbook of Flotation Reagents: Chemistry, Theory and Practice: Volume 1: Flotation of Sulfide Ores*; Elsevier Science: Amsterdam, The Netherlands, 2007; ISBN 978-0-08-047137-2.
3. Liu, G.; Lu, Y.; Zhong, H.; Cao, Z.; Xu, Z. A Novel Approach for Preferential Flotation Recovery of Molybdenite from a Porphyry Copper–Molybdenum Ore. *Miner. Eng.* **2012**, *36–38*, 37–44. [CrossRef]
4. Laskowski, J.S.; Castro, S.; Ramos, O. Effect of Seawater Main Components on Frothability in the Flotation of Cu-Mo Sulfide Ore. *Physicochem. Probl. Miner. Process.* **2013**, *50*, 17–29.
5. Ansari, A.; Pawlik, M. Floatability of Chalcopyrite and Molybdenite in the Presence of Lignosulfonates. Part I. Adsorption Studies. *Miner. Eng.* **2007**, *20*, 600–608. [CrossRef]
6. Moreno, P.A.; Aral, H.; Cuevas, J.; Monardes, A.; Adaro, M.; Norgate, T.; Bruckard, W. The Use of Seawater as Process Water at Las Luces Copper–Molybdenum Beneficiation Plant in Taltal (Chile). *Miner. Eng.* **2011**, *24*, 852–858. [CrossRef]
7. Pearse, M.J. An Overview of the Use of Chemical Reagents in Mineral Processing. *Miner. Eng.* **2005**, *18*, 139–149. [CrossRef]
8. Prasad, M.S. Reagents in the Mineral Industry—Recent Trends and Applications. *Miner. Eng.* **1992**, *5*, 279–294. [CrossRef]
9. Somasundaran, P. *Reagents in Mineral Technology*; CRC Press: Boca Raton, FL, USA, 1987; ISBN 978-0-8247-7715-9.
10. Yin, W.; Zhang, L.; Xie, F. Flotation of Xinhua Molybdenite Using Sodium Sulfide as Modifier. *Trans. Nonferrous Met. Soc. China* **2010**, *20*, 702–706. [CrossRef]
11. Li, M.; Wei, D.; Liu, Q.; Liu, W.; Zheng, J.; Sun, H. Flotation Separation of Copper–Molybdenum Sulfides Using Chitosan as a Selective Depressant. *Miner. Eng.* **2015**, *83*, 217–222. [CrossRef]
12. Kelebek, S.; Yoruk, S.; Smith, G.W. Wetting Behavior of Molybdenite and Talc in Lignosulphonate/Mibc Solutions and Their Separation by Flotation. *Sep. Sci. Technol.* **2001**, *36*, 145–157. [CrossRef]
13. Wang, X.; Zhao, B.; Liu, J.; Zhu, Y.; Han, Y. Dithiouracil, a Highly Efficient Depressant for the Selective Separation of Molybdenite from Chalcopyrite by Flotation: Applications and Mechanism. *Miner. Eng.* **2022**, *175*, 107287. [CrossRef]
14. Wang, C.; Liu, R.; Wu, M.; Xu, Z.; Tian, M.; Yin, Z.; Sun, W.; Zhang, C. Flotation Separation of Molybdenite from Chalcopyrite Using Rhodanine-3-Acetic Acid as a Novel and Effective Depressant. *Miner. Eng.* **2021**, *162*, 106747. [CrossRef]
15. Hirajima, T.; Mori, M.; Ichikawa, O.; Sasaki, K.; Miki, H.; Farahat, M.; Sawada, M. Selective Flotation of Chalcopyrite and Molybdenite with Plasma Pre-Treatment. *Miner. Eng.* **2014**, *66–68*, 102–111. [CrossRef]
16. Miki, H.; Matsuoka, H.; Hirajima, T.; Suyantara, G.P.W.; Sasaki, K. Electrolysis Oxidation of Chalcopyrite and Molybdenite for Selective Flotation. *Mater. Trans.* **2017**, *58*, 761–767. [CrossRef]
17. Hirajima, T.; Miki, H.; Suyantara, G.P.W.; Matsuoka, H.; Elmahdy, A.M.; Sasaki, K.; Imaizumi, Y.; Kuroiwa, S. Selective Flotation of Chalcopyrite and Molybdenite with H2O2 Oxidation. *Miner. Eng.* **2017**, *100*, 83–92. [CrossRef]
18. Suyantara, G.P.W.; Hirajima, T.; Miki, H.; Sasaki, K.; Yamane, M.; Takida, E.; Kuroiwa, S.; Imaizumi, Y. Selective Flotation of Chalcopyrite and Molybdenite Using H2O2 Oxidation Method with the Addition of Ferrous Sulfate. *Miner. Eng.* **2018**, *122*, 312–326. [CrossRef]

19. Suyantara, G.P.W.; Hirajima, T.; Miki, H.; Sasaki, K.; Yamane, M.; Takida, E.; Kuroiwa, S.; Imaizumi, Y. Effect of Fenton-like Oxidation Reagent on Hydrophobicity and Floatability of Chalcopyrite and Molybdenite. *Colloids Surf. A Physicochem. Eng. Asp.* **2018**, *554*, 34–48. [CrossRef]

20. Grano, S.R.; Johnson, N.W.; Ralston, J. Control of the Solution Interaction of Metabisulphite and Ethyl Xanthate in the Flotation of the Hilton Ore of Mount Isa Mines Limited, Australia. *Miner. Eng.* **1997**, *10*, 17–39. [CrossRef]

21. Hassanzadeh, A.; Hasanzadeh, M. Chalcopyrite and Pyrite Floatabilities in the Presence of Sodium Sulfide and Sodium Metabisulfite in a High Pyritic Copper Complex Ore. *J. Dispers. Sci. Technol.* **2017**, *38*, 782–788. [CrossRef]

22. Houot, R.; Duhamet, D. The Use of Sodium Sulphite to Improve the Flotation Selectivity between Chalcopyrite and Galena in a Complex Sulphide Ore. *Miner. Eng.* **1992**, *5*, 343–355. [CrossRef]

23. Khmeleva, T.; Skinner, W.; Beattie, D.A.; Georgiev, T.V. The Effect of Sulphite on the Xanthate-Induced Flotation of Copper-Activated Pyrite. *Physicochem. Probl. Miner. Process.* **2002**, *36*, 185–195.

24. Khmeleva, T.N.; Chapelet, J.K.; Skinner, W.M.; Beattie, D.A. Depression Mechanisms of Sodium Bisulphite in the Xanthate-Induced Flotation of Copper Activated Sphalerite. *Int. J. Miner. Process.* **2006**, *79*, 61–75. [CrossRef]

25. Miki, H.; Hirajima, T.; Muta, Y.; Suyantara, G.P.W.; Sasaki, K. Effect of Sodium Sulfite on Floatability of Chalcopyrite and Molybdenite. *Minerals* **2018**, *8*, 172. [CrossRef]

26. Mu, Y.; Peng, Y. The Role of Sodium Metabisulphite in Depressing Pyrite in Chalcopyrite Flotation Using Saline Water. *Miner. Eng.* **2019**, *142*, 105921. [CrossRef]

27. Petrus, H.T.B.M.; Hirajima, T.; Sasaki, K.; Okamoto, H. Effects of Sodium Thiosulphate on Chalcopyrite and Tennantite: An Insight for Alternative Separation Technique. *Int. J. Miner. Process.* **2012**, *102–103*, 116–123. [CrossRef]

28. Haga, K.; Tongamp, W.; Shibayama, A. Investigation of Flotation Parameters for Copper Recovery from Enargite and Chalcopyrite Mixed Ore. *Mater. Trans.* **2012**, *53*, 707–715. [CrossRef]

29. Hassanzadeh, A.; Karakaş, F. The Kinetics Modeling of Chalcopyrite and Pyrite, and the Contribution of Particle Size and Sodium Metabisulfite to the Flotation of Copper Complex Ores. *Part. Sci. Technol.* **2017**, *35*, 455–461. [CrossRef]

30. Suyantara, G.P.W.; Hirajima, T.; Miki, H.; Sasaki, K.; Kuroiwa, S.; Aoki, Y. Effect of Na_2SO_3 on the Floatability of Chalcopyrite and Enargite. *Miner. Eng.* **2021**, *173*, 107222. [CrossRef]

31. Castro, S. Physico-Chemical Factors in Flotation of Cu-Mo-Fe Ores with Seawater: A Critical Review. *Physicochem. Probl. Miner. Process.* **2018**, *54*, 1223–1236. [CrossRef]

32. Chen, Y.; Chen, X.; Peng, Y. The Depression of Molybdenite Flotation by Sodium Metabisulphite in Fresh Water and Seawater. *Miner. Eng.* **2021**, *168*, 106939. [CrossRef]

33. Shirley, D.A. High-Resolution X-ray Photoemission Spectrum of the Valence Bands of Gold. *Phys. Rev. B* **1972**, *5*, 4709–4714. [CrossRef]

34. Sasaki, K.; Takatsugi, K.; Ishikura, K.; Hirajima, T. Spectroscopic Study on Oxidative Dissolution of Chalcopyrite, Enargite and Tennantite at Different pH Values. *Hydrometallurgy* **2010**, *100*, 144–151. [CrossRef]

35. McIntyre, N.S.; Cook, M.G. X-ray Photoelectron Studies on Some Oxides and Hydroxides of Cobalt, Nickel, and Copper. *Anal. Chem.* **1975**, *47*, 2208–2213. [CrossRef]

36. Ghahremaninezhad, A.; Dixon, D.G.; Asselin, E. Electrochemical and XPS Analysis of Chalcopyrite ($CuFeS_2$) Dissolution in Sulfuric Acid Solution. *Electrochim. Acta* **2013**, *87*, 97–112. [CrossRef]

37. Brion, D. Etude Par Spectroscopie de Photoelectrons de La Degradation Superficielle de FeS_2, $CuFeS_2$, ZnS et PbS a l'air et Dans l'eau. *Appl. Surf. Sci.* **1980**, *5*, 133–152. [CrossRef]

38. Tan, B.J.; Klabunde, K.J.; Sherwood, P.M.A. X-ray Photoelectron Spectroscopy Studies of Solvated Metal Atom Dispersed Catalysts. Monometallic Iron and Bimetallic Iron-Cobalt Particles on Alumina. *Chem. Mater.* **1990**, *2*, 186–191. [CrossRef]

39. Shuxian, Z.; Hall, W.K.; Ertl, G.; Knözinger, H. X-ray Photoemission Study of Oxygen and Nitric Oxide Adsorption on MoS_2. *J. Catal.* **1986**, *100*, 167–175. [CrossRef]

40. Benoist, L.; Gonbeau, D.; Pfister-Guillouzo, G.; Schmidt, E.; Meunier, G.; Levasseur, A. XPS Analysis of Lithium Intercalation in Thin Films of Molybdenum Oxysulphides. *Surf. Interface Anal.* **1994**, *22*, 206–210. [CrossRef]

41. Seifert, G.; Finster, J.; Müller, H. SW Xα Calculations and X-ray Photoelectron Spectra of Molybdenum(II) Chloride Cluster Compounds. *Chem. Phys. Lett.* **1980**, *75*, 373–377. [CrossRef]

42. Kim, K.S.; Baitinger, W.E.; Amy, J.W.; Winograd, N. ESCA Studies of Metal-Oxygen Surfaces Using Argon and Oxygen Ion-Bombardment. *J. Electron Spectrosc. Relat. Phenom.* **1974**, *5*, 351–367. [CrossRef]

43. Stevens, G.C.; Edmonds, T. Electron Spectroscopy for Chemical Analysis Spectra of Molybdenum Sulfides. *J. Catal.* **1975**, *37*, 544–547. [CrossRef]

44. Zingg, D.S.; Makovsky, L.E.; Tischer, R.E.; Brown, F.R.; Hercules, D.M. A Surface Spectroscopic Study of Molybdenum-Alumina Catalysts Using X-ray Photoelectron, Ion-Scattering, and Raman Spectroscopies. *J. Phys. Chem.* **1980**, *84*, 2898–2906. [CrossRef]

45. Patterson, T.A.; Carver, J.C.; Leyden, D.E.; Hercules, D.M. A Surface Study of Cobalt-Molybdena-Alumina Catalysts Using X-ray Photoelectron Spectroscopy. *J. Phys. Chem.* **1976**, *80*, 1700–1708. [CrossRef]

46. Castro, S.; López-Valdivieso, A.; Laskowski, J.S. Review of the Flotation of Molybdenite. Part I: Surface Properties and Floatability. *Int. J. Miner. Process.* **2016**, *148*, 48–58. [CrossRef]

47. Wada, M.; Majima, H. Flotation of Molybdenite. *Flotation* **1962**, *1962*, 9–18. [CrossRef]

48. Suyantara, G.P.W.; Hirajima, T.; Miki, H.; Sasaki, K. Bubble Interactions with Chalcopyrite and Molybdenite Surfaces in Seawater. *Miner. Eng.* **2020**, *157*, 106536. [CrossRef]
49. Song, S.; Zhang, X.; Yang, B.; Lopez-Mendoza, A. Flotation of Molybdenite Fines as Hydrophobic Agglomerates. *Sep. Purif. Technol.* **2012**, *98*, 451–455. [CrossRef]
50. Tanaka, Y.; Miki, H.; Suyantara, G.P.W.; Aoki, Y.; Hirajima, T. Mineralogical Prediction on the Flotation Behavior of Copper and Molybdenum Minerals from Blended Cu–Mo Ores in Seawater. *Minerals* **2021**, *11*, 869. [CrossRef]
51. Suyantara, G.P.W.; Hirajima, T.; Elmahdy, A.M.; Miki, H.; Sasaki, K. Effect of Kerosene Emulsion in $MgCl_2$ Solution on the Kinetics of Bubble Interactions with Molybdenite and Chalcopyrite. *Colloids Surf. A Physicochem. Eng. Asp.* **2016**, *501*, 98–113. [CrossRef]
52. Lin, Q.; Gu, G.; Wang, H.; Liu, Y.; Fu, J.; Wang, C. Flotation Mechanisms of Molybdenite Fines by Neutral Oils. *Int. J. Miner. Metall. Mater.* **2018**, *25*, 1–10. [CrossRef]
53. Suyantara, G.P.W.; Hirajima, T.; Miki, H.; Sasaki, K. Floatability of Molybdenite and Chalcopyrite in Artificial Seawater. *Miner. Eng.* **2018**, *115*, 117–130. [CrossRef]
54. Hornn, V.; Ito, M.; Shimada, H.; Tabelin, C.B.; Jeon, S.; Park, I.; Hiroyoshi, N. Agglomeration-Flotation of Finely Ground Chalcopyrite and Quartz: Effects of Agitation Strength during Agglomeration Using Emulsified Oil on Chalcopyrite. *Minerals* **2020**, *10*, 380. [CrossRef]

Article

Efficient Recovery of Vanadium from High-Chromium Vanadium Slag with Calcium-Roasting Acidic Leaching

Hao Peng [1,2,3], Bing Li [1], Wenbing Shi [1,*] and Zuohua Liu [3,*]

[1] Chongqing Key Laboratory of Inorganic Special Functional Materials, College of Chemistry and Chemical Engineering, Yangtze Normal University, Chongqing 408100, China; cqupenghao@126.com (H.P.); libing@yznu.edu.cn (B.L.)
[2] Chongqing Jiulongyuan High-Tech Industry Group Co., Ltd., Chongqing 400080, China
[3] College of Chemistry and Chemical Engineering, Chongqing University, Chongqing 401133, China
* Correspondence: 20040004@yznu.edu.cn (W.S.); liuzuohua@cqu.edu.cn (Z.L.)

Abstract: High-chromium vanadium slag (HCVS) is an important by-product generated during the smelting process of high-chromium-vanadium-titanium-magnetite. Direct acid leaching and calcium-roasting acid leaching technology were applied to recover vanadium and chromium from HCVS. The effects of experimental parameters on the leaching process, including concentration of H_2SO_4, reaction temperature, reaction time, and liquid-to-solid ratio, were investigated. The XRD and UV-Vis DRS results showed that vanadium and chromium existed in low valence with a spinel structure in the HCVS. The Cr-spinel was too stable to leach out; no more than 8% of the chromium could be leached out both in the direct acid leaching process and calcium-roasting acid-leaching process. Most low valence vanadium could be oxidized to high valence with calcium-roasting technology, and the leaching efficiency could be increased from 33.89% to 89.12% at the selected reaction conditions: concentration of H_2SO_4 at 40 vt.%, reaction temperature of 90 °C, reaction time of 3 h, liquid-to-solid ratio of 4:1 mL/g, and stirring rate of 500 rpm. The kinetics analysis indicated that the leaching behavior of vanadium followed the shrinking core model well, and the leaching process was controlled by the surface chemical reaction, with an Ea of 58.95 kJ/mol and 62.98 kJ/mol for direct acid leaching and roasting acid leaching, respectively.

Keywords: vanadium; calcium roasting; leaching efficiency

Citation: Peng, H.; Li, B.; Shi, W.; Liu, Z. Efficient Recovery of Vanadium from High-Chromium Vanadium Slag with Calcium-Roasting Acidic Leaching. *Minerals* **2022**, *12*, 160. https://doi.org/10.3390/min12020160

Academic Editors: Jean-François Blais, Shuai Wang, Xingjie Wang and Jia Yang

Received: 10 January 2022
Accepted: 25 January 2022
Published: 28 January 2022

Publisher's Note: MDPI stays neutral with regard to jurisdictional claims in published maps and institutional affiliations.

1. Introduction

Vanadium and chromium are strategic transition elements that have been widely used in some fields such as steel-making, energy-storage, catalysts, the petrochemical industry, and green chemistry owing to their excellence hardness, high corrosion resistance, and other excellent physicochemical properties [1–5]. High-chromium vanadium slag (HCVS) is a by-product generated during the smelting process of high-chromium-vanadium-titanium-magnetite, and it is an important vanadium source in China [6–10]. During the smelting process, the vanadium and chromium are reduced into the molten enriched in the spinel, which is hard to destroy directly and restricted the large-scale utilization of HCVS [11]. Thus, some enhancement technologies were needed to recover vanadium and chromium efficiently.

To date, the basic recovery technology for vanadium has been sodium-roasting leaching technology, which was first proposed by Birck in 1912 and is widely used in the Chinese industries since the 1980s [12–14]. The vanadium-containing ores are mixed with the sodium salts (sodium carbonate (800–1000 °C), sodium sulfate (1200–1250 °C), sodium chloride (750–850 °C), and sodium hydroxide (400–800 °C)) at determined mole ratios and then roasted in a vertical kiln under a high temperature atmosphere with O_2 [13,15,16]. The structure of vanadium-containing ores is destroyed in the high temperature and low-valence vanadium is exposed and oxidized to a high valence. The high-valence vanadium is formed as sodium vanadate and can be easily leached out with acid, alkaline, or water-leaching [17–20]. However, some environmental hazards (sulfur dioxide, chlorine, and hydrogen chloride) and large amount of wastewater have limited continuous large-scale industrial application with the high environmental standards of today. To overcome the above problems, a roasting technology called calcium-roasting technology was developed, in which the sodium salts are replaced by calcium salts [21–25]. The roasting process is similar to the sodium-roasting process; the vanadium-containing ores are mixed with lime, limestone, and calcium salts at fixed mole ratios and then roasted at high temperatures, which are higher than with sodium roasting. During the roasting process, the vanadium spinel is decomposed and reacted with calcium salts to form different kinds of calcium-vanadate, which are determined by the mole ratio of the vanadium to calcium salts [26–30]. Usually, some leaching enhancing processes or multiple roasting processes accompany this process to achieve high recovery [16,31–34].

In this paper, direct acid leaching and calcium-roasting acid leaching technology were applied to leach out chromium and vanadium. The effects of experimental parameters including reaction time, liquid-to-solid ratio, reaction temperature, and concentration of H_2SO_4 on the leaching process were investigated. The leaching kinetics were also investigated.

2. Materials and Methods

2.1. Materials

The HCVS was collected from Pangang Group Co. Ltd., Panzhihua City, Sichuan Province, China. It was dried and ground to below 75 μm for further experiments. The elemental composition of HCVS was measured by XRF. The results displayed in Table 1 indicate that the vanadium and chromium were about 5.43 wt.% and 6.84 wt.%, respectively.

Table 1. Elemental accounts of the HCVS (wt.%).

Element	V_2O_5	Cr_2O_3	FeO	CaO	MgO
Percentages (%)	9.7	10.2	24.7	2.8	13.8
Element	SiO_2	Al_2O_3	MnO	TiO_2	
Percentages (%)	25.7	10.3	1.6	2.7	

2.2. Experimental Procedure

The batch experiments were conducted in a 300 mL glass beaker placed in a thermostatic mixing water bath. Firstly, the water bath was heated to a determined temperature. Then, a predetermined concentration of H_2SO_4 solution and a predetermined amount of HCVS or roasting HCVS were added to the beaker. Then, the beaker was placed in the water bath. Finally, the filtrate was collected by vacuum filtration after the required reaction time.

2.3. Analytical Methods

The concentrations of chromium and vanadium in the filtrate were measured by inductively coupled plasma-optical mission spectrometry (ICP-OES, PerkinElmer Optima 6300DV, Kyoto, Japan.) and the leaching efficiency was calculated following Equations (1) and (2):

$$\eta_V = \frac{V \cdot C_V}{m \cdot \omega_V} \times 100\% \tag{1}$$

$$\eta_{Cr} = \frac{V \cdot C_{Cr}}{m \cdot \omega_{Cr}} \times 100\% \tag{2}$$

where C_V and C_{Cr}, are the concentration of chromium and vanadium in the filtrate in g/L; V, is the volume of the filtrate in liters; ω_V, and ω_{Cr}, are the percentages of chromium and vanadium in the HCVS; and m, is the mass of the HCVS used in the batch experiments in grams.

2.4. Characterization

The element percentages of HCVS were measured by XRF (Shimadzu Lab Center XRF-1800, Kyoto, Japan) and the main phases were measured by XRD (Shimadzu Lab Center XRD-6000, Kyoto, Japan). The valences of the vanadium and chromium in the HCVS were detected by UV-Vis DRS (Shimadzu Lab Center, Kyoto, Japan) and XPS (ESCALAB-250Xi, Thermo Fisher Scientific, New York, NY, USA). The thermo-gravimetric analysis was conducted by TG-DSC (Shimadzu Lab Center DSC-60H, Kyoto, Japan) with a heating rate of 10 °C/min from 0 °C to 900 °C.

3. Results

3.1. Characterization of HCVS

The XRD pattern showed in Figure 1a shows that the main crystal structures in the HCVS were Fe_2O_3, $FeCr_2O_4$, $MgFe_2O_4$, Fe_2VO_4 and Fe_2SiO_4 [27,28,30]. The vanadium and chromium mainly existed as spinel structures ($FeCr_2O_4$ and Fe_2VO_4), which are hard to destroy. Therefore, the leaching efficiency of vanadium and chromium may not be high and some enhancing technologies are needed in the further experiments.

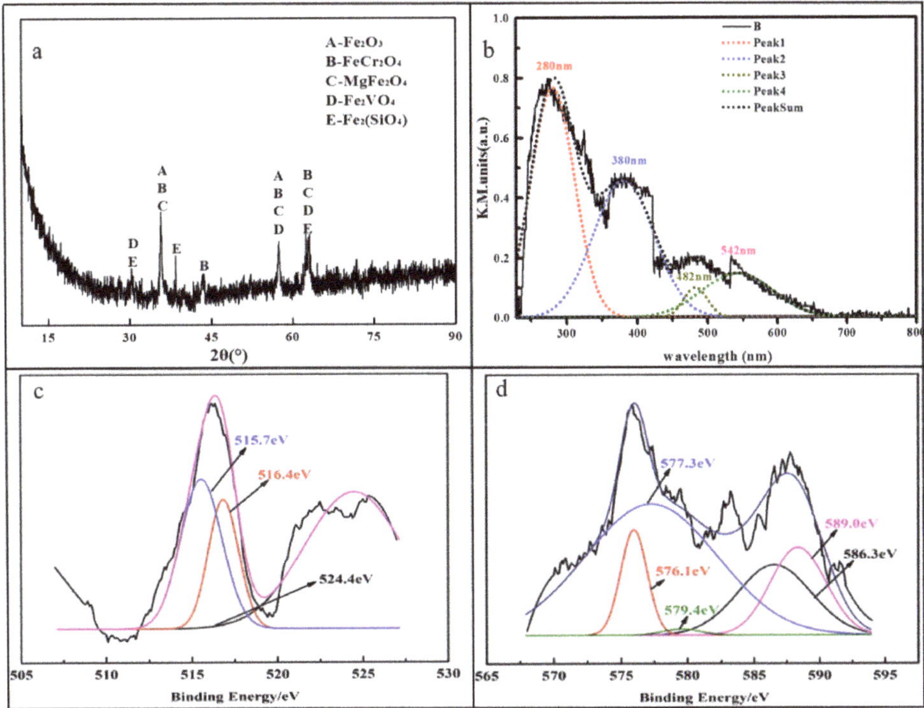

Figure 1. Characterization of HCVS. (**a**) XRD pattern of HCVS; (**b**) UV-Vis DRS of HCVS; (**c**) XPS of vanadium; (**d**) XPS of chromium.

The UV-Vis DRS of HCVS was conducted to help understanding the composition of HCVS; the result is displayed in Figure 1b. The original spectrum signal was analyzed and the peaks were fitted to four main peaks: 280 nm, 380 nm, 482 nm, and 542 nm. The peak at 280 nm was assigned to Fe (III) and confirmed the existence of Fe_2O_3 and $MgFe_2O_4$ [35,36]. The peak at 380 nm was assigned to Cr (III) [37], which corresponds to the $FeCr_2O_4$ phase. The peak at 482 nm was assigned to V=O stretching, which confirmed the existence of V (IV) and V (V) [38,39].

The XPS results showed that most vanadium existed as V(III) and V(IV) (515.7 eV and 516.4 eV), and the Cr (III) accounted for about 82.84% of the HCVS (576.1 eV, 577.3 eV and 586.3 eV), while there were no V (III) and V (V) phases in the XRD pattern. According to our previous study, V(III), V(IV), and V(V) co-exist in HCVS, which means that some V(III) and V(V) compounds in the HCVS exist amorphously and could not detected by XRD [28].

3.2. Direct-Acid-Leaching Process

The direct acid leaching process was conducted to leach out vanadium and chromium from the HCVS. Figure 2 summarizes the effects of the liquid-to-solid ratio, reaction time, concentration of H_2SO_4, and reaction temperature on the leaching process. The leaching efficiencies of vanadium and chromium were relatively low (<35% for vanadium and 8% for chromium), which is consistent with the results of XRD.

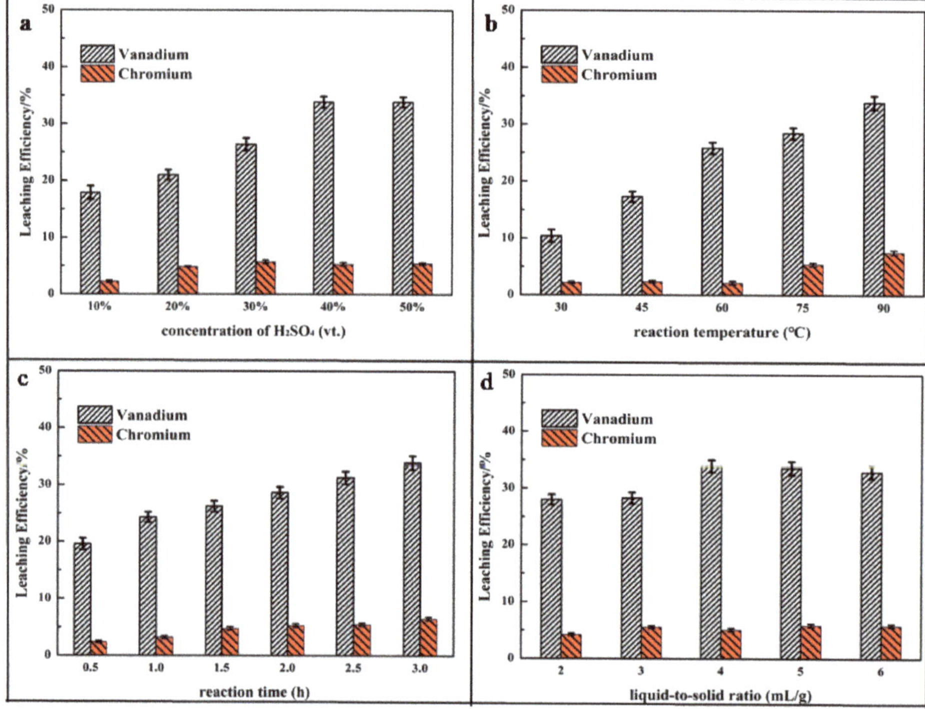

Figure 2. Effect of parameters on leaching efficiency of vanadium and chromium: (**a**) concentration of H_2SO_4; (**b**) reaction temperature; (**c**) reaction time; (**d**) liquid-to-solid ratio.

Figure 2a shows that the leaching efficiency of vanadium increased with the increase of the concentration of H_2SO_4. During the leaching process, the H^+ attacks the spinel and destroys the spinel structure to release vanadium and chromium. With an increasing of concentration of H_2SO_4, the corrosion process of the spinel by the highly concentrated H^+ was intensified and the leaching efficiency of vanadium was increased from 17.87% to 33.90%, as the concentration of H_2SO_4 increased from 10 vt.% to 50 vt.%. As the chromium spinel was more stable than the vanadium spinel, the chromium was harder to leach out, with a leaching efficiency below 8%. Otherwise, the leaching efficiency showed no obvious increase when the concentration of H_2SO_4 increased from 40 vt.% to 50 vt.%, and high concentrations bring high impurities [40]; thus, the concentration of 40 vt.% was selected as optimal for further experiments.

The results showed in Figure 2b indicate that only 10.40% vanadium and 2.18% chromium can be leached out at 30 °C. Higher temperatures enhances the activity of vanadium and chromium ions and further favors the reaction intensity [41]. The leaching efficiency increased to 33.89% for vanadium and 7.56% for chromium at 90 °C. In other words, high reaction temperature was beneficial to the leaching process.

Usually, in order to produce more products, long reaction times are utilized. Figure 2c shows that the leaching efficiency of vanadium and chromium increased with the reaction time, but the increase amplitude was slow. Longer reaction times may not make any contribution to the leaching process; thus, the reaction time of 3 h was selected in the following experiments.

During the leaching process, the HCVS particles was ground fine enough for good contact with the concentrated H_2SO_4 solution. The leaching process was most controlled by parameters such as the reaction temperature and concentration of H_2SO_4, but less by the liquid-to-solid ratio, as the liquid-to-solid ratio had no obvious effect on the leaching efficiency (seen in Figure 2d).

As the spinel structure was hard to destroy directly, the leaching efficiencies of vanadium and chromium were 33.89% and 7.56%, respectively, at the selected optimum conditions: reaction time of 3 h, liquid-to-solid ratio of 4: 1 mL/g, concentration of H_2SO_4 of 40 vt.%, reaction temperature of 90 °C, and stirring rate of 500 rpm.

3.3. Characterization of Roasting HCVS

In order to achieve efficient leaching performance of HCVS, calcium-roasting technology was applied to oxidize the low valence compounds. The obtained TG-DSC curves shown in Figure 3a indicate that there was a dehydration step, with weight loss of 1.23% from 0 °C to 400 °C, and an obvious exothermic peak of the DSC curve at 400 °C was observed, which corresponds to the decomposition of the spinel structure. After the temperature increased to 620 °C, a dramatic mass gain of 6.46% was obtained due to the oxidative decomposition of the vanadium spinel phase. This means that the oxidative roasting of vanadium spinel should be conducted above 620 °C. Thus, the calcium-roasting process was conducted at 650–850 °C and the HCVS was mixed with CaO at a mole ratio of $n(CaO)/n(V_2O_5) = 1.1$

The XRD pattern was used to analyze the phase changes during the calcium-roasting process. The results showed in Figure 3b indicate that some new peaks appeared, corresponding to the new phases of $Ca_2V_2O_5$, $CaFe (Si_2O_6)$, and $Ca_2V_2O_7$. These three new phases appeared at 650 °C, and the crystal structures became more stable as the roasting temperature increased from 650 °C to 850 °C. During the calcium-roasting process, the Fe_2VO_4 decomposed (seen in Equation (3)) to form V_2O_4 at nearly 400 °C according to DG-TSC results, and then reacted with CaO to form $Ca_2V_2O_5$. With the increasing roasting temperature, partial $Ca_2V_2O_5$ was further oxidized to $Ca_2V_2O_7$, which means that in the calcium roasting of HCVS, the V(IV) and V(V) co-existed.

After roasting, some V(III) and V(IV) were oxidized to V(V). The XPS results indicate that only 9.55% V(III) was retained in the roasted HCVS, while Cr(III) still accounted for about 80.32%. As the Cr spinel was more stable than the V spinel [13], the Cr was not oxidized and still existed in $FeCr_2O_4$, according to the XRD results. It was concluded that the chromium was still hard to leach out.

$$2Fe_2VO_4 \rightarrow 4FeO + V_2O_4 \tag{3}$$

$$V_2O_4 + CaO \rightarrow CaV_2O_5 \tag{4}$$

$$2CaV_2O_5 + O_2 + 2CaO \rightarrow 2Ca_2V_2O_7 \tag{5}$$

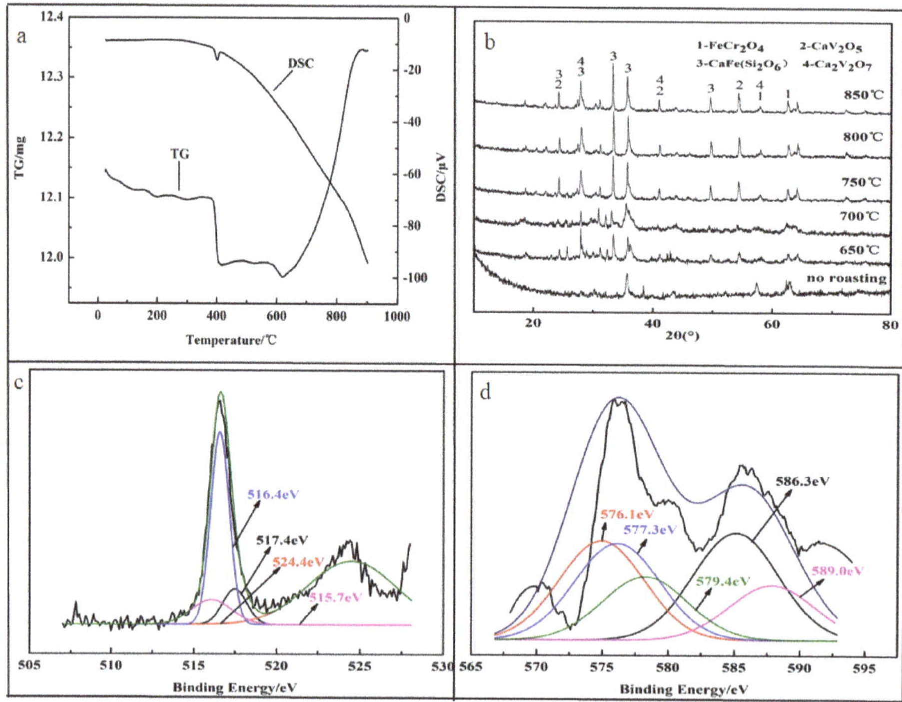

Figure 3. Characterization of roasting HCVS. (**a**) TG and DSC; (**b**) XRD pattern of roasted HCVS; (**c**) XPS of vanadium; (**d**) XPS of chromium.

3.4. Acid Leaching for Roasting HCVS

The acid-leaching experiments with calcium roasting of HCVS were conducted to investigate the effect of calcium roasting on the leaching process under the same reaction conditions as the direct acid leaching process described above. As the Cr spinel was still stable under the roasting temperature, the leaching efficiency of chromium showed no obvious increase compared with the direct acid leaching process; therefore, the leaching behavior of chromium is not discussed in this part.

Figure 4a shows that the calcium roasting made a great contribution to the leaching process. The leaching efficiency of vanadium was increased by nearly 40 percentage (up to 57.54%) after roasting, compared with the direct acidic leaching process (at the concentration of 10 vt.% H_2SO_4). During the roasting process, most V(III) and V(IV) were oxidized to V(V), making a contribution to the great leaching performance of roasting HCVS. The leaching efficiency increased quickly at the beginning and then smoothly with the increase of H_2SO_4 concentration. The highest leaching efficiency was up to 90.12% at a concentration of 50 vt.%, which showed nearly a 58% improvement compared to the direct acid leaching process. Compared to our previous study, the leaching efficiency might be increased more with some enhancing technologies, such as oxidative leaching and electro-oxidative leaching [28]. Otherwise, the formation of the by-product $CaSO_4$, which is a villous particle, might adsorb on the surface of leaching residue and have negative effects on the leaching process [40]; thus, a too high concentration of H_2SO_4 is not suitable for leaching while calcium roasting HCVS. Meanwhile, the leaching efficiency showed little increase as the concentration increased from 40 vt.% to 50 vt.%; thus, a concentration of H_2SO_4 of 40 vt.% was selected for further experiments.

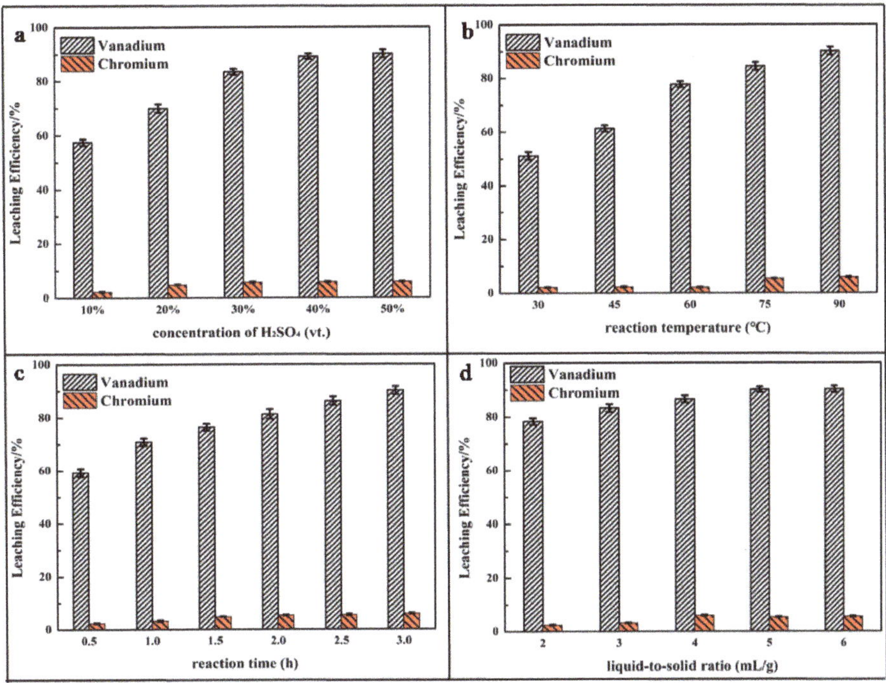

Figure 4. Effect of parameters on leaching efficiency of vanadium and chromium: (**a**) concentration of H$_2$SO$_4$; (**b**) reaction temperature; (**c**) reaction time; (**d**) liquid-to-solid ratio.

The same phenomenon can also be observed in Figure 4b. The leaching process was greatly enhanced by calcium roasting; the leaching efficiency was increased from 10.4% to 89.12% as the reaction temperature increased from 30 °C to 90 °C. Usually, metal ions have high solubility at high temperatures, accompanied by high activity; thus, 90 °C was chose in further experiments.

The results displayed in Figure 4c,d indicate that the liquid-to-solid ratio and reaction time showed similar effects on the leaching process, and a suitable liquid-to-solid ratio and a long reaction time could achieve high leaching efficiency. As can be seen, the calcium-roasting process can oxidize low valence vanadium to high valence vanadium and enhance the leaching process to achieve high leaching efficiency of vanadium, but has no influence on the change trend of leaching efficiency affected by the experimental parameters.

To summarize, low valence vanadium in V spinel was decomposed and oxidized to V(V) during the calcium-roasting process, but Cr spinel was too stable to decompose. For vanadium, 89.12% was leached out under the optimal reaction conditions: reaction time of 3 h, reaction temperature of 90 °C, liquid-to-solid ratio at 4:1 mL/g, concentration of H$_2$SO$_4$ at 40 vt.%, and stirring rate at 500 rpm. Most chromium existing as FeCr$_2$O$_4$ was hard to leach out and was retained in the leaching residue.

3.5. Leaching Kinetics

In order to understand the reaction mechanism, the leaching kinetics of vanadium were analyzed (leaching out chromium was very difficult; thus, it is not analyzed here). Usually, the leaching kinetics followed the shrinking core model described in Equation (6), which was used to describe the liquid-solid reaction [40,42–45]:

$$[(1-\eta)^{-1/3} - 1] + 1/3 \cdot \text{Ln}(1-\eta) = k \cdot t \tag{6}$$

where η is the leaching efficiency of vanadium, in percentage.

The experimental data was fitted to Equation (6) and the results are shown in Figure 5. Based on the fitting results, the Ea for vanadium leached out was calculated following the Arrhenius equations (Equation (7)). Figure 6 shows that the Ea for vanadium leached out was 58.95 kJ/mol for the direct acid leaching process and 62.98 kJ/mol for the calcium-roasting acid leaching process, which indicates that the controlling step for vanadium leaching is the surface chemical reaction [40,43–45]. Compared with the references [40,43,45,46], the Ea was much larger, indicating that the vanadium in the HCVS was hard to leach out by both direct acid leaching technology and calcium-roasting acid leaching technology. In order to improve the leaching efficiency and enhance the leaching process, some more efficient pre-treatment technologies are needed.

$$Lnk = LnA - Ea/(RT) \tag{7}$$

where Ea is the apparent activation energy, A is the pre-exponential factor, and R is the mole gas constant.

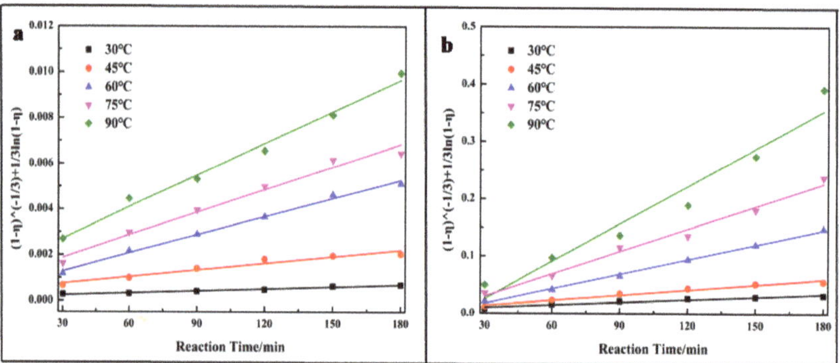

Figure 5. Plot of leaching kinetics of vanadium at various reaction temperatures. (**a**) direct acid leaching process; (**b**) calcium-roasting acid leaching process.

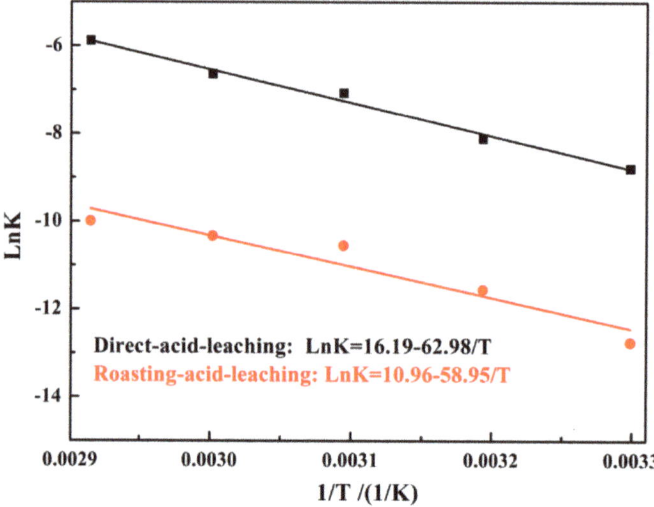

Figure 6. Natural logarithm of reaction rate constant versus reciprocal temperature of vanadium.

4. Conclusions

A direct acid leaching process and a calcium-roasting acid leaching process on HCVS were conducted. The following conclusions were obtained:

(1) The chromium and vanadium existed as spinel structure in the HCVS, which are too stable to destroy directly; only 33.89% of vanadium and 7.56% of chromium could be leached out at the selected conditions during the direct acid leaching process: reaction time of 3 h, liquid-to-solid ratio at 4:1 mL/g, concentration of H_2SO_4 at 40 vt.%, reaction temperature of 90 °C, and stirring rate at 500 rpm. The Ea of the vanadium leached out was 62.98 kJ/mol, which indicates that the vanadium was hard to leach out directly;

(2) Most low valence vanadium could be oxidized to high valence during the calcium-roasting process, and the leaching efficiency could achieve 89.12% under the optimal conditions: reaction time of 3 h, liquid-to-solid ratio at 4:1 mL/g, reaction temperature of 90 °C, concentration of H_2SO_4 at 40 vt.%, and stirring rate at 500 rpm. The leaching behavior followed the shrinking core model well, and the controlling step was the surface chemical reaction, with an Ea of 58.95 kJ/mol for the calcium-roasting acid leaching process.

(3) Chromium was hard to leach out both in the direct acid leaching process and the calcium-roasting acid leaching process; the leaching efficiency was below 8%. Higher roasting temperatures and new additive agents will be needed for efficient chromium recovery in our future works.

Author Contributions: Methodology, B.L.; supervision, W.S. and Z.L.; other contributions were made by H.P. All authors have read and agreed to the published version of the manuscript.

Funding: This work was supported by the Chongqing Science and Technology Commission (No. cstc2021jcyj-msxmX0129) and the Science and Technology Research Program of Chongqing Municipal Education Commission (No. CXQT20026).

Data Availability Statement: Not applicable.

Conflicts of Interest: The authors declare no conflict of interest.

References

1. Zhu, H.; Xiao, X.; Guo, Z.; Han, X.; Liang, Y.; Zhang, Y.; Zhou, C. Adsorption of vanadium (V) on natural kaolinite and montmorillonite: Characteristics and mechanism. *Appl. Clay Sci.* **2018**, *161*, 310–316. [CrossRef]
2. Wei, Z.; Liu, D.; Hsu, C.; Liu, F. All-vanadium redox photoelectrochemical cell: An approach to store solar energy. *Electrochem. Commun.* **2014**, *45*, 79–82. [CrossRef]
3. Nicholas, N.J.; da Silva, G.; Kentish, S.; Stevens, G.W. Use of Vanadium(V) Oxide as a Catalyst for CO2 Hydration in Potassium Carbonate Systems. *Ind. Eng. Chem. Res.* **2014**, *53*, 3029–3039. [CrossRef]
4. Wen, J.; Ning, P.; Cao, H.; Zhao, H.; Sun, Z.; Zhang, Y. Novel method for characterization of aqueous vanadium species: A perspective for the transition metal chemical speciation studies. *J. Hazard. Mater.* **2018**, *364*, 91–99. [CrossRef]
5. Peng, H. A literature review on leaching and recovery of vanadium. *J. Environ. Chem. Eng.* **2019**, *7*, 103313. [CrossRef]
6. Yan, B.; Wang, D.; Wu, L.; Dong, Y. A novel approach for pre-concentrating vanadium from stone coal ore. *Miner. Eng.* **2018**, *125*, 231–238. [CrossRef]
7. Xue, N.; Zhang, Y.; Huang, J.; Liu, T.; Wang, L. Separation of impurities aluminum and iron during pressure acid leaching of va-nadium from stone coal. *J. Clean. Prod.* **2017**, *166*, 1265–1273. [CrossRef]
8. Hu, X.; Yue, Y.; Peng, X. Release kinetics of vanadium from vanadium titano-magnetite: The effects of pH, dissolved oxygen, temperature and foreign ions. *J. Environ. Sci.* **2018**, *64*, 298–305. [CrossRef]
9. Wen, J.; Jiang, T.; Zhou, W.; Gao, H.; Xue, X. A cleaner and efficient process for extraction of vanadium from high chromium vanadium slag: Leaching in (NH4)2SO4-H2SO4 synergistic system and NH4+ recycle. *Sep. Purif. Technol.* **2019**, *216*, 126–135. [CrossRef]
10. Wen, J.; Jiang, T.; Xu, Y.; Liu, J.; Xue, X. Efficient Separation and Extraction of Vanadium and Chromium in High Chromium Vanadium Slag by Selective Two-Stage Roasting–Leaching. *Metall. Mater. Trans. A* **2018**, *49*, 1471–1481. [CrossRef]
11. Gao, H.; Jiang, T.; Xu, Y.; Wen, J.; Xue, X. Change in phase, microstructure, and physical-chemistry properties of high chromium vanadium slag during microwave calcification-roasting process. *Powder Technol.* **2018**, *340*, 520–527. [CrossRef]
12. Zhang, G.; Zhang, Y.; Bao, S. The Effects of Sodium Ions, Phosphorus, and Silicon on the Eco-Friendly Process of Vanadium Precipitation by Hydrothermal Hydrogen Reduction. *Minerals* **2018**, *8*, 294. [CrossRef]

13. Li, H.-Y.; Fang, H.-X.; Wang, K.; Zhou, W.; Yang, Z.; Yan, X.-M.; Ge, W.-S.; Li, Q.-W.; Xie, B. Asynchronous extraction of vanadium and chromium from vanadium slag by stepwise sodium roasting–water leaching. *Hydrometallurgy* **2015**, *156*, 124–135. [CrossRef]

14. Qiu, S.; Wei, C.; Li, M.; Zhou, X.; Li, C.; Deng, Z. Dissolution kinetics of vanadium trioxide at high pressure in sodium hydroxide–oxygen systems. *Hydrometallurgy* **2011**, *105*, 350–354. [CrossRef]

15. Cai, Z.; Zhang, Y. Phase transformations of vanadium recovery from refractory stone coal by novel NaOH molten roasting and water leaching technology. *RSC Adv.* **2017**, *7*, 36917–36922. [CrossRef]

16. Ji, Y.; Shen, S.; Liu, J.; Xue, Y. Cleaner and effective process for extracting vanadium from vanadium slag by using an innovative three-phase roasting reaction. *J. Clean. Prod.* **2017**, *149*, 1068–1078. [CrossRef]

17. Crundwell, F. The mechanism of dissolution of the feldspars: Part II dissolution at conditions close to equilibrium. *Hydrometallurgy* **2015**, *151*, 163–171. [CrossRef]

18. Crundwell, F. The mechanism of dissolution of minerals in acidic and alkaline solutions: Part IV equilibrium and near-equilibrium behaviour. *Hydrometallurgy* **2015**, *153*, 46–57. [CrossRef]

19. Crundwell, F. The mechanism of dissolution of minerals in acidic and alkaline solutions: Part III. Application to oxide, hydroxide and sulfide minerals. *Hydrometallurgy* **2014**, *149*, 71–81. [CrossRef]

20. Crundwell, F. The mechanism of dissolution of minerals in acidic and alkaline solutions: Part II Application of a new theory to silicates, aluminosilicates and quartz. *Hydrometallurgy* **2014**, *149*, 265–275. [CrossRef]

21. Zhang, Y.; Zhang, T.-A.; Dreisinger, D.; Lv, C.; Lv, G.; Zhang, W. Recovery of vanadium from calcification roasted-acid leaching tailing by enhanced acid leaching. *J. Hazard. Mater.* **2019**, *369*, 632–641. [CrossRef] [PubMed]

22. Fang, D.; Liao, X.; Zhang, X.; Teng, A.; Xue, X. A novel resource utilization of the calcium-based semi-dry flue gas desulfurization ash: As a reductant to remove chromium and vanadium from vanadium industrial wastewater. *J. Hazard. Mater.* **2018**, *342*, 436–445. [CrossRef]

23. Zhang, J.; Zhang, W.; Xue, Z. Oxidation Kinetics of Vanadium Slag Roasting in the Presence of Calcium Oxide. *Miner. Process. Extr. Metall. Rev.* **2017**, *38*, 265–273. [CrossRef]

24. Gao, H.; Jiang, T.; Zhou, M.; Wen, J.; Li, X.; Wang, Y.; Xue, X. Effect of microwave irradiation and conventional calcification roasting with calcium hydroxide on the extraction of vanadium and chromium from high-chromium vanadium slag. *Int. J. Miner. Process.* **2017**. [CrossRef]

25. Wen, J.; Jiang, T.; Zhou, M.; Gao, H.; Liu, J.; Xue, X. Roasting and leaching behaviors of vanadium and chromium in calcification roasting–acid leaching of high-chromium vanadium slag. *Int. J. Miner. Process.* **2018**, *25*, 515–526. [CrossRef]

26. Xiaoyong, Z.; Qingjing, P.; Yuzhu, O.; Renguo, T. Research on the Roasting Process with Calcium Compounds for Silica Based Vanadium Ore. *Chin. J. Process Eng.* **2001**, *2*, 189–192.

27. Yue, H.-R.; Xue, X.-X.; Zhang, W.-J. Reaction Mechanism of Calcium Vanadate Formation in V-slag/CaO Diffusion System. *Metall. Mater. Trans. B* **2021**, *52*, 944–955. [CrossRef]

28. Peng, H.; Guo, J.; Zhang, X. Leaching Kinetics of Vanadium from Calcium-Roasting High-Chromium Vanadium Slag Enhanced by Electric Field. *ACS Omega* **2020**, *5*, 17664–17671. [CrossRef]

29. Peiyang, S.; Bo, Z.; Maofa, J. Kinetics of the Carbonate Leaching for Calcium Metavanadate. *Minerals* **2016**, *6*, 102.

30. Juhua, Z.; Wei, Z.; Li, Z.; Songqing, G. Mechanism of vanadium slag roasting with calcium oxide. *Int. J. Miner. Process.* **2015**, *138*, 20–29.

31. Liu, B.; Du, H.; Wang, S.-N.; Zhang, Y.; Zheng, S.-L.; Li, L.-J.; Chen, D.-H. A novel method to extract vanadium and chromium from vanadium slag using molten NaOH-NaNO3 binary system. *AIChE J.* **2012**, *59*, 541–552. [CrossRef]

32. Liu, H.-B.; DU, H.; Wang, D.-W.; Wang, S.-N.; Zheng, S.-L.; Zhang, Y. Kinetics analysis of decomposition of vanadium slag by KOH sub-molten salt method. *Trans. Nonferrous Met. Soc. China* **2013**, *23*, 1489–1500. [CrossRef]

33. Yang, K.; Zhang, X.; Tian, X.; Yang, Y.; Chen, Y. Leaching of vanadium from chromium residue. *Hydrometallurgy* **2010**, *103*, 7–11. [CrossRef]

34. Peng, H.; Yang, L.; Chen, Y.; Guo, J. A Novel Technology for Recovery and Separation of Vanadium and Chromium from Vana-dium-Chromium Reducing Residue. *Appl. Sci.* **2019**, *10*, 198. [CrossRef]

35. Chai, L.; Yang, J.; Zhang, N.; Wu, P.J.; Li, Q.; Wang, Q.; Liu, H.; Yi, H. Structure and spectroscopic study of aqueous Fe(III)-As(V) complexes using UV-Vis, XAS and DFT-TDDFT. *Chemosphere* **2017**, *182*, 595–604. [CrossRef]

36. Bunnag, N.; Kasri, B.; Setwong, W.; Sirisurawong, E.; Chotsawat, M.; Chirawatkul, P.; Saiyasombat, C. Study of Fe ions in aquamarine and the effect of dichroism as seen using UV–Vis, NIR and X-ray. *Radiat. Phys. Chem.* **2020**, *177*, 109107. [CrossRef]

37. Yu, S.; Wang, B.; Pan, Y.; Chen, Z.; Meng, F.; Duan, S.; Cheng, Z.; Wu, L.; Wang, M.; Ma, W. Cleaner production of spherical nanostructure chromium oxide (Cr2O3) via a facile combination membrane and hydrothermal approach. *J. Clean. Prod.* **2018**, *176*, 636–644. [CrossRef]

38. Nagaraju, G.; Chithaiahb, P.; Ashokac, S.; Mahadevaiah, N. Vanadium pentoxide nanobelts: One pot synthesis and its lithium storage behavior. *Cryst. Res. Technol.* **2012**, *47*, 868–875. [CrossRef]

39. Shin, K.-H.; Jin, C.-S.; So, J.-Y.; Park, S.-K.; Kim, D.-H.; Yeon, S.-H. Real-time monitoring of the state of charge (SOC) in vanadium redox-flow batteries using UV–Vis spectroscopy in operando mode. *J. Energy Storage* **2019**, *27*, 101066. [CrossRef]

40. Peng, H.; Guo, J.; Zheng, X.; Liu, Z.; Tao, C. Leaching kinetics of vanadium from calcification roasting converter vanadium slag in acidic medium. *J. Environ. Chem. Eng.* **2018**, *6*, 5119–5124. [CrossRef]

41. Deng, R.; Xie, Z.; Liu, Z.; Tao, C. Leaching kinetics of vanadium catalyzed by electric field coupling with sodium persulfate. *J. Electroanal. Chem.* **2019**, *854*, 113542. [CrossRef]
42. Peng, H.; Liu, Z.; Tao, C. Leaching Kinetics of Vanadium with Electro-oxidation and H2O2 in Alkaline Medium. *Energy Fuels* **2016**, *30*, 7802–7807. [CrossRef]
43. Peng, H.; Yang, L.; Chen, Y.; Guo, J.; Li, B. Recovery and Separation of Vanadium and Chromium by Two-Step Alkaline Leaching Enhanced with an Electric Field and H2O2. *ACS Omega* **2020**, *5*, 5340–5345. [CrossRef] [PubMed]
44. Wang, X.; Jia, Y.; Ma, S.; Zheng, S.; Sun, Q. Effect of mechanical activation on the leaching kinetics of niobium-bearing mineralisation in KOH hydrothermal system. *Hydrometallurgy* **2018**, *181*, 123–129. [CrossRef]
45. Luo, M.-J.; Liu, C.-L.; Xue, J.; Li, P.; Yu, J.-G. Leaching kinetics and mechanism of alunite from alunite tailings in highly concentrated KOH solution. *Hydrometallurgy* **2017**, *174*, 10–20. [CrossRef]
46. Peng, H.; Guo, J.; Huang, H.; Li, B.; Zhang, X. Novel Technology for Vanadium and Chromium Extraction with KMnO4 in an Alkaline Medium. *ACS Omega* **2021**, *6*, 27478–27484. [CrossRef]

Article

Extraction of Potassium from Feldspar by Roasting with CaCl$_2$ Obtained from the Acidic Leaching of Wollastonite-Calcite Ore

Tülay Türk , Zeynep Üçerler , Fırat Burat, Gülay Bulut and Murat Olgaç Kangal *

Mineral Processing Engineering Department, Faculty of Mines, Istanbul Technical University, Maslak, Istanbul 34469, Turkey; turktu@itu.edu.tr (T.T.); ucerler@itu.edu.tr (Z.Ü.); buratf@itu.edu.tr (F.B.); gbulut@itu.edu.tr (G.B.)
* Correspondence: kangal@itu.edu.tr

Abstract: Potassium, which is included in certain contents in the structure of K-feldspar minerals, has a very important function in the growth of plants. Turkey hosts the largest feldspar reserves in the world and is by far the leader in feldspar mining. The production of potassium salts from local natural sources can provide great contributions both socially and economically in the agriculture industry along with glass production, cleaning materials, paint, bleaching powders, and general laboratory purposes. In this study, potassium extraction from K-feldspar ore with an 8.42% K$_2$O content was studied using chloridizing (CaCl$_2$) roasting followed by water leaching. Initially, to produce wollastonite and calcite concentrates, froth flotation tests were conducted on wollastonite-calcite ore after comminution. Thus, wollastonite and calcite concentrates with purities of 99.4% and 91.96% were successfully produced. Then, a calcite concentrate was combined with hydrochloric acid (HCl) under optimal conditions of a 1 mol/L HCl acid concentration, a 60 °C leaching temperature, and a 10 min leaching time to produce CaCl$_2$. To bring out the importance of roasting before the dissolution process, different parameters such as roasting temperature, duration, and feldspar—CaCl$_2$ ratios were tested. Under optimal conditions (a 900 °C roasting temperature, a 60 min duration, and a 1:1.5 feldspar—CaCl$_2$ ratio), 98.6% of the potassium was successfully extracted by the water leaching process described in this article.

Keywords: potassium; calcium chloride; K-feldspar; wollastonite; roasting–leaching

Citation: Türk, T.; Üçerler, Z.; Burat, F.; Bulut, G.; Kangal, M.O. Extraction of Potassium from Feldspar by Roasting with CaCl$_2$ Obtained from the Acidic Leaching of Wollastonite-Calcite Ore. *Minerals* **2021**, *11*, 1369. https://doi.org/10.3390/min11121369

Academic Editors: Xingjie Wang, Jia Yang and Shuai Wang

Received: 26 October 2021
Accepted: 29 November 2021
Published: 3 December 2021

Publisher's Note: MDPI stays neutral with regard to jurisdictional claims in published maps and institutional affiliations.

1. Introduction

Potassium is an element that occupies an important place in human life and is widely used in different industries as a component of potash. Potassium fertilizer is an essential nutrient, particularly for fruit formation and development. It accelerates plant growth and increases crop yields. In addition to the fertilizer industry, large amounts of potassium salts are used annually to produce glass, cleaning materials, paint, and bleaching powders and for general laboratory purposes [1]. The consumption and cost of potassium salts vary annually and are between 25 and 30 million tons and USD 300 and 500, respectively. In Turkey, potassium salts (potassium chloride (KCl) and potassium sulfate (K$_2$SO$_4$)), which are needed by the chemical and fertilizer industries are provided by imports.

Water-soluble potash is used to produce potassium salts; however, the increase in the need for potassium salts and the decrease in water-soluble potash sources have led the producers to find an alternative source such as K-feldspars. Feldspars constitute more than 60% of the earth's crust and contain microcline, albite, and aluminum silicate with varying concentrations (5–12%) of potassium oxide (K$_2$O) [2–4]. Turkey has approximately 14% of the world's high-quality feldspar reserves. While the most valuable sodium feldspar deposits are located in Western Anatolia (Çine, Milas, Yatağan, and Bozdoğan regions), the majority of K-feldspar deposits are found in the Kırşehir massive [5,6]. In addition to large K-feldspar reserves, there is also a significant amount of wollastonite-calcite deposits in the Kırşehir–Buzlukdağı region. About 4 million tonnes of potash are imported for agricultural and industrial activities each year in Turkey. KCl and $CaCl_2$ salts, which are the main raw materials of the fertilizer industry, can be obtained from feldspar and wollastonite minerals, respectively. Obtaining these high value-added products from different industrial raw materials in the same geographical region significantly increases resource use and efficiency while reducing costs.

Wollastonite and calcite minerals, which are used in ceramics, plastics, metallurgy, paint, and other industries are usually found in the same deposits. The most common method for separating these two industrial minerals is flotation. As wollastonite has a needle-like structure, the flotation process can be performed in the presence of collectors such as amines or fatty acids. In a study by Kangal et al., the best result was obtained at pH 6 and 1500 g/t K-oleate dosage. Under these conditions, a calcite concentrate assaying 55.89% CaO, 0.35% SiO_2, 0.03% Fe_2O_3, and 42.30% loss on ignition (LOI) was produced. At the same time, a wollastonite concentrate containing 52.71% SiO_2, 44.65% CaO, 0.44% Fe_2O_3, and 0.60% LOI was obtained [7]. Several investigations have been carried out to separate wollastonite from calcite-rich wollastonite ore [8–10]. Çinku and co-authors indicated that a marketable wollastonite concentrate could be produced by flotation following magnetic separation from wollastonite ore containing 40% CaO and 48% SiO_2. As a result of the flotation tests at pH 9, the optimal flotation conditions were determined as 1000 g/t of collector dosage, 1.7 kg/t of sodium silicate, and 500 g/t of caustic soda. Using oleic acid, a concentrate containing 54.43% SiO_2, 41.44% CaO, 0.59% Fe_2O_3, and 0.97% loss on ignition was obtained. In addition, when oleic acid is used, a concentrate assaying 57.10% SiO_2, 40.88% CaO, 0.30% Fe_2O_3, and 0.13% loss on ignition was produced [9]. In another important study by Ravi et al., a 97% purity wollastonite concentrate was obtained from low-content wollastonite ore through the flotation and magnetic separation methods. In the flotation experiments, a particle size of 100 μm and a sodium oleate concentration of 1 kg/t were found to be optimal, and a wollastonite concentrate (about 95% purity) with 2.64% iron content and 0.4% ignition loss was produced [9]. In the study by Bulut and co-authors, a wollastonite (about 87.9% purity) concentrate containing 0.44% Fe_2O_3 and 52.71% SiO_2 with 0.60% loss on ignition was obtained using 1500 g/t of potassium oleate. In addition, a calcite concentrate containing 99.8% $CaCO_3$ was recovered as a by-product with 85.4% recovery [10].

$CaCl_2$ is an essential component for obtaining potassium salts from feldspar. Although it is not conventional to treat wollastonite with acid to produce $CaCl_2$, acidic digestion of wollastonite can be used as a step in the carbonation pathway for CO_2 sequestration. Zhang et al. [11] studied the leaching of wollastonite at various hydrochloric acid (HCl) concentrations and temperatures. It was reported that 81.9% and 96.1% of calcium were dissolved using 1 and 4 mol/L HCl, respectively. By maintaining a 4 mol/L HCl concentration, constant leaching tests were performed at different temperatures (40, 60, 80, and 90 °C) and the calcium dissolutions values were achieved as 91%, 96.9%, 97.1%, and 97.5%, respectively.

In the literature, many studies have been carried out to produce KCl from run-of-the-mine feldspar ore using pure $CaCl_2$ salt. Using K-feldspar ore and pure $CaCl_2$, roasting tests were carried out in a high-temperature furnace at various temperatures for 1 h. After the roasting was completed, the leaching process was performed under the constant conditions of a 10% solid-in-pulp ratio, a temperature of 60 °C, and a 500 rpm mixing speed [12]. As a result, a 98.3% potassium extraction was obtained with the dissolution process after roasting at 850 °C. In another study by Kangal et al. [13], a 90% dissolution efficiency was achieved at a 15% solid-in-pulp ratio and at a 40 °C temperature.

Yuan et al. [14] reported a 91% potassium dissolution (under the leaching condition of 70 °C, 30 min, and a solid–liquid ratio of 1:50) from a 50–75 μm potassium feldspar fraction with a 13.25% K_2O content by roasting at 900 °C for 40 min with a 1:1.15 $CaCl_2$–potassium feldspar ratio. Serdengeçti et al. [15] investigated the production of KCl from a potassium feldspar ore containing 9.69% K_2O and succeeded in dissolving 99.8% of potassium through water leaching (a 60 °C temperature and a 120 min leaching time) following the roasting process (1:1.5 potassium feldspar—$CaCl_2$ ratio, 850 °C of temperature, and 60 min of roasting time). Samantray et al. [16] investigated the effects of the $CaCl_2$ amount, roasting temperature, leaching time, and temperature on KCl production from feldspar ore containing 11.64% K_2O. After roasting at 900 °C and with a ratio of 1:1 $CaCl_2$—feldspar, more than 80% of potassium was dissolved after 30 min. Tanvar and Dhawan [17] investigated the effects of additives, roasting temperature, and time on the recovery of potash from feldspar containing 9.67% K_2O. Roasting experiments were carried out at 950 °C for 60 min, and $CaCl_2$ was used as the slag material. At the end of citric acid leaching at room temperature, a 1:10 solid–liquid ratio, and 60 min, a 95% potassium dissolution efficiency was achieved. Samantray et al. [18] investigated the extraction of potassium from feldspar ore containing 11.64% K_2O using eggshell powder as a calcium source. Feldspar and eggshells were ground together below 45 μm and roasted at 900 °C for 30 min with a 1:1.8 eggshell—feldspar ratio. Then, this mixture was leached with HCl for 30 min at room temperature and a 99% potassium recovery was obtained.

Today, the main approach of beneficiation and production is based on a circular economy and environment. In this study, potassium recovery from K-feldspar ore was achieved using $CaCl_2$ produced from wollastonite-calcite ore in the same region. In the first part of this study, wollastonite-calcite ore was subjected to flotation tests to produce wollastonite and calcite concentrates. High-grade $CaCl_2$ was produced through a HCl acid treatment using a calcite concentrate obtained as a result of flotation. Second, roasting followed by the water leaching process was applied to K-feldspar with $CaCl_2$. With this original and innovative work, the extraction of potassium using wollastonite—calcite and K-feldspar ores located in a similar mineralization zone will be possible, and these local natural resources can be brought into the economy.

2. Materials and Methods

2.1. Material and Characterization

The wollastonite-calcite and potassium feldspar ores were obtained from the Buzlukdagi Region, Kırşehir, Turkey. The representative ore samples were crushed below 2 mm using jaw, cone, and roller crushers, and wet sieving was performed to determine the particle size distribution. According to the results, the d_{80} size of wollastonite and feldspar samples were found to be 1.2 and 0.5 mm, respectively. The BSE (Back-Scattered Electron) images of samples are given in Figure 1. The chemical contents of the samples (Table 1) were determined by inductively coupled plasma (ICP) at Activation Laboratories Ltd., Hamilton, ON, Canada. To determine the loss on ignition of the samples, the samples were placed in crucibles and weighed. After, they were kept in an oven at 100 °C overnight to remove moisture. Then, the temperature was raised to 550 °C, and the organic materials were completely removed after 4 h. The carbonate structure turned into oxides at 1000 °C after 2 h, and the ratio of the initial and final weight difference to the feed represented the ignition loss.

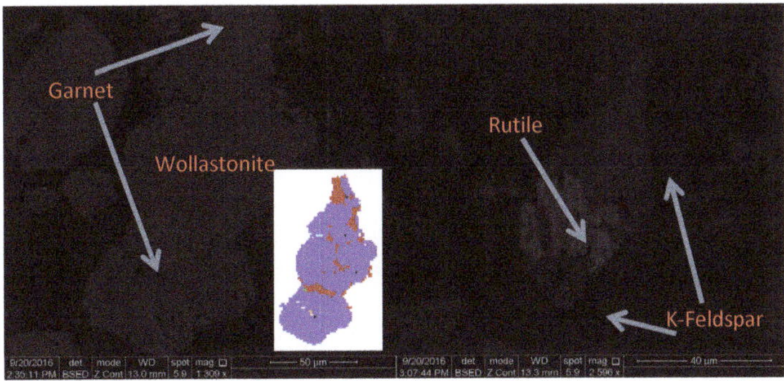

Figure 1. BSE images of wollastonite-calcite and feldspar samples.

Table 1. Chemical analyses of the representative samples.

Compound	Content, %	
	Wollastonite-Calcite	K-Feldspar
SiO_2	28.00	62.30
Al_2O_3	1.91	18.70
Fe_2O_3	0.45	1.89
TiO_2	0.04	0.20
Na_2O	0.47	4.32
K_2O	0.40	8.40
CaO	48.20	2.16
MgO	1.15	0.26
P_2O_5	0.03	0.09
SrO	0.04	0.14
MnO	0.03	0.06
LOI	20.80	1.83

LOI: Loss on ignition.

An X-ray diffraction (XRD) analysis was conducted with a copper X-ray-sourced Panalytical X'Pert Pro diffractometer (Malvern Panalytical Ltd., Malvern, UK). PDF-4/Minerals International Centre of Diffraction Data (ICDD) software (PDF-4/Minerals, ICDD, Delaware County, PA, USA) was used for mineral characterization. The crystals' mineral phase ratio was determined through the Rietveld method. XRD patterns of wollastonite-calcite and feldspar samples are shown in Figures 2 and 3, respectively. According to the results, the secondary mineral was calcite in the wollastonite-calcite sample and the feldspar sample mostly consisted of potassium feldspar.

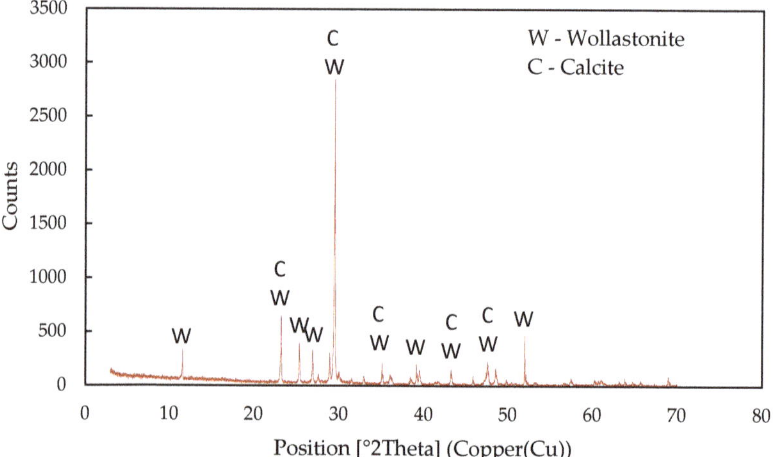

Figure 2. X-ray diffraction patterns of the wollastonite-calcite sample.

The chemical and mineralogical properties of samples were defined with a differential thermal analysis (DTA) and the thermogravimetric analysis (TGA) equipment's STA 449 F3 Jupiter® thermal analyzer (NETZSCH, Selb, Germany). DTA and TGA curves of wollastonite-calcite, calcite concentrate (from flotation), and feldspar samples are illustrated in Figures 4–6, respectively.

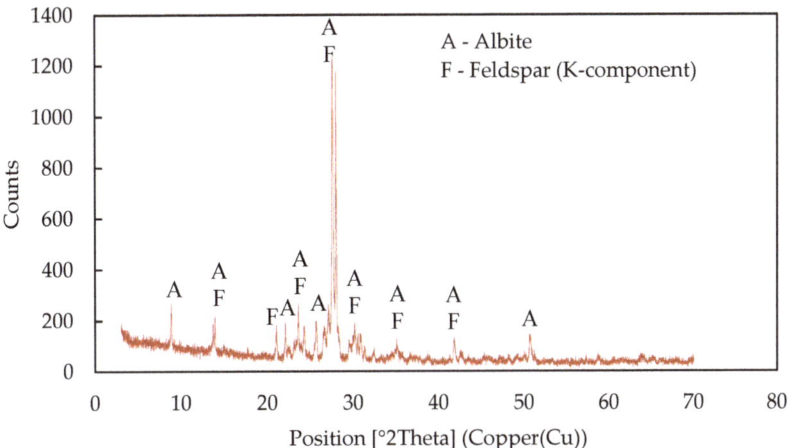

Figure 3. X-ray diffraction patterns of the potassium feldspar sample.

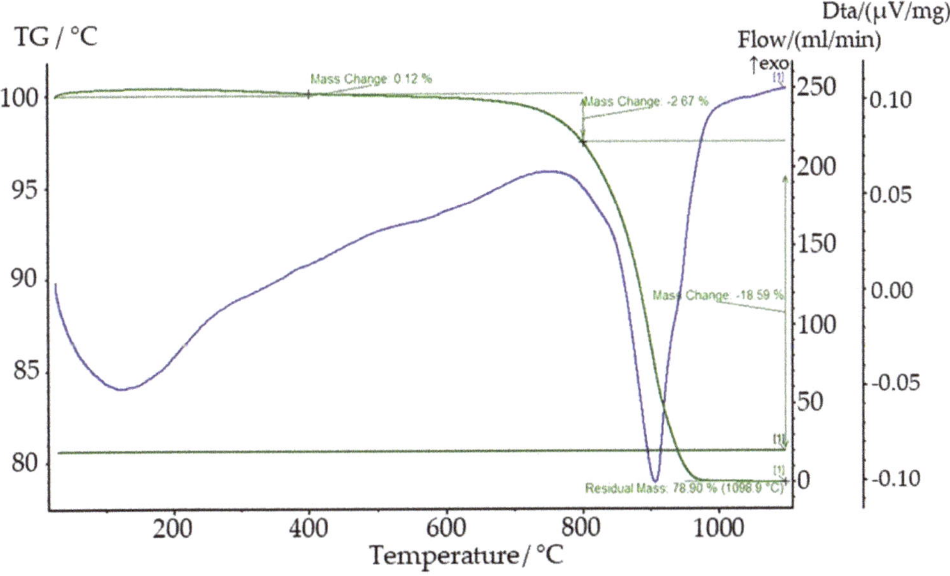

Figure 4. DTA-TGA curves of the wollastonite-calcite sample.

As can be seen in Figure 4, a loss of 21% of weight occurs in the wollastonite ore as the temperature rises from 700 °C to 950 °C. The main reason for this loss is calcite, which constitutes more than 50% of the structure of the ore. The DTA and TGA results shown in Figure 5 indicate that a significant amount of structural deterioration occurred at 900 °C. Only 0.6% of weight loss was determined in the feldspar ore sample between 400 and 800 °C (Figure 6). Partial structural deterioration was encountered at 700–750 °C. This small loss might be explained by probable carbonate components in the ore.

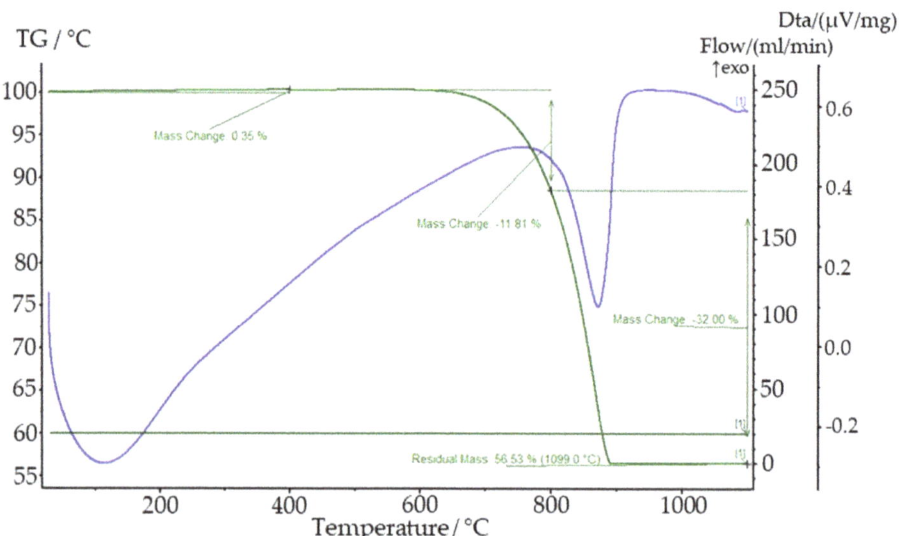

Figure 5. DTA–TGA curves of the calcite concentrate.

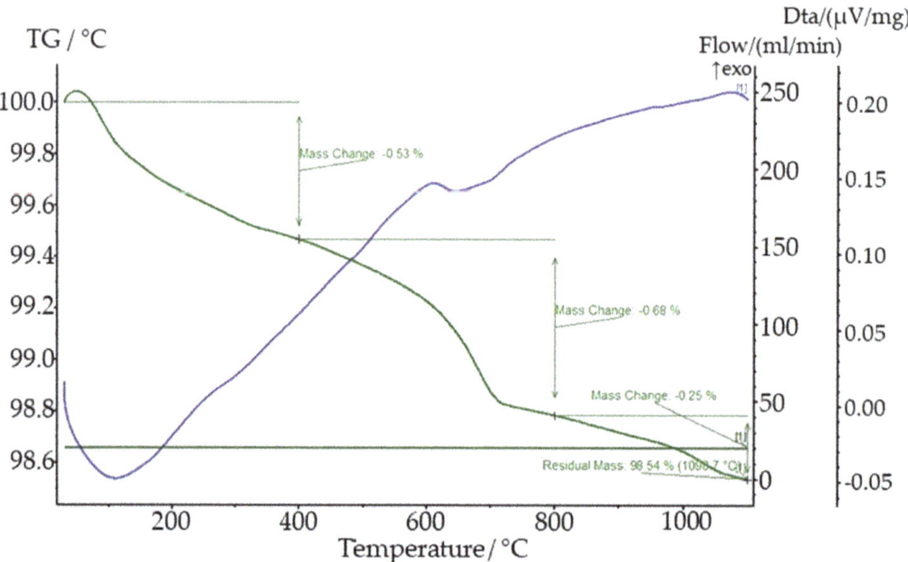

Figure 6. DTA–TGA curves of the potassium feldspar sample.

2.2. Methods

Calcite must be treated with hydrochloric acid to produce $CaCl_2$. Calcite reacts with HCl to form $CaCl_2$ and releases CO_2 and H_2O as by-products (Equation (1)) [19].

$$CaCO_3 + 2HCl \rightarrow CaCl_2 + CO_2 + H_2O \tag{1}$$

The potassium dissolution process can be finished by mixing the $CaCl_2$ obtained in the first equation with K-feldspar ore in certain proportions and by subjecting them to roasting and leaching processes. The mixture of feldspar and $CaCl_2$, which is exposed to a temperature higher than the decomposition temperature during the roasting process, reacts to form KCl salt. On the other hand, quartz and anorthite minerals are produced as by-products (Equation (2)) [14].

Since the KCl salt obtained as a result of this process is easily soluble, it can be taken into a solution by leaching with water.

$$CaCl_2 + 2KAlSi_3O_8 \rightarrow CaAl_2Si_2O_8 + 4SiO_2 + 2KCl \qquad (2)$$

In the first step, wollastonite-calcite ore was subjected to comminution (below 75 μm) and then froth flotation was used to separate calcite and wollastonite. An amount of 300 gr of wollastonite-calcite ore was subjected to self-aerated Denver flotation equipment with a 1.5 L cell volume and a 1200 rpm mixing rate. The conditions of the flotation tests are given in Table 2. While calcite particles were floated with the help of a collector, wollastonite particles were not activated and taken as the sinking product. Afterward, the calcite concentrate was subjected to HCl acid leaching (Figure 7). Different HCl concentrations (0.25, 0.5, 2, 3, 4, and 6 mol/L), temperatures (25, 40, 60, and 80 °C), and leaching times (5, 10, 15, 30, 60, 90, and 120 min) were investigated to determine the optimal treatment conditions with a constant 1:2 solid–liquid ratio. After obtaining the pregnant solution, $CaCl_2$ was precipitated by evaporation. Second, the $CaCl_2$ sample obtained from the calcite concentrate was ground and mixed in an agate mortar with potassium feldspar in a 1:1.5 feldspar–$CaCl_2$ ratio until it had a homogeneous appearance. Afterward, 7.5 g of the mixture was spread in a thin layer on four porcelain crucibles (diameter: 8 cm) and the roasting tests were performed in a Protherm brand PLF 130/6 furnace at elevated temperatures of 800, 850, 900, and 950 °C for 1 h. After roasting, water leaching was performed for 2 h under constant conditions, namely a 10% solid-in-pulp ratio, a 60 °C temperature, and a 500 rpm mixing speed.

Table 2. The conditions of the flotation experiment.

Particle Size	-74 μm
pH	6.0
KOl Dosage	300 + 300 + 300 + 300 + 300 g/t
Conditioning Time	10 + 5 + 5 + 5 + 5 min
Flotation Time	2 + 2 + 1 + 1 + 1 min

After acid leaching, titration was used to determine HCl consumption. A 0.01 mol/L NaOH solution was prepared and transferred into a 10 mL burette. A beaker containing 30 mL of the test solution was placed under the burette. Four drops of phenolphthalein (a color-change indicator) were added to the solution, and the burette faucet was opened enough to allow a single droplet to pass. A 0.01 mol/L NaOH solution was mixed into the solution until it turned pink, at which point the faucet was closed, and the spent NaOH was detected. Lab-scale vacuum equipment with fine filter paper (Sartorius brand 391 code, blue label) was used for liquid–solid separation and the filtered cake was dried at 70 °C for approximately 24 h. The elemental analysis of the liquid was provided by the atomic absorption spectrometer (Varian brand AA240FS, Palo Alto, CA, USA).

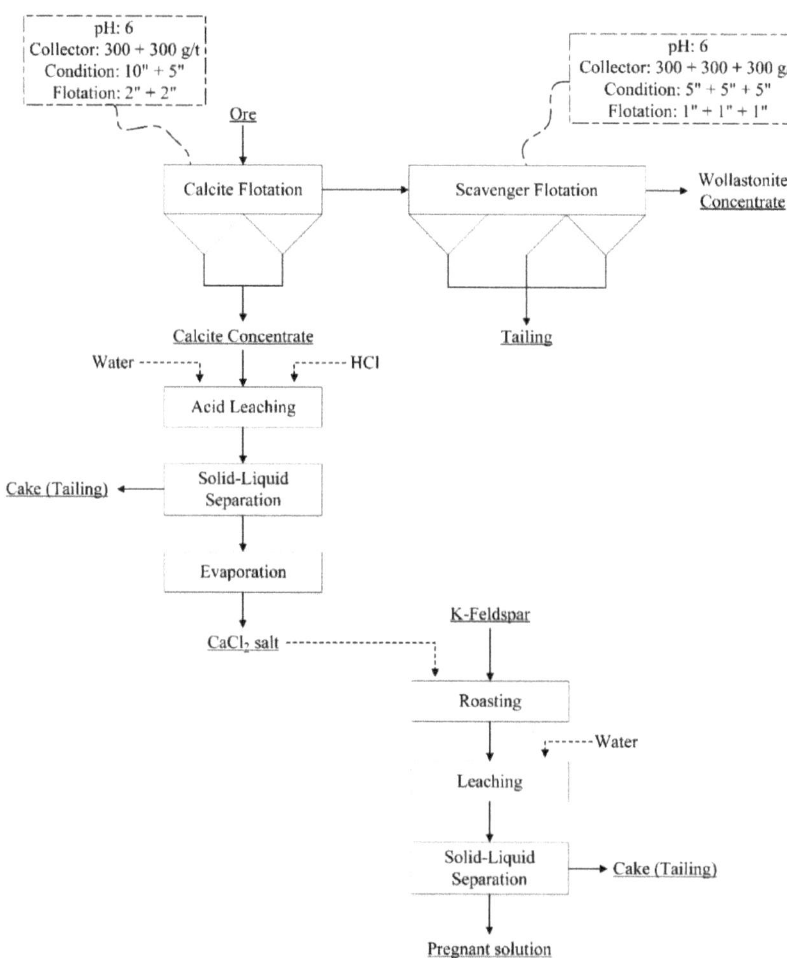

Figure 7. General flowchart of the chemical processing following flotation.

3. Results and Discussion

3.1. Flotation of Wollastonite-Calcite Ore

To produce $CaCl_2$ for the roasting and leaching experiments, previously, wollastonite-calcite ore was subjected to flotation experiments after comminution, and a calcite concentrate with low iron content was obtained in the first two flotation stages. In the following steps, the augite mineral was activated and started to float. On the other hand, wollastonite was not activated and remained in the cell despite the increasing potassium oleate concentration. As seen in Table 3, Float-1 and Float-2 products contained high grades of $CaCO_3$, 91.55% and 92.77%, respectively. Float-3, Float-4, and Float-5 were characterized as tailings, since their Fe_2O_3 contents were high. A marketable wollastonite concentrate having a 99.4% purity was produced with 0.44% Fe_2O_3, 0.95% $CaCO_3$, and 49.70% SiO_2 contents using a 1500 g/t potassium oleate. Calcite concentrates obtained in the first two flotation stages were combined and a final calcite concentrate (91.96% $CaCO_3$) with 0.11% Fe_2O_3 and 2.14% SiO_2 was produced.

Table 3. Results of flotation tests.

Products	Weight (%)	Fe$_2$O$_3$		SiO$_2$		Loss on Ignition		Calcite		Wollastonite	
		C	D	C	D	C	D	P	D	P	D
Float-1 *	31.1	0.09	6.2	1.55	1.8	40.28	59.3	91.55	59.3	3.10	1.8
Float-2 *	15.8	0.15	5.3	3.30	2.0	40.82	30.6	92.77	30.6	6.60	2.0
Float-3	2.9	1.31	8.5	24.20	2.7	36.47	5.1	82.89	5.1	48.40	2.7
Float-4	3.3	2.17	16.0	40.80	5.2	17.52	2.8	39.82	2.8	81.60	5.2
Float-5	3.1	3.09	21.4	45.60	5.4	8.83	1.3	20.07	1.3	91.20	5.4
Sink	43.8	0.44	42.6	49.70	82.9	0.42	0.9	0.95	0.9	99.40	82.9
Total	100.0	0.45	100.0	26.24	100.0	21.08	100.0	47.90	100.0	52.47	100.0

C: Content; D: Distribution; P: Purity. * Combined flotation products.

3.2. CaCl$_2$ Production by Acid Leaching

The final calcite concentrate obtained after the flotation process was treated with HCl to produce CaCl$_2$. The effects of the concentration, temperature, and time on acid leaching were investigated. After the determination of the most suitable condition, CaCl$_2$ was mixed with potassium feldspar in certain proportions, the roasting and dissolution processes were completed, and finally, potassium was successfully extracted.

3.2.1. Effect of HCl Concentration

Based on the formula in Equation (1), the acid treatment was applied to the calcite concentrate. HCl was used as a solvent to investigate the effect of various acid concentrations (0.25, 0.5, 1, 2, 3, 4, and 6 mol/L) on CaCl$_2$ production. The leaching temperature and time were chosen as 60 °C and 10 min, respectively. The HCl-Calcite ratio was adjusted to 2:1 by weight. Calcium dissolution efficiencies versus acid concentration are illustrated in Figure 8.

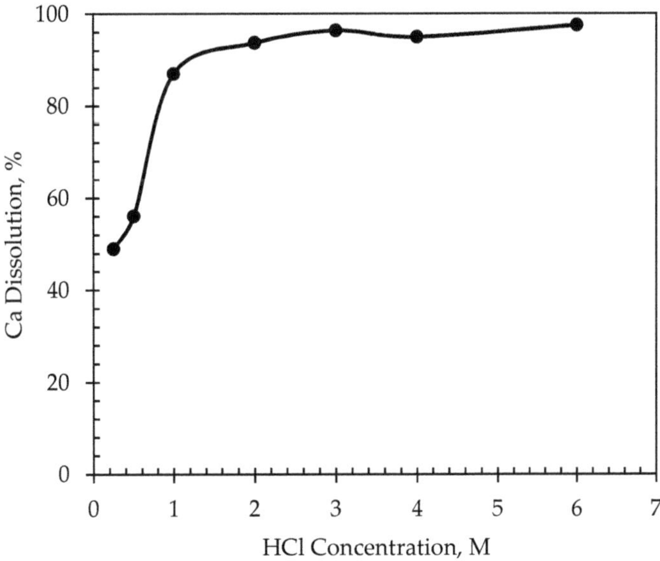

Figure 8. Calcium dissolution curve as a function of acid concentration.

Figure 8 clearly shows that 2 mol/L HCl provides an optimal calcium extraction rate of 93.7%. The 2 mol/L HCl concentration was kept constant in the following experiments, as there was no more than a 3% increase in dissolution efficiency at higher acid concentrations. Similar results were obtained in the study of Zhang et al. [11], who performed the dissolution experiments to produce $CaCl_2$. The highest calcium dissolution rate (96.1%) was obtained at a 4 mol/L HCl concentration, while 1 mol/L HCl led to only 81.9% calcium dissolution. To determine the amount of HCl consumption for our case, the titration method was chosen and the acid consumption was found to be 165 kg/ton for this study.

3.2.2. Effect of Temperature

After a suitable acid concentration was obtained, the effect of temperature on calcium dissolution efficiency was investigated at different temperatures (25, 40, 60, and 80 °C). The HCl–Calcite ratio was kept constant at 2:1, and tests were performed with a 10 min leaching time, a mixing speed of 350 rpm, and 2 different HCl concentrations (1 and 2 mol/L). The calcium dissolution efficiencies are presented in Figure 9.

The results in Figure 9 indicated that the leaching temperature had a considerable effect on the dissolution of calcium. Especially, no significant change was observed after the 2 mol/L HCl concentration at 40 °C. On the other hand, calcium dissolution yields increased continuously with increasing temperature while the dissolution of calcium started to stabilized after the 1 mol/L HCl concentration at 60 °C. Considering the gelation problem caused by the use of a vacuum after the experiment using a 2 mol/L HCl concentration, a 1 mol/L HCl concentration, and a 60 °C temperature were found to be more suitable for the dissolution process. Zhang et al. [11] indicated the importance of temperature on leaching in their study. After 20 min of dissolution at a 4 mol/L HCl concentration, 91.0%, 96.9%, 97.1%, and 97.5% calcium dissolution efficiencies were achieved at 40, 60, 80, and 90 °C, respectively.

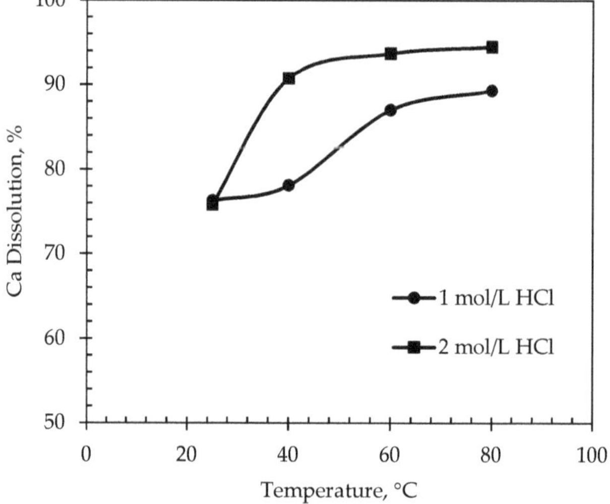

Figure 9. Calcium dissolution efficiency curve in different hydrochloric acid concentrations depending on temperature.

3.2.3. Effect of Leaching Time

The effect of leaching time on the calcium dissolution was investigated using various durations (5, 10, 15, 30, 60, 90, and 120 min). The HCl-Calcite ratio was kept constant at 2:1, and tests were performed at a 1 mol/L HCl concentration, a 60 °C temperature, and a mixing speed of 350 rpm. The results are shown in Figure 10. As seen in Figure 10, although there is not much change in recovery, the recovery decreases as the leaching time increases. A 5 min leaching time is not considered sufficient for the system to balance; therefore, a 10 min leaching time was preferred.

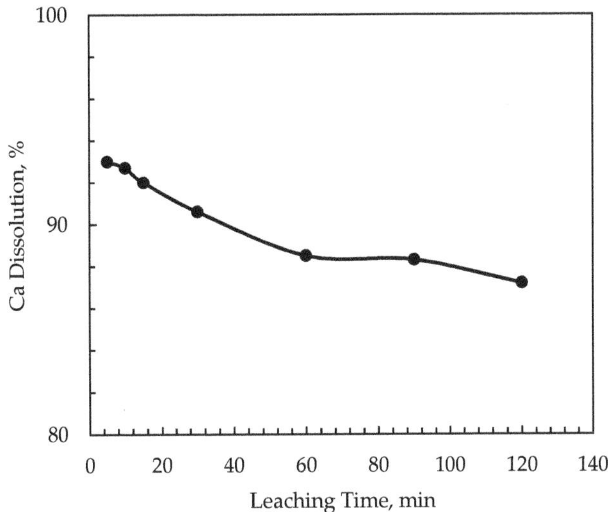

Figure 10. Calcium dissolution curve at different leaching times.

3.3. The Roasting Followed by Leaching

Roasting followed by water leaching was investigated by adopting Equation (2). $CaCl_2$ produced in the previous process was treated with K-feldspar. The aim was to replace calcium ions with potassium at temperatures above the melting point of the feldspar sample. The feldspar—$CaCl_2$ ratio was kept constant at 1:1.5, and the mixture was roasted at different roasting temperatures (800, 850, 900, and 950 °C) for an hour. Water was used as a solvent to dissolve the potassium from the roasted sample. Under constant conditions, a pregnant solution containing potassium ions was obtained. Table 4 shows the results of the roasting process followed by leaching.

Table 4. Potassium dissolution recoveries after roasting with calcium chloride.

Roasting Temperature, °C	Potassium Recovery, %
800	97.1
850	97.1
900	98.6
950	97.3

According to the results, the highest potassium extraction was achieved at a 900 °C roasting time. Since $CaCl_2$ is an expensive additive, producing $CaCl_2$ from local reserves can be helpful in improving both the usage of natural resources and in producing high-value-added products. In light of the data obtained in this study, it is possible to design a process that allows for the recovery of potassium from wollastonite-calcite and K-feldspar ores in the same region.

4. Conclusions

As it is known, mineral processing is very significant in terms of both technological and economic evaluation of raw materials before chemical beneficiation. In this study, using K-feldspar and $CaCl_2$ roasting followed by leaching tests were performed to extract potassium from the natural resources that were located in the same district. For that purpose, the comminuted wollastonite-calcite ore was subjected to flotation tests, and a marketable wollastonite concentrate and a high purity $CaCO_3$ were produced. Calcite concentrate from the flotation test was treated with HCl to produce $CaCl_2$. Under the optimal conditions of a 1 mol/L HCl acid concentration, a 60 °C leaching temperature, and a 10 min leaching time, 92.7% of calcium was extracted. The $CaCl_2$ produced by the evaporation method was mixed with K-feldspar in certain proportions and then subjected to roasting experiments followed by dissolving experiments. The optimal feldspar–$CaCl_2$ ratio was found to be 1:1.5, and 98.6% of the potassium was successfully extracted from the potassium feldspar ore. It is inevitable for economic development to produce KCl, which is very important for the fertilizer industry, from wollastonite-calcite and K-feldspar ores in the same region. Obtaining high-value-added products from different industrial raw materials increases the added value of the raw material considerably and will enable Turkey to use its natural resources more efficiently.

Author Contributions: T.T., Z.Ü., F.B., G.B. and M.O.K. conceived and designed the experiments; T.T., Z.Ü. and M.O.K. performed the experiments and analyzed the data; T.T., Z.Ü., F.B., G.B. and M.O.K. contributed to writing the paper. All authors have read and agreed to the published version of the manuscript.

Funding: This research was funded by Istanbul Technical University Circulating Capital Enterprise R&D, Project ID: 42017.

Data Availability Statement: No applicable.

Acknowledgments: The authors express their sincere thanks and appreciation to Istanbul Technical University Circulating Capital Enterprise R&D and BS Invest Co. for financial support and for kindly providing the feldspar samples.

Conflicts of Interest: The authors declare no conflict of interest.

References

1. Scherer, H.W.; Mengel, K.; Dittmar, H.; Drach, M.; Vosskamp, R.; Trenkel, M.E.; Gutser, R.; Steffens, G.; Czikkely, V.; Niedermaier, T.; et al. Fertilizers. In *Ullmann's Encyclopedia of Industrial Chemistry*; Wiley-VCH Verlag GmbH & Co. KGaA: Weinheim, Germany, 2002.
2. Burat, F.; Kökkiliç., O.; Kangal, O.; Gürkan, V.; Çelik, M.S. Quartz-feldspar separation for glass and ceramic industry. *Miner. Metall. Process.* **2007**, *24*, 75–80. [CrossRef]
3. Kalyoncu, E.; Burat, F. Selective separation of coloring impurities from feldspar ore by innovative single-stage flotation. *Miner. Process. Extr. Metall. Rev.* **2021**, *42*, 1–6. [CrossRef]
4. Rao, J.R.; Nayak, R.; Suryanarayana, A. Feldspar for potassium, fertilisers, catalysts and cement. *Asian J. Chem.* **1998**, *10*, 690–706.
5. Kangal, M.O.; Bulut, G.; Yeşilyurt, Z.; Güven, O.; Burat, F. An alternative source for ceramics and glass augen-gneiss. *Minerals* **2017**, *7*, 70. [CrossRef]
6. Kangal, M.O.; Bulut, G.; Yeşilyurt, Z.; Baştürkcü, H.; Burat, F. Characterization and production of Turkish nepheline syenites for industrial applications. *Physicochem. Probl. Miner. Process.* **2019**, *55*, 605–616.
7. Kangal, M.O.; Bulut, G.; Guven, O. Physicochemical characterization of natural wollastonite and calcite. *Minerals* **2020**, *10*, 228. [CrossRef]
8. Çinku, K.; Ipekoğlu, B.; Tulgarlar, A. Enrichment of Karaköy wollastonite ore. *Ind. Miner. Symp.* **1997**, 86–91.
9. Ravi, B.P.; Krishna, S.J.G.; Patil, M.R.; Kumar, P.S.; Rudrappa, C.; Naganoor, P.C. Beneficiation of a wollastonite from Sirohi, Rajasthan. *Int. J. Innov. Sci. Eng. Technol.* **2014**, *1*, 450–453.
10. Bulut, G.; Akçin, E.S.; Kangal, M.O. Value added industrial minerals production from calcite-rich wollastonite. *Gospod. Surowcami Miner.* **2019**, *35*, 43–58.
11. Zhang, J.; Zhang, R.; Geerlings, H.; Bi, J. A novel indirect wollastonite carbonation route for CO_2 sequestration. *Chem. Eng. Technol.* **2010**, *33*, 1177–1183. [CrossRef]
12. Kangal, M.O.; Bulut, G.; Burat, F.; Üçerler, Z.; Türk, T. *Developing a Process for Obtaining High Value Added KCl with Calcium Chloride to be Produced from Kırşehir High Calcite Content Wollastonite*; R&D Project, Final Report; ITU, Mineral Processing Engineering Department: İstanbul, Turkey, 2019; 39p. (In Turkish)
13. Kangal, M.O.; Burat, F.; Güney, A.; Türk, T.; Şavran, C. *Development of Processes for Obtaining High Value Added Potassium Salts from Turkish Potassium Feldspar Ore*; Final Report; The Scientific and Technological Research Council of Turkey (Project No: 218M107): İstanbul, Turkey, 2019; 53p. (In Turkish)
14. Yuan, B.; Li, C.; Liang, B.; Lü, L.; Yue, H.; Sheng, H.; Xie, H. Extraction of potassium from K-feldspar via the $CaCl_2$ calcination route. *Chin. J. Chem. Eng.* **2015**, *23*, 1557–1564. [CrossRef]
15. Serdengeçti, M.T.; Baştürkcü, H.; Burat, F.; Kangal, M.O. The correlation of roasting conditions in selective potassium extraction from K-feldspar ore. *Minerals* **2019**, *9*, 109. [CrossRef]
16. Samantray, J.; Anand, A.; Dash, B.; Ghosh, M.K.; Behera, A.K. Sustainable process for the extraction of potassium from feldspar using eggshell powder. *ACS Omega* **2020**, *5*, 14990–14998. [CrossRef] [PubMed]

17. Tanvar, H.; Dhawan, N. Recovery of potash values from feldspar. *Sep. Sci. Technol.* **2020**, *55*, 1398–1406. [CrossRef]
18. Samantray, J.; Anand, A.; Dash, B.; Ghosh, M.K.; Behera, A.K. Production of potassium chloride from K-feldspar through roast–leach–solvent extraction route. *Trans. Indian Inst. Met.* **2019**, *72*, 2613–2622. [CrossRef]
19. Lund, K.; Fogler, H.S.; McCune, C.C.; Ault, J.W. Acidization-II. The dissolution of calcite in hydrochloric acid. *Chem. Eng. Sci.* **1975**, *30*, 825–835. [CrossRef]

Article

Hydrometallurgical Treatment of Converter Dust from Secondary Copper Production: A Study of the Lead Cementation from Acetate Solution

Martina Laubertová [1,*], Alexandra Kollová [1], Jarmila Trpčevská [1], Beatrice Plešingerová [1] and Jaroslav Briančin [2]

1   Faculty of Materials, Metallurgy and Recycling, Technical University of Košice, 042 00 Košice, Slovakia; alexandra.kollova@tuke.sk (A.K.); jarmila.trpcevska@tuke.sk (J.T.); beatrice.plesingerova@tuke.sk (B.P.)
2   Institute of Geotechnics SAS, Slovak Academy of Sciences, Watsonova 45, 040 01 Košice, Slovakia; briancin@saske.sk
*   Correspondence: martina.laubertova@tuke.sk

Abstract: The subject of interest in this study was lead cementation with zinc from solution after conventional agitate acidic leaching of converter dust from secondary copper production. The kinetics of lead cementation from an acid solution of lead acetate using zinc powder was studied. The optimal cementation conditions for removing lead from the solution were determined to have a stirring intensity of 300 rpm, a zinc particle size distribution <0.125–0.4> mm and an ambient temperature. Under these conditions, an almost 90% efficiency in removing lead from solution was achieved. The cementation precipitate contains Pb, and a certain amount of Cu. Lead is present in the cementation precipitate in the PbO, Pb_5O_8 and $Pb(Cu_2O_2)$ phases. The solution after cementation was also refined from copper. The solution can be used for further processing in order to obtain a marketable Zn-based product. The resulting cementation precipitate can be further processed and modified to obtain a lead-based product. A kinetic study of the process of lead cementation from solution was also carried out. Based on experimental measurements, the value of apparent activation energy (Ea) which was found to be ~18.66 kJ·mol^{-1}, indicates that this process is diffusion controlled in the temperature range 293–333 K.

Keywords: hydrometallurgy; converter dust; secondary copper; cementation; lead

Citation: Laubertová, M.; Kollová, A.; Trpčevská, J.; Plešingerová, B.; Briančin, J. Hydrometallurgical Treatment of Converter Dust from Secondary Copper Production: A Study of the Lead Cementation from Acetate Solution. *Minerals* **2021**, *11*, 1326. https://doi.org/10.3390/min11121326

Academic Editors: Shuai Wang, Xingjie Wang and Jia Yang

Received: 31 October 2021
Accepted: 23 November 2021
Published: 27 November 2021

Publisher's Note: MDPI stays neutral with regard to jurisdictional claims in published maps and institutional affiliations.

1. Introduction

After iron and aluminium, copper is the third most-used industrial metal [1]. Due to its excellent properties such as ductility, malleability, conductivity of electricity or heat and corrosion resistance, copper is currently irreplaceable in the field of electronics. Moreover, in recent years there has been worldwide growth in sales of electric cars, which entails increased consumption of copper in this branch as well. In 2020 the production of refined copper reached 20.1 million tons [2–4].

Primary copper derived from mining activity is produced mainly (around 84% of total copper production) via the pyrometallurgical route. Alternatively, via the hydrometallurgical route copper is extracted through leaching (solvent extraction) and electrowinning (SX-EW process).

Two of the largest copper producers in 2020 were: Corporación Nacional del Cobre (Codelco), Chile (1,727,000 t Cu) and Glencore, Switzerland (1,258,000 t Cu) [5]. The sole producer of technical grade copper in Slovakia is Kovohuty in Krompachy [6].

Primary copper production consumes much more energy and generates a lot more emissions than the secondary copper recovery processes. Secondary raw materials used for producing copper contain heavy metals such as zinc, tin and lead, which are released from melts in smelting furnaces and converters in the form of dust. The waste from secondary copper production consists of various types of dust: shaft furnace dust, converter dust and anode furnace dust, which are extracted by ventilators and captured in filtering units before they can escape up a chimney. Direct return of these dusts into the copper production process is precluded by their high content of metals such as Zn, Pb, Cd, Sb and Sn. These metals would cause considerable technical problems if the dusts were used in direct processing. They intrude into the production cycle in any case, mostly through the processing of copper scrap from e-waste (discarded electrical and electronic devices) or copper alloys (brasses and bronzes) [7–9]. In 2019 the amount of dust generated in secondary copper production in Slovakia totalled 2370.27 t [10]. Copper smelting flue dusts are defined as "hazardous materials" according to the current European Waste Catalogue and Hazardous Waste List (10 06 03 flue-gas dust & 10 06 06 solid wastes from gas treatment). This classification of the dusts is mainly due to their high content of compounds with the presence of arsenic and lead [11–13].

The potential for dust recycling lies in its hydrometallurgical processing with the aim of recovering commercial products based on Zn, Pb and Sn [14,15]. From the physical point of view, converter dust from secondary copper production consists of a grey-coloured powder particles with principal components of ZnO, Pb_2OCl_2, PbO and some other metal oxides. The lead contained in converter dust from secondary copper production needs to be removed before any further processing of the dust [16].

The authors of individual articles cited here have dealt with hydrometallurgical processing of various types of metal-bearing waste containing lead, for instance dusts [15,17–19], lead paste from batteries [20,21], waste from zinc production [22–24] and residual lead in slags [25,26]. In some cases, mechanical preparation of material was required prior to hydrometallurgical processing. Authors used in mechanical treatment of sample the Falcon gravity concentrator and Knelson gravity concentrator [17]. Mechanical pretreatment was also used on the slag where was ground to less than 1 mm and the samples of the copper smelting dust were dried, ground and sieved, which ensured them to have a size of grains less than 200-mesh [27,28]. The samples were air-dried for 30 days, lightly pulverized with agate mortar and pestle, and dry-sieved to obtain sample with particles passing 106 µm fraction and also the sample was dried and pulverized to—100 mesh (150 µm) [24,29]. The basis for each hydrometallurgical process is leaching. It has been established that the degree of porosity in solid particles influences the rate of diffusion, which in turn affects the efficiency of leaching of prepared material [17]. Hydrometallurgical processing of the types of waste in the above-mentioned studies involved the use either of alkali leaching or acid leaching. The leaching of copper dust was provided in solution of 1–4 M NaOH at temperature of 20–80 °C and the solid to liquid ratio 20. The results obtained were confirmed to exclude Pb in the solution; with the highest extraction 60% Pb at 2 M NaOH after 15 min at 80 °C. Zn extraction 58% was achieved at 4 M NaOH in 15 min at 80 °C [19]. Dust was leached with an alkaline solution ((NH_4)$_2CO_3$ and NH_4OH) to solubilize ZnO and Zn metal where copper, lead, tin and most of the zinc were dissolved in the leach step. Copper, lead and tin were recovered in the cement product [30]. Lead was recovered by cementation from industrial lead sludge solutions of urea acetate using different types of metallic iron [21] and the influence of temperature, sulphuric acid concentration and weight of sample were investigated to recovery zinc and copper from shaft furnace dust [31].

Leaching of lead and zinc from copper shaft furnace dust in NaOH solution is described by following Equations (1)–(3) [19].

$$PbO + NaOH = Na^+ + HPbO_2^- \tag{1}$$

$$PbSO_4 + 2Na^+ + 3OH^- = HPbO_2^- + Na_2SO_4 + H_2O \tag{2}$$

$$ZnO + 2NaOH = ZnO_2^{2-} + 2Na^+ + H_2O \tag{3}$$

Following Equations (4) and (5) occur simultaneously and describes leaching reaction of oxidised copper and lead from slag produced from flash smelting copper concentrates in citric acid solution [27].

$$CuO + C_6H_8O_7 = Cu^{2+} + C_6H_6O_7^{2-} + H_2O \tag{4}$$

$$PbO + C_6H_8O_7 = Cu^{2+} + C_6H_6O_7^{2-} + H_2O \tag{5}$$

Leaching efficiency may be increased through the application of intensification methods during the leaching process, such as using oxidizing agents (hydrogen peroxide H_2O_2 [28], ozone O_3 [32]) or microwave radiation [33,34]. The highest efficiency achieved in leaching lead out of dusts in a citric acid medium was 84.67% [27] and in alkali leaching the highest lead extraction (92.84%) was reached after 30 min using microwave energy for leaching EAF dust in NaOH solution [34]. Leaching is followed by extraction processes involving various methods including electrolysis, cementation, removal of compounds with low solubility (precipitation with chemical agents), salts crystallization, ion exchange, liquids extraction or adsorption [7]. One of the extraction methods with very high efficiency of recovering lead from solution is cementation. Cementation is the process of extracting the metals from solution based on the electrochemical reaction between the cementing metal (zinc, iron) and the ion of the precipitated metal (lead). The electrode potential of the displacing metal must be more negative than that of the displaced one. Cementation is a simple and easy method that has been used for centuries in hydrometallurgy. The method is used for purification of leaching solutions and for recovery of valuable metals as well [20–22,24,25,35–40]. Various factors have been observed influencing the process and efficiency of cementation. In order to achieve high levels of recovery it is necessary to use from two to four times the stoichiometric amount of cementing metal, high temperature (up to 105 °C) or ambient temperature, intensive stirring (110 up to 6000 rpm) or shaking (e.g., 4 cm amplitude, 120 shakes/min frequency). Process thermodynamics are influenced by pH value and redox potential. Modification of pH before cementation can be performed for instance by adding NaOH [41] or H_2SO_4 [37,39,42]. The metals used for cementing in hydrometallurgical processing of waste materials are typically aluminium [22–25,29,41], zinc [37,39,40,43] and iron [35,36], but in one study the researchers used a copper alloy, namely brass [42]. Cementing with iron presents several advantages: removal of heavy metals, simplicity and high speed of the process, the recovery of metals in the form of pure metal.

The general equation of lead cementation with aluminium is described by the Equation (6) [22].

$$3Pb^{2+} + 2Al = 2Al^{3+} + 3Pb \tag{6}$$

The cementation of lead with aluminium from a chloride medium is described by the Equation (7) [25].

$$3PbCl_2 + 2Al = 2AlCl_3 + 3Pb \tag{7}$$

The cementation of cadmium and lead by zinc in a sulphate medium is described by Equations (8) and (9) [40].

$$CdSO_4 + Zn = ZnSO_4 + Cd \tag{8}$$

$$PbSO_4 + Zn = ZnSO_4 + Pb \tag{9}$$

The cementation of copper in the form of $CuSO_4$ by iron is described by the Equation (10) [36].

$$CuSO_4 + Fe = FeSO_4 + Cu \tag{10}$$

The most negative standard potential of those three metals has aluminium, but its impact on process kinetics has been observed [22]. Many studies have focused on the application of various metals in copper cementation, and the consensus seems to be that zinc is a more efficient cementer than aluminium or iron. Although aluminium has lower redox potential than zinc, the protective oxidation layer which forms on the surface of aluminium during cementation slows the process down, thus reducing the amount of target metal recovered [37,39,44]. On the other hand, this passivation layer may be removed by controlling the acidity of the surrounding solution [41]. The selection of cementer metal also depends on the processed material, and if the waste does not contain the given metal, then cementation would lead to contamination and reduced quality of the prepared material. For example, aluminium, zinc or iron can be used for cementation of silver or lead from salt solution, but it has been established that the cementation process runs more slowly with zinc and iron, whereas with aluminium it proceeds faster. Thus, aluminium has been found to be an efficient as well as accessible cementer, and, moreover, less of it is consumed in the process compared with the other two metals [25,29]. The influence of amount and surface morphology of the cementer metal has been observed with regard to the process and result of lead extraction, and the optimal form was found to be powder with micropores [21]. Cementation efficiency is also affected by the presence of specific ions in the solution. For example, cementation of copper using iron proceeded faster in sulphate electrolyte, whereas in iodate solutions the process did not function at all [35]. Regarding results of cementation, almost 100% extraction has been achieved with lead [20,22,29,39,40], copper [36,37,39,42] and iron [37]. Studies of process kinetics have found that in the cementation of copper using zinc [37], lead using aluminium [22] and lead using iron [38], diffusion-controlled phases occurred within certain temperature intervals (22 to 50 °C [37], 50 to 70 °C [22] and 25 to 80 °C [38]). Cementation of lead using aluminium is highly sensitive to temperature change between 40–50 °C, and in this case the process is controlled by the speed of chemical reaction itself [22]. Generally speaking, lead is cemented not only in the form Pb^0, but also PbO [23,24]. In lead cementation moreover, several metals may pass into the precipitate together; for example, in cementing lead using aluminium the precipitate was found to contain aluminium, iron and zinc as well [22]. After processing waste material from zinc production using two-stage leaching in NaOH followed by cementation using powdered zinc, the resulting product contained more than 99% Pb, but silver was also found concentrated within it [23]. The kinetics of cementation processes is generally described as a reaction of the first order, and the step determining the speed of the process is the diffusion-controlled cathode reaction [38,43]. Cementation of copper using zinc is a first-order reaction, and it proceeds with subsequent surface reaction and diffusion-controlled phases [44]. Cementation of copper using aluminium is a first-order reaction as well, and the step determining its speed is the diffusion phase [41].

The insights presented in this theoretical section were applied in the following experimental investigations. The leaching medium used for leaching converter dusts was an acetic acid solution, it was found more effective for metal recovery from the dust. The extraction method selected for extracting lead was cementation, and zinc in metal powder form was used as the cementer. Zinc was chosen as the optimal cementer, since it is contained in the solution after the acid leaching of dusts. The following parameters were monitored for their impact on the speed of the lead cementation process: temperature of the medium, stirring rate and cementer particle size. The aim was to establish the optimal conditions for lead cementation in order to achieve the highest efficiency. The lead was cemented out in the precipitate, and a purified solution was acquired containing zinc for further processing.

2. Materials and Methods

In these experiments we used a representative sample of converter dust from secondary copper production. The chemical composition of the dust (see Table 1) was determined using the atomic absorption spectrometry (AAS) method and a Varian Spectrophotometer AA20+ device (Varian, Belrose, Australia). The contents of zinc 29.9 wt.% and lead 9.33 wt.% made them the principal representatives among the metals in the dust.

Table 1. Chemical analysis of converter dust.

Element	Pb	Zn	Fe	Cu	Ca	Si
Content (wt.%)	9.33	29.90	0.52	7.05	0.23	<LoD [1]

[1] Note: LoD, limit of detection.

Figure 1 presents the distribution of particle sizes in the converter dust. The graphic shows that the particle distribution was bimodal. The predominant portion of the dust occupied the particle size range below 10 μm. Small particle size is in fact advantageous in hydrometallurgical processing. The average unit weight of converter dust is 3.4165 $g \cdot cm^{-3}$ with a standard deviation of 0.0071 $g \cdot cm^{-3}$ [8].

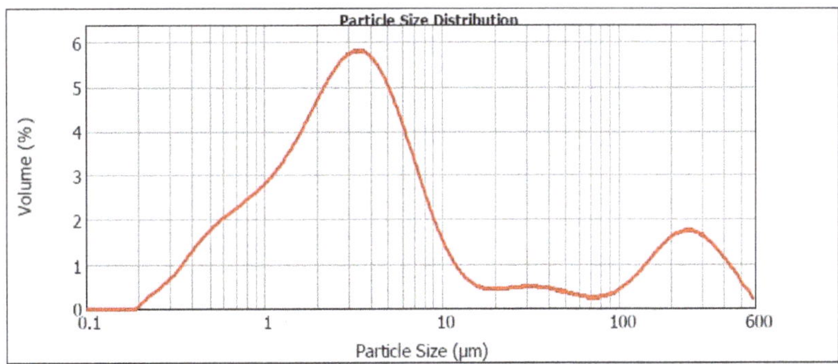

Figure 1. Distribution of particle sizes in our converter dust.

The morphology of the particles under 100× magnifications using a Dino-Lite Pro AM4113T optical stereo microscope (AnMo Electronics Corporation, Hsinch, Taiwan) is presented in Figure 2a. The sample was composed of two fractions: coarse and fine spherical grains, and the larger-size grains were covered in the smaller ones. X-ray diffraction phase analysis (XRD) (PANanalytical, Malvern, UK) produced results shown in Figure 2b, and it was found that zinc was present in the dust in the form of zincite (ZnO). Lead was found in the dust in the mainly form of oxides. Results of leachability testing indicated that this converter dust should be stored only in a hazardous waste dump [8]. Figure 2c,d document the results of SEM and EDX analyses of particles in the dust sample. The EDX analysis shows that the elements predominately represented were Zn, O, Pb, Cu and Cl.

Our thermodynamic calculations applying Van't Hoff's reaction isobar indicated that chemical reactions (11) and (12) had the highest probability of successful course. Figure 3a,b shows the potential–pH diagrams (Pourbaix diagrams) for the systems Pb-C-H_2O and Zn-C-H_2O at 298 K, and it can be seen that lead appears in ion form (Pb^{2+}) in the zone of water stability around 298 K at pH 0 to 2 and 0.0 V to 1.3 V, and in a narrower zone at pH 0 to 4.8 and 0.0 V to 0.2 V. Zinc also appears in ion form (Zn^{2+}) in the water stability zone at temperature 298 K, pH value 0 to 3.4 and 0 V to 1.2V, and in a narrower zone at pH 3.4 to 5.6 and 0 to 0.2 V. PbO dissolves chemically in acetic acid according to Equation (11).

$$2CH_3COOH_{(a)} + PbO = Pb(CH_3COO)_{2\,(a)} + H_2O \quad \Delta G^0_{293K} = -19 \text{ kJ} \cdot mol^{-1} \quad (11)$$

$$2CH_3COOH_{(a)} + ZnO = Zn(CH_3COO)_{2\,(a)} + H_2O \quad \Delta G^0_{293K} = -17.5 \text{ kJ} \cdot mol^{-1} \quad (12)$$

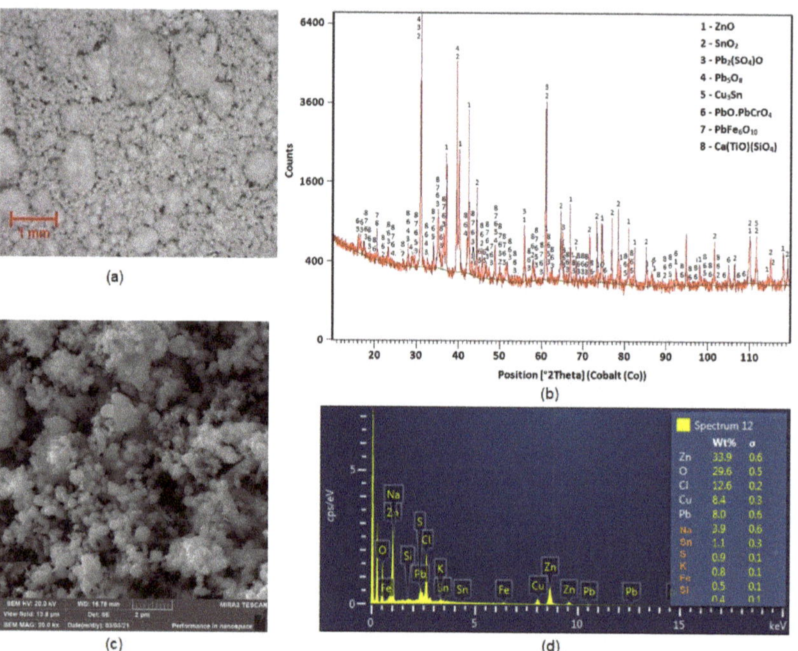

(a)   (b)   (c)   (d)

Figure 2. (**a**) Morphology of converter dust particles under stereo microscopy; (**b**) XRD analysis results; (**c**) SEM analyses, morphology of particles and (**d**) EDX analyses.

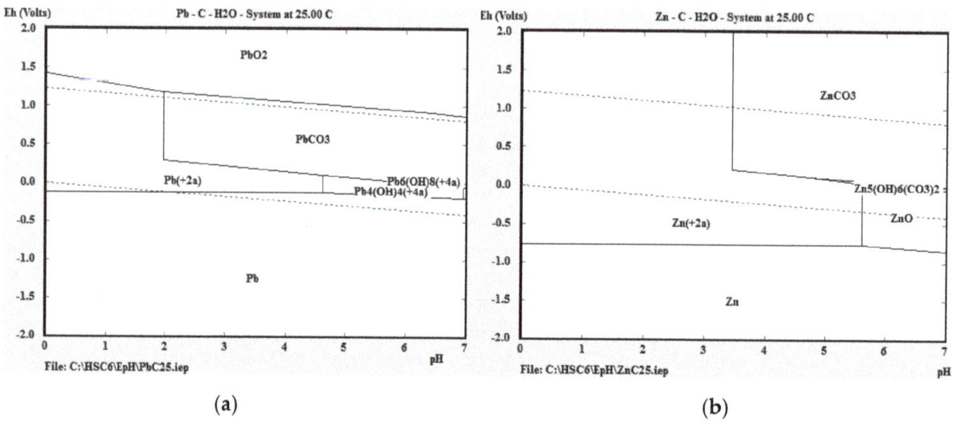

(**a**)   (**b**)

Figure 3. E-pH diagrams of systems (**a**) Pb-C-H_2O and (**b**) Zn-C-H_2O at 298 K, created with HSC Chemistry software, ver. 6 (Outotec, Espoo, Finland).

For ZnO the dissolution zones in the E-pH diagram apply between 0 and 5.7 pH. The aim of leaching was to transfer lead into the solution under optimal conditions, and from the thermodynamic point of view it may be stated that oxides of Pb and Zn will dissolve chemically at temperature 298 K in an acid medium. It is not necessary to raise the temperature further.

In the first phase of the experiments, we carried out leaching of the converter dust in an acetic acid solution under ambient temperature conditions (293 K). The stirring rate was set on the electric stirrer at 400 rpm. Acidity of the medium was measured during the experiments with a hand-held pH meter. We started by pouring 500 mL 0.4 M solution of acetic acid ($C_2H_4O_2$) into a glass vessel, and then added a weighed amount of 12.5 g of converter dust, producing a liquid/solid phase ratio of 40:1. At the following time intervals: 5, 10, 15, 30 and 60 min, we drew off 5 mL samples of the solution for chemical analysis to determine the Pb and Zn contents using the atomic absorption spectrometry (AAS) method on a Varian Spectrophotometer AA20+ type device. At the end of the leaching the resulting sediment was recovered by filtering, and subsequently another 10 mL sample of the solution was taken for final chemical analysis to determine Pb, Zn, Cu and Fe contents by means of AAS method. The solid remnant from filtration was rinsed with 200 mL distilled water, and then dried and weighed. The remaining solution obtained from the leaching process was later re-used in the second phase of our experiments for studying lead cementation. The results of our analyses are summarized in Table 3 (a). During cementation on zinc of lead and copper obtained from leaching in acid solution, the following Equations take place (13) and (14):

$$Zn_{(s)} + Pb(CH_3CO_2)_{2(aq)} = Zn(CH_3COO)_{2(aq)} Pb_{(s)} \quad \Delta G^0_{293K} = -143.155 \text{ kJ·mol}^{-1} \quad (13)$$

$$Zn_{(s)} Cu(CH_3CO_2)_{2(aq)} = Zn(CH_3COO)_{2(aq)} + Cu_{(s)} \quad \Delta G^0_{293K} = -211.878 \text{ kJ·mol}^{-1} \quad (14)$$

The electrode potential of redox vapour Zn^{2+}/Zn^0 is less than that of redox vapours Pb^{2+}/Pb^0 (15) and Cu^{2+}/Cu^0 (16), which is in fact a condition for lead and copper cementation using zinc.

$$E^0_{Zn^{2+}/Zn^0} < E^0_{Pb^{2+}/Pb^0} \quad -0.763 \text{ V} < -0.126 \text{ V} \quad (15)$$

$$E^0_{Zn^{2+}/Zn^0} < E^0_{Cu^{2+}/Cu^0} \quad -0.763 \text{ V} < +0.337 \text{ V} \quad (16)$$

The thermodynamic conditions for cementation of lead and copper using powder zinc in the temperature range 293 to 353 K are described in Equations (13) and (14) presented in Table 2.

Table 2. Thermodynamic conditions for cementing lead and copper using powder zinc.

$Zn_{(s)} + Pb(CH_3CO_2)_{2(aq)} = Zn(CH_3COO)_{2(aq)} + Pb_{(s)}$	
T (K)	**ΔG (kJ)**
293	−143.155
313	−142.094
333	−141.950
353	−142.578

$Zn_{(s)} + Cu(CH_3CO_2)_{2(aq)} = Zn(CH_3COO)_{2(aq)} + Cu_{(s)}$	
T (K)	**ΔG (kJ)**
293	−211.878
313	−212.397
333	−212.903
353	−213.399

In our cementation experiments we focused on investigating the influence of stirring intensity (100, 300 and 400 rpm), solution temperature (293, 313, 333 and 353 K) and cementer particle size on the speed of lead cementation. We used powder zinc as cementer with two particle size distribution: <0.125–0.4> mm and <0.08–0.125> mm. For the purposes of comparing the influence of particulate cementer. We also performed an experiment using a zinc-coated plate with dimensions (20 mm × 35 mm × 1 mm). The cementation procedure was as follows. First, we poured 100 mL of solution resulting from leaching into a glass vessel. Then, we added a stoichiometrically-calculated amount of 0.3 g Zn^0(s) in powder form, representing the excess of the calculated amount of zinc over the amount of lead determined in the solution after leaching. We drew off 5 mL samples for measuring Pb and Cu contents after 5, 10, 15, 30 and 60 min of cementation, or after 0, 5, 10, 15 30 min, if the cementation procedure lasted only 30 min. Once cementation was completed, the precipitate formed in the solution was recovered by filtering, and then dried, and a representative sample was taken for subsequent analysis. The results were processed using standard mathematical and statistical methods with the aid of generally available computer software: Microsoft Office Excel 2007, (Microsoft Corporation, Redmond, WA, USA), and SigmaPlot 10.0 (Systat Software, Erkrath, Germany). The converter dust and the resulting cementation precipitate were analyzed using combinations of these methods: X-ray diffraction phase analysis (XRD) PANanalytical, Malvern, UK), scanning electron microscopy (SEM) (TESCAN, Brno, Czech Republic) and energy-dispersive X-ray analysis (EDX) (Oxford Instruments, Oxford, UK). A MIRA3 FE-SEM scanning electron microscope (TESCAN, USA) was used for investigating particle morphology. EDX provided semi-quantitative element analysis. For qualitative phase analysis we used an X-PANalytical X'Pert PRO MRD (Co-Kα) diffractometer (PANanalytical, Malvern, UK). Particle morphology was examined under 100 $\times$ magnifications with a Dino-Lite ProAM413T optical stereo microscope device (AnMo Electronics Corporation, Hsinch, Taiwan).

3. Results and Discussion

3.1. Influence of Stirring Intensity

Experiments started by studying the influence of this variable, maintaining the temperature of the cementation solution containing powder zinc at 293 K. It has been tried three different stirring rates in separate experiments, namely 100, 300 and 400 revolutions per min (rpm). The solution obtained after converter dust leaching by acetic acid was analyzed using the AAS method. The solution contained 1.570 g·dm^{-3} Pb and 8.756 g·dm^{-3} Zn. The dependence on time of lead concentration in the solution is shown in Figure 4a, while Figure 4b presents a graphical comparison of lead concentrations after 30 min of cementation using powder zinc when the three different stirring rates were applied.

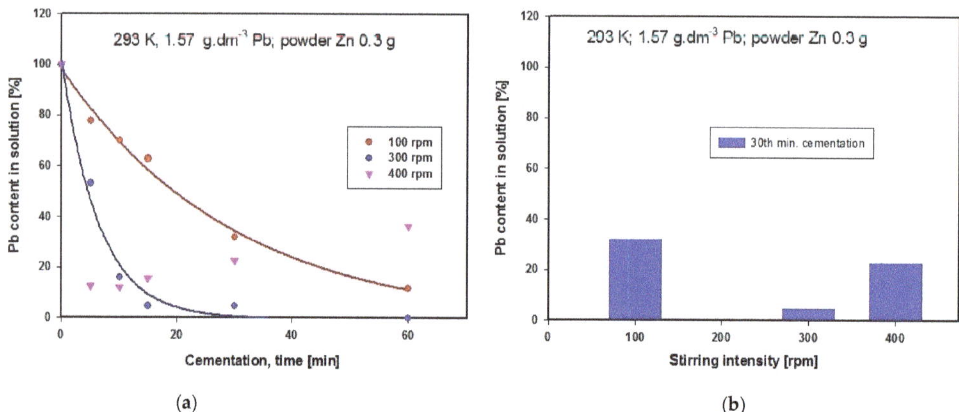

Figure 4. (**a**) Time dependence of Pb concentration in solution after cementation using zinc at different stirring rates at constant temperature; (**b**) Comparison of Pb concentrations in solution after cementation using zinc at different stirring rates at constant temperature.

The graph in Figure 4a suggests that the optimal stirring rate for obtaining the lowest lead concentration in the solution is 300 rpm, given a cementation period of 30 min, and this is confirmed by the results of comparison of the three different stirring rates presented in Figure 4b. The zinc content in the solution after cementation using powder zinc remained unchanged. At the higher stirring rate of 400 rpm, retroactive leaching of lead was observed, while lower revolutions produced lower lead concentration in the final medium. Optimal stirring speed ensured removal of layers of metal deposits (Pb and Cu) from the surface of the cementer (Zn), thus enabling constant contact between the solution ($Pb(CH_3CO_2)_{2(aq)}$; $Cu(CH_3COO)_{2(aq)}$) and cementer ($Zn_{(s)}$). Our further experiments were therefore carried out using the optimal stirring intensity of 300 rpm.

3.2. Influence of Temperature

The influence of temperature was investigated at a constant stirring rate of 300 rpm stirring rate. The initial solution volume was 100 mL, to which we added 0.3 g of powder zinc. Total duration of each experiment was maintained at 30 min. After the end of cementation, a 10 mL sample of solution was drawn off for determination of Pb, Zn and Cu contents. Individual experiments were performed at temperatures of 293 K, 313 K, 333 K and 353 K. Figure 5a is a graphical representation of the influence of temperature on lead content in the solution expressed in percentages, and Figure 5b shows the results of comparing lead concentration after five minutes of cementation at temperatures 293, 313, 333 and 353 K. The graph in Figure 5c shows the changes in concentrations of Pb, Zn and Cu in the solution during cementation using zinc at 293 K. A graphical comparison of cementation efficiency for lead and copper at 293 K is given in Figure 5d, where the efficiency for Cu is seen to be 100%, while that for Pb is 87.13. The reaction occurring during cementation of Cu using Zn (Equation (14)) has more negative value for Gibbs energy than the reaction during cementation of Pb using Zn (Equation (13)) within the temperature range applied in these experiments 293–353 K. It follows from this that cementation of Cu on Zn is more likely to happen than cementation of Pb on Zn, and this is the reason why the efficiency of Cu cementation is higher than that for Pb (see also Figure 5d below).

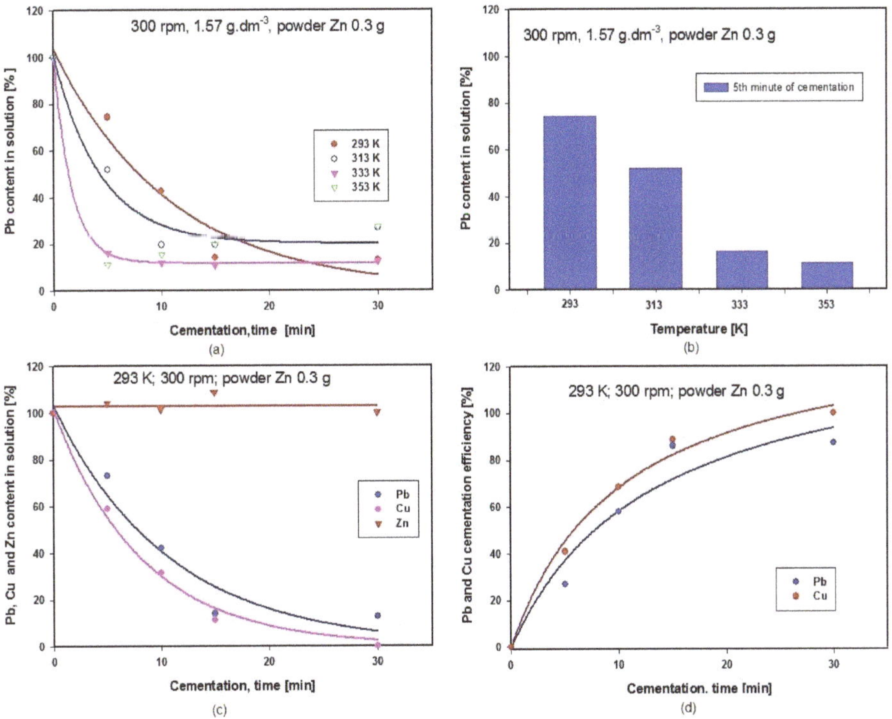

Figure 5. (**a**) Time dependence of Pb content in solution during cementation at different temperatures; (**b**) Comparison of Pb concentrations after 5 min cementation at different temperatures; (**c**) Time dependence of Pb, Cu and Zn concentrations in solution during cementation using zinc; (**d**) Comparison of Pb and Cu cementation efficiencies at 293 K.

The morphology of cementation precipitate particles under $100 \times$ magnification using a Dino-Lite ProAM413T optical stereomicroscope is presented in Figure 6a. The image clearly shows the variety of shapes and sizes of particles ranging from 1 to 1.5 mm. XRD analysis (see image in Figure 6b revealed that lead was contained in the precipitate in phases PbO, Pb_5O_8 and Pb (Cu_2O_2), and copper in the forms Cu, $Cu(OH)_2$ and Cu_4Zn.

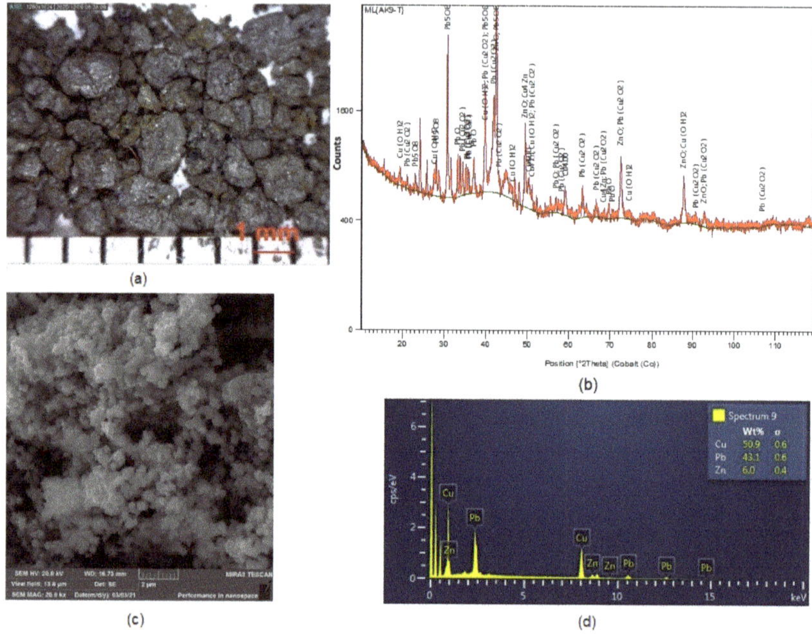

Figure 6. (**a**) Image showing morphology of cementation precipitate particles; (**b**) XRD analysis of precipitate; (**c**) SEM image of precipitate particle morphology; (**d**) EDX analysis after cementation using powder zinc.

Figure 6c,d presents the results of SEM and EDX examinations of particles in the analysed sample of cementation precipitate. It follows from our EDX analysis that Cu, Pb and Zn were present in the precipitate, predominantly Cu ($\approx$33.8 wt.%) and Pb ($\approx$16.89 wt.%). The precipitate also contained unreacted zinc ($\approx$5.25 wt.%). Table 3 (b) summarizes the results of the chemical analysis of the solution after the cementation process using powder zinc. This analysis demonstrates that the cementation succeeded in removing Cu as well as Pb from the solution. The increased amount of Zn in the solution after cementation was caused by the presence of more than the stoichiometric amount of zinc used as the cementer.

Table 3. Chemical analysis of solution (a) before and (b) after cementation.

	Concentration [g/L]	Pb	Zn	Fe	Cu
(a)	Before cementation	1.5706	8.756	0.0104	1.7138
(b)	After cementation	0.0012	10.28	<LoD [1]	0.005

[1] Note: <LoD[2] below limit of detection.

3.3. Influence of Zinc Surface Dimensions

The influence of powder zinc particle size was investigated under ambient temperature conditions (293 K) and stirring rate 300 rpm. For cementation we used weighed amounts of 0.3 g powder zinc with two particle size distribution, namely <0.125–0.4> mm and <0.08–0.125> mm. In both cases the initial and final pH values were the same at 4.2

and 4.4, respectively. We also tried using a zinc-coated plate (20 mm × 35 mm × 1 mm) as cementer under the same experimental conditions as for powder zinc, apart from the stirring rate, which was set at 50 rpm. If more intensive stirring had been applied, excessive swirling would have occurred in the solution, which would not have been conducive to cementation or to the successful technical course of the experiment. The zinc plate was degreased with ethanol and rinsed with distilled water prior to the experiment. A 20 mm length of the plate was submerged in the solution. This form of cementation using a zinc plate proceeded rapidly, and within the first minute a layer of black cementation precipitate started covering the plate. The thickness of the precipitate layer increased with time, and later it began peeling off the plate. After 20 min we observed red particles appearing in the precipitate, which we assumed to be cemented out copper. The graph in Figure 7a shows the influence of powder Zn particle size on the concentration of lead in the solution during cementation at constant temperature 293 K. The most suitable of the particle sizes investigated appeared to be that with distribution <0.125–0.4> mm. After 30 min of cementation using powder zinc with this particle size, the Pb concentration in the solution reached its lowest value compared with the other particle size investigated. The steeper gradient of the curve for cementation using powder zinc with particle size distribution <0.125–0.4> mm compared with the curve produced by using the zinc plate indicates a higher rate of Pb cementation. Figure 7b presents a graphical comparison of the efficiency of Pb cementation after five minutes of cementation using powder zinc with different particle sizes at constant temperature 293 K. The highest efficiency after five minutes of cementation was achieved using powder zinc with particle size <0.08–0.125> mm. At this time point we also measured the lowest Pb concentration in the solution using powder zinc with particle size <0.08–0.125> mm, whereas after 15 min the Pb concentration in the solution was increased. This was probably caused by the zinc particles tending to cluster together (at around 10 min), so that later (at around 15 min) some retro-leaching of the lead started.

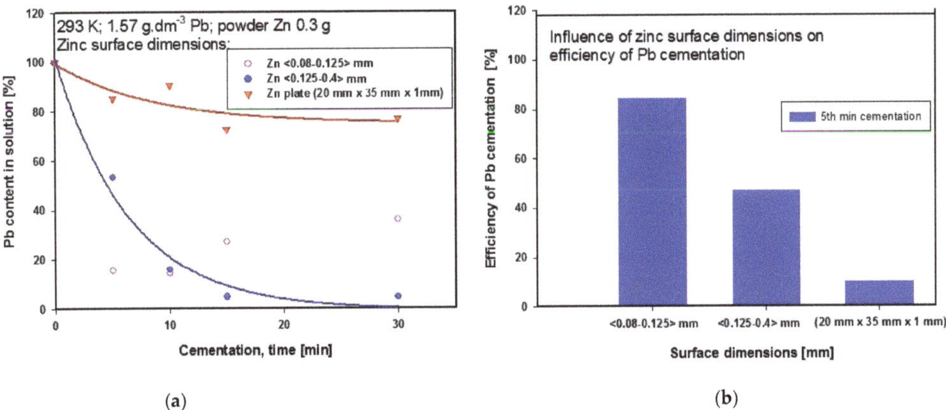

(a) (b)

Figure 7. (a) Influence of cementer Zn particle size on Pb concentration in solution during cementation at 293 K (b) Comparison of Pb cementation efficiencies after 5 min procedure at 293 K using different Zn particle sizes.

3.4. Investigation of Kinetics of Lead Cementation Process in Acid Medium

The average chemical reaction rate (13) and in ion form (15) is expressed through Relation (17),

$$Zn_{(s)} + Pb(CH_3CO_2)_{2(aq)} = Zn(CH_3COO)_{2(aq)} + Pb_{(s)} \quad \Delta G293_K = -143.155 \text{ kJ} \quad (17)$$

$$Zn^0 + Pb^{2+} = Zn^{2+} + Pb^0 \quad (18)$$

$$v = -\frac{\Delta c_{Pb^{2+}}}{\Delta t} \left[mol \cdot dm^{-3} \cdot s^{-1} \right] \tag{19}$$

whereby a negative value represents the change in concentration of lead-based ions over time. The calculated values of chemical reaction rate are given in Figure 8a.

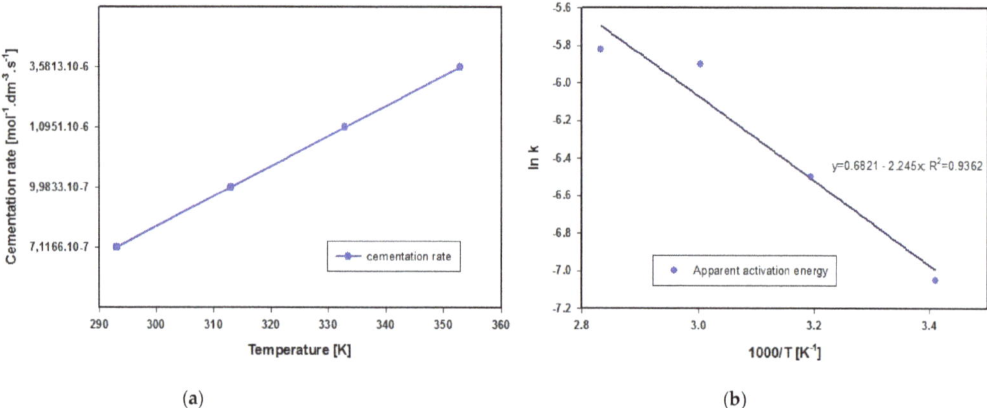

(**a**)　　　　　　　　　　　　　(**b**)

Figure 8. (**a**) Lead cementation rate at different temperatures of solution. (**b**) The Arrhenius equation expresses the interdependence of temperature and rate constant in chemical reactions.

Applying Arrhenius' equation in logarithmic form (17),

$$\ln k = \ln A - \frac{E_a}{RT} \tag{20}$$

where k = rate constant, A = frequency factor, T = temperature in Kelvin (K) and R = universal gas constant, we calculated the value of apparent activation energy as E_a = 18.66 kJ·mol^{-1}, which according to Zelikman et al., 1983 [45] indicates that the process of lead cementation in acetic acid solution proceeds in the diffusion zone in the temperature interval from 293 to 333 K. The course of linear dependence lnk on 1000/T presented in Figure 8b suggests similarly that the mechanism of lead cementation does not change within the temperature interval used here. The apparent activation energy E_a determined using the linear regression method and Equation (20) is given in Figure 8b.

4. Conclusions

Secondary copper production using the pyrometallurgical route generates dust. Due to its heavy metals content and the fineness of its particles, this dust may represent a threat to the environment and to human health, so it is categorized as hazardous waste. According to the results of leachability testing in line with European standard EN 12457-1: 2002 [46], such dust may be stored only in hazardous waste dumps. On the other hand, because of its not inconsiderable non-ferrous metals content, such dust can be seen as a valuable secondary raw material. Research into dust processing currently focuses on use of the hydrometallurgical route for metals recovery.

This study focused on the experimental investigation of lead removal from acid solution after leaching of converter dust. The first step involved leaching of the dust in an acetic acid medium under ambient temperature conditions. Subsequently the process of lead cementation using zinc was studied, the aim of cementation being to remove the lead from the solution. The variables investigated for their influence on the efficiency of the cementation process were temperature of the medium, intensity of stirring and particle size of the powder zinc cementer. The kinetics of the process of lead cementation using powder zinc in acid medium was also studied. The main results of this experimental study are as follows:

(1) The highest efficiency of lead removal from the solution was achieved at ambient temperature (293 K) with stirring rate 300 rpm and zinc particle size <0.08–0.125> mm after five minutes. Under these conditions almost 90% efficiency of lead removal from the solution was achieved. The cementation precipitate was found to contain copper as well as lead, which means that copper was also removed from the solution. Lead was represented in the precipitate in the phases PbO, Pb_5O_8 and $Pb(Cu_2O_2)$. Figure 9 presents suggested flowsheet for lead convert dust recycle process.

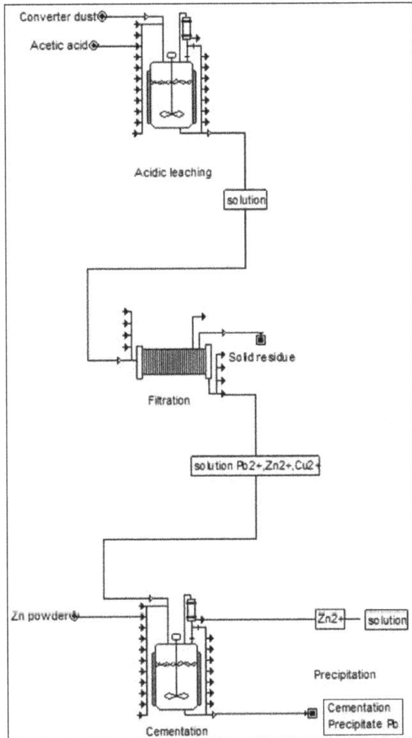

Figure 9. Suggested flowsheet for copper convert dust recycle process.

(2) Experimental measurements revealed the value of apparent activation energy as $E_a = 18.66$ kJ·mol^{-1}, indicating that the process of lead cementation in acetic acid solution proceeds in the diffusion zone in the temperature interval from 293 to 333 K. The course of linear dependence lnk on 1000/T suggests similarly that the mechanism of lead cementation does not change within the temperature interval used here. We established that stirring rate has greater impact on the chemical reaction rate than solution temperature.

Author Contributions: Conceptualization, M.L. and A.K.; methodology, M.L.; software, M.L.; validation, J.T., B.P. and J.B.; formal analysis, J.B. and B.P.; investigation, M.L. and A.K.; resources, A.K.; data curation, M.L.; writing—original draft preparation, M.L. and A.K.; writing—review and editing, M.L.; visualization, M.L.; supervision, J.T.; project administration, J.T.; funding acquisition, J.T. All authors have read and agreed to the published version of the manuscript.

Funding: This research was funded by the Ministry for Education of the Slovak Republic, grant number VEGA 1/0641/20.

Data Availability Statement: The data presented in this study are available on request from the corresponding author.

Acknowledgments: This work was supported by the Ministry for Education of the Slovak Republic under grant MŠ SR VEGA 1/0641/20.

Conflicts of Interest: The authors declare no conflict of interest.

References

1. U.S. Geological Survey. *Mineral Commodity Summaries 2021: Sand and Gravel (Industrial)*; U.S. Geological Survey: Reston, VA, USA, 2021; ISBN 9781411343986.
2. Demand, V. Copper Drives Electric Vehicles. Available online: https://www.copper.org/publications/pub_list/pdf/A6191-ElectricVehicles-Factsheet.pdf (accessed on 29 October 2021).
3. Lynch, J. Copper's Role in Growing Electric Vehicle Production. Available online: https://www.reuters.com/article/sponsored/copper-electric-vehicle (accessed on 20 October 2021).
4. McGuiness, P.J.; Ogrin, R. *Securing Technology-Critical Metals for Britain: Ensuring the United Kingdom's Supply of Strategic Elements & Critical Materials for a Clean Future*; University of Birmingham: Birmingham, UK, 2021; ISBN 9780704429697.
5. Top Ten Global Copper Producers—2020. Available online: https://nma.org/wp-content/uploads/2021/07/Top-Ten-Copper-Producers-2020.pdf (accessed on 22 October 2021).
6. Kovohuty Krompachy. Available online: https://www.kovohuty.sk/home.html (accessed on 29 October 2021).
7. Havlík, T. *Hydrometallurgy Principles and Applications*; Woodhead Publishing Limited: Southston, UK, 2008; ISBN 978-1-84569-407-4.
8. Orac, D.; Laubertova, M.; Piroskova, J.; Klein, D.; Bures, R.; Klimko, J. Characterization of dusts from secondary copper production. *J. Min. Metall. Sect. B Metall.* **2020**, *56*, 221–228. [CrossRef]
9. Forsén, O.; Aromaa, J.; Lundström, M. Primary copper smelter and refinery as a recycling plant—a system integrated approach to estimate secondary raw material tolerance. *Recycling* **2017**, *2*, 19. [CrossRef]
10. Waste Production and Waste Management in the Slovak Republic. Available online: http://cms.enviroportal.sk/odpady/verejne-informacie.php?rok=B-2017&kr=v&kat%5B%5D=100603&kat%5B%5D=100606 (accessed on 20 October 2021).
11. European Waste Catalogue and Hazardous Waste List. Available online: http://www.nwcpo.ie/forms/EWC_code_book.pdf (accessed on 20 October 2021).
12. Balladares, E.; Kelm, U.; Helle, S.; Parra, R.; Araneda, E. Chemical-mineralogical characterization of copper smelting flue dust. *Dyna* **2014**, *81*, 11. [CrossRef]
13. Guo, L.; Lan, J.; Du, Y.; Zhang, T.C.; Du, D. Microwave-enhanced selective leaching of arsenic from copper smelting flue dusts. *J. Hazard. Mater.* **2020**, *386*, 121964. [CrossRef]
14. Kukurugya, F.; Vindt, T.; Havlík, T. Behavior of zinc, iron and calcium from electric arc furnace (EAF) dust in hydrometallurgical processing in sulfuric acid solutions: Thermodynamic and kinetic aspects. *Hydrometallurgy* **2015**, *154*, 20–32. [CrossRef]
15. Trung, Z.H.; Kukurugya, F.; Takacova, Z.; Orac, D.; Laubertova, M.; Miskufova, A.; Havlik, T. Acidic leaching both of zinc and iron from basic oxygen furnace sludge. *J. Hazard. Mater.* **2011**, *192*, 1100–1107. [CrossRef] [PubMed]
16. Laubertová, M.; Pirošková, J.; Dociová, S. The technology of lead production from waste. *World Metall. - ERZMETALL* **2017**, *70*, 47–54.
17. Okanigbe, D.O.; Popoola, A.P.I.; Adeleke, A.A.; Otunniyi, I.O.; Popoola, O.M. Investigating the impact of pretreating a waste copper smelter dust for likely higher recovery of copper. *Procedia Manuf.* **2019**, *35*, 430–435. [CrossRef]
18. Maruskinova, G.; Kukurugya, F.; Parilak, L.; Havlik, T.; Kobialkova, I. Hydrometallurgical treatment of EAF steelmaking dust. *Metall* **2015**, *68*, 522–524.
19. Pirošková, J.; Laubertová, M.; Miškufová, A.; Oráč, D. Hydrometallurgical treatment of copper shaft furnace dust for lead recovery. *World Metall. - ERZMETALL* **2018**, *71*, 37–42.
20. Ghica, V.G.; Buzatu, T. Cemention with iron in solutions of lead acetate in a two-step process. *Rev. Chim.* **2013**, *64*, 580–583.
21. Volpe, M.; Oliveri, D.; Ferrara, G.; Salvaggio, M.; Piazza, S.; Italiano, S.; Sunseri, C. Metallic lead recovery from lead-acid battery paste by urea acetate dissolution and cementation on iron. *Hydrometallurgy* **2009**, *96*, 123–131. [CrossRef]
22. Farahmand, F.; Moradkhani, D.; Sadegh Safarzadeh, M.; Rashchi, F. Optimization and kinetics of the cementation of lead with aluminum powder. *Hydrometallurgy* **2009**, *98*, 81–85. [CrossRef]
23. Siegmund, A.; Shibata, E.; Alam, S.; Grogan, J.; Kerney, U. *PbZn 2020: 9th International Symposium on Lead and Zinc Processing*; Springer Nature: Cham, Switzerland, 2020; Volume 53, ISBN 9788578110796.
24. Silwamba, M.; Ito, M.; Hiroyoshi, N.; Tabelin, C.B. Recovery of Lead and Zinc from Zinc Plant Leach Residues by Concurrent Dissolution-Cementation. *Metals* **2020**, *10*, 531. [CrossRef]
25. Abdollahi, P.; Yoozbashizadeh, H.; Moradkhani, D.; Behnian, D. A Study on Cementation Process of Lead from Brine Leaching Solution by Aluminum Powder. *OALib* **2015**, *2*, 1–6. [CrossRef]
26. Forte, F.; Horckmans, L.; Broos, K.; Kim, E.; Kukurugya, F.; Binnemans, K. Closed-loop solvometallurgical process for recovery of lead from iron-rich secondary lead smelter residues. *RSC Adv.* **2017**, *7*, 49999–50005. [CrossRef]
27. Gargul, K.; Boryczko, B.; Bukowska, A.; Jarosz, P.; Małecki, S. Leaching of lead and copper from flash smelting slag by citric acid. *Arch. Civ. Mech. Eng.* **2019**, *19*, 648–656. [CrossRef]

28. Liu, W.F.; Fu, X.X.; Yang, T.Z.; Zhang, D.C.; Chen, L. Oxidation leaching of copper smelting dust by controlling potential. *Trans. Nonferrous Met. Soc. China Engl. Ed.* **2018**, *28*, 1854–1861. [CrossRef]
29. Raghavan, R.; Mohanan, P.K.; Patnaik, S.C. Innovative processing technique to produce zinc concentrate from zinc leach residue with simultaneous recovery of lead and silver. *Hydrometallurgy* **1998**, *48*, 225–237. [CrossRef]
30. Gabler, R.C.; Jones, J.R. *Metal Recovery from Secondary Copper Converter Dust by Ammoniacal Carbonate Leaching*; Bureau of Mines: Washington, DC, USA, 1988; p. 8.
31. Oráč, D.; Hlucháňová, B.; Havlík, T.; Miškufová, A.; Petrániková, M. Leaching of zinc and copper from blast furnace dust of copper production of secondary raw materials. *Acta Metall. Slovaca* **2009**, *3*, 147–153.
32. Soria-Aguilar, M.J.; Carrillo-Pedroza, F.R.; Preciado-Nuñez, S. Treatment of BF and BOF dusts by oxidative leaching. In Proceedings of the 2008 Global Symposium on Recycling, Waste Treatment and Clean Technology, REWAS, Cancun, Mexico, 12–15 October 2008.
33. Sabzezari, B.; Koleini, S.M.J.; Ghassa, S.; Shahbazi, B.; Chelgani, S.C. Microwave-leaching of copper smelting dust for Cu and Zn extraction. *Materials* **2019**, *12*, 1822. [CrossRef]
34. Laubertova, M.; Havlik, T.; Parilak, L.; Derin, B.; Trpcevska, J. The effects of microwave-assisted leaching on the treatment of electric arc furnace dusts (EAFD). *Arch. Metall. Mater.* **2020**, *65*, 321–328. [CrossRef]
35. Tzaneva, B. Electrochemical investigation of cementation process. *Bulg. Chem. Commun.* **2016**, *48*, 91–95.
36. Shishkin, A.; Mironovs, V.; Vu, H.; Novak, P.; Baronins, J.; Polyakov, A.; Ozolins, J. Cavitation-dispersion method for copper cementation from wastewater by iron powder. *Metals* **2018**, *8*, 920. [CrossRef]
37. Panão, A.S.I.; De Carvalho, J.M.R.; Correia, M.J.N. Copper Removal from Sulphuric Leaching Solutions by Cementation. 2006. Available online: https://www.semanticscholar.org/paper/Copper-Removal-from-Sulphuric-Leaching-Solutions-by-Panao-Carvalho/3a21ea867265ed4638524d9f13e08ced3d6359e1 (accessed on 20 October 2021).
38. Makhloufi, L.; Saidani, B.; Hammache, H. Removal of lead ions from acidic aqueous solutions by cementation on iron. *Water Res.* **2000**, *34*, 2517–2524. [CrossRef]
39. Kazakova, N.; Lucheva, B.; Iliev, P. A study on the cementation process of non-ferrous metals from a brine leaching solution. *J. Chem. Technol. Metall.* **2020**, *55*, 223–227.
40. Klimko, J.; Kobialkova, I.U.; Piroskova, J.; Orac, D.; Havlik, T.; Klein, D. Sustainable development in the tinplate industry: Refining tinplate leachate with cementation. *Physicochem. Probl. Miner. Process.* **2020**, *56*, 219–227. [CrossRef]
41. Demirkran, N.; Künkül, A. Recovering of copper with metallic aluminum. *Trans. Nonferrous Met. Soc. China Engl. Ed.* **2011**, *21*, 2778–2782. [CrossRef]
42. Stankovic, V. Cementation of Copper Onto Brass Particles in a Packed Bed. *J. Min. Metall.* **2004**, *40*, 21–39. [CrossRef]
43. De Oliveira, V.A.A.; Penna, J.M.S.; Magalhães, L.S.; Leão, V.A.; dos Santos, C.G. Kinetics of Copper and Cadmium Cementation By Zinc Powder. *Tecnol. em Metal. Mater. e Min.* **2019**, *16*, 255–262. [CrossRef]
44. Demirkiran, N.; Ekmekyapar, A.; Künkül, A.; Baysar, A. A kinetic study of copper cementation with zinc in aqueous solutions. *Int. J. Miner. Process.* **2007**, *82*, 80–85. [CrossRef]
45. Zelikman, A.H.; Voľdman, G.M.; Beljajevskaja, L.V. *Teoria Gidrometallurgičeskich Processov*; Moskva Metallurgija: Moskva, Russia, 1975; p. 503.
46. European Standard EN 12457-1:2002. *Characterisation of Waste. Leaching. Compliance Test for Leaching of Granular Waste Materials and Sludges*; European Union: Maastricht, The Netherlands, 2006.

Review

Leaching Chalcocite in Chloride Media—A Review

Norman Toro [1,*], Carlos Moraga [2], David Torres [1], Manuel Saldaña [1], Kevin Pérez [3] and Edelmira Gálvez [4]

1   Faculty of Engineering and Architecture, Universidad Arturo Prat, Iquique 1100000, Chile;
    dtorres.albornoz@gmail.com (D.T.); manuel.saldana@ucn.cl (M.S.)
2   Escuela de Ingeniería Civil de Minas, Facultad de Ingeniería, Universidad de Talca, Curicó 3340000, Chile;
    carmoraga@utalca.cl
3   Departamento de Ingeniería Química y Procesos de Minerales, Universidad de Antofagasta,
    Antofagasta 1270300, Chile; kevin.perez.salinas@ua.cl
4   Departamento de Ingeniería Metalúrgica y Minas, Universidad Católica del Norte,
    Antofagasta 1270709, Chile; egalvez@ucn.cl
*   Correspondence: notoro@unap.cl; Tel.: +56-975387250

Abstract: Chalcocite is the most abundant secondary copper sulfide globally, with the highest copper content, and is easily treated by conventional hydrometallurgical processes, making it a very profitable mineral for extraction. Among the various leaching processes to treat chalcocite, chloride media show better results and have a greater industrial boom. Chalcocite dissolution is a two-stage process, the second being much slower than the first. During the second stage, in the first instance, it is possible to oxidize the covellite in a wide range of chloride concentrations or redox potentials (up to 75% extraction of Cu). Subsequently, CuS_2 is formed, which is to be oxidized. It is necessary to work at high concentrations of chloride (>2.5 mol/L) and/or increase the temperature to reach a redox potential of over 650 mV, which in turn decreases the thickness of the elemental sulfur layer on the mineral surface, facilitating chloride ions to generate a better porosity of this. Finally, it is concluded that the most optimal way to extract copper from chalcocite is, during the first stage, to work with high concentrations of chloride (50–100 g/L) and low concentrations of sulfuric acid (0.5 mol/L) at a temperature environment, as other variables become irrelevant during this stage if the concentration of chloride ions in the system is high. While in the second stage, it is necessary to increase the temperature of the system (moderate temperatures) or incorporate a high concentration of some oxidizing agent to avoid the passivation of the mineral.

Keywords: Cu_2S; CuS; dissolution; chloride

Citation: Toro, N.; Moraga, C.; Torres, D.; Saldaña, M.; Pérez, K.; Gálvez, E. Leaching Chalcocite in Chloride Media—A Review. *Minerals* **2021**, *11*, 1197. https://doi.org/ 10.3390/min11111197

Academic Editors: Kyoungkeun Yoo, Shuai Wang, Xingjie Wang and Jia Yang

Received: 26 September 2021
Accepted: 24 October 2021
Published: 28 October 2021

Publisher's Note: MDPI stays neutral with regard to jurisdictional claims in published maps and institutional affiliations.

1. Introduction

Most of copper minerals correspond to sulfide minerals and a minor part to oxidized minerals. Chalcopyrite is the most abundant among the sulfide copper minerals [1–4]. However, this mineral is refractory to conventional leaching processes due to forming a passivating layer that prevents contact between the mineral and the leaching solution [5–7]. Positive results have only been achieved for this mineral when working at medium-high temperatures (over 60 °C); therefore, it has not been possible to implement it on a large scale in industrial heap leaching processes [8].

In the natural mineral, chalcopyrite is commonly associated with secondary sulfides that include chalcocite (Cu_2S), digenite ($Cu_{1.8}S$), and covellite (CuS) [9]. Chalcocite is the most abundant secondary copper sulfide, with the highest copper content, and is easily treated by hydrometallurgical processes, which makes it a very profitable mineral for its extraction [10–14]. Chalcocite has a dark gray color and belongs to the copper-rich mineral family ranging from CuS to Cu_2S (see Table 1), commonly found in the enriched supergenic environment below the oxidized zone of copper porphyry deposits [15–19]. This is formed by oxidation, reduction, dissemination, and migration of primary sulfides such as chalcopyrite [20–22], being also the main component of the tufts of copper and white metal [23–25].

Table 1. Composition, structure, and stability of minerals Cu_xS (Data from [19,26]).

Mineral	Composition	System	Stability
Chalcocite (low T)	$Cu_{1.99-2}S$	Monoclinic	$T < 105\,°C$
Chalcocite (high T)	$Cu_{1.98-2}S$	Hexagonal	$\sim105\,°C < T < \sim425\,°C$
Chalcocite (high P and T)	Cu_2S	Tetragonal	$1\,kbar < P, T < 500\,°C$
Djurleite	$Cu_{1.93-1.96}S$	Monoclinic	$T < 93\,°C$
Digenite (low T)	$Cu_{1.75-1.8}S$	Cubic	Metastable
Digenite (high T)	$Cu_{1.73-2}S$	Cubic	$83\,°C < T$
Anilite	$Cu_{1.75}S$	Orthorhombic	$T < 72\,°C$

Due to the fact that hydrometallurgical processes are more economical and friendly to the environment than pyrometallurgical processes [27], and that it is relatively easy to dissolve copper from chalcocite, several investigations for the leaching of this mineral with the use of multiple additives and in different media have been carried out, such as those that include bioleaching [22,28–36], ferric sulfate solution [12,37–40], chloride medium [13,24,41–45], MnO_2 as an oxidizing agent [46], pressure leaching [47,48], cyanide medium [49,50], ethylenediaminetetraacetic acid (EDTA) [51,52], and synthetic chalcocite (white metal) [23,53,54].

Among the different leaching media to treat secondary copper sulfides, chloride media has had the greatest growth at an industrial level. This is not only due to the good results presented by chloride media in heap leaching processes but also due to freshwater shortage. This is due to three aspects. (i) Chlorinated media have a higher dissolution rate than traditional sulfated systems. This is due to the ability of the chloride ion to stabilize cuprous through the formation of $CuCl_3^{2-}$ [8,13,43,46]. Furthermore, the addition of chloride ions makes it possible to overcome passivation due to the formation of the sulfur layer. The chloride ions increase the redox potential, generating a thinner layer and making it easier for the chloride ions to cause porosity [55]. Figure 1 shows the effect of chloride ions in a heap leaching process at an industrial level (Chilean mine company). For an acid medium without the addition of chloride, extractions of 35% Cu are obtained, while when working at 20 and 50 g/L of chloride, extractions of 50% and 55% Cu are obtained, respectively (for 90 days) [56]. (ii) It is an economic system since the chloride present in seawater (20 g/L Cl^-) or wastewater from desalination plants ($\sim$40 g/L Cl^-) can be used. (iii) Government restrictions on the use of aquifer water in large-scale mining projects. Although mining consumes much less water in its processes than other industries such as agriculture, mining deposits are generally found in arid areas where freshwater is scarce. Therefore, the use of seawater becomes practically a necessity. For example, in Chile (the largest copper producer in the world), it is projected that by 2030 seawater will represent almost 50% of the consumption of water in mining [57].

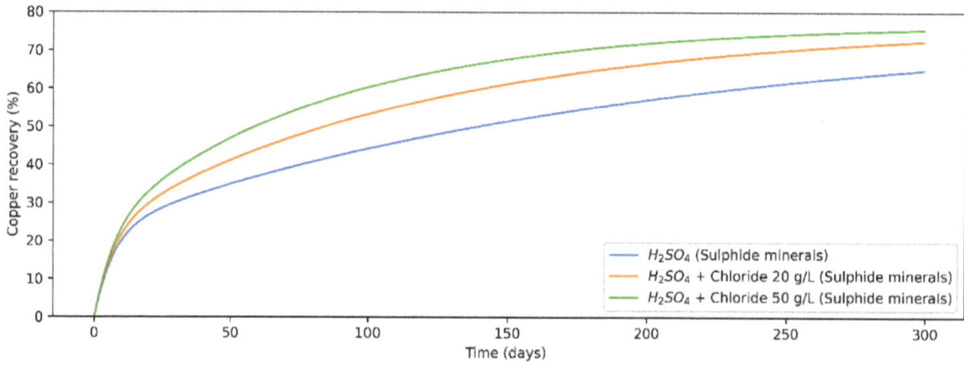

Figure 1. Extraction of copper from sulfide ores in a heap leach using H_2SO_4 and chloride (modified from: [56]).

A bibliographic review based on scientific publications in recent years on chalcocite leaching in chloride media is carried out in the present manuscript. The objective of this work is to evaluate the different operational parameters that influence the dissolution of chalcocite, comparing the impact that each one has on the extraction of copper from it.

2. Fundamentals

The oxidative dissolution of chalcocite by Fe^{3+}, O_2, or Cu^{2+}, either in a sulfate or chloride system, occurs in two stages (reactions 1 and 2), where the chalcocite dissolves, a progressive transformation occurs of this copper sulfide, passing through different stages called polysulfides (digenite $Cu_{1.8}S$; geerite $Cu_{1.6}S$; spionkopite $Cu_{1.4}S$; yarrowite $Cu_{1.1}S$), until reaching the covellite CuS [31,39,58]. The formation of said intermediate polysulfides in the transformation of chalcocite to covellite during the first stage of leaching generates a passivating layer in the mineral particle, which can be seen in Figure 2, where said layers formed are represented. Between 10% to 20% extraction, a thin layer of covellite covers the surface of the mineral, and in the same way, but toward the interior, an intermediate layer is formed with a decreasing proportion of Cu/S. In the second leaching stage, when 49% to 55% of copper has already been extracted, the chalcocite has already been converted to covellite, and a mixture of polysulfide and sulfur is generated on the surface, while in the interior, it remains covellite and as measured. As the leaching progresses, the covellite progressively converts into sulfur and polysulfide (CuS_n) with a decreasing Cu/S ratio [42].

$$Cu_2S + 2\,Fe^{3+} = Cu^{2+} + 2Fe^{2+} + CuS \tag{1}$$

$$CuS + 2\,Fe^{3+} = Cu^{2+} + 2Fe^{2+} + S^0 \tag{2}$$

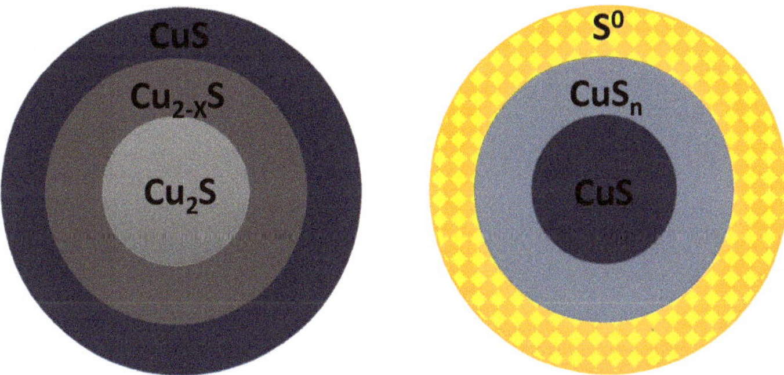

Figure 2. Graphic representation of the dissolution of the two stages of chalcocite leaching (modified from: [42]).

According to Niu et al. [39], in Equation (1) a rapid leaching occurs from chalcocite to covellite due to the low activation energy required (4–25 kJ/mol) in the kinetic model of the unreacted nucleus, being a reaction controlled by the diffusion of the oxidant on the surface of the mineral. Meanwhile, Equation (2) shows leaching is slower and can accelerate as a function of temperature [13]. The researchers Ruan et al. [59] and Miki et al. [38] have concluded that this is because this reaction is controlled by chemical and/or electrochemical reactions under the kinetic model of the unreacted nucleus, requiring an activation energy of around 71.5–72 kJ/mol for the transformation of covellite to dissolved copper. Nicol and Basson [60] suggest that in the oxidation of covellite, an intermediate stage occurs in which it is transformed to a polysulfide CuS_2.

$$Cu_2S_2 = CuS_2 + Cu^{2+} + 2e^- \tag{3}$$

$$CuS_2 = Cu^{2+} + 2S^0 + 2e^- \tag{4}$$

Covellite can be oxidized over a wide range of chloride concentrations or potentials to CuS_2 polysulfide (Equation (4)). Still, the oxidation of CuS_2 (Equation (5)) can only occur under conditions of high chloride concentrations or high potentials (chloride concentrations greater than 2.5 mol/L or potentials greater than 650 mV) [60].

For the dissolution of chalcocite in a sulfated–chloride medium, various investigations have been carried out using different additives and operational conditions (Table 2). There is a consensus in all the investigations regarding the positive effect on the dissolution kinetics of Cu_2S when adding chloride, either synthetic or using seawater. This is because chloride ions promote the formation of long crystals that allow the reagent to penetrate through the passivating layer [43].

Table 2. Comparison of previous investigations for the leaching of chalcocite in a chloride medium.

Investigation	Leaching Agent	Parameters Evaluated	Temperature (°C)	Reference	Max Cu Extraction (%)
The kinetics of leaching chalcocite (synthetic) in acidic oxygenated sulfate–chloride solutions	$NaCl$, H_2SO_4, HCl, HNO_3, and Fe^{3+}	Oxygen flow, stirring speed, temperature, acid concentration, ferric ion concentration, chloride concentration, and particle size	65–94	[24]	97
Leaching kinetics of digenite concentrate in oxygenated chloride media at ambient pressure	$CuCl_2$, HCl, and $NaCl$	Effect of stirring speed, oxygen flow, cupric ion concentration, chloride concentration, acid concentration, and temperature effect	50–100	[61]	95
Leaching of sulfide copper ore in a $NaCl$–H_2SO_4–O_2 media with acid pre-treatment	$NaCl$ and H_2SO_4	Chloride concentration, effect of agitation with compressed air, percentage of solids, and particle size	20	[62]	78
The kinetics of dissolution of synthetic covellite, chalcocite, and digenite in dilute chloride solutions at ambient temperatures	HCl, Cu^{2+}, and Fe^{3+}	Potential redox effect, chloride concentration, acid concentration, temperature, dissolved oxygen, and pyrite effect	35	[38]	98
Leaching of pure chalcocite in a chloride media using seawater and wastewater	$NaCl$, H_2SO_4, and Cl^- from seawater and wastewater	Chloride and acid concentration	25	[13]	68
Modeling the kinetics of chalcocite leaching in acidified cupric chloride media under fully controlled pH and potential	HCl, $CuCl_2$, $NaCl$, KCl, $CaCl_2$, and $MgCl_2$	Chloride concentration, cupric concentration, particle size, and temperature	25–65	[63]	98
Leaching of pure chalcocite in a chloride medium using wastewater at high temperature	H_2SO_4 and Cl^- from wastewater	Temperature effect	65–95	[43]	97
The response of the sulfur chemical state to different leaching conditions in chloride leaching of chalcocite	$FeCl_3$, $CuCl_2$, and HCl	Chloride concentration, temperature, and potential redox effect.	35–55	[42]	88
Leaching of pure chalcocite with reject brine and MnO_2 from manganese nodules	H_2SO_4, MnO_2, and Cl^- from seawater and wastewater	MnO_2, chloride, and acid concentration	25	[46]	71

In a Cu_2S leaching process, adding O_2 to the system at ambient pressure, with H_2SO_4 being the leaching agent, the leaching agents generated during leaching in a Cu^{2+}/Cl^- system are Cu^{2+}, $CuCl^+$, $CuCl_2^-$, and $CuCl^{3-}$. The general reaction being the following:

$$2Cu_2S + O_2 + 4H^+ + 8Cl^- = 4CuCl_2^- + 2H_2O + 2S^0 \tag{5}$$

Although chalcocite leaching reactions occur in two stages, guiding us to Equation (5), where the following occurs:

$$4Cu_2S + O_2 + 4H^+ + 8Cl^- = 4CuCl_2^- + 2H_2O + 4CuS \tag{6}$$

$$4CuS + O_2 + 4H^+ + 8Cl^- = 4CuCl_2^- + 2H_2O + 4S^0 \tag{7}$$

By leaching Cu_2S, the expected resulting products should be soluble copper such as $CuCl_2^-$ and a solid residue of elemental sulfur (S^0) with covellite residues or copper polysulfides (CuS_2) that still contain valuable metals.

$CuCl_2^-$ is the predominant soluble species due to the complexation of Cu (I) with Cl^- at room temperature in a system with high chloride concentrations (greater than 1 mol/L). This $CuCl_2^-$ is stable in a potential range between 0–500 mV and pH < 6–7 (depending on the chloride concentration in the system).

For a sulfated–chloride system where MnO_2 is incorporated as an oxidizing agent, the following reactions are proposed:

$$2Cu_2S + MnO_2 + 4H^+ + 4Cl^- = 2CuCl_2^- + Mn^{2+} + 2CuS + 2H_2O \tag{8}$$

$$2CuS + MnO_2 + 4H^+ + 4Cl^- = 2CuCl_2^- + Mn^{2+} + 2S^0 + 2H_2O \tag{9}$$

During the first leaching stage (Equation (8)), the chalcocite becomes covellite; this reaction being thermodynamically possible with a Gibbs free energy value of -138.59 kJ. The second reaction (Equation (9)) is slower and is also thermodynamically possible ($\Delta G^0 = -84.512$ kJ).

3. Operational Variables

3.1. Effect on Chloride Concentration

Several authors point out that chalcocite leaching in a chloride medium is the best way to dissolve this copper sulfide [8,13,24,38,41,63]. Even if chloride ions are added to a chalcocite leaching with H_2SO_4 or HNO_3, the kinetics increases considerably. As explained by Cheng and Lawson [24], this occurs because in leaching with only sulfate or nitrate ions, a layer of elemental sulfur is formed on the surface of the particles. In this way, an impermeable particle is generated, that is, contact between the particle with the leaching agent is prevented. This implies that the kinetics decrease in the first leaching stage and prevent the reaction in the second stage. However, when chloride ions are found, either alone or associated with sulfate or nitrate, dissolution kinetics increase along with copper extraction, as shown in Figure 3.

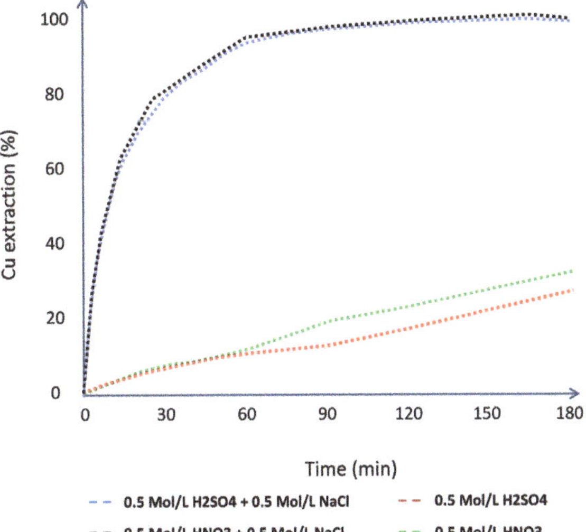

Figure 3. Effect of chloride ions on the acid leaching of chalcocite (T = 85 °C, particle size 31 μm) (Modified from: [24]).

Several studies have shown that working at high chloride concentrations favors the leaching kinetics of secondary sulfides [24,38,44,64]. Chloride ions pass through the sulfur layer and generate a porous layer instead of an amorphous layer formed in the sulfate and nitrate system. The porous layer allows the entry of the leaching solution through said pores, thus allowing contact with the particle, thus accelerating the leaching kinetics in the first stage and making possible the dissolution reaction in the second stage of leaching [24,65,66]. In the study carried out by Toro et al. [13], leaching tests were carried out in stirred reactors for a pure mineral of chalcocite in an acid medium, comparing different concentrations of chloride in the system (20, 40, and 100 g/L). In their results, the authors indicate that the highest Cu extractions are obtained when working at the highest chloride concentrations (see Figure 4). Furthermore, in other studies [43,46] involving the use of seawater (20 g/L Cl$^-$) and wastewater from desalination plants (~40 g/L Cl$^-$) for the dissolution of Cu_2S in an acid medium, the researchers point out that better results are obtained when working with wastewater compared to seawater due to its higher concentration of chloride. Additionally, it is highlighted that the waste generated when working with wastewater is stable (such as elemental sulfur) and non-polluting.

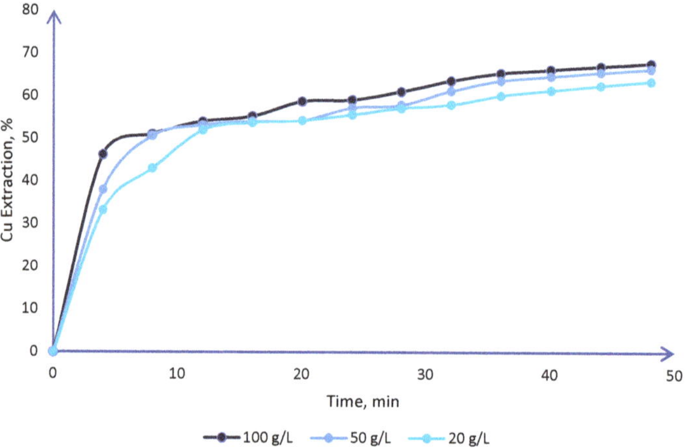

Figure 4. Effect of the chloride concentration in Cu_2S solution (T = 25 °C, H_2SO_4 = 0.5 mol/L) (Modified from: [13]).

The high dissolution rate in the chloride system relative to the sulfated system is attributed to the ability of the chloride ion to stabilize the cuprous ion through the formation of $CuCl_3^{2-}$. In the chloride system, copper can be extracted directly from the chalcocite without causing the oxidation of Cu^+ to Cu^{2+}. On the other hand, in the sulfated system, Cu+ must be oxidized to Cu^{2+} on the surface of the particles before copper is released into the solution [8,13,41,64]. The addition of chloride ions allows breaking the passivated sulfur layer since an increase in the concentration of chloride ions implies an increase in the redox potential [42], and a higher redox potential generates a thinner layer that makes it easier for chloride ions to generate porosity [13].

3.2. Effect on Stirring Speed

The agitation speed in a reactor leaching system decreases the thickness of the boundary layer and maximizes the gas–liquid interface area [67]. This variable is not very significant in copper extraction for tests of the dissolution of Cu_2S in an acid–chloride medium. There is a consensus on the part of different authors in previous research [24,38,61,62,68] where it is stated that it is only necessary to stir at a sufficient speed to keep all the chalcocite particles in suspension within the reactor. Additionally, it is important to note that of the various agitation systems used in these investigations it is advisable to work with mechanical agitation since with other systems anomalous results are obtained, for example, in the study carried out by Herreros and Viñals [62] the authors indicate that in their air agitation tests the results were superior under the same operational conditions compared to mechanical agitation tests. This occurred because the air increased the extraction of copper. After all, the oxygen reacted with the CuCl (solid) formed during the leaching process, favoring the formation of $CuCl^+$. On the other hand, Velásquez-Yévenes [68], in his study, mentions that when working with the use of magnetic agitation the mineral is reduced in size due to the abrasion that is generated when it passes under the rotating magnet, generating an increase in the dissolution of chalcocite.

3.3. Effect on Acid Concentration

Regarding the acid concentration in a sulfate–chloride system, the findings presented by Dutrizac [69], Cheng and Lawson [24], Senanayake [45], Toro et al. [13], Saldaña et al. [44], and Torres et al. [46] confirm that the concentration of chloride ions in the system is the variable that most influences the kinetics of the dissolution of chalcocite at room temperature, making other operational variables, such as acid concentration, particle size, stirring speed, etc., less relevant. These same results were obtained for other copper sulfides such as covellite [64] and chalcopyrite [70].

Toro et al. [13] performed statistical analysis (ANOVA) for the dissolution of Cu_2S in a chloride medium in stirred reactors. For this, the copper extraction was evaluated through the effect of the independent variables with the response surface optimization method (See Table 3). In their results, the researchers indicate that, although sulfuric acid helps to improve the dissolution kinetics of the mineral, the chloride concentration in the system has much more impact on copper extraction, as shown in Figure 5. These results are consistent with those presented by Cheng and Lawson [24], where the researchers mention that a low concentration of H_2SO_4 (0.02 mol/L) is sufficient to dissolve chalcocite and later phases of it such as djurleite and digenite. However, it is essential to maintain a high concentration of chloride ions since in its absence the dissolution kinetics considerably decrease (first stage) and later the covellite is not dissolved (second stage). On the other hand, a recent study by Torres et al. [46] worked with wastewater at different concentrations of sulfuric acid to dissolve a pure chalcocite mineral. The authors mention that the same results were obtained in their results, even in short periods of time at H_2SO_4 concentration ranges between 0.1 and 1 mol/L (see Figure 6).

Table 3. Experimental parameters used in statistical analysis [13].

Experimental Parameters	Low	Medium	High
Time (h)	4	8	12
Cl^- concentration (g/L)	20	50	100
H_2SO_4 (mol/L)	0.5	1	2

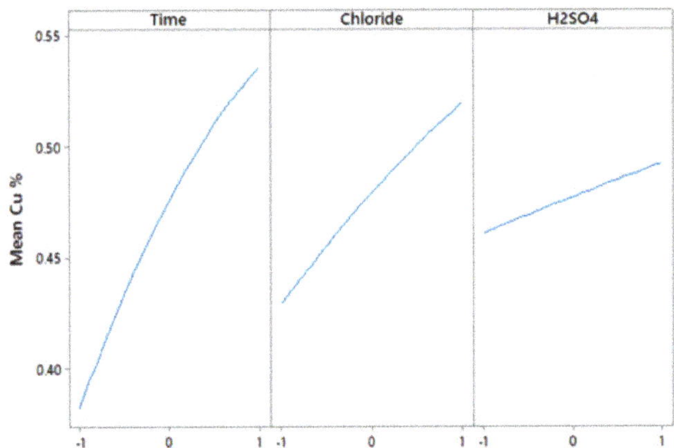

Figure 5. Linear effect graph for the extraction of Cu from chalcocite in a chloride medium [13].

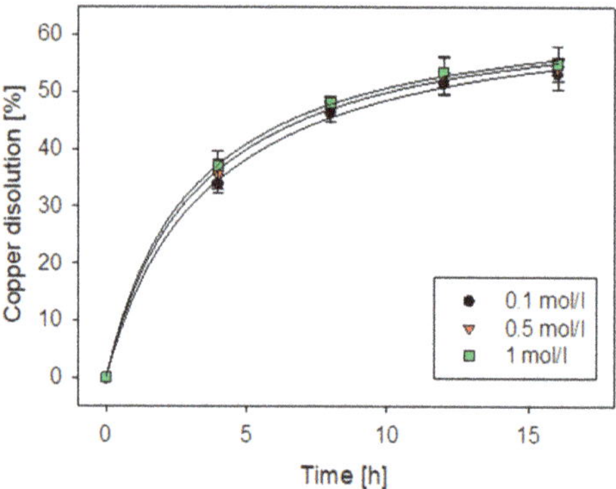

Figure 6. Effect on the H_2SO_4 concentration in the Cu_2S solution with the use of wastewater ($\sim$40 g/L Cl$^-$) [46].

3.4. Particle Size Effect

The effect of particle size on chalcocite leaching has been studied by different authors; however, these studies have been carried out with relatively small particle sizes: 25 to 4 mm [71]; 4 mm to 12 µm [62]; 4 to 0.054 mm [12]; 11 to 63 µm [24]; 150 to 75 µm [63]; 150 to 106 µm [72]. These authors agree that a smaller particle size implies an increase in the dissolution kinetics and the extraction rate in the first leaching stage. But the effect decreases significantly in the second stage. Naderi et al. [71] reported that for fine particle sizes the first stage is controlled by diffusion through the liquid film. In the second stage, the accumulation of the elemental sulfur layer in the solid product, accompanied by a jarosite precipitate, transformed the control mechanism into solid diffusion. Phyo et al. [12] studied the effect on the dissolution kinetics of Cu_2S in stirred reactors using an acid medium. In their results, as can be seen in Figure 7a, a significant effect of the particle size is observed in the dissolution of copper, especially in the size of $-0.074 + 0.054$ mm, which in 2.5 h had already reached 45% recovery compared to the almost 17 hours it took to achieve the same recovery with $-4 + 2$ mm particles. In Figure 7b, the researchers observed a turning point of around 75% copper dissolution at different times depending on the granulometry and divided the second stage into two sub-stages, indicating that the first sub-stage has a dissolution speed 20 times faster than the second sub-stage.

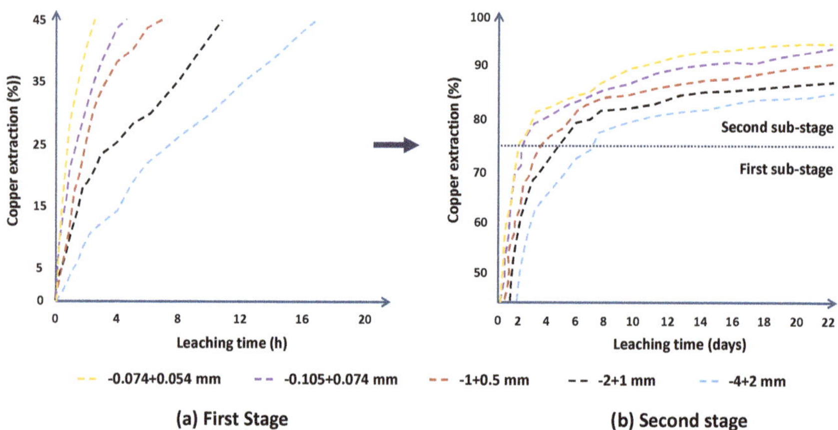

Figure 7. Cu_2S dissolution at different particle sizes in two different stages: (**a**) first stage, (**b**) second stage ([Fe^{3+}] = 10 g/dm^3, pH = 1.00–1.50, Eh = 750 mV, temperature = 45 °C) (Modified from: [12]).

3.5. Effect of Temperature

It is known that temperature is the operational variable that most influences the dissolution of copper sulfide minerals [43,73,74]. For the specific case of chalcocite, it becomes a critical parameter during the second stage, being much slower, and can be accelerated with temperature, which indicates that the process is controlled by chemical and/or electrochemical reactions [43,59]. Miki et al. [38] pointed out that the dissolution rate of CuS is largely independent of the concentration of chloride and HCl in the ranges of 0.2 to 2.5 mol/L and 0.1 to 1 mol/L, reporting activation energy values of 71.5 kJ/mol. Therefore, it can be concluded that the process is controlled by a chemical or electrochemical reaction on the surface of the mineral.

In the results presented by Pérez et al. [43] for the dissolution of a pure chalcocite mineral in a chloride medium in a stirred reactor at different temperatures, the authors point out that at temperatures above 65 °C extractions of copper were close to 40% in short periods of time (15 min) (see Figure 7), which, the researchers concluded, is due to the phase change that governs the first stage of leaching from chalcocite to covellite, which requires low activation energy. More energy is necessary for the second stage to become a copper polysulfide, which requires more demanding conditions to achieve its complete dissolution. In addition, Pérez et al. [43] mentioned that there is good synergy between the chloride concentration in the system and the temperature, since, in their research, they achieved copper extractions of 97% in 3 hours under the conditions operations that are presented in Figure 8. The research carried out by Ruiz et al. [54] investigated the dissolution of white metal (chalcocite and djurleite) working under similar operational conditions. Without chloride in the system, 55% extractions were obtained in a time of 5 hours at a temperature of 105 °C.

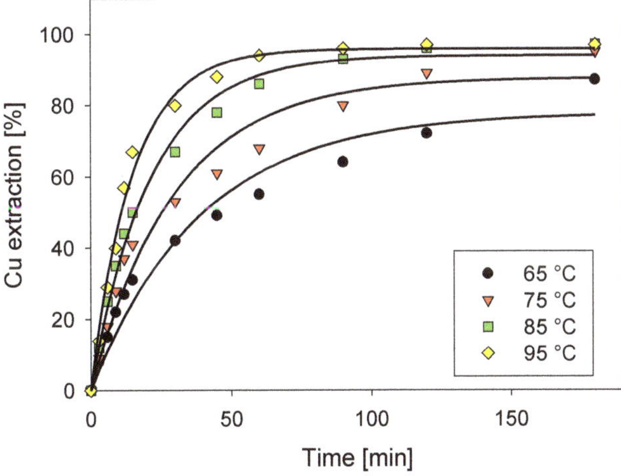

Figure 8. Cu_2S dissolution as a function of temperature (0.5 mol/L H_2SO_4 and 100 g/L of Cl^-) [43].

3.6. Effect of Redox Potential

Miki et al. [38] studied the effect of the redox potential in a Cu_2S solution (synthetic) with the use of a 0.2 M HCl solution, 0.2 g/L of Cu (II), and 2 g/L of Fe (III)/Fe (II) at a temperature of 35 °C. The researchers, in their findings, reported that the dissolution of Cu_2S occurs rapidly at a potential of 500 mV but then stops when 45% copper is removed (end of the first stage). For potentials of 550 mV, there is then an increase in the dissolution until reaching 50% extraction of copper. Subsequently, the copper mineral present is mainly covering, which requires potentials of at least 600 mV to be able to dissolve. However, Niu et al. [39] point out that these results were not determined in a range of industrial redox potentials. In their experiments for the dissolution of Cu_2S, Niu et al. [39] worked in mini glass columns (30 cm long and 6 cm in diameter), adding $Fe_2(SO_4)_3$ as a leaching agent. In their results, the researchers note that the dissolution rate of the second stage of Cu_2S leaching was insensitive to the redox potential at moderate temperatures (30–40 °C) in the industrial range of 650–800 mV. In the study conducted by Hashemzadeh et al. [63], the researchers modeled the dissolution kinetics of Cu_2S in chloride media using leaching data obtained under fully controlled temperature, pH, and solution potential. In their results, the researchers mentioned that an increase in the chloride concentration and temperature generated an increase in the redox potential, increasing from 680 to 830 mV with the addition of 0.1 chlorides and 3 mol/L of NaCl, respectively, and consequently higher dissolution kinetics, mainly in the second leaching stage.

The results obtained from the aforementioned studies are directly related to the formation of a layer of elemental sulfur on the surface of the covellite in the second stage, which decreases with the increase in the potential of the solution; however, the layer in the solid surface is a mixture of sulfur and polysulfides (CuS_n) [42], where these could be responsible for a slow reaction during this stage.

3.7. Effect of Oxidizing Agents
3.7.1. Air

In another study carried out by Cheng and Lawson [24], the effect of aeration on the leaching of chalcocite was seen, but in this case, agitating leaching was carried out, showing that there is no greater contribution. Moreover, by adding a greater flow of air to the system slightly lower recoveries are obtained. This is because the particles adhere to the bubbles generated and are dragged toward the walls of the reactor. On the other hand, in the research carried out by Liu and Granata [75], the effect of aeration was studied by analyzing historical data in two leaching piles of chalcocite as the main mineral present, one with and the other without aeration. In the results, they observed that in leaching in an aerated heap, better results were obtained, but there was only a comparative change after 200 days of leaching with respect to the non-aerated heap. For the aerated pile, from 200 days onward, there was no significant change in copper recovery. This is because aeration is no longer beneficial at this point. After all, we are in the second stage of the dissolution of chalcocite, which is a controlled chemical reaction. Furthermore, the authors mentioned that the total acid consumption per ton of ore processed was higher in the case of the aerated pile, but the net acid consumption per ton of copper produced was the same in both cases.

3.7.2. Ferric Ions

The effect of the concentration of ferric ions in the leaching of chalcocite has also been studied. For the first reaction stage, there is a positive effect on the leaching kinetics with an increase in the concentration of ferric ions. When the concentration is lowered, the leaching rate of chalcocite is considerably lower. For the second stage, the ferric ions are not as noticeable as in the first stage. Still, better reaction rates are obtained by increasing their concentration because of the increase generated in the redox potential of the leaching solution [24,65,72].

3.7.3. MnO_2

The use of MnO_2 as an oxidizing agent in chloride media has recently been studied for secondary and primary sulfides [46,70,76–78] where positive results have been obtained in the dissolution of copper. For example, for a mineral refractory to conventional processes such as chalcopyrite, in the study carried out by Toro et al. [70] it was possible to extract 77% of copper at room temperature when working at high concentrations of MnO_2 (4/1 and 5/1) and chloride (~40 g/L), which allowed maintaining of the redox potential values between 580 and 650 mV. For the specific case of chalcocite, Torres et al. [46] worked under the same operational conditions as Toro et al. [13] (see Table 4). In their results, Torres et al. (2020a) showed that incorporating MnO_2 at low concentrations significantly improves the dissolution of chalcocite in short periods, which is important in continuous leaching operations.

Table 4. Comparison between studies for the dissolution of chalcocite in chloride media, with and without the addition of MnO_2.

Experimental Conditions and Results	[13]	[46]
Temperature (°C)	25	25
Particle size of Cu_2S (μm)	$-147 + 104$	$-147 + 104$
H_2SO_4 concentration (mol/L)	0.5	0.5
MnO_2/Cu_2S ratio (w/w)	-	0.25/1
Dissolution in seawater after 4 h (%)	32.8	35.6
Dissolution in reject brine after 4 h (%)	36	40
Dissolution in seawater after 48 h (%)	63.4	64.7
Dissolution in reject brine after 48 h (%)	64.6	66.2

4. Conclusions

Among the various leaching processes to treat chalcocite, chloride media show better results and have greater industrial relevance. This is because of the positive results in copper extraction, low cost, and the possibility of working with seawater. However, chalcocite leaching is a process that occurs in two stages, which must be evaluated individually according to the different operational parameters that can be tested in the process.

In general:

Working in chloride media favors the dissolution of Cu_2S, accelerating the leaching kinetics in the first stage and making possible the dissolution reaction in the second stage. This is mainly due to two reasons: (i) the chloride ions in the system allow the cuprous ions to be stabilized through the formation of $CuCl_3^{2-}$, allowing the copper to be extracted directly from Cu_2S without prior oxidation of Cu^+ to Cu^{2+}; (ii) the chloride ions promote the formation of long crystals that allow the penetration of the reagent through the passivating layer. Furthermore, the concentration of chloride ions is the variable that most influences the dissolution kinetics of Cu_2S at room temperature, making other operational variables, such as acid concentration, particle size, stirring rate, and addition of other oxidizing agents (air, ferric ions, etc.), less relevant.

Evaluating by stage:

The first stage of leaching occurs quickly, requiring low activation energy (4–25 kJ/mol) in the unreacted core model, via a controlled reaction by diffusion of the oxidant on the mineral surface, while reaction 2 is much slower and requires higher activation energy (71.5–72 kJ/mol), being a stage controlled by chemical reaction. During the second stage, in the first instance, it is possible to oxidize the covellite in a wide range of chloride concentrations or redox potentials (up to 75% extraction of Cu). Subsequently, CuS_2 is formed, which to be oxidized it is necessary to work at high concentrations of chloride (>2.5 mol/L) and/or increase the system's temperature. This is because to dissolve covellite it is necessary to increase the redox potential of the system (>650 mV), which in turn decreases the thickness of the elemental sulfur layer on the mineral surface, facilitating chloride ions to generate a better porosity of this. Furthermore, it is important to note good synergy between the chloride concentration in the system and the temperature. The operational parameters impact differently in each of the stages, as can be seen below in Table 5:

Table 5. Impact of the different operational parameters on the dissolution of Cu_2S.

Parameters	First Stage	Second Stage
Chloride concentration	Increases dissolution kinetics	Help prevent passivation
Stirring rate	It is not relevant	It is not relevant
Acid concentration	A low concentration of H_2SO_4 (0.02 mol/L) is sufficient to dissolve the mineral. The same results are obtained between 0.1 and 1 mol/L of H_2SO_4.	Increases dissolution kinetics
Particle size	Increase in dissolution kinetics.	Slight increase in dissolution kinetics
Temperature	Significantly accelerates dissolution	Significantly accelerates dissolution Helps prevent passivation
Redox potential	Low redox potential is required ($\geq$500 mV)	High redox potential values are required (>650 mV)
Oxidizing agents (air, Fe^{3+}, MnO_2)	Increases dissolution kinetics by adding low concentrations	Increases dissolution kinetics, but only at high concentrations

Finally, it is concluded that the most optimal way to extract copper from chalcocite is, during the first stage, to work at high concentrations of chloride (50–100 g/L) and low concentrations of sulfuric acid (0.5 mol/L) at a temperature environment. Other variables become irrelevant during this stage if the concentration of chloride ions in the system is high. In the second stage, it is necessary to increase the temperature of the system (moderate temperatures) or incorporate a high concentration of some oxidizing agent to avoid the passivation of the mineral.

Author Contributions: N.T., D.T. and K.P. contributed in research and wrote paper, C.M., M.S. and E.G. contributed with research, review and editing. All authors have read and agreed to the published version of the manuscript.

Funding: This research received no external funding.

Acknowledgments: Kevin Pérez acknowledges the infrastructure and support of Doctorado en Ingeniería de Procesos de Minerales of the Universidad de Antofagasta. Carlos Moraga acknowledges the support of Centro Tecnológico de Conversión de Energía of the University of Talca.

Conflicts of Interest: The authors declare no conflict of interest.

References

1. Bogdanović, G.D.; Petrović, S.; Sokić, M.; Antonijević, M.M. Chalcopyrite leaching in acid media: A review. *Metall. Mater. Eng.* **2020**, *26*, 177–198. [CrossRef]
2. Phuong Thao, N.T.; Tsuji, S.; Jeon, S.; Park, I.; Tabelin, C.B.; Ito, M.; Hiroyoshi, N. Redox potential-dependent chalcopyrite leaching in acidic ferric chloride solutions: Leaching experiments. *Hydrometallurgy* **2020**, *194*, 105299. [CrossRef]
3. Taboada, M.E.; Hernández, P.C.; Padilla, A.P.; Jamett, N.E.; Graber, T.A. Effects of Fe+2 and Fe+3 in Pretreatment and Leaching on a Mixed Copper Ore in Chloride Media. *Metals* **2021**, *11*, 866. [CrossRef]
4. Turan, M.D.; Sarı, Z.A.; Nizamoğlu, H. Pressure leaching of chalcopyrite with oxalic acid and hydrogen peroxide. *J. Taiwan Inst. Chem. Eng.* **2021**, *118*, 112–120. [CrossRef]
5. Dutrizac, J. Elemental sulphur formation during the ferric chloride leaching of chalcopyrite. *Hydrometallurgy* **1990**, *23*, 153–176. [CrossRef]
6. O'Connor, G.M.; Eksteen, J.J. A critical review of the passivation and semiconductor mechanisms of chalcopyrite leaching. *Miner. Eng.* **2020**, *154*, 106401. [CrossRef]
7. Ram, R.; Coyle, V.E.; Bond, A.M.; Chen, M.; Bhargava, S.K.; Jones, L.A. A scanning electrochemical microscopy (SECM) study of the interfacial solution chemistry at polarised chalcopyrite (CuFeS2) and chalcocite (Cu2S). *Electrochem. Commun.* **2020**, *115*, 106730. [CrossRef]
8. Torres Albornoz, D.A. Copper and Manganese Extraction Through Leaching Processes, Universidad Politécnica de Cartagena. 2021. Available online: https://repositorio.upct.es/bitstream/handle/10317/9373/data_C.pdf?sequence=1&isAllowed=y (accessed on 1 October 2021).
9. Hidalgo, T.; Verrall, M.; Beinlich, A.; Kuhar, L.; Putnis, A. Replacement reactions of copper sulphides at moderate temperature in acidic solutions. *Ore Geol. Rev.* **2020**, *123*, 103569. [CrossRef]

10. Domic, E.M. A Review of the Development and Current Status of Copper Bioleaching Operations in Chile: 25 Years of Successful Commercial Implementation. In *Biomining*; Springer: Berlin/Heidelberg, Germany, 2007; pp. 81–95.
11. Ghorbani, Y.; Becker, M.; Mainza, A.; Franzidis, J.-P.; Petersen, J. Large particle effects in chemical/biochemical heap leach processes—A review. *Miner. Eng.* **2011**, *24*, 1172–1184. [CrossRef]
12. Phyo, H.A.; Jia, Y.; Tan, Q.; Zhao, S.; Liang, X.; Ruan, R.; Niu, X. Effect of particle size on chalcocite dissolution kinetics in column leaching under controlled Eh and its implications. *Physicochem. Probl. Miner. Process.* **2020**, *56*, 676–692. [CrossRef]
13. Toro, N.; Briceño, W.; Pérez, K.; Cánovas, M.; Trigueros, E.; Sepúlveda, R.; Hernández, P. Leaching of pure chalcocite in a chloride media using sea water and waste water. *Metals* **2019**, *9*, 780. [CrossRef]
14. Watling, H.R. The bioleaching of sulphide minerals with emphasis on copper sulphides—A review. *Hydrometallurgy* **2006**, *84*, 81–108. [CrossRef]
15. Dunn, J.G.; Ginting, A.R.; O'Connor, B. A thermoanalytical study of the oxidation of chalcocite. *J. Therm. Anal.* **1994**, *41*, 671–686. [CrossRef]
16. Evans, H.T. The crystal structures of low chalcocite and djurleite. *Zeitschrift Krist.-Cryst. Mater.* **1979**, *150*, 299–320. [CrossRef]
17. Reddy, S.L.; Fayazuddin, M.; Frost, R.L.; Endo, T. Electron paramagnetic resonance and optical absorption spectral studies on chalcocite. *Spectrochim. Acta Part A Mol. Biomol. Spectrosc.* **2007**, *68*, 420–423. [CrossRef]
18. Tanda, B.C.; Eksteen, J.J.; Oraby, E.A. Kinetics of chalcocite leaching in oxygenated alkaline glycine solutions. *Hydrometallurgy* **2018**, *178*, 264–273. [CrossRef]
19. Wang, L.-W. High Chalcocite Cu2S: A Solid-Liquid Hybrid Phase. *Phys. Rev. Lett.* **2012**, *108*, 085703. [CrossRef] [PubMed]
20. Kang, J.; Lee, J.Y. Production of chalcocite by selective chlorination of chalcopyrite using cuprous chloride. *Miner. Metall. Process.* **2017**, *34*, 76–83. [CrossRef]
21. Parikh, R.S.; Liddell, K.C. Mechanism of anodic dissolution of chalcocite in hydrochloric acid solution. *Ind. Eng. Chem. Res.* **1990**, *29*, 187–193. [CrossRef]
22. Wu, B.; Yang, X.; Wen, J.; Wang, D. Semiconductor-Microbial Mechanism of Selective Dissolution of Chalcocite in Bioleaching. *ACS Omega* **2019**, *4*, 18279–18288. [CrossRef]
23. Castillo, J.; Sepúlveda, R.; Araya, G.; Guzmán, D.; Toro, N.; Pérez, K.; Rodríguez, M.; Navarra, A. Leaching of white metal in a NaCl-H2SO4 system under environmental conditions. *Minerals* **2019**, *9*, 319. [CrossRef]
24. Cheng, C.Y.; Lawson, F. The kinetics of leaching chalcocite in acidic oxygenated sulphate-chloride solutions. *Hydrometallurgy* **1991**, *27*, 249–268. [CrossRef]
25. Dunn, J.G.; Ginting, A.; O'Connor, B.H. Quantitative determination of phases present in oxidised chalcocite. *J. Therm. Anal.* **1997**, *50*, 51–62. [CrossRef]
26. Posfai, M.; Buseck, P.R. Djurleite, digenite, and chalcocite: Intergrowths and transformations. *Am. Mineral.* **1994**, *79*, 308–315.
27. Saldaña, M.; Neira, P.; Flores, V.; Robles, P.; Moraga, C. A Decision Support System for Changes in Operation Modes of the Copper Heap Leaching Process. *Metals* **2021**, *11*, 1025. [CrossRef]
28. Bampole, D.L.; Luis, P.; Mulaba-Bafubiandi, A.F. Sustainable copper extraction from mixed chalcopyrite–chalcocite using biomass. *Trans. Nonferrous Met. Soc. China* **2019**, *29*, 2170–2182. [CrossRef]
29. Bostelmann, H.; Southam, G. The biogeochemical reactivity of phosphate during bioleaching of bornite-chalcocite ore. *Appl. Geochem.* **2019**, *104*, 193–201. [CrossRef]
30. Kang, J.; Qiu, G.; Gao, J.; Wang, H.; Wu, X.; Ding, J. Bioleaching of chalcocite by mixed microorganisms subjected to mutation. *J. Cent. South Univ. Technol.* **2009**, *16*, 218–222. [CrossRef]
31. Leahy, M.J.; Davidson, M.R.; Schwarz, M.P. A model for heap bioleaching of chalcocite with heat balance: Bacterial temperature dependence. *Miner. Eng.* **2005**, *18*, 1239–1252. [CrossRef]
32. Leahy, M.J.; Davidson, M.R.; Schwarz, M.P. A model for heap bioleaching of chalcocite with heat balance: Mesophiles and moderate thermophiles. *Hydrometallurgy* **2007**, *85*, 24–41. [CrossRef]
33. Petersen, J.; Dixon, D.G. The dynamics of chalcocite heap bioleaching. In *Hydrometallurgy*; Young, C., Alfantazi, A., Anderson, C., James, A., Dreisinger, D.B., Harris, B., Eds.; In Proceedings of the 5th International Symposium Honoring Professor Ian M. Ritchie; TMS Publishers: Vancouver, BC, Canada, 2003.
34. Petersen, J.; Dixon, D.G. Modeling and Optimization of Heap Bioleach Processes. In *Biomining*; Springer: Berlin/Heidelberg, Germany, 2007; pp. 153–176.
35. Wu, B.; Wen, J.-K.; Chen, B.-W.; Yao, G.-C.; Wang, D.-Z. Control of redox potential by oxygen limitation in selective bioleaching of chalcocite and pyrite. *Rare Met.* **2014**, *33*, 622–627. [CrossRef]
36. Zhang, Y.; Zhao, H.; Zhang, Y.; Liu, H.; Yin, H.; Deng, J.; Qiu, G. Interaction mechanism between marmatite and chalcocite in acidic (microbial) environments. *Hydrometallurgy* **2020**, *191*, 105217. [CrossRef]
37. Bolorunduro, S.A. Kinetics of Leaching of Chalcocite in Acid Ferric Sulfate Media: Chemical and Bacterial Leaching, University of British Columbia. 1999. Available online: https://open.library.ubc.ca/soa/cIRcle/collections/ubctheses/831/items/1.0078730 (accessed on 1 October 2021).
38. Miki, H.; Nicol, M.; Velásquez-Yévenes, L. The kinetics of dissolution of synthetic covellite, chalcocite and digenite in dilute chloride solutions at ambient temperatures. *Hydrometallurgy* **2011**, *105*, 321–327. [CrossRef]
39. Niu, X.; Ruan, R.; Tan, Q.; Jia, Y.; Sun, H. Study on the second stage of chalcocite leaching in column with redox potential control and its implications. *Hydrometallurgy* **2015**, *155*, 141–152. [CrossRef]

40. Palencia, I.; Romero, R.; Mazuelos, A.; Carranza, F. Treatment of secondary copper sulphides (chalcocite and covellite) by the BRISA process. *Hydrometallurgy* **2002**, *66*, 85–93. [CrossRef]

41. Fisher, W.W.; Flores, F.A.; Henderson, J.A. Comparison of chalcocite dissolution in the oxygenated, aqueous sulfate and chloride systems. *Miner. Eng.* **1992**, *5*, 817–834. [CrossRef]

42. Hashemzadeh, M.; Liu, W. The response of sulfur chemical state to different leaching conditions in chloride leaching of chalcocite. *Hydrometallurgy* **2020**, *192*, 105245. [CrossRef]

43. Pérez, K.; Jeldres, R.; Nieto, S.; Salinas-Rodríguez, E.; Robles, P.; Quezada, V.; Hernández-Ávila, J.; Toro, N. Leaching of pure chalcocite in a chloride media using waste water at high temperature. *Metals* **2020**, *10*, 1–9. [CrossRef]

44. Saldaña, M.; Rodríguez, F.; Rojas, A.; Pérez, K.; Angulo, P. Development of an empirical model for copper extraction from chalcocite in chloride media. *Hem. Ind.* **2020**, *74*, 285–292. [CrossRef]

45. Senanayake, G. Chloride assisted leaching of chalcocite by oxygenated sulphuric acid via Cu(II)-OH-Cl. *Miner. Eng.* **2007**, *20*, 1075–1088. [CrossRef]

46. Torres, D.; Trigueros, E.; Robles, P.; Leiva, W.H.; Jeldres, R.I.; Toledo, P.G.; Toro, N. Leaching of Pure Chalcocite with Reject Brine and MnO2 from Manganese Nodules. *Metals* **2020**, *10*, 1426. [CrossRef]

47. Muszer, A.; Wódka, J.; Chmielewski, T.; Matuska, S. Covellinisation of copper sulphide minerals under pressure leaching conditions. *Hydrometallurgy* **2013**, *137*, 1–7. [CrossRef]

48. Veltman, H.; Pellegrini, S.; Mackiw, V.N. Direct acid pressure leaching of chalcocite concentrate. *JOM* **1967**, *19*, 21–25. [CrossRef]

49. Kyle, J.; Aye, K.T.; Breuer, P.; Meakin, R. The dissolution of covellite and chalcocite in cyanide solutions. The dissolution of covellite and chalcocite in cyanide solutions. In Proceedings of the World Gold 2011: 50th Annual Conference of Metallurgists of CIM, Montreal, QC, Canada, 2–5 October 2011; 2011.

50. Shantz, R.; Fisher, W.W. The kinetics of the dissolution of chalcocite in alkaline cyanide solution. *Metall. Trans. B* **1977**, *8*, 253–260. [CrossRef]

51. Konishi, Y.; Katoh, M.; Asai, S. Leaching kinetics of copper from natural chalcocite in alkaline Na4EDTA solutions. *Metall. Trans. B* **1991**, *22*, 295–303. [CrossRef]

52. Tomášek, J.; Neumann, L. Dissolution of secondary copper sulphides using complex-forming agents (EDTA, EDA). Part II: Chalcocite dissolution in EDTA and EDA. *Int. J. Miner. Process.* **1982**, *9*, 41–57. [CrossRef]

53. Herreros, O.; Quiroz, R.; Viñals, J. Dissolution kinetics of copper, white metal and natural chalcocite in Cl2/Cl− media. *Hydrometallurgy* **1999**, *51*, 345–357. [CrossRef]

54. Ruiz, M.C.; Abarzúa, E.; Padilla, R. Oxygen pressure leaching of white metal. *Hydrometallurgy* **2007**, *86*, 131–139. [CrossRef]

55. Moraga, G.A.; Jamett, N.E.; Hernández, P.C.; Graber, T.A.; Taboada, M.E. Chalcopyrite Leaching with Hydrogen Peroxide and Iodine Species in Acidic Chloride Media at Room Temperature: Technical and Economic Evaluation. *Metals* **2021**, *11*, 1567. [CrossRef]

56. Saldaña, M.; Toro, N.; Castillo, J.; Hernández, P.; Navarra, A. Optimization of the heap leaching process through changes in modes of operation and discrete event simulation. *Minerals* **2019**, *9*, 1–13. [CrossRef]

57. COCHILCO. *Proyeccion Agua Mineria del Cobre 2019–2030*; COCHILCO: Santiago, Chile, 2020; Available online: https://www.cochilco.cl/Listado%20Temtico/proyeccion%20agua%20mineria%20del%20cobre%202019-2030%20VF.pdf (accessed on 1 October 2021).

58. Whiteside, L.S.; Goble, R.J. Structural and compositional changes in copper sulfides during leaching and dissolution. *Can. Mineral.* **1986**, *24*, 247–258.

59. Ruan, R.; Zou, G.; Zhong, S.; Wu, Z.; Chan, B.; Wang, D. Why Zijinshan copper bioheapleaching plant works efficiently at low microbial activity—Study on leaching kinetics of copper sulfides and its implications. *Miner. Eng.* **2013**, *48*, 36–43. [CrossRef]

60. Nicol, M.; Basson, P. The anodic behaviour of covellite in chloride solutions. *Hydrometallurgy* **2017**, *172*, 60–68. [CrossRef]

61. Ruiz, M.C.; Honores, S.; Padilla, R. Leaching kinetics of digenite concentrate in oxygenated chloride media at ambient pressure. *Metall. Mater. Trans. B* **1998**, *29*, 961–969. [CrossRef]

62. Herreros, O.; Viñals, J. Leaching of sulfide copper ore in a NaCl–H2SO4–O2 media with acid pre-treatment. *Hydrometallurgy* **2007**, *89*, 260–268. [CrossRef]

63. Hashemzadeh, M.; Dixon, D.G.; Liu, W. Modelling the kinetics of chalcocite leaching in acidified cupric chloride media under fully controlled pH and potential. *Hydrometallurgy* **2019**, *189*, 105114. [CrossRef]

64. Pérez, K.; Toro, N.; Saldaña, M.; Salinas-Rodríguez, E.; Robles, P.; Torres, D.; Jeldres, R.I. Statistical Study for Leaching of Covellite in a Chloride Media. *Metals* **2020**, *10*, 477. [CrossRef]

65. Hashemzadeh, M.; Dixon, D.G.; Liu, W. Modelling the kinetics of chalcocite leaching in acidified ferric chloride media under fully controlled pH and potential. *Hydrometallurgy* **2019**, *186*, 275–283. [CrossRef]

66. Hernández, P. Estudio del Equilibrio Sólido-Líquido de Sistemas Acuosos de Minerales de Cobre Con Agua de Mar, Aplicado A Procesos de Lixiviación, Universidad de Antofagasta. 2013. Available online: https://intranetua.uantof.cl/docs/carreras/tesis_pia_hernandez.pdf (accessed on 1 October 2021).

67. Schlesinger, M.; King, M.; Sole, K.; Davenport, W. *Extractive Metallurgy of Copper*, 5th ed.; Elsevier: Amsterdam, The Netherlands, 2011; ISBN 9780080967899.

68. Velásquez-Yévenes, L. The Kinetics of the Dissolution of Chalcopyrite in Chloride Media, Murdoch University. 2009. Available online: https://researchrepository.murdoch.edu.au/id/eprint/475/ (accessed on 1 October 2021).

69. Dutrizac, J.E. The leaching of sulphide minerals in chloride media. *Hydrometallurgy* **1992**, *29*, 1–45. [CrossRef]
70. Toro, N.; Pérez, K.; Saldaña, M.; Jeldres, R.I.; Jeldres, M.; Cánovas, M. Dissolution of pure chalcopyrite with manganese nodules and waste water. *J. Mater. Res. Technol.* **2019**, *9*, 798–805. [CrossRef]
71. Naderi, H.; Abdollahy, M.; Mostoufi, N. Kinetics of chemical leaching of chalcocite from low-grade copper ore: Size-distribution behavior. *J. Min. Environ.* **2015**, *6*, 109–118. [CrossRef]
72. Toro Villarroel, N.R. Optimización de Parámetros Para la Extracción de Elementos Desde Minerales en Medios Ácido, Universidad Politécnica de Cartagena. 2020. Available online: https://repositorio.upct.es/handle/10317/8504 (accessed on 1 October 2021).
73. Chang, K.; Zhang, Y.; Zhang, J.; Li, T.; Wang, J.; Qin, W. Effect of temperature-induced phase transitions on bioleaching of chalcopyrite. *Trans. Nonferrous Met. Soc. China* **2019**, *29*, 2183–2191. [CrossRef]
74. Li, Y.; Kawashima, N.; Li, J.; Chandra, A.P.; Gerson, A.R. A review of the structure, and fundamental mechanisms and kinetics of the leaching of chalcopyrite. *Adv. Colloid Interface Sci.* **2013**, *197*, 1–32. [CrossRef]
75. Liu, W.; Granata, G. The Effect of Aeration on Chalcocite Heap Leaching. In *Extraction*; Davis, B., Moats, M., Wang, S., Gregurek, D., Kapusta, J., Battle, T., Schlesinger, M., Alvear, G., Jak, E., Goodall, G., Eds.; The Minerals, Metals & Materials Series; Springer: Cham, Switzerland, 2018. [CrossRef]
76. Devi, N.B.; Madhuchhanda, M.; Rao, K.S.; Rath, P.C.; Paramguru, R.K. Oxidation of chalcopyrite in the presence of manganese dioxide in hydrochloric acid medium. *Hydrometallurgy* **2000**, *57*, 57–76. [CrossRef]
77. Kowalczuk, P.B.; Manaig, D.O.; Drivenes, K.; Snook, B.; Aasly, K.; Kleiv, R.A. Galvanic leaching of seafloor massive sulphides using MnO2 in H2SO4-NaCl media. *Minerals* **2018**, *8*, 235. [CrossRef]
78. Torres, D.; Ayala, L.; Jeldres, R.I.; Cerecedo-Sáenz, E.; Salinas-Rodríguez, E.; Robles, P.; Toro, N. Leaching Chalcopyrite with High MnO2 and Chloride Concentrations. *Metals* **2020**, *10*, 107. [CrossRef]

Article

Maintenance of the Metastable State and Induced Precipitation of Dissolved Neodymium (III) in an Na_2CO_3 Solution

Youming Yang [1,2], Xiaolin Zhang [1], Kaizhong Li [1,*], Li Wang [3], Fei Niu [1,2], Donghui Liu [1] and Yuning Meng [1]

1   Faculty of Materials Metallurgy and Chemistry, Jiangxi University of Science and Technology, Ganzhou 341000, China; Yanguming@126.com (Y.Y.); userlin116@126.com (X.Z.); niufeijxust@foxmail.com (F.N.); Fiboy@163.com (D.L.); ynmeng88@126.com (Y.M.)
2   Institute of Engineering Research, Jiangxi University of Science and Technology, Ganzhou 341000, China
3   College of Vanadium and Titanium, Panzhihua University, Panzhihua 617000, China; wl003@126.com
*   Correspondence: likaizhong1105@163.com

Abstract: Rare earths dissolved in carbonate solutions exhibit a metastable state. During the period of metastability, rare earths dissolve stably without precipitation. In this paper, neodymium was chosen as a representative rare earth element. The effects of additional NaCl and CO_2 on the metastable state were investigated. The metastable state can be controlled by adding NaCl to the Na_2CO_3 solution. Molecular dynamics studies indicated that the Cl^- provided by the additional NaCl partially occupied the coordination layer of Nd^{3+}, causing the delayed formation of neodymium carbonate precipitation. In addition, the additional NaCl decreased the concentration of free carbonate in the solution, thereby reducing the behavior of free contact between carbonate and Nd, as well as resulting in the delay of Nd precipitate formation. Consequently, the period of the metastable state was prolonged in the case of introduction of NaCl. However, changing the solution environment by introducing CO_2 can destroy the metastable state rapidly. Introduction of CO_2 gas significantly decreased the CO_3^{2-} content in the solution and increased its activity, resulting in an increase of the free CO_3^{2-} concentration of the solution in the opposite direction. As a result, the precipitation process was accelerated and the metastable state was destroyed. It was possible to obtain a large amount of rare earth carbonate precipitation in a short term by introducing CO_2 into the solution with dissolved rare earths in the metastable state to achieve rapid separation of rare earths without introducing other precipitants during the process.

Keywords: neodymium; metastable state; maintenance; induced precipitation

Citation: Yang, Y.; Zhang, X.; Li, K.; Wang, L.; Niu, F.; Liu, D.; Meng, Y. Maintenance of the Metastable State and Induced Precipitation of Dissolved Neodymium (III) in an Na_2CO_3 Solution. *Minerals* **2021**, *11*, 952. https://doi.org/10.3390/min11090952

Academic Editors: Shuai Wang, Xingjie Wang, Jia Yang and Kenneth N. Han

Received: 19 July 2021
Accepted: 27 August 2021
Published: 31 August 2021

Publisher's Note: MDPI stays neutral with regard to jurisdictional claims in published maps and institutional affiliations.

1. Introduction

Rare earths are strategic metal resources that are used in a wide range of industries. For example, they can be found in the development of high-tech advanced materials for permanent magnets, luminescence, catalysis and hydrogen storage, as well as in basic industries such as metallurgy, machinery and petrochemicals in general [1–5].

Rare earth carbonate is a barely soluble substance, with a solubility in water of only 10^{-5}–10^{-7} mol·L^{-1} [6,7]. However, when rare earth ions are added to a higher concentration of alkali metal carbonate solution, there occurs the phenomenon of rare earth dissolution in the carbonate solution. The amount of rare earth dissolution increases with increased carbonate concentration. As early as 1963, Taketatsu [8,9] found that when a certain amount of rare earth chloride solution was gradually added to a concentrated K_2CO_3 solution, sediment of the rare earth carbonate was generated first and then dissolved again with the passage of reaction time. The dissolution amount of rare earth increased with the increase of CO_3^{2-} concentration and the atomic number of the rare earth (except for Ce and Y). Restricted by the situation of the industry at that time, rare earth resources were not as scarce as nowadays, but comparatively abundant. Therefore, the discovery of the regularity of dissolution of rare earths in carbonate solution did not attract the attention of the rare earth separation industry.

Vasconcellos et al. [10] carried out a feasibility study on selective dissolution, separation and enrichment of a rare earth in carbonate solution based on Taketatsu's regularity [8,9]. Low-Ce rare earth carbonate concentrates were selectively dissolved and successfully enriched with yttrium by using $NH_4HCO_3/(NH_4)_2CO_3$ solution as a carrier. The grade of yttrium increased from 2.4% to 81.0%. In addition, it was also found that the concentration of NH_4^+ influenced the dissolution behavior of rare earth in the solution system in that a higher concentration of NH_4^+ could enhance the solubility of rare earths. Reference to research on the equilibrium hydrochemical behavior of neodymium in a Na^+-Cl^--CO_3^{2-}-HCO_3^- solution system [11], shows that there is a relationship between the solubility of rare earth in a carbonate solution and the concentration of the NaCl salt as an impurity, and the dissolved amount of rare earth increases with increasing concentration of NaCl in the solution. High concentration of NaCl results in high ionic strength of the solution. In addition, according to research data on rare earth adsorption in a water-bearing sand layer [12], at higher ionic strength of the solution, the more significant was the rare earth adsorption in the sand layer. In other words, a greater amount of rare earth loss was caused by dissolution. Thus, when the concentration of NH_4^+ in the solution increased, this resulted in a greater amount of dissolved rare earths, as observed by Vasconcellos [10], which can be attributed to the influence of the ionic strength of the solution [13].

Nowadays, a number of techniques have been developed for the separation of rare earths, such as solvent extraction [14], ion exchange, membrane separation and ionic liquids [15], as well as other methods. Among them, the most widely used is the traditional solvent extraction technique. The other methods are less used because of high cost. The core component of solvent extraction is the extractant. Currently, a number of high-performance extractants have been developed [16], but toxicity and loss of extractants are always the key weaknesses limiting the development of the technology. In addition, due to the increasingly stringent requirements of environmental protection, the high salt wastewater generated during the separation of rare earths remains a problem [17], and is also a bottleneck in the solvent extraction separation process [18]. For the healthy development of the rare earth industry, it is necessary and urgent to develop a new type of highly efficient and environmentally friendly separation technology.

A good and feasible green method to separate rare earths is by using the metastable state of the carbonate solution which dissolves them. In our previous study [19], a series of experiments on metastable states was carried out by choosing neodymium as an example of rare earth elements. Our results indicated that neodymium dissolved in a sodium carbonate solution exhibited some metastable properties. Among them, the most important one was that there is a limit to the dissolution of neodymium in a certain concentration of sodium carbonate solution, after which there is instantaneous saturated solubility. When the dissolved neodymium in the solution does not exceed its solubility, it is stable in the solution for a period (metastable period) without precipitating neodymium carbonate. Our previous study was not very comprehensive and limited by the length of the paper, so a follow-up study of metastable solution-induced precipitation was not carried out.

Now, combined with the idea of the solid-liquid separation of rare earths, we are continuing to consider the potential value of the metastable state. Rare earths are dissolved and enriched in the metastable period and precipitated and separated after exceeding this period. No other impurities are introduced in this process. In addition, the carbonate solution can be recycled. Therefore, this may be a potential method for green separation of rare earths. Hence, how to manually control the metastable state and the precipitation of rare earth carbonate is the core content of the present study.

In this study, the artificial control of metastable states is discussed in detail. Neodymium was chosen as a representative of rare earth elements. The effect of changing the solution environment, such as ion concentration, on the metastable state was studied, and the effective conditions for maintaining and destroying the metastable state were discovered.

2. Experiment

2.1. Raw Materials and Equipment

The rare earth material used in the experiment was a 10 g·L^{-1} dilute $NdCl_3$ solution obtained by diluting high purity $NdCl_3$ solution with deionized water. The high purity $NdCl_3$ solution was purchased from the rare earth smelting & separating plant in Longnan, Jiangxi Province, and its distribution is shown in Table 1. Solutions of Na_2CO_3, Na_2CO_3/NaCl with different concentration gradients and dilute hydrochloric acid for acidification were obtained by dissolving AR-grade Na_2CO_3, solid NaCl and HCl solution with deionized water.

Table 1. Content of the high purity solution of $NdCl_3$.

Concentration of Nd^{3+} (mol·L^{-1})			Concentration of H^+ (mol·L^{-1})		Specific Gravity (g·mL^{-1})	
1.3568			<0.10		1.326	
Non-rare earths impurities (µg·mL^{-1})						
Fe_2O_3			SiO_2		CaO	
<0.50			2.49		7.3	
Rare Earth Impurities/REO (µg·mL^{-1})						
La_2O_3	CeO_2	Pr_6O_{11}	Sm_2O_3	Eu_2O_3	Gd_2O_3	Tb_2O_3
<100	<100	500	<100	<100	<100	<100
Dy_2O_3	Ho_2O_3	Er_2O_3	Tm_2O_3	Yb_2O_3	Lu_2O_3	Y_2O_3
<100	<100	<100	<100	<100	<100	<100

An experiment in which CO_2 gas was used to induce precipitation of a metastable state solution was carried out by using an autoclave with a CO_2 high-pressure cylinder, as shown in Figure 1. Other equipment used in experiments is shown in Table 2.

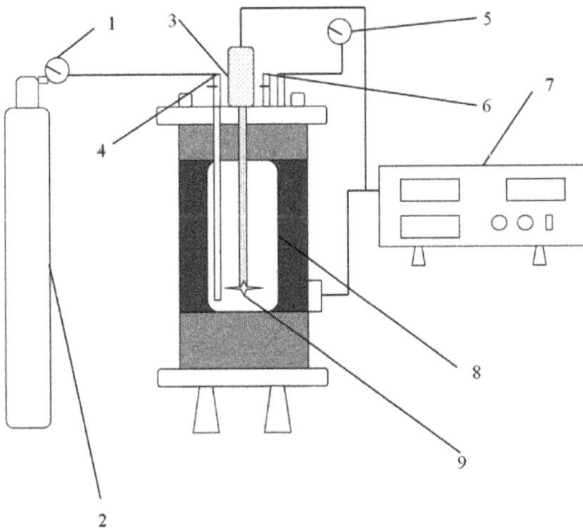

Figure 1. Schematic diagram of autoclave ventilation experiment: 1—CO_2 pressure reducing valve; 2—high pressure cylinder; 3—rotating motor; 4—air inlet; 5—safety valve; 6—air outlet; 7—controller; 8—polytetrafluoroethylene tank; 9—agitator.

Table 2. Information of equipment.

Equipment	Model	Manufacturers
High-speed centrifuge	TGL16MS	Yancheng Anxin Experimental Instrument Co., Ltd. (Yancheng, China)
Computing server	IBM System X3850 X5	International Business Machines Corporation (Armonk, NY, USA)
UV-Visible Spectrophotometer (UV-vis)	UV-5500PC	Shanghai yoke instrument Co., Ltd. (Shanghai, China)
Fourier transform infrared spectrometer (FTIR)	ALPHA	Bruker Corporation (Billerica, MA, USA)
inductively coupled plasma-optical emission spectroscopy (ICP-OES)	ULTIMA2	HORIBA Jobin Yvon (Newark, NJ, USA)
Kang's oscillator	KS	Changzhou Putian Instrument Manufacturing Company (Changzhou, China)

2.2. Maintenance and Mechanism of Metastable State of the Neodymium Dissolved in Na$_2$CO$_3$ Solution

2.2.1. Maintaining Metastable State by NaCl

Following on from our previous research [19], the effect of additional NaCl on the maintenance of the metastable state of a solution of dissolved Nd^{3+} was investigated. The neodymium concentration in 2 mol·L^{-1} Na$_2$CO$_3$ solution was controlled at 2.621 g·L^{-1}. The ionic strength of the solution was controlled by adding NaCl to Na$_2$CO$_3$ solution to create a mixed electrolyte NaCl/Na$_2$CO$_3$ solution. The concentration of the additional NaCl ranged from 0 to 0.5 mol·L^{-1}. During the experiment, the volume of the Na$_2$CO$_3$ (or mixed electrolyte NaCl/Na$_2$CO$_3$) solution was fixed at 25 mL, and the NdCl$_3$ solution was added to it drop by drop with oscillation. The time was set from 0 to 480 min. When the experiment finished, the solution containing precipitates was further centrifuged at 6000 rpm for 5 min. After that, the obtained supernatant was split into two parts; one was selected as aqueous sample for the further testing, and the another was completely acidulated by using dilute hydrochloric acid, after which complexometric titration was used to determine the concentration of Nd^{3+}.

2.2.2. Effect of NaCl on Neodymium Coordination and Solid Phase Precipitates

The metastable state solution in each representative stationary period was scanned by ultraviolet-visible light (UV-vis) full-wavelength scanning. In order to provide an experimental comparison, a blank solution (Na$_2$CO$_3$ and NdCl$_3$ solution only addition with water) was also scanned by UV-vis full wavelength. The precipitates were collected as solid samples and detected by Fourier transform infrared spectroscopy (FTIR). Because drying could cause the sample to decompose [20], producing errors in the results, the samples were all stored in deionized water. Before determination, the water was filtered, the samples dried with filter paper, and the analysis carried out immediately.

2.2.3. Mechanism of Maintaining the Metastable State by NaCl

In order to find the maintenance mechanism of the additional NaCl on the metastable state, a molecular dynamics (MD) calculation was carried out using Materials Studio 8.0 [21] software. The solution model was established by using an Amorphous Cell module, and the model was geometry optimized using a Forcite module. Finally, MD calculation and radial distribution function (RDF) analysis [22] were carried out to reveal the relationship between the RDF and the coordination number of each component in the solution. The average coordination number of each component was calculated by Equation (1).

$$N(L) = \int_0^L g(r)\, \rho\, 4\pi r^2 dr \tag{1}$$

where N(L) refers to the number of coordination atoms(molecules) in the $0-L$ spherical shell around the target atom, ρ refers to the number density of coordination atoms (molecules), where the value is the ratio of the number of atoms (molecules) to the volume of space, g(r) refers to the RDF value, and indicates the probability of the occurrence of coordination atoms(molecules) within a certain distance, and r refers to the cutoff radius.

2.3. Induced Precipitation of Neodymium Carbonates in Metastable State Solution

The precipitation process of neodymium carbonates by introducing CO_2 gas into the metastable state solution was studied. The procedure of the dissolution of Nd^{3+} in the solution was as previously described. Further, the dissolved Nd^{3+} solution was transferred to an autoclave for introducing CO_2 gas. The solution contained a large amount of halogen Cl^-, so a corrosion-resistant polytetrafluoroethylene tank was selected as the inner tank of the autoclave. During the experiment, the input pressure of CO_2 was uniformly controlled at 0.2 Mpa. The time was set from 0–60 min. When the set time was reached, the solution containing precipitate was centrifuged at 6000 r·min^{-1} for 5 min, then the supernatant obtained after centrifugation was split in two parts. One was acidified, and the concentration of neodymium in the supernatant was determined by inductively coupled plasma emission spectrometer (ICP-OES). The other was analyzed by CO_3^{2-} and HCO_3^- acid-base titration to determine the concentration of CO_3^{2-} and HCO_3^-.

3. Results and Discussion

3.1. Maintenance and Mechanism of the Metastable State of Neodymium Dissolved in Na₂CO₃ Solution

3.1.1. Maintaining Metastable State by NaCl

The metastable period of the solution with the addition of NaCl was greater than that of the solution without NaCl due to neodymium being dissolved stably in solution for a longer time. As shown in Figure 2a, the metastable period was sustained only for about 120 min without the addition of NaCl. However, the metastable period reached 240 min when the concentration of the additional NaCl was 0.2 mol·L^{-1}. Moreover, the amount of dissolved neodymium in the solution reached 2.578 g·L^{-1}, which was only 1.64% lower than the initial amount of 2.621 g·L^{-1}. Surprisingly, when the additional concentration of sodium chloride in the solution reached 0.5 mol·L^{-1}, the metastable period was extended to 480 min, i.e., twice as long as before.

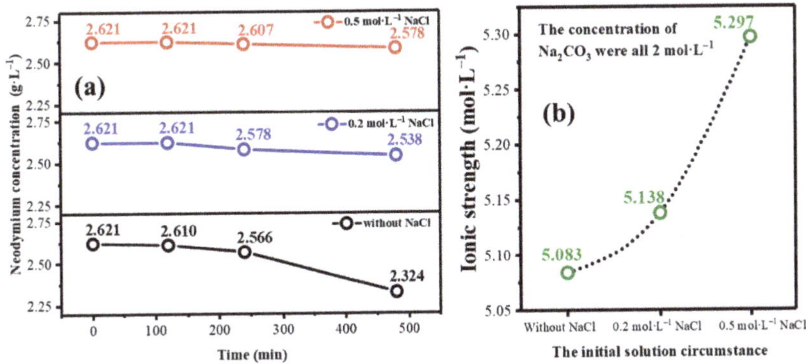

Figure 2. The effect of the concentration of the additional NaCl in (**a**) and Ionic strength of solution in (**b**).

The molar concentration in the above solution was converted into mass molar concentration according to Equation (2). Then, the ionic strength of each solution was calculated by using Equation (3). The results are shown in Figure 2b which shows that the addition of NaCl effectively improved the ionic strength of the solution.

$$C = bB \cdot \rho \tag{2}$$

$$I = \frac{1}{2} \sum C_i bB_i^2 \tag{3}$$

bB refers to the molar concentration, ρ refers to the density of the solution, and I refers to the ionic strength of the solution.

In the case of the equivalent concentration of Na_2CO_3, the addition of NaCl resulted in neodymium dissolving stably in Na_2CO_3 solution for a longer time. The higher the concentration of NaCl in solution, and the stronger the ionic strength, resulting in a longer metastable period of the solution. Hence, the additional NaCl maintained the metastable state effectively.

3.1.2. Effect of NaCl on Neodymium Coordination and Solid Phase Precipitates

To confirm whether the coordination reaction between Nd^{3+} and CO_3^{2-} still occurred in the mixed electrolyte solution of $NaCl/Na_2CO_3$, the aqueous samples in each period of the above experiments were collected and scanned by UV-vis with full wavelength. As shown in Figure 3, the characteristic peak of neodymium was not found in the UV-vis spectra of the sample of the blank $NaCl/Na_2CO_3$ mixed electrolyte solution with only added water. Characteristic peaks of neodymium at the 340–370 nm and 500–620 nm wavebands [23] were observed in the spectra of the blank $NdCl_3$ solution with added water only.

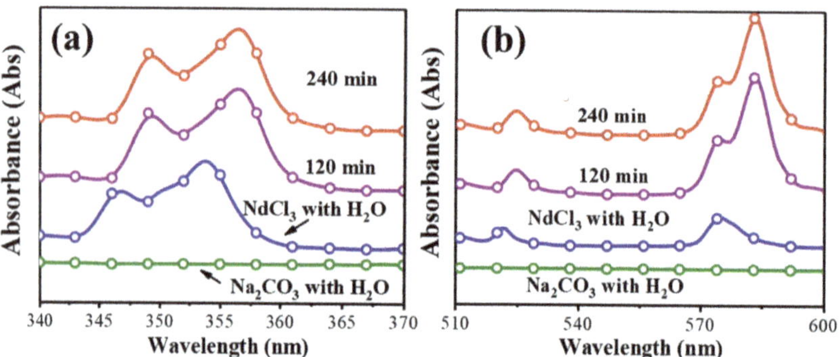

Figure 3. UV-vis spectrum of $NdCl_3/Na_2CO_3$ solution with the addition of 0.5M NaCl, wavelength in 340 to 370 nm in (**a**), and 510 to 600 nm in (**b**).

As shown in Figure 3, characteristic peaks of neodymium were obtained at 349 and 357 nm wavelengths in the spectra of the $NaCl/Na_2CO_3$ mixed electrolyte solution with dissolved neodymium. Compared with the blank neodymium solution spectrum, it is worth noting that the characteristic peaks of neodymium obtained from the $NaCl/Na_2CO_3$ solutions with dissolved neodymium were slightly red-shifted from the initial 347 and 354 nm wavelengths due to the high alkalinity of the $NaCl/Na_2CO_3$ mixed electrolyte solution.

There were also neodymium characteristic peaks at 524 and 575 nm wavelengths in the UV-vis spectrum of the neodymium-dissolving solution, as in our previously results [19]. In addition, a new peak with higher intensity was observed at 583 nm. This indicates that neodymium could still coordinate with CO_3^{2-} in the Na_2CO_3 solution with the addition of NaCl. The precipitate sample generated from the solution after 480 min was collected and analyzed by FTIR. The results shown in Figure 4.

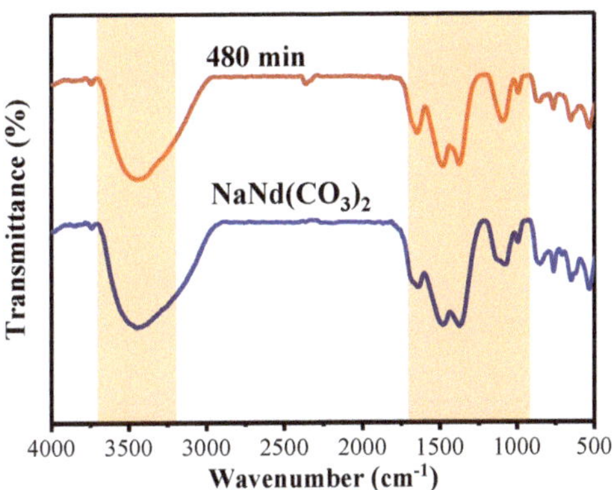

Figure 4. FTIR pattern of the neodymium precipitate, red line refers to the precipitate obtained in the presence of 0.5 mol NaCl, and blue line refers to the Comparison sample.

Figure 4 shows that the characteristic infrared peak position of the precipitate generated from the NaCl/Na$_2$CO$_3$ solution with dissolved neodymium was consistent with that of the blank sample NaNd(CO$_3$)$_2$ solid phase. This was consistent with results reported in previous studies [11,24]. These reports and our results confirm that in the presence of additional NaCl in a highly concentrated CO$_3^{2-}$ solution, the insoluble rare earth existed only in the form of an NaNd(CO$_3$)$_2$ double salt, while Nd$_2$(CO$_3$)$_3$ was almost non-existent.

3.1.3. Mechanism of Maintaining the Metastable State by NaCl

In order to investigate the mechanism of the maintenance of the metastable state by the additional NaCl, molecular dynamic (MD) calculations were carried out. The construction and optimization of the solution components were consistent with our previous study [19]. The model of the solution with the addition of 0.5 mol·L^{-1} NaCl (with the better metastable condition), and the corresponding blank solution (Na$_2$CO$_3$/NaCl solution with the only additional water) was also established by using the Amorphous Cell module. After that, geometry optimization and the MD calculation were carried out. The components in the model and calculation parameters are shown in Table 3. The energy changed during the geometry optimization process is presented in Figure 5a.

Table 3. Modeling parameters of the solutions.

Components	The Metastable State Solution ρ: 1.164 g·L^{-1}		Corresponding Blank Solution ρ: 1.148 g·L^{-1}	
	Number	Mass Fraction (%)	Number	Mass Fraction (%)
H$_2$O	10,590	84.7	10,590	85.0
Na$^+$	630	6.4	630	6.5
CO$_3^{2-}$	280	7.5	280	7.5
Nd^{3+}	3	0.2	0	0.0
Cl$^-$	79	1.2	70	1.1

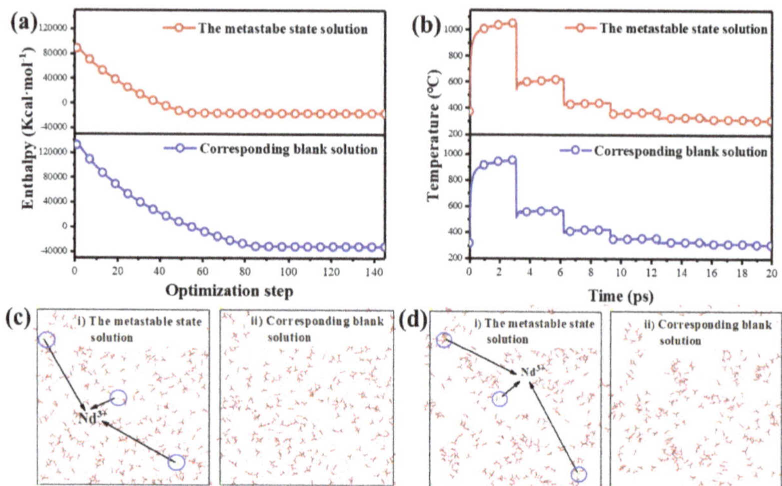

Figure 5. Geometry optimization process of the solution model (**a**). Temperature change during the MD calculation in (**b**). Geometry optimization and after MD calculation of the solution model (**c**,**d**). In the figures, (**i**) refers to the metastable state solution and (**ii**) is the corresponding blank solution.

As shown in Figure 5a, the overall energy decreased gradually with increase in the number of optimization steps without large energy disturbances. At the end of optimization, the energy tended to be relatively minimized and achieved convergence. Change of the temperature is presented in Figure 5b, which shows that the temperature of the models decreased gradually with the increase of simulation time. At the end of the calculation, it was stable at about 298 K $\pm$ 10% and there was no significant disturbance. This result is very reliable. The optimized models are presented in Figure 5c, which shows that after the geometry optimization step, the Na^+, CO_3^{2-}, Nd^{3+}, Cl^- ions in each solution model were randomly and uniformly distributed in the model box, and no agglomeration existed.

The models after the MD calculation are presented in Figure 5d, which shows that the solution was generally homogeneous. However, in the local region, the components of the solution model had different degrees of agglomeration due to the interaction between ions (or molecules). Among them, the CO_3^{2-} distribution exhibited local agglomeration. A large number of Na^+ ions were distributed around the CO_3^{2-}. This can be attributed to the incomplete dissociation of Na^+ and CO_3^{2-} at the high concentration of the $NaCl/Na_2CO_3$ mixed electrolyte solution. We speculate that the concentration of carbonate that could move freely (called free CO_3^{2-}) in the solution was limited and at a low level.

Moreover, Nd^{3+} was also almost surrounded by CO_3^{2-} in that Nd^{3+} was coordinated with about three CO_3^{2-}. This shows that Nd^{3+} can coordinate with carbonate despite the addition of NaCl to Na_2CO_3 solution. In addition, all kinds of complex ions in the form of $Nd_n(CO_3)_m^{3n-2m}$ (m $\geq$ 2) existed, but in different proportions. As shown in Figure 6a, there was a specific Cl^- ion around some Nd^{3+} ions, indicating that, unlike in previous studies, Cl^- might also coordinate with Nd^{3+}. Previous studies of steady-state dissolution of rare earths in CO_3^{2-} solution [11,24], showed that the coordination between Cl^- and Nd^{3+} could almost be ignored under the condition of steady-state solution equilibrium because of weak Cl^- coordination ability, which does not agree with our study. The difference could be ascribed to the difference between metastable and steady states. Besides, it is reasonable to speculate that Cl^- occupied the coordination layer of Nd^{3+}, delaying the formation of carbonate precipitation. Therefore, the addition of NaCl maintained the metastable state.

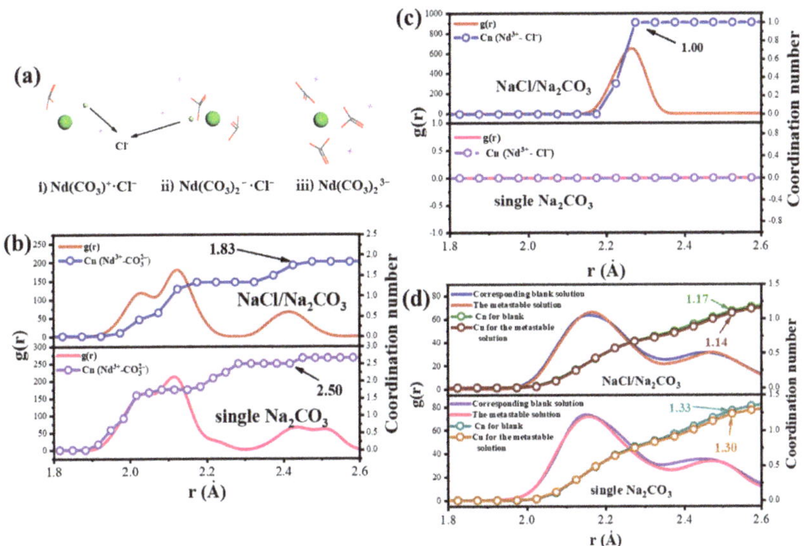

Figure 6. Coordination between Nd^{3+}, Cl^- & CO_3^{2-} (a). RDF of the ion pairs Nd^{3+}-CO_3^{2-}, Nd^{3+}-Cl^- & Na^+-CO_3^{2-} and their coordination number (b–d), respectively.

In order to further quantify the interaction from the microscopic level, the main ion pairs of Nd^{3+}-CO_3^{2-}, Na^+-CO_3^{2-} and Nd^{3+}-Cl^- in the solution were analyzed by radial distribution function (RDF), and their coordination numbers were calculated. Figure 6b shows that the RDF peak position of Nd^{3+} and CO_3^{2-} in the $NaCl/Na_2CO_3$ mixed electrolyte solution was almost the same as that in the single Na_2CO_3 solution in the chemical bond range (r < 2.6 Å) [25]. This directly proves that Nd^{3+} can coordinate with CO_3^{2-} even in the presence of NaCl. At the same time, the RDF peak intensity of Nd^{3+}-CO_3^{2-} in the presence of NaCl was slightly lower than that in the single Na_2CO_3 solution. The difference of RDF peak intensity indicated that the additional NaCl had an influence on the interaction between Nd^{3+} and CO_3^{2-}. From the point of view of coordination number, with the presence of the additional NaCl the average coordination number around Nd^{3+} was about 1.83 CO_3^{2-}, which was lower than the average coordination number of 2.50 in the single Na_2CO_3 solution. The reason for the decrease of coordination number may be that part of Cl^- occupied the coordination layer of Nd^{3+} in the metastable period, causing the decrease in the average coordination number between Nd^{3+} and CO_3^{2-}. This is consistent with Figure 6a.

To quantitatively explain the coordination between Nd^{3+} and Cl^- in the mixed electrolyte, the RDF and the coordination number were further analyzed and calculated. Figure 6c shows that the RDF peak at 2.275 Å within the range of chemical bond (<2.6 Å) was clearly observed from the spectrum of the Nd^{3+}-Cl^- ion pair in the mixed electrolyte solution of $NaCl/Na_2CO_3$. One Nd^{3+} was coordinated with about one Cl^-. In contrast to the single Na_2CO_3 solution, there was insignificant evidence of an interaction existing between Cl^- and Nd^{3+}. Hence, it was proved that that Cl^- could coordinate with Nd^{3+} in the presence of NaCl during the metastable period.

It is worth noting that the RDF peak position of Nd^{3+}-CO_3^{2-} was earlier than that of Nd^{3+}-Cl^-. The result indicates that CO_3^{2-} tends to occupy the coordination layer of Nd^{3+} and reacted with it first, then Cl^- entered the coordination layer of Nd^{3+}, although, previous studies [11,24] showed that in an environment of high concentration of CO_3^{2-} solution, Cl^- did not coordinate with Nd^{3+} in a steady state. However, our result does not conflict with these previous studies because of the difference between metastable and steady states. In addition, when the placement time exceeded the metastable period, the precipitate was still consistent with that obtained in the steady state. Therefore, there is sufficient reason to speculate that in the metastable state, Cl^- participated in the coordination reaction and temporarily occupied the coordination layer of Nd^{3+}, and then was re-released into the solution with the passage of time. When the metastable period ended, the Cl^- in the coordination layer of Nd^{3+} had been exhausted. A diagram of Nd^{3+}, Cl^- and CO_3^{2-} coordination in the metastable state is shown in Figure 7.

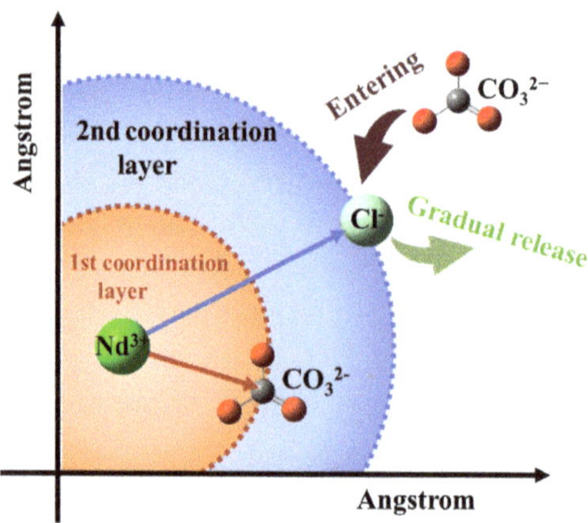

Figure 7. Coordination process of Nd^{3+} in the $NaCl/Na_2CO_3$ mixed electrolyte solution.

The concentration of free CO_3^{2-} in solution was another key factor affecting the existence of metastable states. To explore the interaction between Na^+ and CO_3^{2-} in the solution in the presence of additional NaCl, the RDF of the Na^+-CO_3^{2-} ion pair and its average coordination number were also analyzed and calculated. The results in Figure 6d show that there was interaction between Na^+ and CO_3^{2-} in the mixed electrolyte solution, and the position of the RDF peak was basically the same as that in single Na_2CO_3 solution. The average coordination number was about 1.14 CO_3^{2-} around a Na^+ ion in the solution with additional NaCl, which was not much different from the corresponding blank solution number of 1.17. However, the value was lower than the value of 1.30 in a single Na_2CO_3 solution.

The existence of this difference does not mean that the degree of dissociation of Na^+-CO_3^{2-} ion pairs in the mixed electrolyte solution was higher because of the introduction of Na^+ via the addition of NaCl. The introduced Na^+ would also tend to interact with CO_3^{2-}, reducing the average coordination number. In this regard, the conversion calculation of the number of Na^+ ions around the CO_3^{2-} is a better illustration of the problem, as listed in Table 4.

Table 4. Distribution of Na^+ around CO_3^{2-} in the solution.

r (Å)	Average Coordination Number (Cn)	
	NaCl/Na$_2$CO$_3$ Mixed Electrolyte Solution	Single Na$_2$CO$_3$ Solution
2.225	1.29	1.28
2.275	1.51	1.50
2.325	1.64	1.67
2.375	1.79	1.82
2.425	1.99	2.02
2.475	2.24	2.26
2.525	2.44	2.46
2.575	2.57	2.57
2.625	2.63	2.63

Table 4 shows that the distribution of Na^+ around CO_3^{2-} in mixed electrolyte and single Na_2CO_3 solutions was the same at a cutoff distance of 2.575 Å (within chemical bond range), which was 2.57 Na^+ around CO_3^{2-}. The total number of Na^+ was 630 in the mixed electrolyte, higher than that of 580 in a single Na_2CO_3 solution, due to the existence of additional NaCl. Even so, the distribution of Na^+ around CO_3^{2-} was almost the same. This means that the dissociation degree of the Na^+-CO_3^{2-} ion pairs with the additional NaCl should be much lower than that in the single Na_2CO_3 solution. Therefore, the additional NaCl could also affect the interaction of Na^+-CO_3^{2-} ion pairs, further reducing the concentration of free CO_3^{2-}. Thus, the delayed the formation of neodymium carbonates was delayed and the metastable period was extended.

3.2. Induced Precipitation of Neodymium Carbonates in Metastable State Solution

In theory, if the system dominated by CO_3^{2-} in the solution could be rapidly transformed into the coexistence of HCO_3^- and CO_3^{2-}, a rapid destruction of metastable state could be achieved and most of the dissolved neodymium could be separated from the solution via a self-precipitation. Theoretically, introduction of acidic CO_2 gas into the solution, as shown in Equation (4), would neutralize the original alkaline Na_2CO_3 solution so that the induction of neodymium carbonate from the metastable solution might be more quickly achieved.

$$Na_2CO_{3(aq)} + CO_{2(g)} + H_2O_{(l)} = 2NaHCO_{3(aq)} \tag{4}$$

The specific experimental results in Figure 8a show that the metastable state was rapidly terminated after the introduction of CO_2 gas. The concentration of dissolved Nd^{3+} in the solution fell rapidly, and neodymium carbonate precipitate were generated and separated from the solution. After gassing with CO_2 for only 5 min, the Nd^{3+} concentration in the solution decreased from 2.621 to 1.793 $g·L^{-1}$, and the precipitation rate reached 31.60%. When the ventilation time was extended to 30 min, only 0.239 $g·L^{-1}$ was left in the solution, and the precipitation rate of Nd^{3+} reached 90.87%. When the ventilation time reached 60 min, the neodymium concentration in the solution further decreased from 0.239 $g·L^{-1}$ at 30 min to 0.138 $g·L^{-1}$, and the precipitation rate slowly increased from 90.78% to almost 95%. The results indicate that Nd^{3+} in the solution entered the insoluble solid phase and was separated from the solution via self-precipitation.

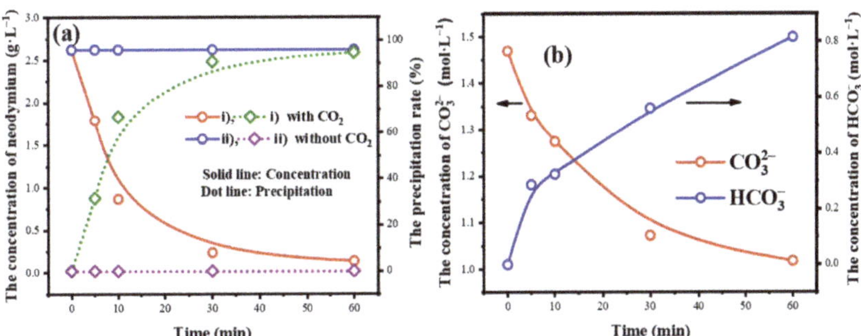

Figure 8. The concentration of dissolved Nd^{3+} in the solution and the precipitation rate with the ventilation time of CO_2 (**a**), and the concentration of CO_3^{2-}/HCO_3^- with the ventilation time of CO_2 (**b**).

Acid-base titration analysis was carried out to determine the concentration of CO_3^{2-} and HCO_3^- in the solution after CO_2 was injected. Figure 8b shows that the concentration of CO_3^{2-} in the solution decreased with increasing CO_2 introduction time. The concentration of HCO_3^- showed an upward trend during the introduction of CO_2, and the system gradually changed from a single CO_3^{2-}-dominated to a CO_3^{2-} and HCO_3^--dominated system.

When the CO_2 introduction time reached 5 min, the concentration of HCO_3^- in the solution approached $0.286\ mol \cdot L^{-1}$ and the corresponding concentration of CO_3^{2-} was $1.331\ mol \cdot L^{-1}$, which was much lower than the initial value of $2\ mol \cdot L^{-1}$. After 60 min, the concentration of HCO_3^- in the solution increased than several times and reached about $0.816\ mol \cdot L^{-1}$. Besides, the corresponding concentration of CO_3^{2-} was reduced to $1.018\ mol \cdot L^{-1}$, and the solution system could no longer be considered to be dominated by single CO_3^{2-}, but both CO_3^{2-} and HCO_3^-. The reason was that the solution of $NaCl/Na_2CO_3$ mixed electrolyte with higher basicity spontaneously absorbed acidic CO_2 gas, resulting in the conversion of CO_3^{2-} to HCO_3^- in a short time.

After introducing CO_2, the concentration of CO_3^{2-} in the solution decreased, but the solution volume was almost unchanged. In other words, it was equivalent to diluting the solution. It is well known that ion pairs dissociate more completely in dilute solution. Thus, there is no doubt that the concentration of free CO_3^{2-} in the solution was increased, and it is important that CO_2 could induce the precipitation of neodymium carbonate in the metastable state solution. In addition, the concentration of CO_3^{2-} decreased as CO_2 introduction time and the precipitation rate of neodymium carbonate increased. Therefore, the introduction of carbon dioxide into metastable solutions is a new potential method for the separation of rare earths. This is because carbon dioxide is nonpolluting and does not introduce other ion impurities as with the use of precipitants (e.g., ammonium carbon, which is commonly used in industry, introduces ammonium ions). In addition, the separation is achieved by the rapid production of rare earth precipitation in a short time after its entry into the solution.

4. Conclusions

We chose neodymium as an example of a rare earth element and studied the maintenance of the metamorphic state of Na_2CO_3 solution with dissolved Nd^{3+}. The results show that the metastable state can be successfully prolonged by adding sodium chloride. In addition, the introduction of carbon dioxide is an effective way to terminate the metastable state and generate neodymium carbonate.

The main conclusions are as follows. (1) The higher the additional NaCl concentration, the longer the metastable period, because the additional NaCl affects the interaction of Na^+-CO_3^{2-}-ion pairs and influences the concentration of free CO_3^{2-} in the solution. (2) The Cl^- introduced by the high concentration of NaCl can occupy the coordination layer of Nd^{3+} temporarily and delay the formation of rare earth carbonate precipitation. (3) After introduction of CO_2 gas, the existing environment of the solution directly changes in a short time from a single CO_3^{2-}-dominated system to a predominantly CO_3^{2-} and HCO^{3-}-dominated system. As a result, the metamorphic state of the solution is quickly terminated and the precipitation of Nd carbonate is advanced.

Author Contributions: Conceptualization, K.L.; data curation, X.Z.; formal analysis, Y.M.; funding acquisition, Y.Y.; investigation, K.L.; methodology, K.L.; project administration, Y.Y.; validation, F.N.; visualization, K.L.; writing—original draft, K.L.; writing—review & editing, L.W. and D.L. All authors have read and agreed to the published version of the manuscript.

Funding: This research was funded by National Natural Science Foundation of China, grant number 51774155, and The APC was funded by 51774155.

Data Availability Statement: Not Applicable.

Acknowledgments: The authors gratefully acknowledge the financial supports of the Program of National Natural Science Foundation of China (51774155).

Conflicts of Interest: The authors declare no conflict of interest.

References

1.   Takagi, K.; Hirayama, Y.; Okada, S.; Yamaguchi, W.; Ozaki, K. Novel powder processing technologies for production of rare-earth permanent magnets. *Sci. Technol. Adv. Mater.* **2021**, *22*, 150–159. [CrossRef] [PubMed]

2. Prokofev, P.A.; Kolchugina, N.B.; Skotnicova, K.; Burkhanov, G.S.; Kursa, M.; Zheleznyi, M.V.; Dormidontov, N.A.; Cegan, T.; Bakulina, A.S.; Koshkidko, Y.S.; et al. Blending Powder Process for Recycling Sintered Nd-Fe-B Magnets. *Materials* **2020**, *13*, 3049. [CrossRef] [PubMed]

3. Nelson, J.J.M.; Schelter, E.J. Sustainable Inorganic Chemistry: Metal Separations for Recycling. *Inorg. Chem.* **2019**, *58*, 979–990. [CrossRef] [PubMed]

4. Lixandru, A.; Venkatesan, P.; Jönsson, C.; Poenaru, I.; Hall, B.; Yang, Y.; Walton, A.; Güth, K.; Gauß, R.; Gutfleisch, O. Identification and recovery of rare-earth permanent magnets from waste electrical and electronic equipment. *Waste Manag.* **2017**, *68*, 482–489. [CrossRef] [PubMed]

5. Kuz'Min, M.D.; Skokov, K.; Jian, H.; Radulov, I.; Gutfleisch, O. Towards high-performance permanent magnets without rare earths. *J. Phys. Condens. Matter* **2014**, *26*, 64205. [CrossRef] [PubMed]

6. Bingqian, W. *Rare Earth Metallurgy*; Central South University of Technology Press: Changsha, China, 1997.

7. Firsching, F.H.; Mohammadzadei, J. Solubility products of the rare-earth carbonates. *J. Chem. Eng. Data* **1986**, *31*, 40–42. [CrossRef]

8. Taketatsu, T. The solubilities and anion-exchange behavior of rare earth elements in potassium carbonate solutions. *Anal. Chim. Acta* **1965**, *32*, 40–45. [CrossRef]

9. Taketatsu, T. The Dissolution and Anion Exchange Behavior of Rare Earth and Other Metallic Elements in Potassium Bicar-bonate, Potassium Carbonate and Ammonium Carbonate Solutions. *Bull. Chem. Soc. Jpn.* **1963**, *36*, 549–553. [CrossRef]

10. de Vasconcellos, M.E.; da Rocha, S.; Pedreira, W.; Queiroz, C.A.D.S.; Abrão, A. Solubility behavior of rare earths with ammonium carbonate and ammonium carbonate plus ammonium hydroxide: Precipitation of their peroxicarbonates. *J. Alloy. Compd.* **2008**, *451*, 426–428. [CrossRef]

11. Rao, L.; Rai, D.; Felmy, A.R.; Novak, C.F. Solubility of NaNd(CO$_3$) 6H$_2$O (c) in Mixed Electrolyte (Na-Cl-CO$_3$-HCO$_3$) and Synthetic Brine Solutions. In *Actinide Speciation in High Ionic Strength Media*; Springer: Berlin/Heidelberg, Germany, 1999; pp. 153–169.

12. Tang, J.; Johannesson, K.H. Rare earth elements adsorption onto Carrizo sand: Influence of strong solution complexation. *Chem. Geol.* **2010**, *279*, 120–133. [CrossRef]

13. Thakur, P.; Xiong, Y.; Borkowski, M. An improved thermodynamic model for the complexation of trivalent actinides and lanthanide with oxalic acid valid to high ionic strength. *Chem. Geol.* **2015**, *413*, 7–17. [CrossRef]

14. Jing, Y.; Chen, J.; Chen, L.; Su, W.; Liu, Y.; Li, D. Extraction Behaviors of Heavy Rare Earths with Organophosphoric Extractants: The Contribution of Extractant Dimer Dissociation, Acid Ionization, and Complexation. A Quantum Chemistry Study. *J. Phys. Chem. A* **2017**, *121*, 2531–2543. [CrossRef] [PubMed]

15. Pavón, S.; Fortuny, A.; Coll, M.; Sastre, A.M. Rare earths separation from fluorescent lamp wastes using ionic liquids as extractant agents. *Waste Manag.* **2018**, *82*, 241–248. [CrossRef] [PubMed]

16. Qiu, L.; Pan, Y.; Zhang, W.; Gong, A. Application of a functionalized ionic liquid extractant tributylmethylammonium dibutyldigly-colamate ([A336][BDGA]) in light rare earth extraction and separation. *PLoS ONE* **2018**, *13*, e0201405. [CrossRef]

17. Sun, P.; Huang, K.; Liu, H. The nature of salt effect in enhancing the extraction of rare earths by non-functional ionic liquids: Synergism of salt anion complexation and Hofmeister bias. *J. Colloid Interface Sci.* **2019**, *539*, 214–222. [CrossRef]

18. Li, C.; Zhuang, Z.; Huang, F.; Wu, Z.; Hong, Y.; Lin, Z. Recycling Rare Earth Elements from Industrial Wastewater with Flowerlike Nano-Mg(OH)2. *ACS Appl. Mater. Interfaces* **2013**, *5*, 9719–9725. [CrossRef] [PubMed]

19. Yang, Y.; Zhang, X.; Li, L.; Wei, T.; Li, K. Metastable Dissolution Regularity of Nd3+ in Na2CO3 Solution and Mechanism. *ACS Omega* **2019**, *4*, 9160–9168. [CrossRef]

20. Fannin, C.; Edwards, R.; Pearce, J.; Kelly, E. A Study on the Effects of Drying Conditions on the Stability of NaNd (CO$_3$)$_2$·6H$_2$O and NaEu (CO$_3$)$_2$·6H$_2$O. *Appl. Geochem.* **2002**, *17*, 1305–1312. [CrossRef]

21. Biovia, D.S. *Materials Studio 8.0*; Dassault Systèmes: San Diego, CA, USA, 2014.

22. Allen, M.P.; Tildesley, D.J. *Computer Simulation of Liquids*; Oxford University Press: Oxford, UK, 2017; ISBN 0192524704.

23. Anggraeni, A.; Arianto, F.; Mutalib, A.; Pratomo, U.; Bahti, H.H. Fast and simultaneously determination of light and heavy rare earth elements in monazite using combination of ultraviolet-visible spectrophotometry and multivariate analysis. In *AIP Conference Proceedings*; AIP Publishing LLC: Bandung, Indonesia, 2017; Volume 1848, p. 30004. [CrossRef]

24. Rao, L.; Rai, D.; Felmy, A.R.; Fulton, R.W.; Novak, C.F. Solubility of NaNd (CO$_3$)$_2$ 6 H$_2$O (c) in Concentrated Na$_2$CO$_3$ and NaHCO$_3$ Solutions. *Radiochim. Acta* **1996**, *75*, 141–148.

25. Jianfeng, Z. Dynamics Simulation of Interaction between Impurity Inhibitors and Aluminum Impurities. In *Ionic Rare Earth Ores No Title*; Jiangxi University of Science and Technology: Jiangxi, China, 2015.

 *minerals*

Article

Performance of TiB$_2$ Wettable Cathode Coating

Bo Yang [1,2,†], Ruzhen Peng [1,3,†], Dan Zhao [1,†], Ni Yang [1,4,†], Yanqing Hou [1,*] and Gang Xie [1,4,*]

[1]   Faculty of Metallurgy and Energy Engineering, Kunming University of Science and Technology, Kunming 650093, China; yangbo@stu.kust.edu.cn (B.Y.); yxp@stu.kust.edu.cn (R.P.); zhaoddan@stu.kust.edu.cn (D.Z.); yqh@stu.kust.edu.cn (N.Y.)

[2]   School of Materials and Civil Engineering, Guizhou Normal University, Guiyang 550014, China

[3]   Department of Materials and Metallurgy Chemistry, Jiangxi University of Science and Technology, Ganzhou 341000, China

[4]   Kunming Metallurgical Research Institute, Kunming 650093, China

*   Correspondence: hyqsimu@kust.edu.cn (Y.H.); wqs@stu.kust.edu.cn (G.X.)

†   These authors contributed equally to this work.

Abstract: A TiB$_2$ wettable cathode coating was deposited on a graphite carbon cathode material via atmospheric plasma spraying (APS). The microstructure and phase composition of the TiB$_2$ coating were analyzed via scanning electron microscopy (SEM), X-ray diffraction (XRD), and energy-dispersive X-ray spectroscopy (EDS). The wettability and corrosion resistance of the coating were studied in a molten-aluminum electrolytic system. The results showed that the surface of the TiB$_2$ coating prepared via plasma spraying was flat and that the main phase of the coating was TiB$_2$. The wettability between the TiB$_2$ coating and liquid aluminum was better than that between graphite cathode carbon block and liquid aluminum. The abilities of the TiB$_2$ coating and graphite cathode carbon block to resist sodium (Na) penetration and prevent molten salt corrosion were compared through a corrosion test. The TiB$_2$ coating was found to have better resistance to Na penetration and better refractory cryolite corrosion resistance than graphite cathode carbon block.

Keywords: plasma spraying; TiB$_2$ wettable cathode coating; wettability; corrosion resistance

Citation: Yang, B.; Peng, R.; Zhao, D.; Yang, N.; Hou, Y.; Xie, G. Performance of TiB$_2$ Wettable Cathode Coating. *Minerals* **2022**, *12*, 27. https://doi.org/10.3390/min12010027

Academic Editors: Kenneth N. Han, Shuai Wang, Xingjie Wang and Jia Yang

Received: 4 November 2021
Accepted: 22 December 2021
Published: 24 December 2021

Publisher's Note: MDPI stays neutral with regard to jurisdictional claims in published maps and institutional affiliations.

1. Introduction

Currently, cryolite molten salt electrolysis is the only method for smelting aluminum [1]. During the electrolysis of molten cryolite, the cathode carbon block exhibits some corrosion properties; especially, sodium (Na) permeates into the cathode carbon block and causes the expansion of the carbon block, which is the main cause of the cathode damage and failure [2].

Many studies on cathodic carbon blocks have focused on wettable cathodes, among which TiB$_2$ has become a research hotspot, because it is characterized by good wettability to liquid aluminum, low resolution, resistance to electrolyte erosion, and good electrical conductivity [3]. At present, many domestic and foreign studies on TiB$_2$ and its composites mainly focus on TiB$_2$ ceramic cathodes [4,5], TiB$_2$-carbon adhesive coating [6–8], TiB$_2$ layer prepared by plating in molten salt [9–11], TiB$_2$ non-carbon adhesive coating [12], TiB$_2$ coating prepared via chemical vapor deposition [13], TiB$_2$ coating prepared via self-spreading high-temperature synthesis TiB$_2$ coating [14], and TiB$_2$ coating prepared via plasma spraying [15].

In recent years, our research group has conducted numerous studies on the TiB$_2$ coating preparation via atmospheric plasma spraying (APS) and has achieved certain results. Through several orthogonal experiments on spraying parameters, the best process conditions for the coating preparation have been obtained [16]. The influence of mechanical properties has been studied [17]. The current study mainly investigates the microstructure of TiB$_2$ coating prepared via APS under optimal process conditions and the wettability and corrosion resistance of TiB$_2$ coating in a molten-salt electrolytic aluminum system.

2. Materials and Methods

2.1. Materials

The TiB_2 powder used in the experiment was obtained from a certain chemical industry research institute in Shenyang ($TiB_2 \geq 98.5\%$), with a particle size of $-325-+400$ mesh (38–45 µm). For aluminum electrolysis, a graphite cathode carbon block was adopted as the matrix (the specification was 100 mm × 100 mm × 30 mm); it was provided by an aluminum factory in Yunnan and needed to be polished and sandblasted in advance.

2.2. Preparation of Plasma-Sprayed TiB_2 Coating

During the TiB_2 coating preparation, first, the carbon block and TiB_2 powder were placed into the drying box and heated up to 100 °C for 4 h to remove the moisture, improve the TiB_2 powder fluidity, and improve the bonding strength between the carbon block and TiB_2 coating. Furthermore, TiB_2 powder was deposited on the graphite cathode carbon block using the DH-2080 plasma spray gun(Shanghai Ruifa Spraying Machinery Co., Ltd., Shanghai, China), with argon as the main gas, hydrogen as the secondary gas, and argon and hydrogen as the powder carrier gases. Before the preparation of the TiB_2 wettable cathode coating via spraying, the DH-2080 plasma spray gun was used to preheat the surface of the roughened and clean carbon block. The preheating temperature was maintained between 100 °C and 200 °C to shorten the temperature difference between the substrate and the TiB_2 powder and reduce the residual stress between the coating and the carbon block substrate. After the preheating, the plasma-spraying process parameters (Table 1) were adjusted, and the TiB_2 coating was sprayed on the carbon block substrate to obtain a TiB_2 coating with ~1000 µm thickness.

Table 1. Parameters of TiB_2 coating prepared via atmospheric plasma spraying (APS).

Name	Voltage	Current	Distance	Main Air Flow	Powder Feeding Rate	Ar Air Flow	Particle Diameter	Spray Gun Moving Speed
Parameter	72 V	550 A	90 mm	1800 L·h^{-1}	27.34 g·min^{-1}	120 L·h^{-1}	(d_{50}) $\leq$ 37.4 µm	100 mm·s^{-1}

2.3. TiB_2 Coating Characterization

The TiB_2 coating prepared via plasma spraying was peeled off from the base carbon block and then ground. The phase composition of the prepared TiB_2 coating powder was analyzed using the X'pert 3 powder-type diffractometer produced by PANalytical (Malvern, UK). The parameters were as follows: Cu target, acceleration voltage of 40 kV, current of 40 mA, and sweep speed of $8°$ min^{-1}. After the TiB_2 cathode coating subjected to different treatments was polished with sandpaper, the coating surface was observed under a QUANTA600 scanning electron microscope to detect the microscopic internal structure of the coating. Moreover, energy spectrum analysis was performed using the NORAN SYSTEM SIX energy spectrometer (SelectScience, Bath, UK).

2.4. Method for Wettability Measurement

In this study, the wetting angle between the plasma-sprayed TiB_2 inert cathode coating and molten aluminum was not directly measured; only the wettability between the TiB_2 inert cathode coating and molten aluminum was qualitatively characterized. First, an appropriate number of aluminum ingots were placed in the silicon carbide crucible, and the silicon carbide crucible was placed in a resistance furnace, model RF-15-10, produced by Changsha Gongtai Experimental Electric Furnace Co., Ltd (Changsha, China), and the heating system was set to 750 °C. When the aluminum ingot in the silicon carbide crucible became molten, the plasma-sprayed TiB_2 cathode coating sample and the aluminum electrolytic cathode carbon block preheated at the same temperature were simultaneously immersed in the molten aluminum; the immersion time was 5 s. Subsequently, the sample was taken out from the molten aluminum and placed on a steel plate at 45 °C with the horizontal ground. After the molten aluminum on the surface of the sample was cooled, aluminum was found to occur on the TiB_2 inert cathode coating and the carbon block. The spreading situations on the cathode and the wettability of the TiB_2 coating and carbon block were qualitatively evaluated.

2.5. Corrosion Resistance Test Method

Determination of the dissolution loss of TiB_2 coating in molten aluminum: The TiB_2 coating sample was soaked in high-temperature liquid aluminum at 960 °C for 48 h. After the molten aluminum sample was cooled, the change in the Ti content in the molten aluminum was analyzed, and then the dissolution loss of the TiB_2 coating in the molten aluminum was evaluated. The ARL841OX XRF was used to determine the titanium content in the liquid aluminum.

Static molten salt corrosion test of TiB_2 coating: The TiB_2 coating sample was placed on the bottom of a 0.5 L graphite crucible, and the prepared electrolyte (including Na_3AlF_6, NaF, Al_2O_3, and CaF_2 in mass percentages of 71.5%, 14.5%, 9.0%, and 5.0%, respectively) was added into a graphite crucible. The crucible was placed into a 960 °C atmosphere furnace and kept for 100 h. After the corrosion was over, the crucible was cooled to room temperature. Then, the TiB_2 coating sample at the bottom of the electrolytic cell was removed, and the residual electrolyte on the surface was cleaned. The sample was then cut along the central axis. Scanning electron microscopy (SEM) and energy-dispersive X-ray spectroscopy (EDS) were used to characterize the TiB_2 coating after static corrosion.

Static molten salt corrosion test of TiB_2 coating: The TiB_2 coating sample was placed on the bottom of a 0.5 L graphite crucible, and the prepared electrolyte (including Na_3AlF_6, NaF, Al_2O_3, and CaF_2 in mass percentages of 71.5%, 14.5%, 9.0%, and 5.0%, respectively) was added into the graphite crucible. The crucible was then placed into a 980 °C pit furnace. The graphite anode embedded in cast iron was inserted into the electrolytic cell. The area of the graphite anode was 28 cm^2. After the electrolyte was melted, the current density was kept at 0.7 $A \cdot cm^{-2}$, and electrolysis was conducted for 4 h. The electrolyte was added to the electrolytic cell every 15 min. The subsequent treatment after electrolysis was the same as that in the static test.

3. Results and Discussion

3.1. TiB_2 Coating Microstructure

Figure 1a,b are the SEM images of the TiB_2 coating surface prepared via APS at two different magnifications. Figure 1a shows that the coating internal structure was relatively dense and uniformly distributed. The coating mainly showed two colors of dark gray and white. Moreover, small cracks occurred in the coating. The formation of such microcracks depends on the melting of TiB_2, the volume shrinkage of the particles, and the TiB_2 coating cooling process. Figure 1b shows that after the TiB_2 powder was remelted and recrystallized, the particle size of the coating was not uniform. The distribution of such particles led to the generation of pores in the coating, and the increase in porosity will reduce the coating hardness and the bonding strength between the coating and the substrate; moreover, it will destroy the conductive network of the TiB_2 coating and increase the coating resistivity.

Figure 2a,b are the cross-sectional topography images of the TiB_2 coating under two different magnifications. The figure shows a clear boundary between the substrate and the coating. The gray area is the TiB_2 coating, and the black area is the carbon block substrate. The coating was formed by injecting TiB_2 powder into a high-temperature plasma jet and then transporting it to the substrate. The temperature at the center of the plasma flame was as high as 32,000 K, and the temperature at the nozzle outlet was still up to 20,000 K. The TiB_2 powder particles were instantly heated to a molten or semi-melted state in the high-temperature plasma jet. When the particles hit the carbon block substrate surface, they spread into a flat liquid covering and clung to the concave and convex points on the surface of the carbon block substrate. It shrank and bit the anchor point during condensation. Therefore, the combination of the TiB_2 coating and the carbon block substrate was mainly mechanical embedded-type. Because the TiB_2 coating was formed by solidifying a single particle as a unit to the substrate surface in a layered accumulation, the difference in particle size between the droplets resulted in some voids in the coating. Moreover, it can be seen from the cross section of the coating that the substrate and the coating. The bonding interface of the layer was clear, smooth, and tortuous, which shows that the coating was tightly bonded.

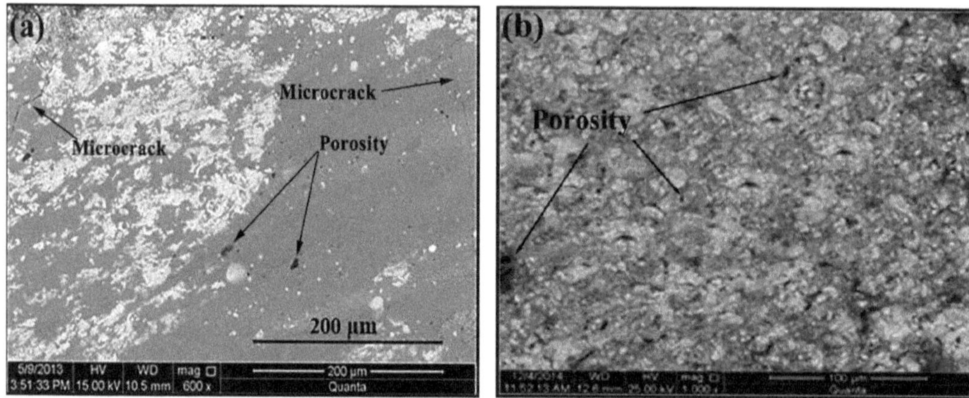

Figure 1. Scanning electron microscopy (SEM) image of TiB$_2$ coating surface morphology. (**a**) SEM of coating under 600 times, (**b**) SEM of coating under 1000 times.

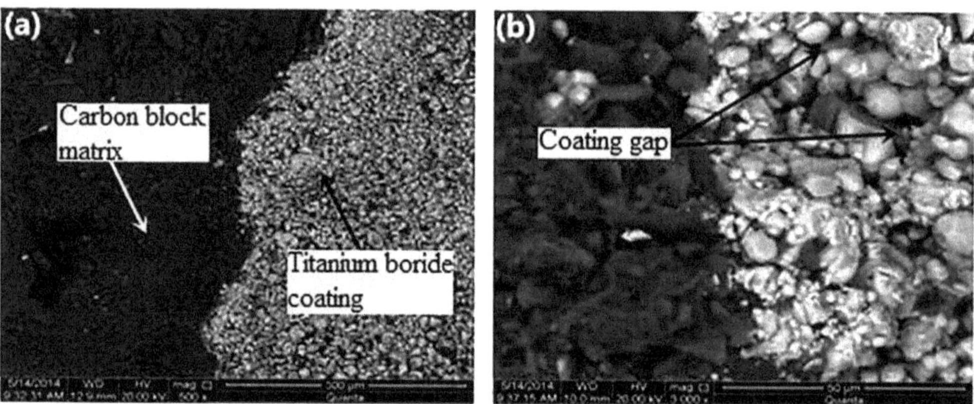

Figure 2. TiB$_2$ coatings sectional SEM micrographs. (**a**) SEM of coating under 600 times, (**b**) SEM of coating under 1000 times.

3.2. X-ray Diffraction Pattern of TiB$_2$ Coating

Figure 3 is the X-ray diffraction (XRD) analysis result of the TiB$_2$ coating prepared via APS. The figure shows that the coating was mainly composed of TiB$_2$, TiO$_2$, and a small amount of B$_2$O$_3$, indicating that TiB$_2$ was oxidized during the spraying process. This was mainly due to the air involved in the plasma jet and the molten TiB$_2$ particles during the plasma-spraying process. Contact causes TiB$_2$ particles to be oxidized to form oxidation products such as TiO$_2$ and B$_2$O$_3$.

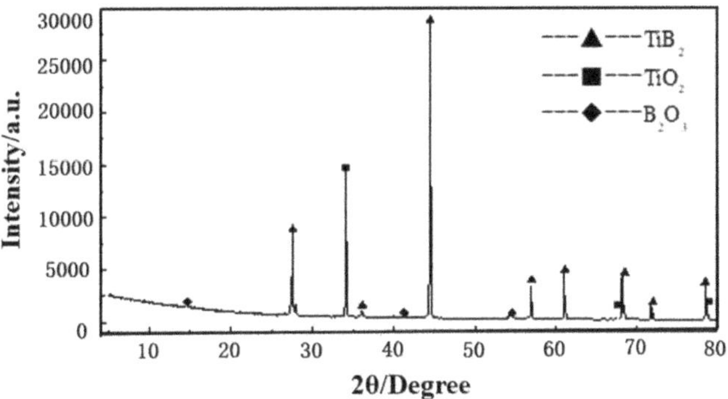

Figure 3. X-ray diffraction (XRD) patterns of plasma-sprayed TiB_2 coating.

3.3. Wettability of TiB_2 Coating

Figure 4 compares the peeling of molten aluminum from two samples. Figure 4c shows that the liquid aluminum spreading on the surface of the carbon block cathode was easily peeled off from the substrate and will not peel off with the carbon block material. Figure 4d shows that it is desired to spread the It is very difficult for the liquid aluminum on the surface of the layer to peel off from the TiB_2 substrate. The adhesion of the liquid aluminum to the coating is strong, and during the peeling of the coating, part of the coating materials peel off with the substrate; this also shows that the coating prepared by TiB_2 inert cathode in this study has good wettability.

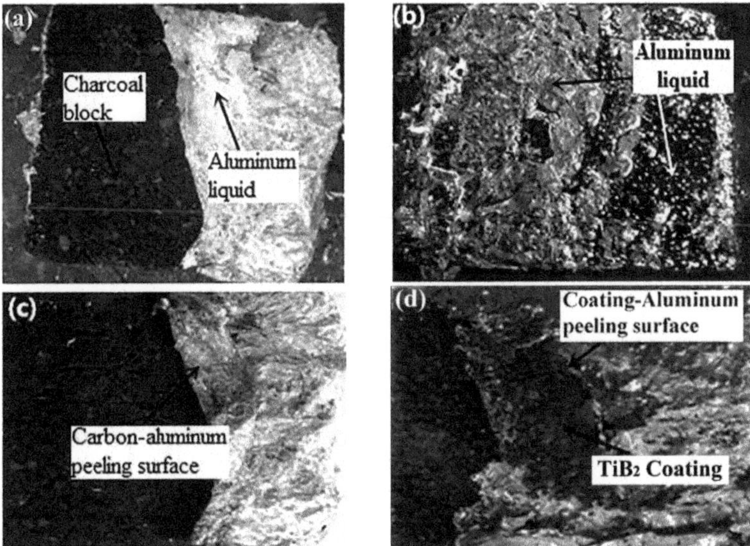

Figure 4. Characterization of wettability of TiB_2 coating prepared using cathode carbon block and plasma spraying to liquid aluminum. (**a**) SEM image wetting effect diagram of cathode carbon block on molten aluminum, (**b**) SEM image wetting effect diagram of TiB2 coating on molten aluminum, and comparative characterization of wettability of liquid aluminum after peeling, (**c**) SEM image of the peeling of the molten aluminum from the cathode carbon block, (**d**) liquid aluminum peeling from TiB_2 coating.

3.4. Corrosion Resistance of TiB₂ Coating Analysis of Solubility Loss of TiB₂ Coating in Molten Aluminum

Figure 5 depicts the change curve of the TiB₂ coating prepared via plasma spraying in molten aluminum. The figure shows that the quality of the TiB₂ coating remained unchanged and basically stable within 48 h of corrosion in high-temperature molten aluminum, demonstrating a good resistance to corrosion by molten aluminum. This is mainly because the TiB₂ material had good wettability to liquid aluminum and low solubility in liquid aluminum. Moreover, the TiB₂ coating prepared via plasma spraying was uniform and dense, demonstrating the good resistance of the TiB₂ material to liquid aluminum erosion.

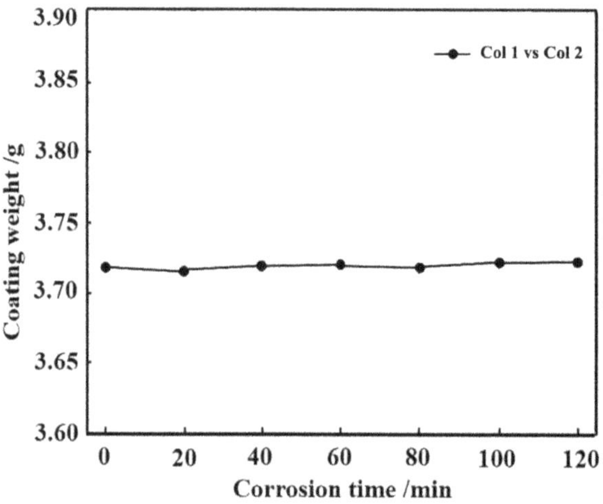

Figure 5. Variation curve of TiB₂ coating quality with erosion time after aluminum alloy erosion.

According to the XRD analysis, after the TiB₂ coating was corroded in high-temperature liquid aluminum (960 °C) for 120 min, no Ti component was detected, and the main component in the liquid aluminum was Al (Figure 6). The TiB₂ coating prepared via plasma spraying could be well wetted by molten aluminum and had excellent corrosion resistance to molten aluminum. Moreover, the elemental analysis method was used to detect the Ti content in the molten aluminum. The Ti content of the molten aluminum was 0.0042%, which was only 0.0016% higher than that of the original aluminum (0.0026%). It was found that the titanium content in the original aluminum was 0.0026% (mass percent), and after the TiB₂ coating was immersed in the aluminum solution for 48 h, the titanium content in the analyzed aluminum was 0.0042% (mass percent), which was only 0.0016% higher than the original aluminum content of 0.0026%. According to this calculation, the industrial tank was coated with 1 mm. The service life of thick pure TiB₂ coating should be over four years.

3.5. Analysis of Static Corrosion Results of TiB₂ Coating

Resistance to Na penetration is an important indicator of wettable cathode materials. Figure 7a is a low-power-SEM image of the surface of the sample after the static corrosion of the carbon block, and Figure 7b is a low-power-SEM image of the TiB₂ coating surface after static corrosion. The figure shows that after cryolite corrosion under the same conditions, the surface of the carbon block cathode was loose and porous, the area of the carbon block was significantly reduced, and the electrolyte was more likely to penetrate into the carbon block. Figure 8a is the surface micro-topography image of the TiB₂ coating after corrosion, and Figure 8b is the cross-sectional micro-topography image of the TiB₂ coating after corrosion. Figure 8a shows that after the coating surface was corroded, the TiB₂ particles still maintained the original crystal structure state and were not corroded by cryolite melt.

Figure 8b shows that the TiB$_2$ coating cross section also maintained the original complete structure; the coating structure was still dense. After 100 h of static corrosion, the TiB$_2$ coating internal structure was dense, and the corrosion resistance was good. Figure 9 shows the EDS spectrum of the TiB$_2$ coating after corrosion. The figure shows that the main component element in the coating was Ti, and the metal Na$^+$ ions did not penetrate into the coating.

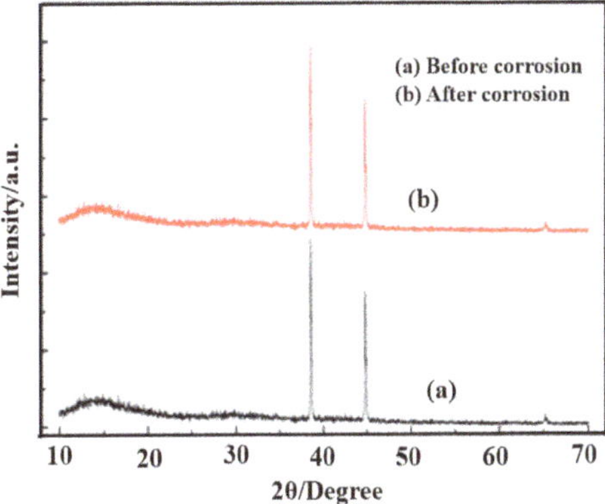

Figure 6. XRD analysis results of TiB$_2$ coatings before and after erosion in liquid aluminum for around 120 min. (**a**) Before corrosion. (**b**) After corrosion.

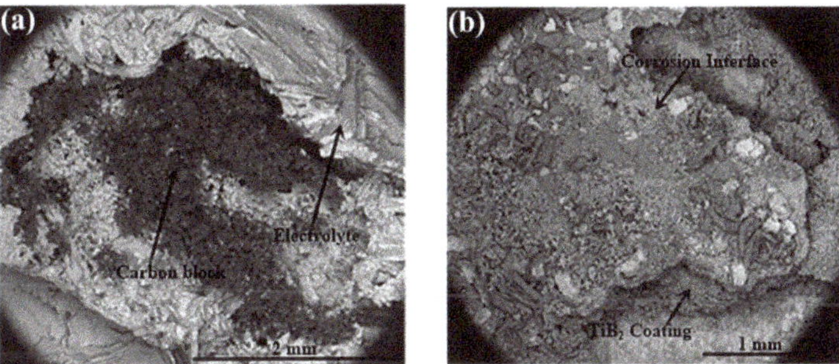

Figure 7. Charcoal surface and the coating surface after etching contrast. (**a**) Low magnification SEM image of sample surface after carbon block static corrosion, and (**b**) low magnification SEM image of TiB$_2$ coating surface after static corrosion.

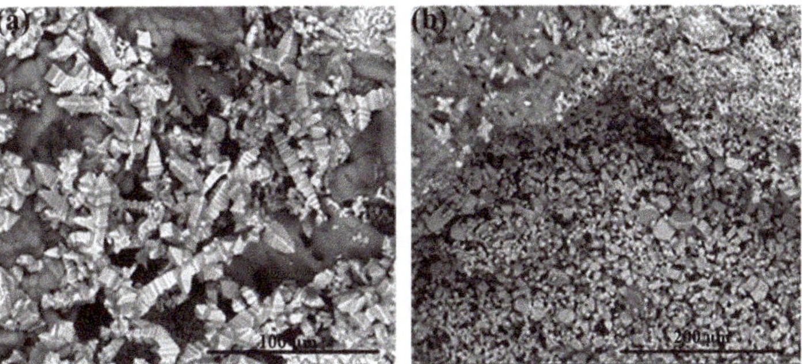

Figure 8. (**a**) Surface and (**b**) cross section after TiB$_2$ coating corrosion.

Figure 9. Energy-dispersive X-ray spectroscopy (EDS) analysis after corrosion of TiB$_2$ coating surface.

3.6. Analysis of Dynamic Corrosion Results of TiB$_2$ Coating

Figure 10 is a cross-sectional view of the TiB$_2$ coating after dynamic corrosion, and Figure 11 is the cross section of the TiB$_2$ coating after the dynamic corrosion experiment and the EDS energy spectrum of points taken near the TiB$_2$ coating surface.

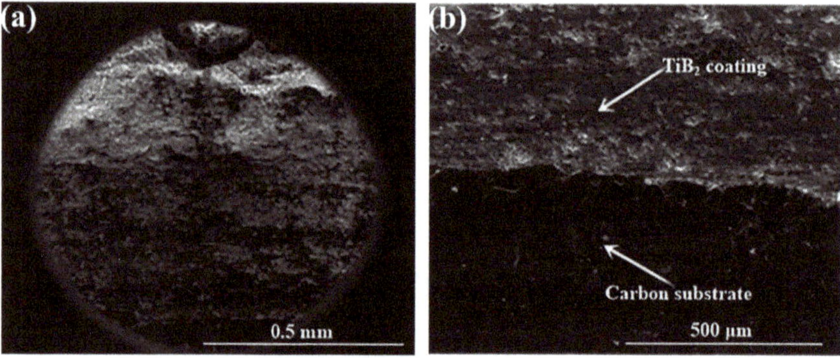

Figure 10. The section after the corrosion of the TiB$_2$ coating. (**a**) Low-magnification figure and (**b**) Micro-topography.

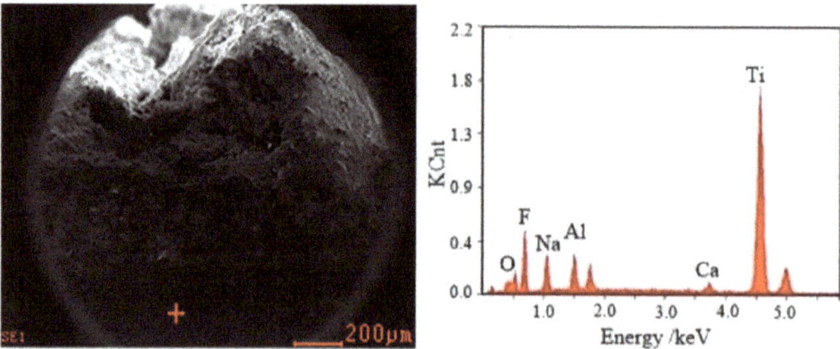

Figure 11. Cross section and EDS spectrum of coating after corrosion.

Figure 10a shows that the TiB_2 coating after electrolysis maintained a compact structure. Figure 10b shows that the bonding surfaces of the TiB_2 coating and the carbon block substrate were in good contact, without peeling or porosity of the coating. The TiB_2 coating had good bonding strength with the carbon block substrate. From the analysis of the EDS results, a small amount of cryolite components such as F, Na, Al, and Ca occurred in this area. The main component of the coating was Ti; this indicates that a small amount of cryolite penetrated the coating surface after the electrolysis experiment; the penetration depth was not deep, indicating that the TiB_2 coating obtained via plasma spraying has improved resistance to cryolite corrosion. The TiB_2 wettable cathode coating prepared through plasma spraying technology does not contain carbon elements, which avoids the defect that the TiB_2/C carbon glue coating is easily corroded by the electrolyte due to carbon elements.

4. Conclusions

In this study, under optimal parameters and the optimal particle size, a TiB_2 coating was prepared on a carbon cathode surface using plasma-spraying equipment, and the phase and microstructure of the coating were analyzed via SEM, XRD, and other characterization methods. After the analysis, the wettability and corrosion resistance of the coating were measured, and the following conclusions were drawn:

The TiB_2 coating prepared via APS had a smooth surface free of peeling and cracking. The TiB_2 coating internal structure was dense and uniform. The large particles were in a semi-melted state, forming a disc-shaped structure embedded in the coating, and the small particles were completely melted; they connected the large particles to fill the pores between the large particles. In the plasma-spraying process, the air drawn into the plasma jet contacted the molten TiB_2 particles, causing the TiB_2 particles to be oxidized and form oxidation products such as TiO_2 and B_2O_3. TiB_2 is still the most important phase component in the coating.

The wettability between the TiB_2 wettable cathode coating and molten aluminum was significantly better than that between graphite cathode carbon block and molten aluminum. Through static corrosion experiments, the abilities of the TiB_2 coating and graphite cathode carbon block to resist Na penetration and to prevent molten cryolite corrosion were compared. The TiB_2 coating resistance to Na penetration and corrosion resistance to molten cryolite were better than those of the graphite cathode carbon block. Moreover, after the TiB_2 coating was dynamically corroded for 4 h, only a small amount of F, Na, and Al penetrated into the TiB_2 coating inner surface. Given the results, the TiB_2 coating prepared via plasma spraying has good liquid aluminum wetting ability and good resistance to Na penetration.

5. Future Prospective

The oxidation behavior of TiB2 in the plasma spraying process is very complex, the process is very much influenced by the transfer (momentum transfer, heat transfer and mass transfer), especially the transfer phenomenon of gas phase B2O3 under high temperature conditions determines the TiB2 oxidation behavior, and the plasma spraying TiB2 coating is carried out under high temperature conditions, and metallurgical bonding is observed through experiments, however, the bonding mechanism and the effect on TiB2 coating properties are still unclear; finally the introduction of multi-component TiB2 base can further improve the properties of the coating such as corrosion resistance and wettability. Therefore, on the basis of this thesis study, the following studies are proposed to be carried out in the future:

(1) Study the momentum transfer, heat transfer and mass transfer behaviors of TiB2 during the movement to the surface of carbon block during the spraying process, analyze the time-varying laws of flow field and temperature field, and reveal the oxidation behavior of TiB2 and the symmetry mechanism of multi-physical field;

and study in depth the mechanism of pore formation of TiB2 coating under high temperature conditions and study the transfer phenomenon of gas phase B2O3.

(2) To study the metallurgical bonding behavior of TiB2 coating process for electrolytic aluminum by plasma spraying, to study the generation of new substances and formation of new chemical bonds by quantum chemical methods, and to reveal the possible chemical reactions by searching for transition states, and to propose the mechanism of the influence of metallurgical bonding on the performance of TiB2 coating through the study.

(3) On the basis of TiB2 coating research, we study multi-component TiB2 based coating materials for aluminum electrolysis, add rare earth group elements to TiB2 raw materials, study the mechanism of its addition on the performance of coating materials, and explore new high-performance TiB2-based coating materials for aluminum electrolysis.

Author Contributions: For research articles with several authors, Conceptualization, B.Y. and Y.H.; methodology, R.P.; validation, N.Y. and D.Z.; formal analysis, B.Y. and R.P.; investigation, Y.H.; resources, G.X.; writing—original draft preparation, B.Y.; writing—review and editing, B.Y.; supervision, Y.H.; funding acquisition, Y.H. and G.X. All authors have read and agreed to the published version of the manuscript.

Funding: This research received no external funding.

Data Availability Statement: Data sharing not applicable. No new data were created or analyzed in this study. Data sharing is not applicable to this article.

Acknowledgments: This work was financially supported by the National Natural Science Foundation of China (22168019, 52074141). The authors are grateful to NSFC, China for their support.

Conflicts of Interest: The authors declare no conflict of interest.

References

1. Wang, X.; Leng, H.; Han, B.; Wang, X.; Hu, B.; Luo, H. Solidified microstructures and elastic modulus of hypo-eutectic and hyper-eutectic TiB2-reinforced high-modulus steel. *Acta Mater.* **2019**, *176*, 84–95. [CrossRef]
2. Kang, S.H.; Kim, D.J.; Kang, E.S.; Baek, S.S. Pressureless sintering and properties of titanium diboride ceramics containing chromium and iron. *J. Am. Ceram. Soc.* **2001**, *84*, 893–895. [CrossRef]
3. Lee, D.B.; Lee, Y.C.; Kim, D.J. The oxidation of TiB2 ceramics containing Cr and Fe. *Oxid. Met.* **2001**, *56*, 177–189. [CrossRef]
4. Luo, X.; Wang, Z.; Hu, X.; Shi, Z.; Gao, B.; Wang, C.; Chen, G.; Tu, G. Influence of Metallic Additives on Densification Behaviour of Hot-Pressed TiB2. In *Light Metals*; The Minerals, Metals & Materials Society: Warrendale, PA, USA, 2009; pp. 1151–1155.
5. Liao, N.; Xu, L.; Jin, Y.; Xiao, Y.; Grasso, S.; Hu, C. Synthesis, microstructure, and mechanical properties of TiC-SiC-Ti3SiC2 composites prepared by in situ reactive hot pressing. *Int. J. Appl. Ceram. Technol.* **2020**, *17*, 1601–1607. [CrossRef]
6. Liu, W.; Hao, C.-Y.; Zhao, X.-D.; Wang, X.-J.; Shi, G.-L. Strengthening mechanisms and tribological properties of ultra-hard AlMgB14 based composite reinforced with TiB2. *Sci. Adv. Mater.* **2020**, *12*, 866–872. [CrossRef]
7. Fang, Z.; Wu, X.-L.; Yu, J.; Li, L.-B.; Zhu, J. Penetrative and migratory behavior of alkali metal in different binder based TiB2–C composite cathodes. *Trans. Nonferrous Met. Soc. China* **2014**, *24*, 1220–1230. [CrossRef]
8. Fan, X.; Huang, W.; Zhou, X.; Zou, B. Preparation and characterization of NiAl–TiC–TiB2 intermetallic matrix composite coatings by atmospheric plasma spraying of SHS powders. *Ceram. Int.* **2020**, *46*, 10512–10520. [CrossRef]
9. Zhang, T.; Song, S.; Xie, C.; He, G.; Xing, B.; Li, R.; Zhen, Q. Preparation of highly-dense TiB2 ceramic with excellent mechanical properties by spark plasma sintering using hexagonal TiB2 plates. *Mater. Res. Express* **2019**, *6*, 125055. [CrossRef]
10. Islak, B.Y.; Candar, D. Synthesis and properties of TiB2/Ti3SiC2 composites. *Ceram. Int.* **2021**, *47*, 1439–1446. [CrossRef]
11. Zhuang, W.; Yang, H.; Yang, W.; Cui, J.; Huang, L.; Wu, C.; Liu, J.; Sun, Y.; Meng, C. Microstructure, tensile properties, and wear resistance of in situ TiB2/6061 composites prepared by high energy ball milling and stir casting. *J. Mater. Eng. Perform.* **2021**, *30*, 7730–7740. [CrossRef]
12. Xiao, P.; Gao, Y.; Xu, F.; Yang, S.; Li, B.; Li, Y.; Huang, Z.; Zheng, Q. An investigation on grain refinement mechanism of TiB2 particulate reinforced AZ91 composites and its effect on mechanical properties. *J. Alloys Compd.* **2019**, *780*, 237–244. [CrossRef]
13. Liu, J.H.; Blanpain, B.; Wollants, P. A XPS study of atmospheric plasma sprayed TiB2 coatings. *Key Eng. Mater.* **2008**, *368–372*, 1347–1350. [CrossRef]
14. Gu, J.; Lee, S.-H.; Vu, V.H.; Yang, J.; Lee, H.-S.; Kim, J.-S. Fast fabrication of SiC particulate-reinforced SiC composites by modified PIP process using spark plasma sintering—Effects of green density and heating rate. *J. Eur. Ceram. Soc.* **2021**, *41*, 4037–4047. [CrossRef]

15. Wang, R.; Wang, S.; Li, D.; Xing, A.; Zhang, J.; Li, W.; Zhang, C. Theoretical characterization of the temperature dependence of the contact mechanical properties of the particulate-reinforced ultra-high temperature ceramic matrix composites in Hertzian contact. *Int. J. Solids Struct.* **2021**, *214–215*, 35–44. [CrossRef]
16. Song, B.; Yang, W.; Liu, X.; Chen, H.; Akhlaghi, M. Microstructural characterization of TiB2–SiC–BN ceramics prepared by hot pressing. *Ceram. Int.* **2021**, *47*, 29174–29182. [CrossRef]
17. Liu, Y.; Li, Z.; Peng, Y.; Huang, Y.; Huang, Z.; Zhang, D. Effect of sintering temperature and TiB2 content on the grain size of B4C-TiB2 composites. *Mater. Today Commun.* **2020**, *23*, 100875. [CrossRef]

Article

Preparation of Antimony Sulfide and Enrichment of Gold by Sulfuration–Volatilization from Electrodeposited Antimony

Wei Wang [1,2], Shuai Wang [3], Jia Yang [1,2,4,*], Chengsong Cao [1], Kanwen Hou [1,2], Lixin Xia [1,2], Jun Zhang [1,2], Baoqiang Xu [1,2,4] and Bin Yang [1,2,4]

[1] Faculty of Metallurgical and Energy Engineering, Kunming University of Science and Technology, Kunming 650093, China; 20202202039@stu.kust.edu.cn (W.W.); 201710201260@stu.kust.edu.cn (C.C.); hkw1347@stu.kust.edu.cn (K.H.); xialixin7620@163.com (L.X.); 15979099078@163.com (J.Z.); kmxbq@kmust.edu.cn (B.X.); kgyb2005@126.com (B.Y.)

[2] National Engineering Laboratory for Vacuum Metallurgy, Kunming University of Science and Technology, Kunming 650093, China

[3] College of Chemistry and Chemical Engineering, Central South University, Changsha 410083, China; wangshuai@csu.edu.cn

[4] State Key Laboratory of Complex Nonferrous Metal Resources Clear Use, Kunming University of Science and Technology, Kunming 650093, China

* Correspondence: yangjia0603@kust.edu.cn; Tel.: +86-13033351342

Abstract: Electrodeposited antimony can be treated with sulfuration–volatilization technology, which causes antimony to volatilize in the form of antimony sulfide. During this process, gold is enriched in the residue, thereby realizing the value-added use of antimony and the recovery of gold. In this study, the thermodynamic conditions of antimony sulfide were analyzed by the Clausius–Clapeyron equation. Moreover, the volatilization behavior of antimony sulfide and the enrichment law of gold were studied by heat volatilization experiments. The effects of the sulfide temperature and volatilization pressure on the separation efficiency of antimony and gold enrichment were investigated. The results demonstrate that the sulfuration rate was the highest, namely 96.06%, when the molar ratio of sulfur to antimony was 3:1, the sulfur source temperature was 400 °C, the antimony source temperature was 550 °C, and the sulfuration time was 30 min. Antimony sulfide prepared under these conditions was volatilized at 800 °C over 2 h at an evaporation pressure of 0.2 atm, and the volatilization rate was the highest, namely 92.81%. Antimony sulfide with a stibnite structure obtained from the sulfuration–volatilization treatment of electrodeposited antimony meets the ideal stoichiometric ratio of sulfur and antimony in Sb_2S_3 (3:2), and gold is enriched in the residue.

Keywords: electrodeposited antimony; sulfuration–volatilization; antimony sulfide; enriched gold

Citation: Wang, W.; Wang, S.; Yang, J.; Cao, C.; Hou, K.; Xia, L.; Zhang, J.; Xu, B.; Yang, B. Preparation of Antimony Sulfide and Enrichment of Gold by Sulfuration–Volatilization from Electrodeposited Antimony. *Minerals* **2022**, *12*, 264. https://doi.org/10.3390/min12020264

Academic Editor: Benjamin P. Wilson

Received: 31 December 2021
Accepted: 17 February 2022
Published: 19 February 2022

Publisher's Note: MDPI stays neutral with regard to jurisdictional claims in published maps and institutional affiliations.

1. Introduction

Antimony is a rare metal, and its average abundance in the earth's crust is only 0.2–0.5% [1]. In 2014, the European Union classified antimony as a key raw material [2]. The conductivity of antimony is between conductor and insulator, it is not easily oxidized at room temperature, and it has corrosion resistance. In addition, antimony compounds also have the excellent characteristics of high heat resistance and low resistance, and are often used to manufacture laser-guided bombs, various missile seekers, and friction materials. Sb and its compounds were first only used in fireworks, firewood, and other daily life fields. With the rapid development of science and technology, antimony and its compounds have become widely used in the nuclear industry, batteries, semiconductors, catalysis, enamel, and alloys, as well as in the medical and pharmaceutical fields, the military, etc. Antimony is also characterized by low substitution and high military demand, and a single supply source mainly obtained via primary mining [3–5].

Antimony is a traditionally dominant mineral resource in China. According to survey data released by the United States Geological Survey (USGS), China's antimony ore resource reserves account for 25% of the world's total reserves, with antimony ore reserves ranking first in the world. Moreover, China's antimony ore mining accounted for 52% of the world's antimony mining in 2020, with extraction of antimony ores also ranking first in the world [6]. With the increasing development of mining technology, the production process of antimony mineral resources, mainly jamesonite and stibnite, has become increasingly mature. At present, the research on antimony production in China has gradually shifted to investigations of complex polymetallic symbiotic ores, represented by antimony–gold deposits [7]. The traditional treatment method of antimony minerals is the smoke method [8,9]; this is because antimony–gold ore contains gold and other minerals, and gold is often associated with sulfides, such as pyrite, antimonite, and arsenopyrite. In addition to natural gold and continuous gold, which represent the free existence of raw gold, most gold is embedded in sulfide and oxide minerals, which belong to the category of refractory gold mines [10–14]. Considering the enrichment and extraction of precious metals, sodium sulfide leaching–sodium thioantimonate solution electrodeposition is a commonly used technology for the treatment of antimony–gold ore.

Antimony–gold ore can be treated by a sodium sulfide leaching–sodium thioantimonate solution electrowinning process. However, during this treatment, up to 10.3% of the gold in the ore will be dissolved into the leaching solution, and this gold will be precipitated during electrowinning before the antimony in the leaching solution is reduced; the antimony produced in this solution system contains not only a large amount of sodium salt, but also gold [15–18]. Therefore, there is an urgent need to explore processes that can simultaneously purify electrodeposited antimony and achieve gold enrichment. The recovery and value-added use of valuable metals from metallurgical by-products can be achieved by sulfidation [19] and vacuum distillation [20,21]. Zhang [22] purified electrodeposited antimony by vacuum distillation and enriched gold in the residue. The present research is based on the problem that electrodeposited antimony contains precious gold, the smelting product of antimony–gold ore. By using the basic principle whereby the saturated vapor pressure difference between Sb, Sb_2S_3, and Au is significant, a two-stage process of sulfuration–volatilization is used to directly sulfurate the electrodeposited antimony. Antimony sulfide can be produced and purified by distillation via a novel process for the treatment of electrodeposited antimony. In this process, antimony is volatilized in the form of sulfide antimony, while gold is enriched in the residue, thereby realizing the value-added use of the main metal antimony and the recovery of gold.

2. Materials and Methods

2.1. Materials and Device

The experimental raw materials used in this study were electrodeposited antimony produced by Shandong Hengbang Group (Yantai, China) and sulfur powder produced by Tianjin Fengchuan Chemical Reagent Technology Co., Ltd. (Tianjin, China). Chemical analysis and gravimetric methods were used to quantitatively determine the chemical composition of the electrodeposited antimony and sulfur powder. The chemical composition of antimony electrodeposited from raw materials was obtained via chemical analysis and X-ray fluorescence, and the results are reported in Table 1.

Table 1. Chemical composition of electrodeposited antimony.

Element	Na	Fe	As	Se	Sb	Au [1]
Content/wt%	0.400	0.170	0.044	0.033	93.690	38

[1] The unit of Au content was g/t.

In this experiment, to fully react the electrodeposited antimony, excess sulfur powder was needed; thus, the molar ratio of S: Sb was set to 3:1. The tube furnace was filled with argon. One of the dual-temperature-zone tube furnaces was the sulfuration temperature zone with an antimony crucible, the other was the sulfur volatilization temperature zone with a sulfur crucible, and the temperature of the sulfuration temperature zone was set to 400–650 °C. The temperature gradient was set to 50 °C, and the heating rate was 10 °C·min^{-1}. The volatilization system was set to 400 °C, the heating rate was 8 °C·min^{-1}, the pressure was 1 atm, and the temperature was maintained for 30 min. A furnace box of the tube furnace was used for the volatilization test, and graphite paper was placed at both ends to collect volatiles. The samples obtained under optimal vulcanization conditions were volatilized at 800 °C. The atmospheric pressure in the volatilization zone was respectively set to 0.2, 0.4, 0.6, 0.8, and 1.0 atm, and the temperature was maintained for 120 min.

Reactions in the system:

$$2Sb + 3S \rightarrow Sb_2S_3 \tag{1}$$

The reaction flowchart is presented in Figure 1.

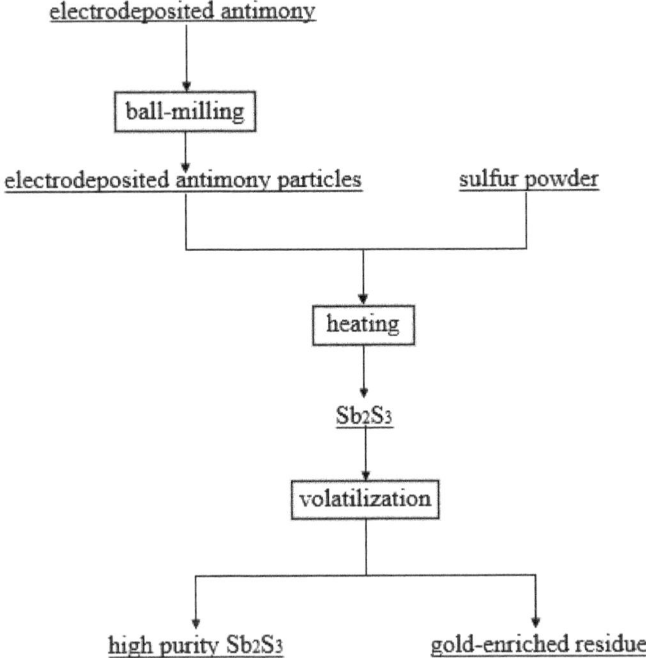

Figure 1. Process flow chart of electrodeposited antimony sulfide volatilization and enrichment of gold.

The experimental device was a dual-temperature-zone tube furnace, produced by Shanghai Yifeng Electric Furnace Co., Ltd. (Shanghai, China), as shown in Figure 2. The shell of the tube furnace was welded with a thin steel plate at the edge, and the furnace lining was made of refractory fiber material. The spiral heating element was made of an iron-chromium-aluminum alloy electric heating wire and inserted into the furnace lining. The tube furnace was equipped with a temperature controller and an Ni-Cr-Ni-Si thermocouple, which could measure, indicate, and automatically control the furnace temperature. Thermocouples for temperature measurement and control were inserted through the thermocouple holes, and the gaps between the holes and thermocouples were filled with cotton fiber.

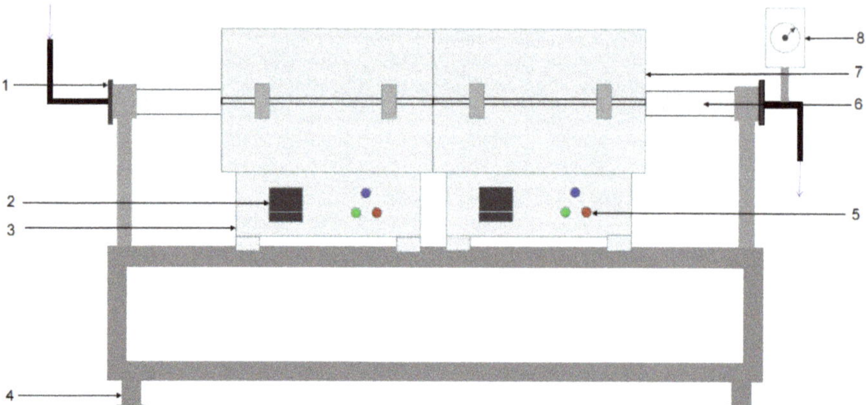

Figure 2. Structure chart of double temperature zone tube furnace: 1—flange connector at inlet, 2—numerical control display, 3—furnace body, 4—holder, 5—control switch, 6—quartz tube, 7—resistance furnace box, and 8—vaccum gauge.

2.2. Methods

Scanning electron microscopy (SEM, VEGA3 TESCAN, Brno, Czech Republic) was employed to observe the microscopic morphology of antimony sulfide. Raman spectroscopy was performed with a Renishaw (London, England). Raman microscope equipped with a 514 nm laser, integrated switchable gratings with 600 or 1800 lines/mm, and a CCD detector. An electron probe micro-analyzer (EPMA-1720H, Shimadzu, Shimane, Japan) was used to scan the elemental distributions of the samples. Energy-dispersive spectroscopy (TM3030Plus, HITACHI, Tokyo, Japan) was employed for the elemental analysis of the micro-samples. X-ray photoelectron spectroscopy (XPS, PHI5000 Versaprobe-II, ULVAC-PHI, Chigasaki, Japan) was carried out to measure the binding energy of the electrons to identify the chemical properties and composition of the volatiles.

The sulfuration rate of electrodeposited antimony was calculated by the following formula:

$$\text{Sulfuration rate} = (m_1 - m_0)/m, \tag{2}$$

where m_0 is the mass (g) of the empty crucible before the experiment; m_1 is the total mass of the crucible containing the electrodeposited antimony after the reaction; and m is calculated by the chemical reaction equation of the reaction between elemental sulfur and elemental antimony.

The volatilization rate of samples after vulcanization was calculated by the following formula:

$$\text{Volatilization rate} = (m_2 - m_3)/m_2, \tag{3}$$

where m_2 is the mass of the sample prepared under the optimal vulcanization conditions before volatilization in the crucible, and m_3 is the mass of the residue in the crucible after volatilization.

3. Theoretical Calculation

The theoretical basis for vacuum metallurgical separation, purification, and refining is based on the different saturated vapor pressures of various substances. The relationship between pressure and temperature when a pure substance reaches two-phase equilibrium can be expressed by the Clausius–Clapeyron equation:

$$dp/dT = H/T(V_g - V_l) \tag{4}$$

In the process of substance sublimation and evaporation, p represents the saturated vapor pressure of the substance at temperature T, Pa; T represents temperature, K; H represents the latent heat of vaporization of the substance, J/mol; V_g represents the molar volume of the substance in the gas phase volume, m³/mol; and V_l indicates the molar volume of the substance in the liquid phase, m³/mol.

For gas–liquid equilibrium, the equilibrium pressure was the vapor pressure of the liquid phase. Since the molar volume V_g when the substance was in the gas phase was much larger than the molar volume V_l when the substance was in the liquid phase, V_l can be ignored, namely:

$$V_g - V_l = V_g \tag{5}$$

Therefore, the Formula (4) was simplified to:

$$dp/dT = H_l/T\,V_g \tag{6}$$

The liquid phase has a low saturated vapor pressure and can be treated as an ideal gas. The gas phase follows the ideal gas law:

$$p\,V_g = R\,T \tag{7}$$

Substituting Formula (7) into Formula (6), we can get:

$$dp/dT = H_l\,p/R\,T^2 \tag{8}$$

Owing to dp/p = p lnp, Formula (8) becomes:

$$dp/dT = H_l\,p/R\,T^2 \tag{9}$$

When the temperature changes, the change in latent heat of evaporation cannot be ignored, then:

$$H_l = H_0 + aT + bT^2 + cT^3 + \ldots\ldots \tag{10}$$

Substituting Formula (10) into Formula (9), we can get:

$$\ln p = -L_0/RT + a/R\,\ln T + b/R\,T + \ldots\ldots + c \tag{11}$$

To further simplify:

$$\lg p = AT^{-1} + B\,\lg T + CT + D \tag{12}$$

From Equation (12), it can be seen that the image is a curve, representing a non-linear relationship.

The constants A, B, C, and D in Formula (12) are evaporation constants, which can be obtained by consulting related thermodynamic manuals.

From the literature [23,24]:

For metal Sb:

$$\lg p = 8.00 - 6060/T \quad T \le 1573\ ^{\circ}\text{C} \tag{13}$$

$$\lg p = 9.15 - 7880/T \quad T > 1573\ ^{\circ}\text{C} \tag{14}$$

For Sb_2S_3:

$$\lg p = 14.67 - 11{,}200/T \quad 673\ ^{\circ}\text{C} \le T < 773\ ^{\circ}\text{C} \tag{15}$$

$$\lg p = 9.92 - 7068/T \quad 773\ ^{\circ}\text{C} \le T < 1223\ ^{\circ}\text{C} \tag{16}$$

For Au:

$$\lg p = (14.50 - 19{,}280/T) - 1.01 \times \lg T \tag{17}$$

Therefore, the relationship between saturated vapor pressure and temperature can be obtained as shown in Table 2 and Figure 3:

Table 2. Saturated vapor pressure of Sb_2S_3 and Au.

Substance	Saturation Vapour Pressure/Pa				
	773 K	873 K	973 K	1073 K	1173 K
Sb	1.45	11.44	59.13	225.05	674.15
Sb_2S_3	5.98	66.65	447.58	2127.46	7752.19
Au	4.38×10^{-14}	2.79×10^{-11}	4.65×10^{-9}	2.96×10^{-7}	9.19×10^{-6}

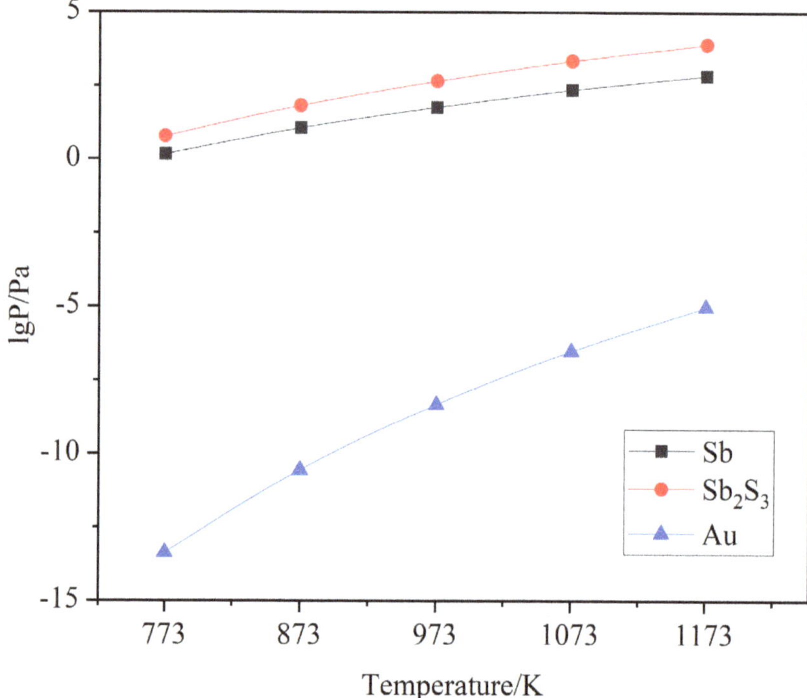

Figure 3. Saturated vapor pressure curves of Sb, Sb_2S_3 and Au.

Table 2 and Figure 3 show that in the temperature range of 400 °C to 900 °C, the saturated vapor pressure of Au was much less than that of Sb and Sb_2S_3, while the saturated vapor pressure of Sb was much lower than that of Sb_2S_3, and varied with increasing temperature. The difference in saturated vapor pressure between Sb and Sb_2S_3 also gradually increases.

In summary, the saturated vapor pressures of Sb_2S_3 and Sb were much greater than that of Au, and the saturated vapor pressure of Sb_2S_3 was much greater than that of Sb, indicating that Au was less volatile and Sb_2S_3 was more volatile than Sb. In this experiment, sulfur vapor was first used to vulcanize Sb to produce Sb_2S_3, and then to volatilize the vulcanized Sb_2S_3. Recovery of antimony sulfide from electrodeposited antimony and enrichment of gold.

4. Experimental Analysis

4.1. Sulfuration Reaction

4.1.1. Sulfuration Rate Calculation

During the sulfuration process, a small portion of the generated antimony sulfide will volatilize. The volatile Sb_2S_3 was condensed in the non-heating zone and adhered to the wall of the quartz tube. A portion of the antimony sulfide generated during the sulfuration process was volatilized, resulting in the calculated sulfuration rate being lower than the actual sulfuration rate. At 400 and 450 °C, the electrodeposited antimony was not fully vulcanized, and the sulfuration rates were only 79.57% and 79.21%, respectively. The sulfuration rate reached 89.78% at 500 °C. A good sulfuration rate of 96.06% was achieved at 550 °C. However, the calculated sulfuration rates at 600 and 650 °C were only 88.89% and 89.78%, respectively. The reason for this is that a portion of the antimony will be in the melting–recrystallization zone, as shown in Figure 4.

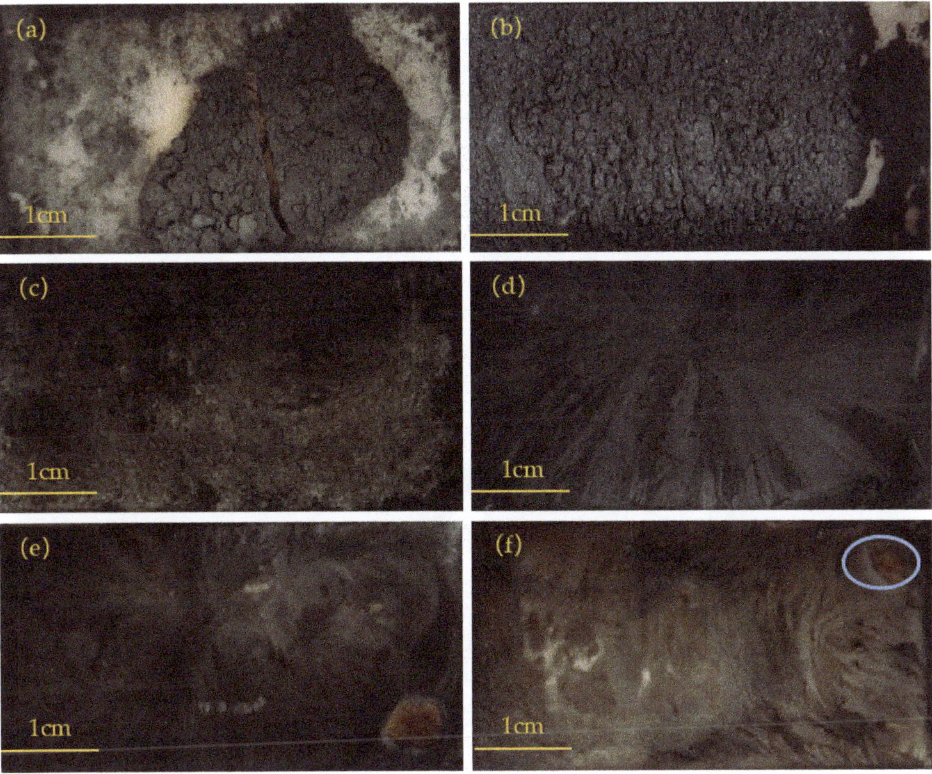

Figure 4. Sulfide sample pictures: (**a**) 400 °C, (**b**) 450 °C, (**c**) 500 °C, (**d**) 550 °C, (**e**) 600 °C, and (**f**) 650 °C.

4.1.2. SEM Analysis

Figure 5 displays the SEM images of the samples after sulfuration at constant atmospheric pressure and a constant holding time. The SEM images (Figure 5a, b) reveal that a large number of bright particles inside the electrodeposited antimony were not vulcanized at 400 and 450 °C. At 500 °C, the sulfuration rate increased to 89.78%, and the unsulfurated electrodeposited antimony particles were relatively reduced as compared to those at 400 and 450 °C; however, there remained a considerable amount of unsulfurated electrodeposited antimony. At 550 °C, the number of bright electrodeposited antimony particles in the visible field of view was very small, and they had been vulcanized to form Sb_2S_3. At 600 and 650 °C, the number of brightly electrodeposited antimony particles in the electron microscopic view decreased significantly as compared with that in the low-temperature reaction.

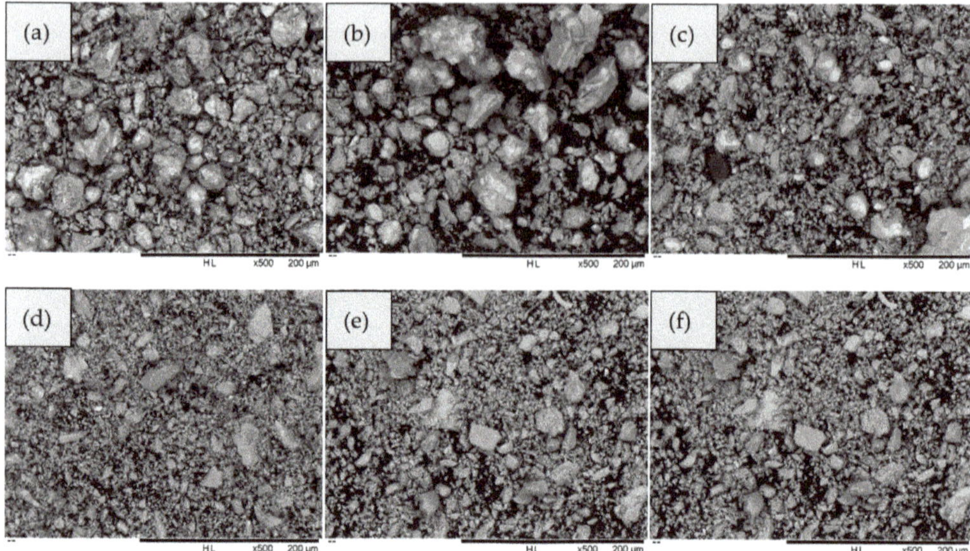

Figure 5. SEM images of antimony sulfide synthesized by electrodeposited antimony sulfide and electrodeposited antimony at different temperatures: (**a**) 400 °C, (**b**) 450 °C, (**c**) 500 °C, (**d**) 550 °C, (**e**) 600 °C, and (**f**) 650 °C.

4.1.3. EPMA Analysis

When a certain temperature was reached, the sulfuration rate decreased with the increase of the sulfuration temperature; this issue was further investigated. After reacting at 600 and 650 °C, melting and recrystallized regions of antimony appeared in the crucible (Figure 4). The samples sulfurized at 650 °C were analyzed by electron probe microanalysis (EPMA). Figure 6 presents the surface scanning images of the antimony sulfide area at 650 °C. The images of elemental Sb and elemental S reveal that their distributions were basically consistent. However, as indicated in the image of elemental Sb, there were some areas where the Sb content was very low, but in the image of elemental Na, the area of concentrated Na was the empty area in the image of elemental Sb.

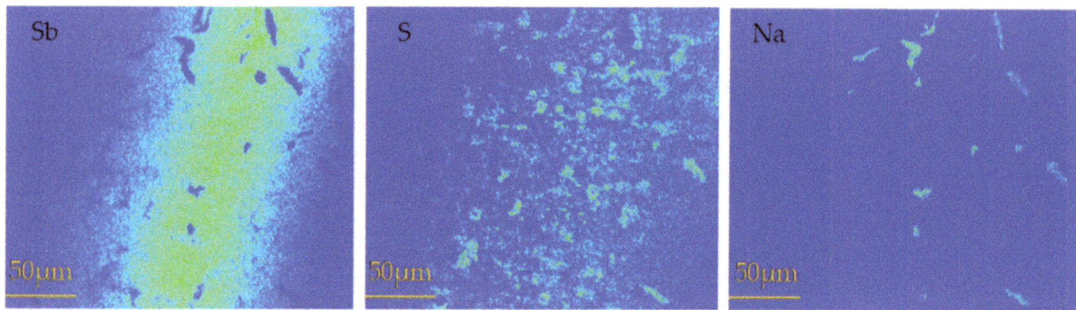

Figure 6. EPMA scanning maps of vulcanized sample at 650 °C.

Figure 7 presents the surface scanning images of the melting–recrystallization zone of antimony at 650 °C (the area circled in Figure 4). The figure reveals that the recrystallization zone was mainly unsulfurated metal Sb, resulting in a decrease in the vulcanization rate, and some areas were enriched with Na.

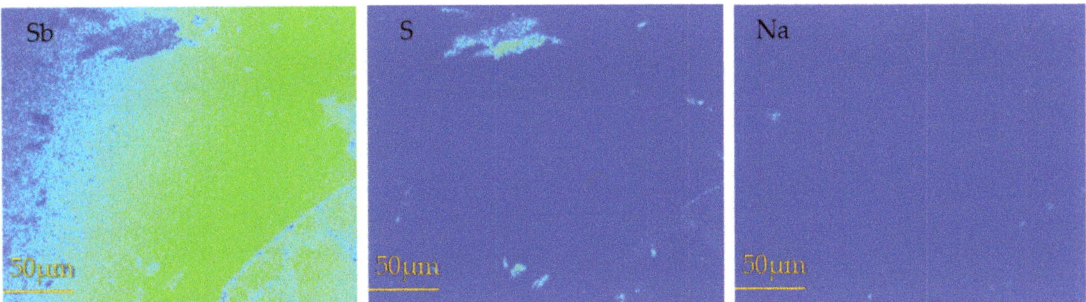

Figure 7. EPMA scanning maps of recrystallization zone in metal antimony melting at 650 °C.

In comprehensive consideration of the images of the sulfide samples presented in Figure 4 and the EPMA results, the main reasons for the occurrence of the melting–recrystallization zone at 600 and 650 °C are the following. (1) The sulfur source temperature remained constant at 400 °C, so the sulfur partial pressure remained certain. With the increase of the antimony source temperature, the amount of sulfur vapor diffused to the antimony end gradually decreased, and the collision between sulfur and antimony at the upper end of the antimony source decreased, resulting in the decrease of the vulcanization rate. (2) With the increase of the reaction temperature, the pressure of the reaction system increases, and the sulfur pressure in the antimony reaction zone becomes higher than that in the low-temperature zone; thus, the gas moves to the sulfur end. (3) When the pressure of the reaction system is high, the gas reaction proceeds in the direction of volume shrinkage, and the gaseous sulfur particles collide with each other to form macromolecular liquid sulfur, resulting in the reduction of sulfur involved in the vulcanization reaction.

4.1.4. EDS Analysis

The chemical formula of antimony sulfide is expressed as Sb_2S_3, the S/Sb ratio of which is 1.5. At 400 and 450 °C, the highest S/Sb ratio calculated by the obtained EDS data was about 1.32, indicating that the vulcanization at 400 and 450 °C was insufficient. At 500 °C, the S/Sb ratio was higher than those at 400 and 450 °C, and the calculated value was 1.45, which was close to the ideal value of 1.50. However, there remained an unsulfurated area, indicating that the vulcanization effect at 500 °C was still not ideal. At 600 and 650 °C, the obtained S/Sb values were distributed in the range of 0.9–1.3. However, as reported in Section 4.1.1, the melting–recrystallization zone of metal antimony appeared at 600 and 650 °C, which led directly to a reduction in the vulcanization rate and ultimately caused the vulcanization effect to be less than ideal. At 550 °C, an S/Sb ratio of 1.46 was obtained, and the sample was well vulcanized; compared with the S/Sb value under other temperature conditions, this was the optimal value. The EDS data also confirm that the best vulcanization effect was achieved at 550 °C.

4.1.5. Raman Spectroscopy Analysis

The experimental results revealed that strong Raman peaks appeared at 71, 99, 135, 178, 241, and 290 cm^{-1}; according to the existing literature [25–28], these are characteristic peaks of antimony sulfide (Figure 8). Figure 8 also shows that at 550 °C, the characteristic peaks of Sb_2S_3 corresponding to 135, 178, and 241 cm^{-1} changed from broad and scattered peaks to narrow and sharp peaks, indicating a better degree of crystallinity of the sulfide formed at this reaction temperature.

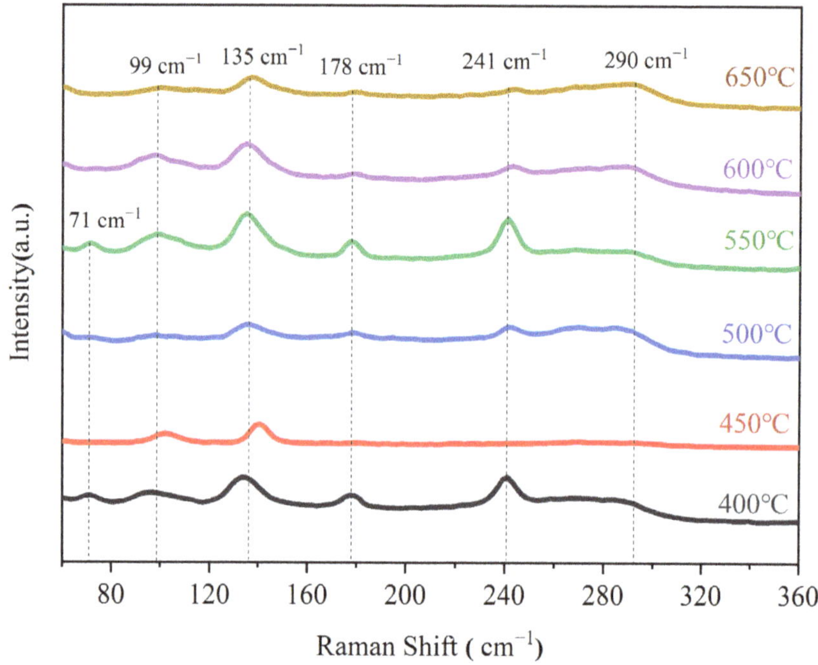

Figure 8. Raman spectra of vulcanized samples at different temperatures.

In summary, the sulfuration effect was the best under conditions of standard atmospheric pressure, a temperature of 550 °C, and a holding time of 30 min. The number of bright electrodeposited antimony particles in the SEM scanning field of sulfide at 550 °C was very small, and these particles had been basically vulcanized to form Sb_2S_3 with a sulfuration rate of 96.06%. As determined by the EDS data, the best S/Sb ratio of 1.46 was achieved at 550 °C, and was the closest to the theoretical value of S/Sb = 1.5.

4.2. Volatilization Reaction

4.2.1. XPS Analysis

The optimal sulfuration conditions of standard atmospheric pressure, a temperature of 550 °C, and a heat preservation time of 30 min were adopted to prepare sulfuration samples as the raw material for the volatilization experiment. In the experiment, the holding time was set to 2 h, the volatilization temperature zone was set to 800 °C, the volatilization pressure gradient was set to 0.2 atm, and the pressure values were 0.2, 0.4, 0.6, 0.8, and 1.0 atm, respectively.

According to the experimental and calculation results, the volatilization rate of antimony sulfide at 0.2 atm was 92.81%, and the contents of Na, Fe, and Se were lower than those under other pressure conditions. To characterize the chemical state of the product, XPS analysis was performed on the volatiles and residues under the conditions of 0.2 atm, 800 °C, and 2 h. The results are presented in Figures 9 and 10, which respectively display the representative XPS spectra of the volatiles and residues. The binding energy of C 1s was designated as the standard value of 284.8 eV [29], and a correction was carried out; the corrected values of charge displacement were respectively 2.50 and 2.40 eV.

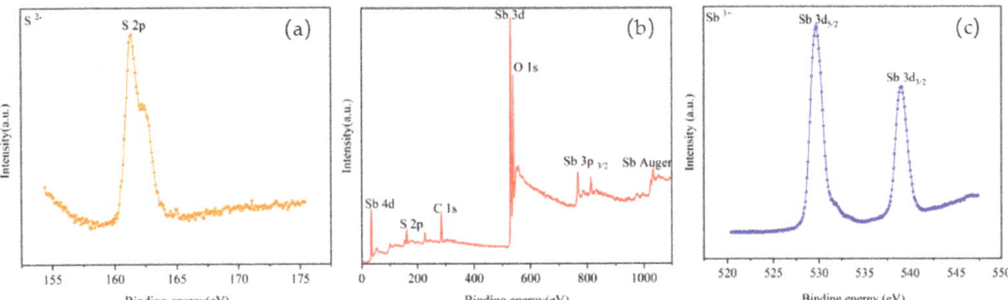

Figure 9. XPS spectra of volatiles: (**a**) S 2p core level, (**b**) Typical XPS survey spectrum, and (**c**) Sb 3d core level.

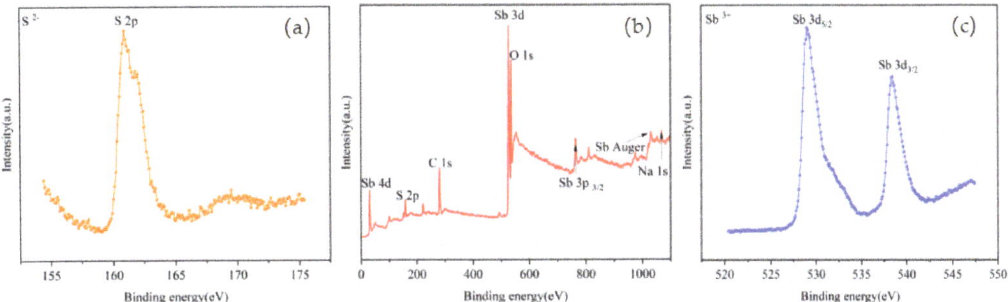

Figure 10. XPS spectra of residue: (**a**) S 2p core level, (**b**) typical XPS survey spectrum, and (**c**) Sb 3d core level.

The high-resolution XPS spectrum of S 2p is presented in Figure 9a, while the high-resolution XPS spectrum of Sb 3d is exhibited in Figure 9c. The double peaks of Sb $3d_{5/2}$ and Sb $3d_{3/2}$ related to the Sb-S bond were observed around 529.80 and 539.05 eV, respectively; these peaks are both characteristic of the Sb^{3+} state. The S 2p peak was observed near 161.50 eV. The peak positions of Sb and S were similar to those reported in the existing literature [30–33], indicating that the phase representing the volatile was Sb_2S_3.

The high-resolution XPS spectrum of S 2p is presented in Figure 10a, and the S 2p peak was observed near 160.90 eV. The high-resolution XPS spectrum of Sb 3d is presented in Figure 10c, and the double peaks of Sb $3d_{5/2}$ and Sb $3d_{3/2}$ related to the Sb-S bond were observed around 529.00 and 538.40 eV, respectively; these peaks are both characteristic of the Sb^{3+} state. The peak positions of Sb and S were similar to those reported in the existing literature [25–28], indicating that the phase of the residue was Sb_2S_3. The Na 1s peak was clearly observed near 1071.32 eV, and the binding energy of Na 1s in the sodium compounds was found to be between 1071.00 and 1071.50 eV based on a database [34]. However, no obvious Na was observed in the XPS spectra of the volatiles; there was no obvious Na 1s peak, indicating that most of the elemental Na remained in the residue.

4.2.2. Chemical Composition Analysis

The chemical compositions of the volatiles and residues under the optimal conditions of a temperature of 800 °C, a pressure of 0.2 atm, and a holding time of 2 h were analyzed. As reported in Table 3, the Na and Fe elements were well separated; compared with their contents in the raw materials, their contents in the volatiles were reduced by about 7.4 and 58.6 times, respectively. Elemental Au was not detected in the volatile species, and it was enriched in the residue by about 1.6 times. These findings demonstrate that this experimental method can realize the enrichment of all the Au in the electrodeposited antimony in the residue, volatilization to obtain the antimony sulfide, and the effective removal of the Na and Fe elements in the antimony sulfide volatiles. Excluding As and Se, other impurities are basically concentrated in the residue. The experimental results reveal that 0.2 atm was the best volatilization pressure within the experimental pressure range. Under this condition, the gold in the electrodeposited antimony was effectively enriched and antimony sulfide was formed simultaneously.

Table 3. Chemical composition of volatiles and residues at 800 °C and 0.2 atm.

Type	Content/wt%					
	Na	**Fe**	**As**	**Se**	**Sb**	**Au** [1]
Volatile	0.054	0.003	0.037	0.017	73.180	0
Remains	2.500	0.150	0.000	0.018	68.950	60

[1] The unit of Au content was g/t.

5. Conclusions

In this study, an electrodeposited antimony sulfuration–volatilization method was proposed to prepare antimony sulfide while enriching gold. The feasibility of the process was demonstrated on a laboratory scale, thereby providing theoretical guidance for the value-added use of the electrodeposited antimony obtained via the alkaline leaching of antimony–gold ore. In the experimental temperature range, as the temperature gradually increased, the S/Sb value of the vulcanized sample was also found to gradually increase. The best S/Sb ratio of 1.46 achieved at 550 °C was close to the theoretical S/Sb ratio of 1.50 at 550 °C. The results of SEM and Raman analyses proved that the reaction conditions were the best curing conditions in the experimental range. Under these conditions, the sulfuration rate was the best, namely 96.06%. Moreover, as compared to those under the other pressure conditions, the S/Sb value of the volatile substance under the condition of 0.2 atm was closer to the theoretical value of S/Sb = 1.5. Compared with their contents in the raw materials, the contents of Na and Fe in the volatile matter were reduced by about 7.4 and 58.6 times, respectively. Elemental Au was not detected in the volatile species, and it was enriched in the residue by about 1.6 times. In summary, all the Au in the electrodeposited antimony was basically enriched in the residue, and the effective removal of the Na and Fe elements in the antimony sulfide volatiles was simultaneously realized. The best volatilization pressure within the experimental pressure range was found to be 0.2 atm, and the best volatilization conditions were found to be a pressure of 0.2 atm, a temperature of 800 °C, and a heat preservation time of 2 h. Electrodeposited antimony was found to enrich the gold in the residue via sulfuration–volatilization treatment and generate antimony sulfide, which effectively increased the added value of these products. Antimony sulfide has good volatility and high commercial value. This method increases the product value and reduces the energy consumption of the reaction. Furthermore, this technology has no toxic and harmful by-products, which can reduce environmental pollution and realize clean and effective metallurgy.

Author Contributions: In this joint work, each author was in charge of their expertise and capability: conceptualization, W.W. and J.Y.; methodology, W.W., S.W., J.Y., C.C., K.H., L.X., J.Z., B.X. and B.Y.; theoretical basis, W.W. and C.C.; formal analysis, W.W.; investigation, K.H.; J.Z. and L.X.; resources, B.X. and B.Y.; data curation, W.W., J.Y. and C.C.; writing—original draft, W.W.; writing—review and editing, W.W. and S.W.; visualization, L.X.; supervision, J.Y.; project administration, J.Y.; funding acquisition, J.Y. All authors have read and agreed to the published version of the manuscript.

Funding: This research was funded by National Natural Science Foundation of China (No. 52104350), the Natural Science Foundation of Yunnan Province (No. 2018FD037, 202001AT070045), Personnel Training Project of Kunming University of Science and Technology (No. KKSY201663027). Independent Project of Innovation Center for Complex Nonferrous Metal Resources Clear Use in Yunnan Province (No. KKPT201663012). The Central Government guides local funds for scientific and

Technological Development (Yunnan) (No. KKST202052004). Analysis and Measurement Center of Kunming University of Science and Technology (2017T20160012, 2020M20192202047).

Data Availability Statement: The data presented in this study are available on request from the corresponding author. The data are not publicly available due to unfinished related ongoing further studies.

References

1. Majzlan, J. Primary and secondary minerals of antimony. *Antimony* **2021**, 17–47. [CrossRef]
2. Dupont, D.; Arnout, S.; Jones, P.T.; Binnemans, K. Antimony Recovery from End-of-Life Products and Industrial Process Residues: A Critical Review. *J. Sustain. Metall.* **2016**, *2*, 79–103. [CrossRef]
3. Zhou, Y.J. Antimony Demand Forecast and Supply-Demand Pattern Analysis. Master's Thesis, China University of Geosciences, Beijing, China, 2015.
4. Anderson, C.G. The metallurgy of antimony. *Geochemistry* **2012**, *72*, 3–8. [CrossRef]
5. Grund, S.C.; Hanusch, K.; Breunig, H.J.; Wolf, H.U. Antimony and antimony compounds. In *Ullmann's Encyclopedia of Industrial Chemistry*; Wiley-VCH Verlag GmbH & Co. KGaA: Weinheim, Germany, 2000. [CrossRef]
6. Proenza, J.A.; Torró, L.; Nelson, C.-E. Mineral deposits of Latin America and the Caribbean. Preface. *Boletín De La Soc. Geológica* **2020**, *72*, P250820. [CrossRef]
7. Yuan, B.; Fan, J.T.; Yu, L.H. Situation Analysis and Countermeasures of Antimony Resources in China. *China Land Resour. Dly.* **2011**, *24*, 47–49.
8. Guo, X.Y.; Xin, Y.T.; Wang, H.; Tian, Q.H. Mineralogical characterization and pretreatment for antimony extraction by ozone of antimony-bearing refractory gold concentrates. *Trans. Nonferrous Met. Soc. China* **2017**, *27*, 1888–1895. [CrossRef]
9. Guo, X.Y.; Xin, Y.T.; Wang, H.; Tian, Q.H. Leaching kinetics of antimony-bearing complex sulfides ore in hydrochloric acid solution with ozone. *Trans. Nonferrous Met. Soc. China* **2017**, *27*, 2073–2081. [CrossRef]
10. Chen, J.; Huang, Z.L.; Yang, R.D.; Du, L.J.; Liao, M.Y. Gold and antimony metallogenic relations and ore-forming process of Qinglong Sb (Au) deposit in Youjiang basin, SW China: Sulfide trace elements and sulfur isotopes. *Geosci. Front.* **2021**, *12*, 605–623. [CrossRef]
11. Zhang, L.; Yang, L.Q.; Groves, D.I.; Sun, S.C.; Liu, Y.; Wang, J.Y.; Li, R.H.; Wu, S.G.; Gao, L.; Guo, J.L.; et al. An overview of timing and structural geometry of gold, gold-antimony and antimony mineralization in the Jiangnan Orogen, southern China. *Ore Geol. Rev.* **2019**, *115*, 103173. [CrossRef]
12. Yang, T.Z.; Xie, B.Y.; Liu, W.F.; Zhang, D.C.; Chen, L. Enrichment of Gold in Antimony Matte by Direct Smelting of Refractory Gold Concentrate. *JOM* **2018**, *70*, 1017–1023. [CrossRef]
13. Rusalev, R.E.; Rogozhnikov, D.A.; Naboichenko, S.S. Investigation of complex treatment of the gold-bearing antimony flotation concentrate. *Solid State Phenom.* **2018**, *284*, 863–869. [CrossRef]
14. Guo, X.D.; Wang, X.J.; Zhao, Y.L.; Tang, M.G.; Xiao, L. Structural characteristics and ore-controlling analysis of Jinlongshan gold-antimony ore belt in Shanxi Province. *Depos. Geol.* **2014**, *33*, 387–388. [CrossRef]
15. Ubaldini, S.; Veglio, F.; Fornari, P.; Abbruzzese, C. Process flow-sheet for gold and antimony recovery from stibnite. *Hydrometallurgy* **2000**, *57*, 187–199. [CrossRef]
16. Yang, T.Z.; Rao, S.; Liu, W.F.; Zhang, D.C.; Chen, L. A selective process for extracting antimony from refractory gold ore. *Hydrometallurgy* **2017**, *169*, 571–575. [CrossRef]
17. Celep, O.; Alp, I.; Deveci, H. Improved gold and silver extraction from a refractory antimony ore by pretreatment with alkaline sulphide leach. *Hydrometallurgy* **2011**, *105*, 234–239. [CrossRef]
18. Zhang, D.C.; Xiao, Q.K.; Liu, W.F.; Chen, L.; Yang, T.Z.; Liu, Y.N. Pressure oxidation of sodium thioantimonite solution to prepare sodium pyroantimonate. *Hydrometallurgy* **2015**, *151*, 91–97. [CrossRef]
19. Zhang, X.F.; Huang, D.X.; Jiang, W.L.; Zha, G.Z.; Deng, J.H.; Deng, P.; Kong, X.F.; Liu, D.C. Selective separation and recovery of rare metals by vulcanization-vacuum distillation of cadmium telluride waste. *Sep. Purif. Technol.* **2020**, *230*, 115864. [CrossRef]
20. Zhang, X.F.; Liu, D.C.; Jiang, W.L.; Xu, W.J.; Deng, P.; Deng, J.H.; Yang, B. Application of multi-stage vacuum distillation for secondary resource recovery: Potential recovery method of cadmium telluride photovoltaic waste. *J. Mater. Res. Technol.* **2020**, *9*, 6977–6986. [CrossRef]
21. Li, H.L.; Wu, X.Y.; Wang, M.X.; Wang, J.; Wu, S.k.; Yao, X.L.; Li, L. Separation of elemental sulfur from zinc concentrate direct leaching residue by vacuum distillation. *Sep. Purif. Technol.* **2014**, *138*, 41–46. [CrossRef]
22. Zhang, F. Study on Purification of Electrodeposited Antimony Enriched Gold by Vacuum Distillation. Master's Thesis, Kunming University of Science and Technology, Kunming, China, 2021.
23. Dai, Y.J. *China Antimony Industry*; Metallurgical Industry Press: Beijing, China, 2014.
24. Ouyang, Z. Study on Clean Extraction Process of Reduction and Sulfur Fixation Roasting of Antimony Sulfide Concentrate. Master's Thesis, Hunan University of Technology, Zhuzhou, China, 2020.

25. Parise, R.; Katerski, A.; Gromyko, I.; Rapenne, L.; Roussel, H.; Karber, E.; Appert, E.; Krunks, M.; Consonni, V. ZnO/Ti-O$_2$/Sb$_2$S$_3$ Core-Shell Nanowire Heterostructure for Extremely Thin Absorber So-lar Cells. *J. Phys. Chem. C* **2017**, *121*, 9672–9680. [CrossRef]
26. Yoshioka, A.; Nagata, K. Raman spectrum of sulfur under high pressure. *J. Phys. Chem. Solids* **1995**, *56*, 581–584. [CrossRef]
27. Escorcia-Garcia, J.; Becerra, D.; Nair, M.-T.-S.; Nair, P.-K. Heterojunction CdS/Sb$_2$S$_3$ solar cells using antimony sulfide thin films prepared by thermal evaporation. *Thin Solid Film.* **2014**, *569*, 28–34. [CrossRef]
28. Tang, A.; Yang, Q.; Qian, Y.-T. Formation of crystalline stibnite bundles of rods by thermolysis of an antimony (III) diethyldithio-carbamate complex in ethylene glycol. *Inorg. Chem.* **2003**, *42*, 8081–8086. [CrossRef]
29. Briggs, D.; Beamson, G. Primary and secondary oxygen-induced C1S binding-energy shifts in X-Ray photoelectron-spectroscopy of polymers. *Anal. Chem.* **1992**, *64*, 1729–1736. [CrossRef]
30. Salinas-Estevane, P.; Sanchez, E.-M. Preparation of Sb$_2$S$_3$ nanostructures by the ionic Liquid-Assisted low power Sonochemical method. *Mater Lett.* **2010**, *64*, 2627–2630. [CrossRef]
31. Mittal, V.K.; Bera, S.; Narasimhan, S.V.; Velmurugan, S. Sorption of Sb(III) on carbon steel surface in presence of molybdate and selenite in citric acid medium. *Appl. Surf. Sci.* **2011**, *258*, 1525–1530. [CrossRef]
32. Hanafi, Z.M.; Ismail, F.M. Colour Problem of Antimony Trisulphide: IV-X-Ray Photoelectron and Diffuse Reflectance. *Z. Für Phys. Chem.* **1987**, *268*, 573–577. [CrossRef]
33. Gheorghiu, A.; Lampre, I.; Dupont, S.; Senemaud, C.; Raghni, M.E.I.; Lippens, P.E.; Olivier-Fourcade, J. Electronic structure of chalcogenide compounds from the system Tl$_2$S-Sb$_2$S$_3$ studied by XPS and XES. *J. Alloys Compd.* **1995**, *28*, 143–147. [CrossRef]
34. Kohiki, S.; Ohmura, T.; Kusao, K. Appraisal of a new charge correction method in X-ray photoelectron spectroscopy. *J. Electron. Spectrosc. Relat. Phenom.* **1983**, *31*, 85–90. [CrossRef]

Review

Recent Advancements in Metallurgical Processing of Marine Minerals

Katarzyna Ochromowicz [1,*], **Kurt Aasly** [2] and **Przemyslaw B. Kowalczuk** [2,*]

1 Department of Analytical Chemistry and Chemical Metallurgy, Faculty of Chemistry, Wroclaw University of Science and Technology, 50-370 Wroclaw, Poland

2 Department of Geoscience and Petroleum, Norwegian University of Science and Technology, S. P. Andersens Veg 15a, 7031 Trondheim, Norway; kurt.aasly@ntnu.no

* Correspondence: katarzyna.ochromowicz@pwr.edu.pl (K.O.); przemyslaw.kowalczuk@ntnu.no (P.B.K.)

Abstract: Polymetallic manganese nodules (PMN), cobalt-rich manganese crusts (CRC) and seafloor massive sulfides (SMS) have been identified as important resources of economically valuable metals and critical raw materials. The currently proposed mineral processing operations are based on metallurgical approaches applied for land resources. Thus far, significant endeavors have been carried out to describe the extraction of metals from PMN; however, to the best of the authors' knowledge, it lacks a thorough review on recent developments in processing of CRC and SMS. This paper begins with an overview of each marine mineral. It is followed by a systematic review of common methods used for extraction of metals from marine mineral deposits. In this review, we update the information published so far in peer-reviewed and technical literature, and briefly provide the future perspectives for processing of marine mineral deposits.

Keywords: deep-sea mining; marine minerals; seafloor massive sulfides; polymetallic nodules; cobalt-rich crusts; mineral processing; hydrometallurgy; pyrometallurgy; metals; extraction

Citation: Ochromowicz, K.; Aasly, K.; Kowalczuk, P.B. Recent Advancements in Metallurgical Processing of Marine Minerals. *Minerals* **2021**, *11*, 1437. https://doi.org/10.3390/min11121437

Academic Editors: Shuai Wang, Xingjie Wang and Jia Yang

Received: 18 November 2021
Accepted: 15 December 2021
Published: 19 December 2021

Publisher's Note: MDPI stays neutral with regard to jurisdictional claims in published maps and institutional affiliations.

1. Introduction

The discovery of deep-sea concretions, later known as polymetallic nodules (PMN), in the 1870s, during the HMS Challenger expedition [1], opened a perspective for new, alternative-to-terrestrial, rich resources of many valuable metals (Cu, Ni, Co, Mn, Ag, REE, etc.). The results from the HMS Challenger expedition initiated numerous investigations aiming at the acquisition of mineral deposits on the seafloor, and subsequent recovery of metals. At present, three main types of marine mineral deposits have been discovered: polymetallic manganese nodules (PMN), cobalt-rich manganese crusts (CRC) and seafloor massive sulfides (SMS). These mineral deposits have gained increasing attention due to the significant content of economically valuable metals and critical raw materials (CRM) [2]. Although the accurate assessment of the total amounts of metals and CRM in the marine mineral deposits is difficult, the estimated hypothetical and speculative abundance is of the order of a million tons, compared to the identified terrestrial deposits [3].

A major challenge for scientists has been to develop a technology for deep sea-floor exploration and extraction, enabling the collection of mineralized material from the seabed, followed by the efficient extraction of metals. Currently, mining technologies are just emerging to exploit and extract these mineral deposits. Based on this rough characteristic, different research projects have started aiming for implementation of various concepts, including solutions known from terrestrial applications. However, until now, with an enormous number of research projects, with access to modern and high-tech equipment, there is no full-scale operation [4]. The technological problems, probably caused by the great diversity of the sea-bed materials, or difficulties with the beneficiation of marine mineral deposits, are not the only ones. A lot of discussion held so far concerns the legitimacy of taking such a strong interference in the environment [5–7]. The main concern arises from the fact that large quantities of very fine sediments from the ocean floor can be discharged into the surface water, together with cold, nutrient-rich seawater during dredging operations. Furthermore, the potential effects on the ecology of oceans are unpredictable and will depend on the type of mining systems used for future large-scale production [8]. One of the most important and least understood threats is the sediment plume that is expected to travel in the ocean column away from the mine site [9].

Each of the marine mineral deposits exhibits unique properties such as mineralogical assemblage, bulk chemical composition, density, porosity, surface area and hardness, each of which are crucial for the development of a beneficiation process ending in an effective recovery of metals. Based on numerous investigations dedicated to defining mineralogy of the marine deposits, it is known that PMN and CRC share similar mineralogical compositions (the valuable metals are mostly found incorporated in either oxide or oxyhydroxide forms), whereas the SMS group includes sulfide minerals containing the metals [10]. Additionally, the chemical composition of numerous samples enabled us to determine the main metallic content in each type of marine deposit. PMN are the main source of Mn and Fe, together with Cu, Ni and Co. CRC are rich in cobalt, but also contain variable amounts of vanadium, titanium, tantalum, tungsten, and REEs. SMS contain predominantly iron, copper, zinc, and many minor elements. Moreover, some elements such as manganese, cobalt, tellurium, and yttrium are more abundant in the marine deposits in comparison to land resources.

Currently proposed mineral processing operations are based on metallurgical approaches applied to terrestrial (land) deposits; however, development of marine minerals processing technology is underway. Recent extensive reviews on processing of PMN can be found elsewhere [11]; however, to the best of the authors' knowledge, there is a lack of thorough review on advancements in processing of CRC and SMS. It should also be noted that the number of publications on PMN processing is quite impressive, compared to the manuscripts that describe the methods for CRC and SMS (Figure 1). We also point out that in this review we focus only on articles on the physicochemical and metallurgical processing of marine minerals, discarding those relating to either technical, ecological, or economic issues.

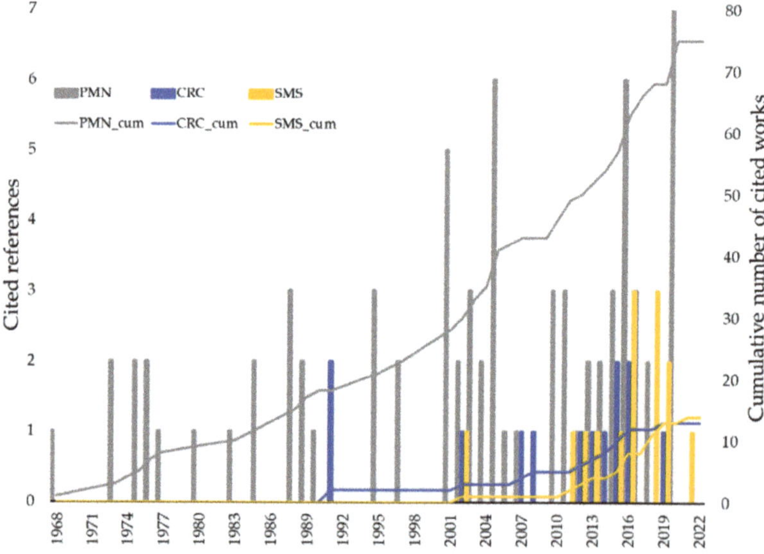

Figure 1. Distribution and cumulative share of reference papers from present work (from 1968 to 2021).

In this paper, we respond to the increasing interest in the processing of marine mineral deposits by collecting information from peer-reviewed and other technical literature on historical and recent developments in processing of PMN, CRC, and SMS. We update the information published so far and provide future perspectives for the recovery of valuable metals from the marine deposits.

2. Brief Characteristics of Marine Resources

Mineral deposits in the world oceans can be divided into marine mineral deposits and deep-marine mineral deposits [2]. Figure 2 shows the distribution of the different types and their geological settings. The first type is found on the continental shelf and comprises deposits accumulated from weathered terrestrial rocks transported and deposited in the ocean through alluvial processes. The latter comprises deposits formed at or in the deep-ocean floor. There are three main types of deep-marine mineral resources currently known from the world's oceans [1]. These are typically referred to as polymetallic manganese nodules (PMN), cobalt-rich manganese crust (CRC), and sea-floor massive sulfides (SMS). They occur in different geological settings on the sea floor and their formation differs significantly. A short overview of each of the three types is given in the following sections.

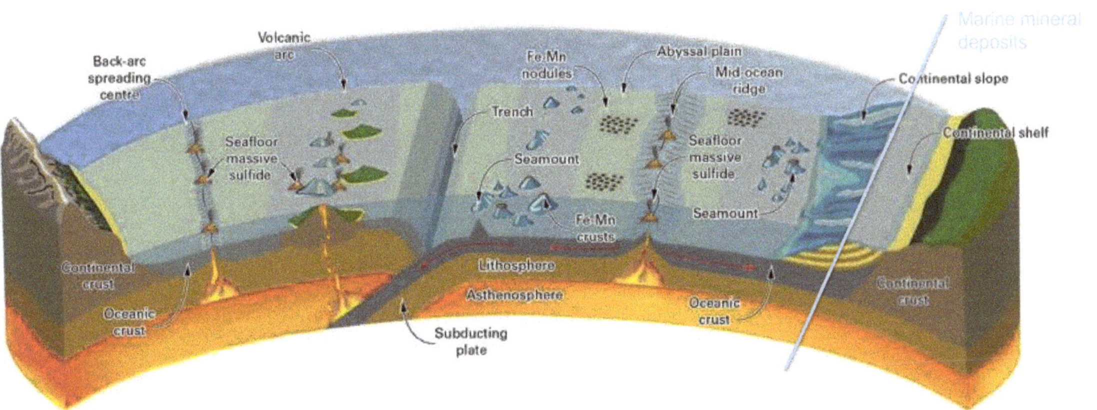

Figure 2. Geological and geographical settings of different types of mineral deposits in the ocean, including the distribution of the different deep-marine minerals. Reprinted with permission from [12]. Copyright 2018 British Geological Survey materials © UKRI 2018 under the Creative Commons Attribution CC-BY 3.0 License.

2.1. Polymetallic Manganese Nodules (PMN)

Three main types of polymetallic manganese nodules (PMN) have been described and summarized elsewhere [1,10]. These are (i) hydrogenetic nodules formed by the direct deposition of manganese and cobalt, nickel, and copper from seawater; (ii) diagenetic nodules formed as a result of remobilization of manganese in the sediment column; (iii) hydrothermal nodules formed as a result of massive discharge of hydrothermal fluids at the seafloor at hotspots and divergent plate margins [1].

The deposition of PMN starts precipitation of metals from the ambient sea water onto some sort of nuclei. This nuclei is typically an older nodule fragment, shark tooth, plankton shell, or rock fragments [10].

The abundance of PMN on the seafloor is mainly controlled by the sedimentation rate [1]. PMN occur mainly on the great abyssal plains at depths ranging from 3000 m to 6000 m and can typically be 1 cm to 12 cm diameter. The most common size in the Clarion–Clipperton Zone (CCZ) is the range 1–5 mm. The growth rate of PMN is controlled by the deposition environment but is generally less than 10 mm/million year for hydrogenetic nodules and more than a 100 mm/million years for diagenetic nodules. A greater diagenetic component of the nodules results in a faster growth rate [13]. However, a combination of the two growth mechanisms is typical, and hence also the average growth rate.

PMN differ significantly in physical properties and mineralogical composition from the known terrestrial deposits. A unique characteristic of PMN compared to the terrestrial resources is the presence of multiple elements in one deposit. A polymetallic manganese nodule contains a range of valuable metals, i.e., Mn, Fe, Cu, Zn, Ni, Co, Mo, and also minor amounts of 22 other elements, including rare earth elements (REE) [14,15]. According to Hein et al. [15], PMN from CCZ might contain a greater tonnage of Mn, Ni, Co, Tl, and Y, and a similar tonnage of As, as the entire "global terrestrial reserve base". It should, however, be noted that numbers from the CCZ do not represent a reserve base but rather a compilation of data from projects in various stages of resource classifications, none of these classified as mineral reserves. The average chemical and mineralogical composition of PMN are given in Tables 1 and 2. Figures 3 and 4 represent PMN samples and their growth structures, respectively. The chemical and mineralogical composition of PMN depends mainly on the processes controlling the deposition of nodules as well as the geographic location for their formation. The most common manganese minerals detected in nodules are todorokite, birnessite, and delta manganese dioxide. Goethite has been determined to be the most common iron-bearing mineral. Nodules are built up of nanometer- scale manganese oxides and iron oxyhydroxides. Fe/Mn ratios typically vary similarly with the type of nodules [16].

Table 1. Average composition of PMN [10,14,15,17].

Cu	Mn	Fe	Ni	Co	Mo	Al	Moisture
				(wt%)			
0.74	26.0	8.9	1.0	0.19	0.05	2.0	13.8

Table 2. Average mineralogical composition of PMN [18,19].

Mn Minerals	(1) Todorokite: oxides of manganese, magnesium, calcium, sodium, and potassium which may be chemically stated as $(Ca, Na, Mn^{2+}, K)(Mn^{4+}, Mn^{2+}, Mg)_6O_{12} \cdot 3H_2O$ (2) Buserite or 10 Å manganite: a sodium manganese oxide hydrate $Na_4Mn_{14}O_{27} \cdot 21H_2O$ (3) Birnessite or 7 Å manganite: $(Na_7Ca_3)Mn_7O_{140} \cdot 28H_2O$ (4) Vernadite $(Mn^{4+}, Fe^{3+}, Ca, Na)(O, OH)_2 \cdot nH_2O$ or MnO
Fe Minerals	Goethite α-FeOOH Feroxyhyte δ-$Fe^{3+}O(OH)$

Figure 3. PMN samples occurring in the eastern German license area within CCZ (**A–C**)—nodules with different proportions of hydrogenetic and diagenetic layer growth structures, (**D**)—cross section of a typical PMN) [20].

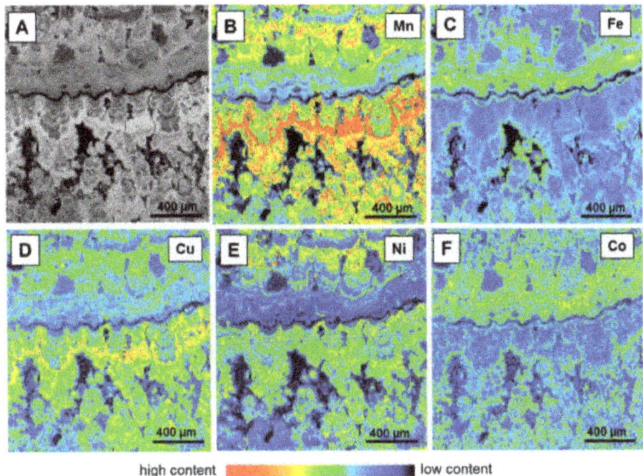

Figure 4. Backscattered electron (BSE) image (**A**) and element distribution maps (**B–F**) of PMN from CCZ [20].

2.2. Cobalt-Rich Manganese Crusts (CRC)

There are still not enough geochemical data, and little is known about the abundance of cobalt-rich manganese crusts (CRC) in most areas of the ocean. There are two main types of manganese crusts according to [1]; (i) cobalt-rich manganese crust, which is a hydrogenous manganese crust with more than 1% (wt%) Co, and (ii) hydrothermal manganese crust, which is the least abundant manganese deposit in the world's oceans. The cobalt-rich manganese crusts are the ones of interest from an economic perspective. CRC occur on sea mounds in any part of the world's oceans, but are typically restricted to ancient seamounts in relatively large depths and are seen to be more frequent in the Pacific Ocean [1]. Criteria for the formation of CRC is that bottom conditions have resulted in minimal sedimentation and provided the substrate, i.e., the host rock or the original rock of the sea mound, free from sediments. Crusts are only formed on sediment-free surfaces. The formation of crusts typically takes place at depths between 400 m and 4000 m, and the most prospective, cobalt-rich manganese crusts are found on large seamounts at less than 1000 m to 1500 m water depth and older than 20 million years [21].

Hydrogenous CRC exhibit strong similarities with hydrogenetic nodules [13], and thus have been attracting investment in exploration for higher concentrations of various metals. The average chemical and mineralogical composition of CRC are given in Tables 3 and 4. The main metals of economic interest are nickel, copper, cobalt, and possibly manganese and titanium [10]. There are also traces of other valuable metals, such as molybdenum, REE, and lithium. The characteristic properties of CRC are very high porosity (60%), high surface area (300 m^2/cm^3 of crust), and extremely slow rates of growth (1–6 mm/Ma) [22]. The CRC consist of a very fine-grained mixture of ferruginous vernadite (mainly δ-$MnO_2 \times H_2O$), X-ray amorphous Fe-oxyhydroxide, aluminosilicate phases, carbonate-fluorapatite (secondary in the older crust generation), minor admixtures of fine-grained, detrital quartz, and feldspar as well as residual biogenetic phases. Figure 5A,B represent CRC samples and their growth structures, respectively.

Table 3. Average content of some main metals in CRC from the mid-Pacific mountains [1].

Mn	Fe	Co	Ni	Cu	Pt
		(wt%)			(ppm)
28.4	14.3	1.18	0.5	0.03	0.5

Table 4. Typical mineralogy of CRC [13].

Mn-Minerals:	Vernadite $(Mn^{4+},Fe^{3+},Ca,Na)(O,OH)_2 \cdot nH_2O$
Fe-Minerals:	Amorphous Fe-oxyhydroxides; Ferroxyhyte δ-$Fe^{3+}O(OH)$; Ferrihydrite $(Fe^{3+})_2O_3 \cdot 0.5H_2O$; Goethite α-$FeO(OH)$
Others:	Quartz SiO_2; Feldspars $(KAlSi_3O_8$-$NaAlSi_3O_8$-$CaAl_2Si_2O_8)$; Phosphates; Carbonates

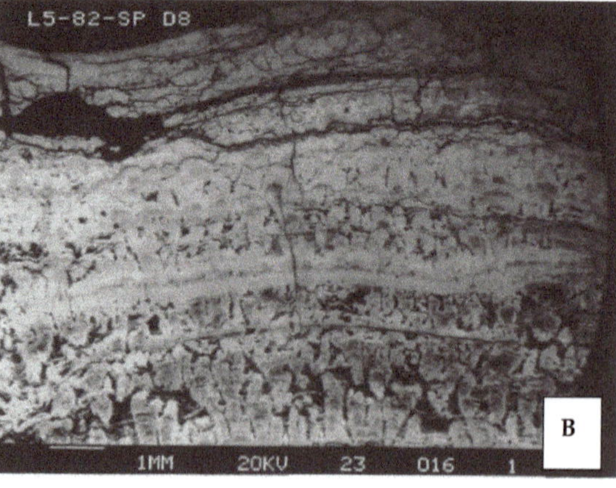

Figure 5. (**A**) CRC as a black capping layer on the top of brownish substrate and (**B**) complex internal structures seen in CRC [23,24].

2.3. Seafloor Massive Sulfides (SMS)

Seafloor massive sulfides (SMS) are derived from the fluid/rock interaction within the oceanic crust [25]. See Figure 6 for a schematic overview of the processes leading to the formation of SMS. Oceanic water penetrates the crust along fractures and cracks and a heat source provides heating of fluids towards the depth. The heat also mobilizes fluids trapped in the crust and the circulating water leaches metals from the host rocks. At a point towards depth, circulating waters reach the point where the water starts to rise along fractures in the feeder system to the black smoker system. The rising fluids bring metals to the feeder zone and ultimately release them to the cold ambient water near the seafloor, where the dissolved metals start to precipitate and create a plume. Precipitation is forced by the cold temperatures in the water on the seafloor as well as by reduced pressure as the fluids travel upwards. The SMS deposits typically occur at water depths down to 4000 m [26], and they are found in a variety of tectonic settings, mainly located to plate boundaries, at the modern seafloor including mid-ocean ridges,

back-arc rifts, and seamounts [27]. The composition of hydrothermal sulfide deposits can vary significantly according to the geodynamic environment, the nature of basement rocks affected by hydrothermal circulation, the water depth, the phase separation processes, and the maturity of deposits. SMS share some mineralogical and chemical characteristics with classic volcanogenic massive sulfides (VMS). The major minerals forming SMS deposits include iron sulfides, such as pyrite and marcasite, as well as the minerals of the most economic interest—chalcopyrite, isocubanite (being copper sulfides) and sphalerite (zinc sulfide). All other minerals of SMS deposits are considered as minor (by-product) ones. The precious metals gold and silver mainly occur in native form, and their Au and Ag grades in SMS deposits are significantly higher than in PMN and CRC deposits. Tables 5 and 6 show the chemical and mineralogical compositions of SMS samples. Figures 7 and 8 represent SMS samples and their growth structures, respectively.

Table 5. Average metal concentration in SMS deposits as related to their tectonic settings. "*N*" = number of deposits included in the calculations. Concentrations in wt%, except Au and Ag reported in parts per million. Data from [10].

Setting	*N*	Cu	Zn	Pb	Fe	Au	Ag
			(wt%)			(ppm)	
Sediment-free MOR	51	4.5	8.3	0.2	27	1.3	94
Ultramafic-hosted MOR	12	13.4	7.2	<0.1	24.8	6.9	69
Sediment-hosted MOR	3	0.8	2.7	0.4	18.6	0.4	64
Intraoceanic back arc	36	2.7	17	0.7	15.5	4.9	202
Transitional back-arcs	13	6.8	17.5	1.5	8.8	13.2	326
Intracontinental rifted arc	5	2.8	14.6	9.7	5.5	4.1	1260
Volcanic arcs	17	4.5	9.5	2	9.2	10.2	197

Among the physical properties of SMS, important from a metallurgical point of view, are their density, water content, and grain size. The bulk density of black and white smoker samples ranged between 1.9 and 3.0, the porosity between 19.4% and 38.8% and it changed with the maturation stage of SMS deposits [28]. The observed trend showed that less mature sulfide samples (the ones closer to the surface) are more porous.

Basics of a hydrothermal vent - a Black Smoker

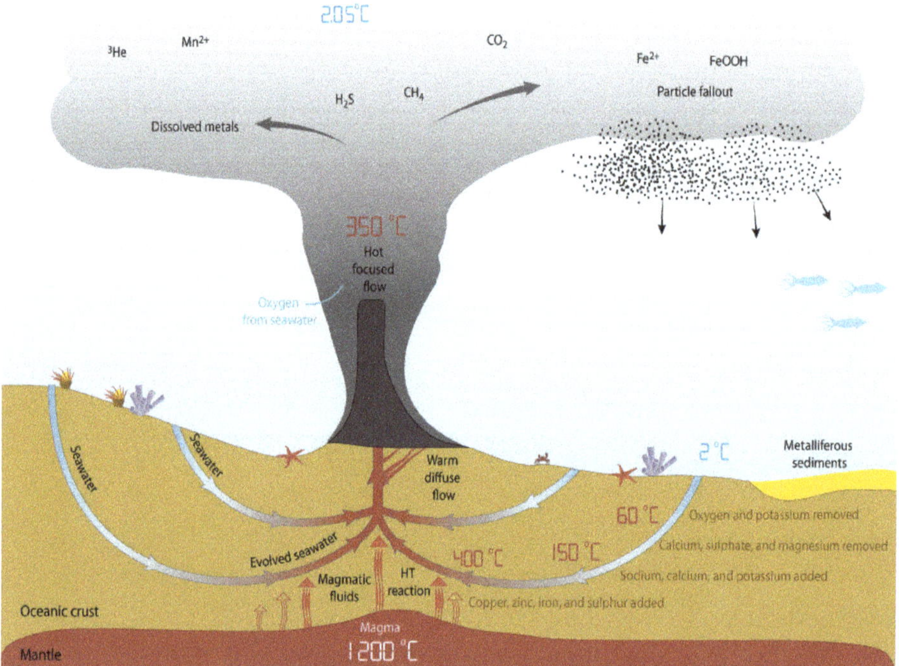

Figure 6. Illustration showing the hydrothermal process resulting in formation of SMS [29].

Figure 7. SMS samples from the Loki's Castle hydrothermal vent on the Mohn's Ridge; with (**a**) possible conduit structures implying fluid channels and (**b**) rust implying Fe derived from sulphides [30].

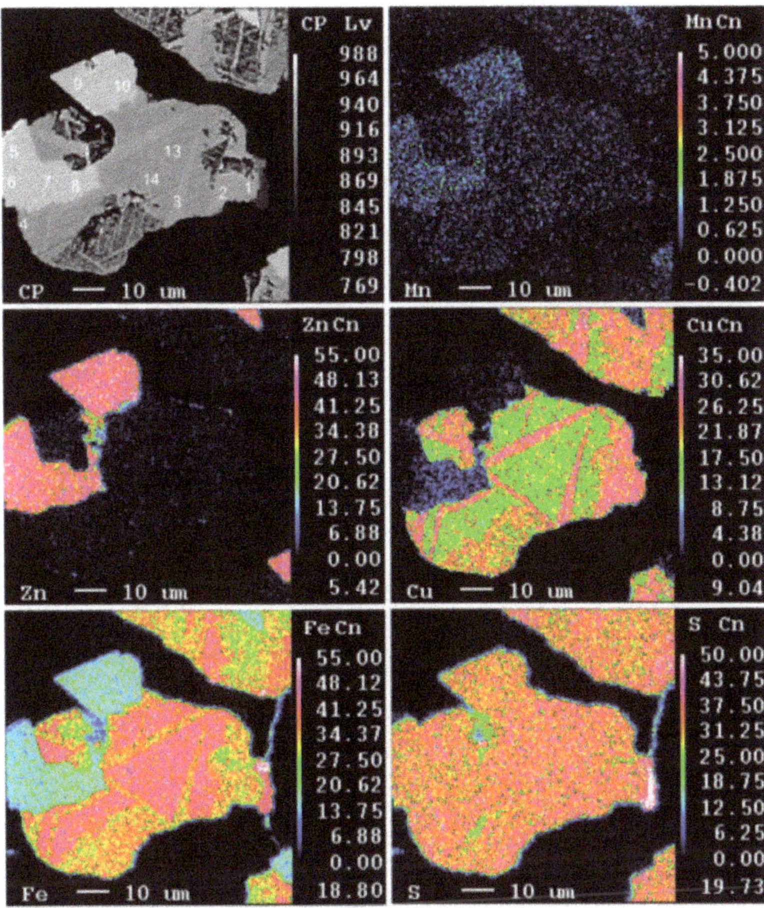

Figure 8. Backscattered electron (BSE) and element distribution maps of SMS from the Loki's Castle hydrothermal vent on the Mohn's Ridge [30].

Table 6. Typical mineralogy of SMS [30,31].

Value Minerals:	Chalcopyrite $CuFeS_2$; Isocubanite $CuFe_2S_3$; Sphalerite ZnS; Wurtzite $(Zn,Fe)S$; Chalcocite Cu_2S
Gangue Minerals:	Pyrite/marcasite FeS_2; Pyrrhotite $Fe_{1-x}S$ (x = 0 to 0.2)
	Baryte $BaSO_4$; Anhydrite $CaSO_4$; Quartz SiO_2; Aragonite/calcite $CaCO_3$

2.4. Comparison of Different Ore Types

The in situ estimated tonnages of discovered nodules and crusts of the CCZ and PPCZ (Table 7) are significant, but apart from Mn, the total tonnages are not more than required to be regarded as a supplement to the land-based reserve base [13]. However, especially CRC deposits are known in the deep oceans also in the Atlantic Ocean, e.g., within the Norwegian Exclusive Economic Zone (EEZ) [32] and within Japanese EEZ [33]. Hence, the tonnages for CRC in Table 7 may be upgraded, but too little is yet known.

Table 7. Comparison of in situ-discovered nodules and crusts in the Clarion–Clipperton Zone (CCZ) and the Prime Pacific Crust Zone (PPCZ), respectively. Table modified after [13], and references therein. * REO = rare earth oxides.

	CCZ	Global Land-Based Reserves	PPCZ
		10^6 metric tons	
Mn	5929	5200	1718
Cu	224	1300	7.4
Ti	59	900	87
Zn	29	480	5
REO *	17	150	20
Ni	278	150	32
Zr	6	57	4.1
Mo	12	19	3.5
Li	2.7	14	0.02
Co	42	13	50
W	1.3	6.3	0.67
Nb	0.4	3.0	0.4
Bi	–	0.68	0.32
Y	1.9	0.48	1.7
Te	0.07	0.05 (0.022)	0.45

When it comes to SMS deposits compared to the land-based reserve base, the amount of work performed on SMS seems not yet to have focused on the available tonnage estimates for different metals. However, several authors have estimated the size of single deposits and their typical grades. For example, Hannington et al. [34], estimated the around 600 Mt of massive sulfides along 89,000 km of Mid-Ocean Ridges with 5% combined Cu + Zn + Pb. These numbers suggest something in the range just higher than what is estimated for the PPCZ.

3. Processing of Marine Minerals

Processing involves separating an ore from a waste and transforming it into a product (e.g., metal). The number and type of steps involved in a particular process may vary significantly depending on the physical, chemical, and mineralogical properties of the processed ore. For terrestrial (land) ores, the processing takes place on land and includes a wide range of techniques. Marine minerals (PMN, CRC, SMS), however, due to their complex mineralogy, high porosity and water content, differ significantly from the land resources, and thus their processing will be different as well. Recovery of metals from the marine minerals might take place either on land or the seabed; however, the processing routes have not been elaborated yet. The currently applied extraction techniques of marine minerals are based on the terrestrial ore practices and can mainly be divided into three major categories: (i) conventional mineral processing, (ii) hydrometallurgical, and (iii) pyrometallurgical treatment. In mineral processing, several technique—from relatively straightforward mechanical operations to complex physicochemical procedures, are employed to prepare a material for further processing and/or to separate the ore from the waste.

The hydrometallurgical treatment mainly includes leaching with various lixiviants and reducing agents, while pyrometallurgy involves smelting, chlorination, and segregation processes. Tables 8–10 summarize the main routes applied so far in the processing of marine minerals, which are discussed in this section.

Table 8. Main routes in the processing of polymetallic manganese nodules (PMN).

	Method	Concept	Results and Main Conclusions	Ref.
PYROMETALLURGY	Reduction smelting	Prereduction at 1000 °C for 1 h; smelting with pure graphite powder → Fe-Cu-Co-Ni alloy + Mn-rich slag	4 wt% graphite at MnO/SiO_2 ratio = 1.6, 1350 °C	[35]
	Segregation roasting	Roasting at 800–1000 °C, 1 h with coke + chlorinating agents (solid chlorides of Na, Mg, NH_4, Li, Cs, Ca)	46–88% Cu, 22.5–32% Ni, 7–21.5% Co, 0–9% Mn, 0–9% Fe. Best temp for Cu–850 °C, for Ni and Co–1050	[17]
	Reduction with hydrogen	Reduction at 130–500 °C, 1 h, or at 400 °C for 8 h	Reduction of PMN in H_2 proceeds in 4 stages: 1st—loss of water (up to 130 °C); 2nd—decomposition of ferric oxyhydroxide (up to 320 °C), 2nd and 3rd—reduction of oxides and hydroxides of Cu, Ni and Co; 4th—reduction of α-Fe_2O_3 to metallic Fe	[36]
	Reduction roasting	Reduction at 1000–1150 °C with anthracite and additives (CaF_2, SiO_2, FeS) + magnetic separation	1100 °C, 2.5 h, 4% CaF_2, 7% anthracite, 5% SiO_2, 6% FeS metals in concentrates: 86.48% Ni, 86.74% Co, 5.63% Mn, 83.91 Cu, 91.46% Fe	[37]
	Reductive smelting	Zero-waste 2-step smelting → Cu-Co-Ni alloy + HC FeMn	Smelting at 1400 °C with 9.4% SiO_2, yielding over 90 and up to 100% for Cu, Co, Mo, and Ni, 97% of Mn in final slag	[38]
PYROMETALLURGY + HYDROMETALLURGY	Reduction roasting + ammoniacal leaching	Reduction at 750–1150 K with wood charcoal and natural gas; Leaching in 1 M $(NH_4)_2CO_3$ in 10% NH_3	Roasting at 1073 K, 2 h, 6% reduction agent, leaching for 210 min at 318 K: for Co roasting temp. 1123 K 90% Ni, >70% Cu, >60% Mo, 90% Co	[39]
		Reduction at 700 °C, 2 h with 10% low-sulfur fuel oil; precipitation of Fe and Mn before leaching; leaching with NH_3 + CO_2	3.5 h leaching time 10% Cu, 22% Ni, 62% Co	[40]
		Reduction at 650–800 °C with coal; leaching (2 stages) with ammonium salt + ammonium hydroxide sol.	1st leaching step (0.05–1 M NH_3) at RT for Cu recovery, 2nd leaching step (up to 2 M NH_3) at ~50 °C for Ni dissolution in residue. Leaching time 0.5–4 h.	[41]
		Reduction at 800 °C with coal; preconditioning with NH_3 + $(NH4)_2CO_3$ + surfactant solution; precipitation of Fe and Mn by air purging; residue leaching in NH_3 + $(NH_4)_2CO_3$	95% Cu, 94% Ni, 80% Co	[42]
	Pyrolysis + acidic leaching	Reduction at 300–500 °C with sawdust ground <1mm under N_2; Leaching with 1 M H_2SO_4 at 60 °C for 1 h	10% sawdust, reduction temp. 500 °C, reduction time 6 min. 96.1% Mn, 91.7% Cu, 92.5% Co, 94.4% Ni	[43]
	Reduction and smelting + chlorine leach	Reduction at 900 °C with coal + SiO_2 + CaO for 2 h; Smelting at 1400 °C for 2 h; Leaching of sulfided Cu-Co-Ni alloy with chlorine gas; SX + EW for Fe, Cu, Co, and Ni	Silicomanganese obtained from slag phase chlorine leach is preferable to an oxygen-pressure leach. 99% Cu, Co, Ni in 3 h	[44]

Table 8. *Cont.*

	Method	Concept	Results and Main Conclusions	Ref.
HYDROMETALLURGY	Hydrochlorination + water leach	Hydrochlorination with HCl gas at 550 °C and water vapour at 300 °C; Leaching of dissolved chloride products with water; Separation of Fe_2O_3 precipitate; SX of Cu, Co, and Ni from PLS; electrolysis.	US Patent	[45]
	Reduction + smelting + acidic pressure leaching	Reduction with fuel oil + air at 1000 °C; smelting in electric furnace; oxidative pressure leaching of matte; slag treatment for Mn recovery;	Ground matte leached at 1 MPa, 110 °C, 2 h, 100 g/L H_2SO_4 99% Cu, Co, Ni, only 0.01 g/L Fe (after sulfidization of matte)	[46]
	Reduction + smelting + POX	Reduction with fuel oil + air at 1000 °C; smelting in electric furnace; oxidative pressure leaching of FeNiCoCu alloy (no conversion to matte) with H_2SO_4 + $CuSO_4$; FeOOH precipitation; SX + EW Of Cu, Co, and Ni	Addition of $CuSO_4$ prevents H_2 formation during leaching, Cu is cemented by less noble metals and leached by sulfuric acid. 1.5 excess of acid, 2–3 excess of $CuSO_4$, 10 bar, 6 h, solid conc. 25–45 g/L	[47]
	Baking + water leaching	Baking with conc. H_2SO_4; water leaching of Cu, Co, Ni, Mn soluble sulfates	N/A	[48]
Acidic	Pressure leaching micellar mediated	Pressure leaching of ground nodules (<100 μm) with H_2SO_4 and surfactants: CTAB, SDS, Triton X 100, Tween 80; conditions: 110–160 °C, S/L 1/10, 2 h.	CTAB, 160 °C, 10% pulp density, 2 h, 5% H_2SO_4 99% Mn, Cu, Co, Ni	[49]
	Pressure leaching/+charcoal	Leaching of ground nodules with H_2SO_4 at 150 °C and 0,55 MPa, 4 h Charcoal addition to remove Fe dissolve MnO_2	150 °C, 0.66 g H_2SO_4 per g of nodule, pO_2 = 0.55 MPa, 4 h or the same conditions + 0.05 g charcoal/g of nodule 77%Cu, 99,8% Ni, 88% Co, 99,8% Mn, 4,5%Fe	[50]
	Atmospheric/Pressure leaching	Comparative leaching with H_2SO_4 at 100 °C and 200 °C	200 °C, 3 h, 0.3 g H_2SO_4/g of nodules, 90% Ni, 91% Cu, 44% Co, 6% Mn, 2% Fe Higher leaching at 100 °C for Co (70%) and Fe (65%)	[51]
	Atmospheric leaching	Leaching with H_2SO_4 + $FeSO_4·7H_2O$ at 80 °C, 90 °C	90 °C, 1.6 excess of H_2SO_4, L/S 7–15. Solution contains $FeSO_4$ in stoichiometric amount to MnO_2. >90% Ni, Cu and Mn, 85% Co	[52]
	Atmospheric leaching + amines	Leaching with H_2SO_4 and aromatic amines (as reductants) at ambient temp. aniline, o-phenylene diamine, o-aminobenzoic acid, o-nitroaniline, p-amino toluene, p-aminobenzene sulfonic acid, 1-naphtylamine	84–99.6% Mn, 23–97.7% Cu, 74–99.3% Ni, 89–99.7% Co	[53]
	Atmospheric leaching + phenols	Leaching with H_2SO_4 and phenols (as reductants) at ambient temp. hydroxybenzene, o-dihydroxybenzene, m-dihydroxybenzene, p-dihydroxybenzene, o-trihydroxybenzene and m-trihydroxybenzene	95% Mn, Cu, Ni, Co	[54]

Table 8. *Cont.*

	Method	Concept	Results and Main Conclusions	Ref.
	Atmospheric leaching	Leaching of ground nodules with H_2SO_3 (dilute aq. solution of SO_2) at ambient temp.	SO_2 ratio to the total weight of nodules (g): $0.94 \cdot 10^{-2}$ —$1.25 \cdot 10^{-2}$. Concentration of 6–8% SO_2 in water is satisfactory. Temp 25 °C, p = 1 atm. >90% Ni, Co, Mn in ~10 min.	[55]
	Atmospheric leaching	Leaching of ground nodules with SO_2 or SO_2 + H_2SO_4 at 30 °C	Particle size −150: +76 µm. 30–40 °C, leaching with SO_2 (only) 89% Mn, 60% Cu, 82,5% Ni, 90% Co, 75% Zn	[56]
	Atmospheric leaching	Leaching of REE in H_2SO_4, 500rpm, 30 °C, 2 h	3 M H_2SO_4 90 °C >90% REE but high co-extraction of Fe, Co, Ni, Cu or 0.2 M H_2SO_4 at 45 °C total extraction of REE 58% low co-extraction (0.3% Mn, 4.63% Fe, 23.7% Cu, 0.2% Co, 31.8% Ni).	[57]
	Atmospheric leaching	Leaching of ground nodules with HCl, at 90–100 °C	1–1.5 M HCl, Grain size ~35 µm. Non-selective towards Fe. >80%Ni, Cu and Zn with 30–35% Fe, Mn, and Co < 20%.	[56]
	Atmospheric leaching + SX	Leaching of powdered nodules in HCl; Solvent extraction of Cu, Co, and Ni	4 M HCl; Cyanex 923 and Cyanex 301 at 25 °C A/O = 1 >90% Cu, Co, Ni	[58]
	Cathodic electroleaching + adsorption	Electrolytic reduction of Cu, Mn, Co, and Ni from acidic slurry sltn on Pt electrodes at 30 °C; Adsorption of metals from lean electrolyte on nodules	Copper leached and deposited on a cathode, MnO_2 deposited on the anode. Adsorption: 1g of nodules mix with 100mL of sltn Cu, Ni, Co, Mn (single or grouped), size fraction −75 and +53 µm 100% Cu, Co, Ni. 50% Mn	[59]
	Slurry electrolysis	Electrolysis in HCl-NaCl medium cathodic reduction at the cathode; anodic oxidation and deposition of MnO_2	Anode: Ti/MnO_2 strip; Cathode: graphite stick, diaphragm; 120 g/L NaCl, 40–70 g/L Mn, 70 °C, pH 0.5–1.5, 200 min, current density: 200 A/m^2 Cu 96–99%, Co 99%, Ni 98–99%, Fe 54–79%, Mn 96–99%	[60]
Basic	Pressure leaching + SX-EW Medium-scale plant	Ammoniacal leaching with reductants: SO_2, CO, Fe(II), Mn(II), thiosulfate, glucose, carbon, Demanganisation step (prec. MnO_2), ammonia stripping and recycling; Cu SX-EW Sulfides precipitation of Co, Ni, and minor impurities (Cu, Zn, Fe), dissolution in H_2SO_4; Co-Ni SX-EW	5 m^3 autoclave, medium temp and pressure. Scale: 500 kg/day avg. 85% Cu, 90% Ni, 80% Co	[61]

Table 8. *Cont.*

	Method	Concept	Results and Main Conclusions	Ref.
Microorganisms assisted	Liquid phase oxidation + Atmospheric leaching	Molten KOH + air to oxidize MnO_2 in nodules and for dissociation of nodules structure; conversion of K_2MnO_4 to $KMnO_4$ and MnO_2 Pure MnO_2 from $KMnO_4$ decomposition; separation of Fe_2O_3 through gravity classification, reductive leaching with $(NH_4)_2SO_3$	50 g/L residue conc., 200 rpm, 100 g/L NH_3, 70 °C 95%Cu, 65% Co, 84% Ni	[62]
	Bacterial leaching	Leaching of REE from nodules with thiobacillus ferroxidans	~100% for Cu and Ni (2 weeks), <5% Fe and Mn, 50% Co	[63]
	Bacterial leaching	Leaching of ground PMN with thermophile Acidianus brierleyi at 65 °C or mesophile Thiobacillus species at 30 °C	A. brierleyi more effective; 100% Cu, Zn (4 days) and 85% Ni, 70% Co, 55% Mn (10 days)	[64]
	Bacterial leaching + pyrite	Thiobacillus ferroxidans + pyrite at 30 °C pyrite as reductant	pH 2, pulp 10%, 3 days leaching, pyrite:nodules ratio 1:1 95% Co, 94% Ni, 97% Mn, 80% Cu Higher leaching rate at anaerobic conditions	[65]
	Bioleaching with marine bacterium isolate	Comparison of acidic leaching and bioleaching; 2.5 M H_2SO_4 + $Na_2S_2O_3$ or 2.5 M HCl + glucose or 2.5 M HNO_3 vs. marine isolate; 30 °C	30–50% Co (HCl), 85% Cu, 85% Ni (HCl), 80%Mn. Bioleahing with marine isolate was much less efficient < 45% Co, ~30% Cu and Ni	[66]
	Electrobioleaching/galvanic leaching	Thiobacillus ferrooxidans, Thiobacillus thiooxidans, 30 °C; Galvanic leaching with pyrite/pyrolusite (MnO_2)	voltage range −600:-1400 mV, 4–5 h; −75 to +53 µm size fraction; galvanic leachingat nodule:pyrite ratio = 2:10 100% Cu, Ni, Co	[67,68]
	Leaching with Fe-reducing bacterium	Decomposition of nodules with Shewanella putrefaciens and NaCl solution	0.5 M NaCl, pH 7, necessary daily addition of 1mmol sodium lactate, leaching of REE with 0,01M HCl	[69]
	Bioleaching with bacteria consortia and reductants	Anaerobic leaching with bacteria consortia and glucose or sodium acetate as reductants, 30 °C, no agitation	Glucose (30% recovery of Mn) 90 days only 42 ppm of Fe was leached, only 30% of Cu, and 30% of Ni	[70]
	Bioleaching with fungi	*Aspergillus niger* (fungal culture) realeses organic acids such as oxalic or citric acid which help reduce host metal oxides/hydroxides in nodules	Activation: 10 min, size <10 µm; Leaching with A. niger, 15 days, 35 °C. (25 days for not-activated material) 95% Cu, Ni, and Co	[71]
			Bioleaching more effective than chemical leaching by carboxylic acids or by fungal metabolites. 97%Cu, 98% Ni, 86% Co, 91% Mn, 36% Fe, 30 days, initial pH 4.5, 35°C, 5% pulp density, particle size <300 µm	[72]
		Aspergillus niger and *Trichoderma* sp.	11 days with A. Niger >80% Mn, Cu, Ni, 70% Co, 30% Fe	[73]

Table 9. Main routes in the processing of cobalt-rich manganese crusts (CRC).

HYDROMETALLURGY	Acidic	Atmospheric leaching	2.75 M HCl with the addition of 18.5 mL of ethanol (reductant) initial pH: 1.5	Mixed diagenetic/hydrogenetic crust shows lower recovery than lower diagenetic or pure hydrogenetic crusts. Mn 75–81%, Fe 49–58%, Co 63–108%, Ni 53–85%, Cu 50–74%, V 58–85%	[74]
			1st stage—leaching with H2SO4 at 80–90 °C; 2nd stage—leaching of residue with HNO_3	50 g of sample + H_2SO_4 (20–25%), S/L = 1/4 or HNO_3 (10–30%) 74–85% Mn, Co, Ni, Cu, Zn, Y, HREE, U, and Hf >90% of elements extracted from the residue	[75]
			Beneficiation of crust sample by froth flotation and magnetic separation(separation from the substrate); leaching with H_2SO_4-H_2O_2; precipitation of Fe with CaO; precipitation with H_2S under pressure removal of Co-Ni mixed sulfides Mn recovery by carbonation at neutral pH precipitation of $MnCO_3$	25 °C, 1 h, 13% solids, 5.9% H_2SO_4, 1.2% H_2O_2 96% MNm 43% Fe, 95% Co, 91%Ni	[76]

Table 10. Main routes in the processing of seafloor massive sulfides (SMS).

PYROMETALLURGY		Zero-waste process: 2-stage reductive smelting	Modification of INCO process; The slag phase from 1st smelting step was directed to the next stage to increase Mn recovery in the form of high carbon ferromanganese (HC FeMn).	$1400\ ^\circ C$ with 9.4% SiO_2, yielding 90–100% for Cu, Co, Mo, and Ni. The final slag: 97% of Mn and low concentrations of Cr, Cu, V, and Ni.	[38]
HYDROMETALLURGY	**Acidic**	Atmospheric leaching	Leaching with HNO_3	10% HNO_3 $90\ ^\circ C$, 2 h, S/L 1/10 >90% Cu, Zn, Fe	[77]
			Galvanic leaching using MnO_2-H_2SO_4-NaCl media.	24 h, temp 30–$80\ ^\circ C$, 0–1.5 M H_2SO_4, 0–1 M NaCl, 0–19.5 g/L MnO_2	[78]
			Simultaneous leaching of SMS and PMN-pure or at different ratios	1 M H_2SO_4 and 1 M NaCl, 700 rpm, $80\ ^\circ C$, 48 h, S/L 50 g/L PMN dosage from 30–100% Cu, Mn, Ni ~100, Zn ~85%	[79]
			Artificial seawater leaching	$12\ ^\circ C$, 0.6–1 g SMS to 500 mL of seawater ppb levels for Cu and Pb	[80]
CONVENTIONAL PROCESSING			Application of ball mill grinding and column flotation to SMS processing; LIBS technology applied for in situ measurement of the metal grade of ore particles	Water-filled grinding at high pressure had an almost comparable grinding performance to wet grinding at the atmospheric pressure; concentrates of Cu and Zn obtained in column flotation	[81]
			Flotation of SMS to separate chalkopyrite and galena as froth, and sphalerite, pyrite, and remaining gangue minerals as tailings.	Flotability of sphalerite increases in the presence of Pb minerals (PbS, $PbSO_4$) and soluble compounds: Cu^{2+}, Zn^{2+}, Pb^{2+}, and $Fe^{2+/3+}$ High separation of chalkopyrite and sphalerite is possible through the combination of surface cleaning with EDTA and depression of lead-activated sphalerite by zinc sulfate	[82]
			SMS grinding and flotation of Cu-minerals (mainly chalkopyrite)	>25% Cu concentrates (~85–90% Cu recovery) ready for Cu smelter ~25% gold recovered in Cu concentrate ~65–70% of Au can be recovered into a pyrite concentrate. Extraction of Au by the conventional technologies of roasting/cyanidation or pressure oxidation/cyanidation	[83]

3.1. Polymetallic Manganese Nodules (PMN)

To obtain high recoveries of valuable metals from PMN, it is necessary to release them by breaking the crystal lattices of the manganese oxides. Therefore, most of the research works propose the reduction of tetravalent manganese to a divalent state [19,84]. This can be accomplished either pyrometallurgically by smelting with gaseous, liquid, or solid reducing agents, or with hydrometallurgy, where the reduction is carried out either before or during a leaching operation, or by a combination of both (Figure 9).

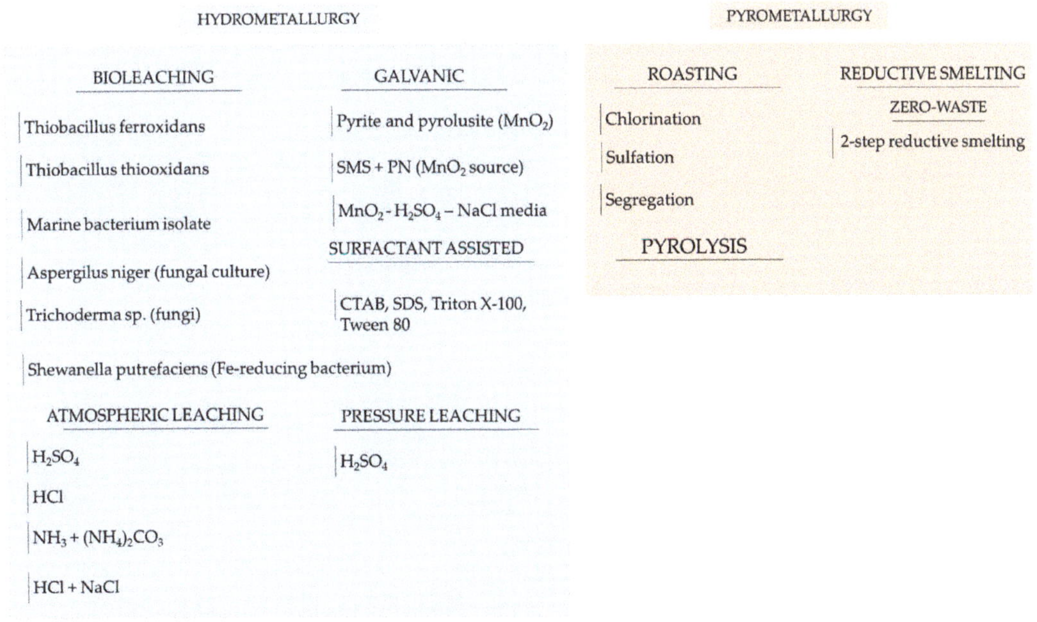

HYDROMETALLURGY

PYROMETALLURGY

BIOLEACHING

Thiobacillus ferroxidans

Thiobacillus thiooxidans

Marine bacterium isolate

Aspergilus niger (fungal culture)

Trichoderma sp. (fungi)

Shewanella putrefaciens (Fe-reducing bacterium)

GALVANIC

Pyrite and pyrolusite (MnO_2)

SMS + PN (MnO_2 source)

MnO_2 - H_2SO_4 – NaCl media

SURFACTANT ASSISTED

CTAB, SDS, Triton X-100, Tween 80

ROASTING

Chlorination

Sulfation

Segregation

PYROLYSIS

REDUCTIVE SMELTING

ZERO-WASTE

2-step reductive smelting

ATMOSPHERIC LEACHING

H_2SO_4

HCl

NH_3 + $(NH_4)_2CO_3$

HCl + NaCl

PRESSURE LEACHING

H_2SO_4

Figure 9. Hydrometallurgical and pyrometallurgical processing of PMN [17,35–73].

Regardless of the approach used for metals recovery, the most common method of PMN pretreatment used to be air drying followed by crushing and grinding to reduce the size of nodules. Usually, the powdered nodules size fractions are less than 100 μm [45,49,55]. When samples are processed pyrometallurgically at high temperatures, the powder is mixed with the flux and reducing agents, and in the case of hydrometallurgical processing, the ground nodules are sieved and leached under specified conditions.

PMN are relatively easy to grind with the Bond-index of about 7 kilowatt hours per ton [84]. The size mineral phases in PMN ranges from below 1 to ca. 5 μm. Due to the high surface area (ca. 200 m^2/g) and porosity (60%) nodules were considered for their use as adsorbents or catalysts [85–89]. In a few works, due to the high concentrations of MnO_2, PMN served as an oxidizing agent [90–92]. High porosities, with pore size diameters in the range 0.01 μm to 0.1 μm, result in a high moisture content (30–40%). This is a major disadvantage in high-temperature metallurgical treatment because it forces the use of a drying operation, and thus it is energetically inefficient. The complex oxidic mineral composition of PMN (a very fine-grained admixture) makes the application of methods of physical beneficiation such as gravity, electrostatic and magnetic separation or flotation to produce concentrates of the valuable metals economically inefficient; instead, either hydro- or pyrometallurgical processing has to be used. Physical separation techniques might be applied in screening for removal of such debris as bones, sharks' teeth, etc. [11].

Pyrometallurgy aims at the reduction of metals in PMN to metallic forms, which can be further recovered. A lot of research works have been dedicated to finding the most effective reducing agent. Many inorganic and organic compounds were tested in this role. The hydrometallurgical treatment has been described in several process options, among which the most popular is either acidic or ammonia leaching under either atmospheric or elevated pressures, with an addition of various kinds of additives such as reductants, surfactants, or microorganisms (biohydrometallurgy) (Figure 9).

The most important metals of economic interest, found in manganese nodules, are Cu, Ni, Co, and Mn and their recovery might be based on the classical smelting processes, dedicated for the copper and nickel metallurgy, known from the terrestrial applications. The most applied pyrometallurgical methods are listed in Table 8. At first, nodules are either dried or calcined at various conditions, then ground, and introduced into a furnace for reduction (at temperatures from 130 °C to over 1400 °C) [15,20,37,40,41,43,85]. The reduction process with various reducing agents, such as hydrogen chloride, ferrous ion, sulfur dioxide, carbon, and many organic compounds [18,36,42,89,93], leads to a manganese-rich slag and an iron-nickel-copper-cobalt alloy. The alloy is then subjected to a converting operation where during oxidation most of the remaining Mn and Fe are removed. In the next step, the obtained Ni-Cu-Co matte might be treated by several methods. The Mn-Fe slag phase can be fed to a furnace to produce the ferro-silico-manganese alloy. A simplified pyrometallurgical route for manganese nodules treatment, created based on the data provided elsewhere [94], is presented in Figure 10.

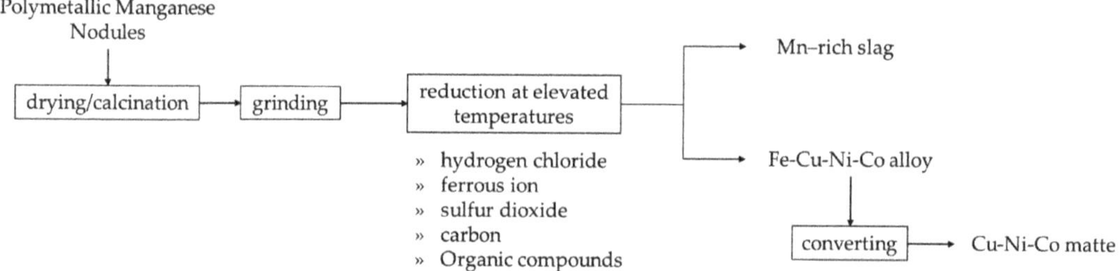

Figure 10. Simplified pyrometallurgical route for manganese nodules treatment.

Two approaches have been known for the manganese recovery from the slag. It can be recovered as either silicomanganese or ferromanganese. Both are marketable products with some limitations to the contents of Mn, Si, C, or S, specified by the American Society for Testing and Materials (ASTM) [46,76], Active Standard A99 and A483/A483M-ASTM. A typical Mn recovery exceeds 95%, but slags still contain a small amount of Cu, Co, and Ni. Sommerfeld et al. proposed two-step reductive smelting of polymetallic nodules resulting in a "zero-waste" process [38]. Their concept was based on the well-known INCO process described elsewhere [94]. Briefly, the idea was to use fluxes (Al_2O_3, TiO_2, FeO, $Na_2B_4O_7$, and SiO_2) in an additional smelting step, designed for the slag phase obtained in the first one to increase the Mn recovery in the form of high carbon ferromanganese (HC FeMn). The optimal conditions were: smelting at 1400 °C with 9.4% SiO_2, yielding over 90% and up to 100% for Cu, Co, Mo, and Ni. The final slag contained 97% of Mn and low concentrations of Cr, Cu, V, and Ni.

Most of the research works dedicated to the treatment of PMN relate to the application of a leaching operation. These works can be generally grouped into two categories based on either the leaching type (acidic/basic or atmospheric/pressure) or leaching mechanism (electroleaching, electrobioleaching, galvanic, or surfactant mediated). Table 8 and Figure 11 summarize the applied methods for the hydrometallurgical treatment of PMN. Hydrometallurgical processing very often is conducted on a previously reduced feed, i.e., under a reducing atmosphere of H_2, CO [46], or SO_2, or by mixing nodules with solid reductants such as coal [45,87]. Another option is to apply reduction leaching, where manganese is reduced to Mn (II) by various reagents, such as SO_2, [61] CO, HCl [20], $FeSO_4$ [52], aromatic amines [53], phenols [54], glucose [66,70], or surfactants [49]. Then, the only pretreatment operation is the comminution of nodules. Some of the most commonly used techniques are shortly described here.

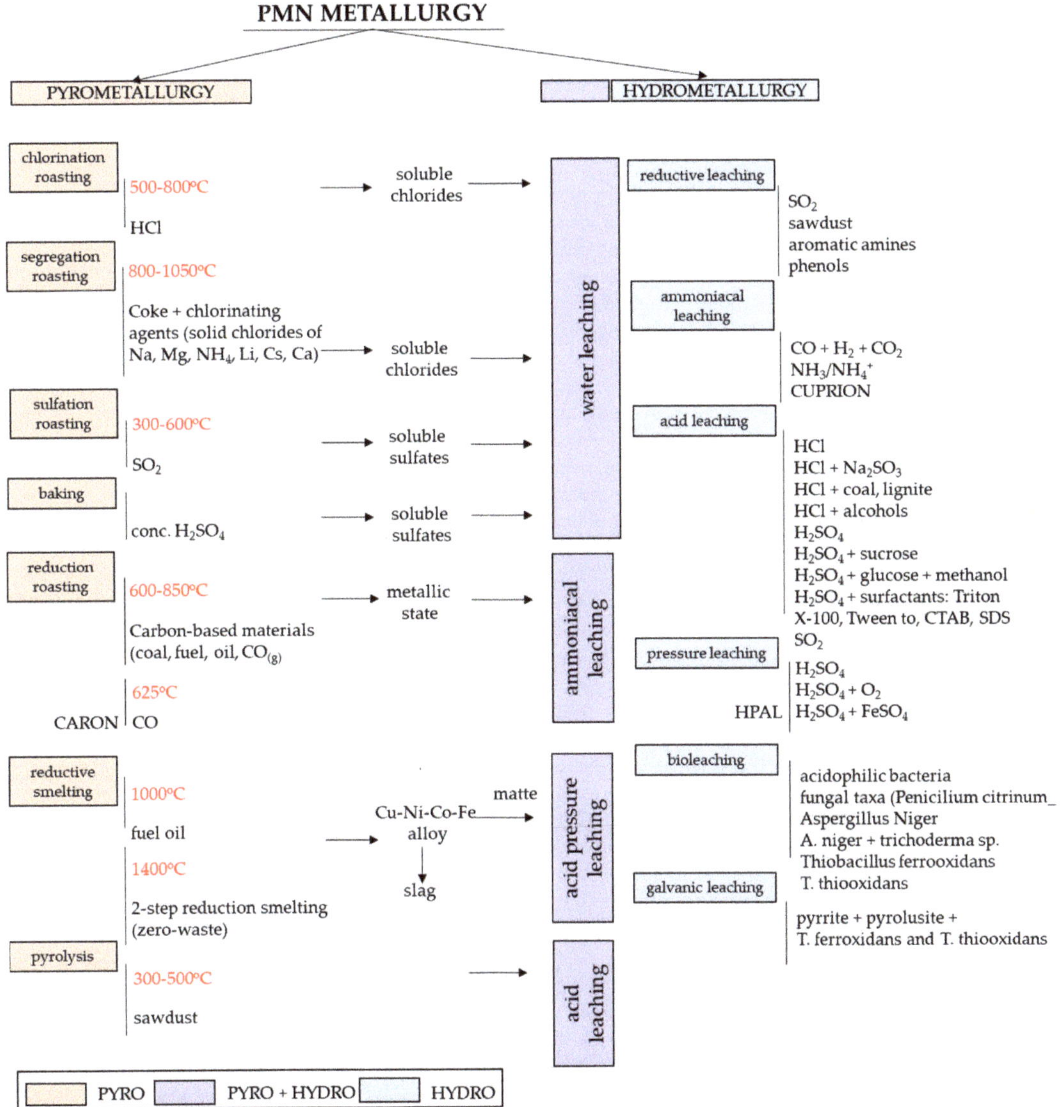

Figure 11. Metallurgical processes for manganese nodules [39,42,43,49,53,54,57,59,63,65,68,71,94–96].

Among all applied techniques, the Cuprion process developed by Kennecott Copper Corporation (KCC), and further improved by several researchers, has been found as the most promising method for recovery of metals from PMN. The process involves a reductive ammonia–ammonium carbonate leach in the presence of the reducing gas containing carbon monoxide, hydrogen, and small amounts of carbon dioxide, oxygen, and nitrogen. The most important step is the reduction of manganese, which helps in breaking the manganese matrix; thus the rest of metals such as copper, nickel, cobalt can easily react with a leaching agent. The carbon monoxide reduces cupric ion to the cuprous state, which is stabilized by the ammonia–ammonium carbonate solution composition. Then, the cuprous ion reacts with manganese dioxide to form the manganous ion, which rapidly precipitates as manganese carbonate. Nickel, copper and cobalt contained in PMN are leached into the solution as the manganese dioxide is converted to manganese carbonate. The manganese dioxide converted to manganese carbonate and iron are rejected as a residue. The main advantages of this process are its mild operating conditions, that is, low energy consumption, low toxicity, acceptable environmental impact, and most importantly high selectivity. However, low cobalt recovery, low pulp density, the use of carbon monoxide as a reducing agent, as well as the removal of manganese are the main disadvantages of the process.

Other methods applied for recovery of metals from PMN are combined pyro- and hydrometallurgical processes. Special attention has been paid to the International Nickel Company (INCO) process described in detail elsewhere [94]. It requires drying the nodules, and thus substantial energy input. Dried PMN are crushed and ground, and then reduced and subjected to smelting in an electric furnace to produce an alloy containing copper, nickel and cobalt, and residual Fe and Mn, as well as the manganese and iron-rich slag. The slag is then treated to produce ferro-manganese. However, the alloy is sulfurized to produce a matte containing copper, nickel, and cobalt, and the residual iron and manganese are removed as an oxide slag. Then, the matte is leached with sulfuric acid, and the leach liquor is processed by solvent extraction and electrowinning for copper, and the raffinate from this process is again processed for nickel recovery. Finally, cobalt is reduced from the solution with hydrogen to produce cobalt powder.

Another option for extraction of metals from PMN is bioleaching utilizing anaerobic Mn-reducing bacteria such as *Thiobacillus ferrooxidans* [63–65], *Thiobacillus thiooxidans* [64,68], *Aspergillus niger* [72], or *Acidianus brierleyi* [64].

Abramovski et al. [97] made a comparative economic analysis of three technologies proposed in the literature: (i) pyrometallurgical treatment with the use of SO_2 as the reductant; (ii) a combined pyro–hydro method, and (iii) hydrometallurgical treatment based on high-pressure acid leaching. The evaluation of scaling up selected lab-test solutions to a pilot-plant level was based on the extraction efficiency, supported by the possibilities for performance, the actual prices of metals, environmental pollution, and other factors. Although authors made no final statement defining the best technology, it can be seen from their analysis that smelting is rather ineffective, due to significant losses of valuable metals (i.e., Co, Ni, V, REE, etc.) to a slag phase.

It seems that the most popular and the most sensible method of ocean nodules treatment is to combine pyro- and hydro-methods. Metals that form either alloys or slag phases after roasting can be extracted with different leaching agents. In this context, different mineral acids, bases, or organic acid solutions were investigated.

3.2. Cobalt-Rich Manganese Crusts (CRC)

Future scenarios for the extraction of CRC from the seabed may cause the substrate or upper layer to stick to the mined crust. Hence, conventional mineral processing operations such as comminution and separation (i.e., gravity, magnetic, flotation) are necessary to remove the substrate before the metallurgical treatment. Among the available literature, the work from 1991 on the flotation of Co-rich crusts [98] deserves attention. This research describes the experiments aiming the beneficiation, i.e., separation of crusts from the metal-barren substrate by the froth flotation technique. Other investigators tried to beneficiate crust samples by using Jig concentrators, allowing for further gravity separation of particles within the ore body [99]. In this case, crust samples were crushed to obtain various particle fractions. In the end, the flowsheet was proposed showing that fractions from 0.5 to 4 mm could be concentrated by Jig separation, while the finer fractions by froth flotation.

In 1991, other researchers presented their study on the leaching of metals from CRC in the H_2SO_4-H_2O_2 media [76]. Crust samples were air dried and beneficiated before the leaching operation. The advantages of their method were leaching at ambient temperatures and that water and oxygen were only used. For the beneficiation, froth flotation (collectors: diesel fuel or kerosene with fatty acids and a lesser amount of sulfonate), and magnetic separation were proposed. For metals recovery from leach solution, authors applied three-step precipitation: first, iron was precipitated with CaO, then Co and Ni were removed as mixed sulfides by the precipitation with H_2S under pressure, and finally, Mn was recovered by carbonation at neutral pH, as $MnCO_3$.

Since that time, the literature on the recovery of metals from CRC has been limited. Except for a few propositions aiming at leaching of crusts samples in HCl [74] and H_2SO_4 [75], there have been no works on that subject. Even if the title indicates the interest in metals recovery from the crust material, in the majority of works it is a rather general consideration of pros and cons. Since CRC, similarly to PMN, exhibit complex oxidic mineral composition, it might be assumed the respective ore processing techniques of crusts will be very similar to nodule extraction discussed in the previous subchapter and summarized in Table 8 and Figure 11.

3.3. Seafloor Massive Sulfides (SMS)

Similarly to CRC, the literature on the processing of seafloor massive sulfides is quite poor—only a few research works concern the extraction of metals from this type of marine deposit. This is due to the limited number of mined samples to be tested.

The Solwara 1 project (the first deep-sea mining project at the international level was approved but failed before the extraction phase) planned to mine mineral-rich hydrothermal vents in the Bismark Sea [83]. However, so far, no commercial extraction and processing has been performed. From 1998 to 2012, several tests on the characterization of samples and processing were performed. The first reported processing test was conducted in 1998 on two samples with different compositions: rich in zinc and poor in copper, and rich in copper and poorer in zinc. High recovery of copper and zinc was obtained. The authors of the presented report claimed that it was similar to terrestrial mining operations, but no flotation conditions were provided. It has to be noted that for all Solwara processing tests, the flotation conditions were not provided. The next reported test was conducted on the chimney sample in 2005, and the tests resulted in high recoveries of copper, silver, gold, lead, and zinc. The final concentrate contained over 26% of copper, however the concentration of metals in the feed was not provided. In 2008 the grindability and floatability tests were performed. The flotation concentrates with a copper grade of 28% and recovery of 90% were produced. In 2012 the final processing tests were performed, achieving the concentrate enriched in gold and degraded in copper. Since then, no other enrichment tests for the Solwara 1 project have been reported.

The idea of direct application of conventional mineral processing technologies to seafloor mineral processing was discussed in work [81]. The presented concept was to implement ball mill grinding and column flotation directly on the seabed before lifting the ores. Before experiments, the ore samples were ground into particles with the size of 88 to 106 µm. Both water-filled grinding and column flotation under elevated pressure had almost comparable performance to wet grinding and flotation at the atmospheric pressure. Moreover, the measurement by Laser-Induced Breakdown Spectroscopy (LIBS) was proposed to apply to the in situ measurement of a metal grade of ore particles.

Recently, Aikawa et al. (2021) [82] presented results on flotation of SMS samples, which were chemically pretreated using ethylene diamine tetra acetic acid (EDTA) to remove anglesite. The authors claimed that application of surface cleaning before flotation and depression of Pb-activated sphalerite could achieve the highest separation efficiencies of chalcopyrite and sphalerite.

Kowalczuk et al. [77] studied the extraction of copper and zinc from SMS rock samples from Loki's Castle hydrothermal vent on the Mohn's Ridge. The mineralogical analysis revealed that SMS samples are totally different than massive sulfides occurring on land. Sphalerite, chalcopyrite and isocubanite showed complex intergrowth texture on the nano scale. Chalcopyrite occurs as 1-micron lamellas in the structure of isocubanite and sphalerite. Sphalerite contained app. 20% of iron. This complex mineralogy provideed challenges to conventional mineral processing methods such as liberation by comminution and upgrading by flotation. The high iron content in sphalerite was a challenge to upgrade a sulfide concentrate by removing pyrite. Thus, nitric acid was proposed as an appropriate leaching agent. The authors found that leaching with HNO_3, at an elevated temperature yielded more than 90% extraction of Cu, Zn, and Fe.

An alternative method was proposed in experiments aiming at copper and silver recovery [78], conducted at the material of the same chemical and mineralogical composition (silica, barite, quartz, galena, pyrite, and marcasite were identified as gangue minerals). The silver phases were not detected due to their low content (15 ppm), thus it was assumed that Ag can occur as an admixture in the crystallographic lattice of sulfides or as micro inclusions. Here, the SMS sample was leached in MnO_2-H_2SO_4-$NaCl$ media, providing the galvanic mechanism of the leaching reactions. It is well known that galvanic effects occur when two sulfide minerals, differing in rest potentials, are coupled together in a solution that acts as an electrolyte [100–103].

In the mentioned study, it was found that copper and silver only started to dissolve in the presence of MnO_2, thus confirming the galvanic interactions between Cu and Ag-phases and MnO_2. The probable galvanic couples formed between primary sulfide minerals of the SMS ore (chalcopyrite, isocubanite, sphalerite, pyrite/marcasite), and MnO_2, acting as a cathode and anode, respectively. The observed leaching rates were dependent on the addition and dosage of MnO_2, but reached only 80 and 35% after 24h, for Cu and Ag, respectively.

Another idea of Kowalczuk's team was presented in the work [79], aiming at simultaneous leaching of massive sulfides and polymetallic nodules. Leaching was performed with the use of H_2SO_4 and $NaCl$, at 80 °C, and different PMN/SMS ratios. The PMN sample was used as a potential source of MnO_2 (27 wt% of Mn)—an oxidant for metals leaching from the SMS sample. The results indicated that leaching of the two types of marine resources enabled high dissolution rates of metals; copper, manganese, and nickel were almost completely extracted, whereas zinc extraction yielded around 85% after 48 h of leaching. In addition to that, the authors showed a few concepts for utilization of leaching residue, i.e., in ceramic production, oil and gas drilling, or metal adsorbents.

In all of the above-discussed works, the common procedure on the material preparation before leaching operation was drying at room temperature (for the SMS sample) or 60 °C (for the PMN sample), and crushing followed by sieving, to obtain the particle size fraction <50 µm. Such a procedure enables the liberation of sulfides from mineral intergrowths.

Other available research works on leaching of SMS deposits are rather scarce. The investigation presented in [80] simulates natural leaching of fine particulate SMS materials suspended in the seawater, during in situ mining, at the stage of returning the material to the ocean after ship-board processing. In this study, the SMS sample from the Trans-Atlantic Geotraverse (TAG) active mound was leached in artificial seawater at 12 °C, to find the effect of copper, iron, and lead leachability. The main mineral phases indicated were: pyrite/marcasite, chalcopyrite, covellite, Fe oxides/oxyhydroxides, and quartz. Before leaching, the sample was crushed and sieved to receive fractions below 45 µm. The concentrations of Cu and Pb in the leachate were at ppb levels, and leached Fe underwent the hydrolysis and formed Fe oxyhydroxides. Additionally, it was stated that the factors that control the leaching process are mineralogy and/or galvanic interactions formed between mineral phases in a seawater environment.

The pyrometallurgical approach to recovery of metals from SMS deposits was presented in [104]. In this work, investigators proposed a two-stage process involving smelting operation and further converting of the produced matte phase. Both operations were run at the temperature of 1350 °C. The SMS sample used in this study had mineralogy typical for seafloor sulfides (bornite, chalcopyrite, chalcocite, marcasite, pyrite, sphalerite) and contained around 39 wt% and 21 wt% of Cu and Fe, respectively. The produced converting matte phase contained ca. 99 wt% of Cu, and the overall copper recovery for both process steps was ca. 96%.

4. Downstream Processing

Many of the hydrometallurgical methods proposed so far concentrate on the extraction of all metals from polymetallic nodules, and only a few are concerned with the selective removal of impurities [18,50,51]. This is an important step because the presence of impurities complicates the process of recovering metals in a pure form.

The common impurity in pregnant leach solutions (PLS), posing a huge problem in metallurgical circuits, is iron. Its extraction is typically between 30% and 90%, sometimes ending in complete dissolution [56,60,72,94]. It is especially frequent when Fe salts are introduced as reducing agents or in bioleaching methods. However, the latter can also effectively reject iron [70], but usually, it is compensated by long extraction times (i.e., 90 days) or low recovery of coexisting metals.

In [44] authors compared the efficiency, operating, and capital costs of five selected processes described in the literature: (a) reduction and ammoniacal leaching, (b) CUPRION ammoniacal leaching, (c) high-temperature and high-pressure sulfuric acid leaching, (d) reduction and hydrochloric acid leaching, and (e) smelting and sulfuric acid leaching. In general, pyrometallurgical processes were found to be less beneficial than hydrometallurgical processes, due to high energy consumption (due to high moisture content). However, the smelting process can be advantageous, because valuable metals could be effectively concentrated in the alloy phase and separated from the impurities, which remain in the slag phase. It also turned out that it was better to combine smelting with leaching in HCl acid instead of H_2SO_4 because the former allows higher recoveries and recycling of chlorine gas and sulfur.

The great challenge arising in the processing of seafloor materials is the handling of multi-component pregnant leach solutions. Usually, downstream processing uses either precipitation or solvent extraction (SX) and electrowinning (EW) technology (Table 11). Precipitation is mainly destined for iron and manganese removal from PLS before extraction of valuable metals, commonly with aid of basic compounds or by purging with air. In many cases, Co and Ni are precipitated and recovered in the form of mixed sulfides, which after redissolving can be directed to the SX stage. Solvent extraction is useful in the separation of the valuable metals from impurities and accompanying metals present in PLS. The selectivity is usually ensured either by the selection of proper extractant or by applying specified conditions (pH, temperature) for selective stripping. Clean solutions of metallic compounds obtained in the SX step are directed to electrowinning to recover pure metal.

Table 11. Downstream processing methods for marine deposits—examples.

Method	Conditions	Results/Main Conclusions	Ref.
Ammonia leaching with SO_2 → precipitation of Mn(aq), recovery and recycle of ammonia → Cu SXandEW → precipitation of Ni, Co sulfides from Cu raffinate → dissolution of Ni(Co) sulfides in H_2SO_4, O_2 → Ni, Co SXandEW	Cu SX with LIX 84I, Ni SX with D2EHPA Co SX with PC-88A	Medium scale demonstration plant 500 kg/day Metals recoveries: Cu: 85% Ni: 90% Co: 80%	[61]
Leaching with 50% H_2SO_4 → precipitation of Fe and Mn at pH 4.5 → co-extraction of Cu and Ni → selectie stripping of Ni and Cu	Leaching at 80–90 °C; SX: Organic phase: LIX 984N + kerosene Acorga M5640 + kerosene, 5 min., A/O = 1	Quantitative and selective stripping of Cu and Ni; Co in raffinate	[105]
Leach liquor → precipitation of Fe → SX of REE	Fe precipit. with $Ca(OH)_2$, at pH = 3.95 SX: 0,1M sols of D2EHPA, PC88A and Cyanex 272, 5 min, A/O = 1; Stripping with 2M HCl	Highly selective extraction of REE with D2EHPA, in the presence of base metals (Cu, Ni, Co, Mn)	[106]

Table 11. *Cont.*

Method	Conditions	Results/Main Conclusions	Ref.
PMN leach liqor → Fe precipitation → Cu SX → Zn SX → Mn SX, Co scrubbing from loaded organic → Ni SX	Fe(II) oxidation with 0.5% H_2O_2, precipit. With $Ca(OH)_2$; Cu SX: 10% LIX84I + kerosene, A/O = 6/1; Zn SX: 0.02 MD2EHPA + kerosene, A/O = 1 Mn SX: 1M NaD2EHPA + kerosene Ni SX: 0.15 M NaD2EHPA + kerosene 30oC, A/O = 1, 5 min; stripping with H_2SO_4	Metals recoveries: Cu: ~100% Zn(II): 99.6% Mn(II): 99.9% Ni(II): 99.3%	[107]
Synthetic solution (equivalent to sea nodule leach liquor) → Mo SX → crystallization → thermal decomposition	Mo SX: 10% *v/v* Alamine 304–1 + kerosene, 5 min., 25 °C, stripping with NH_4OH + $(NH_4)_2CO_3$; Crystallization of $(NH_4)_4Mo_2O_6$; decomposition at 400 °C to MoO_3	purity of $(NH_4)_4Mo_2O_6$ and MoO_3—99.9%	[108]
PMN leaching liquor → Precipitation of Fe → * precipitation of Co, Ni sulfides → dissolution in H_2SO_4 → Co, Ni SX → Co precipitation → roasting → Co_2O_3 → Ni recovery → $NiSO_4·7H_2O$ precipitation of Mn from * → dissolution in 0.5 M H_2SO_4 and crystallization of $MnSO_4·H_2O$	Impurities (Mn, Fe, Cu, Zn) SX: 15% D2EHPA + kerosene Co, Ni SX: 25% P507 (neutralized) kerosene, 27 °C, Co precipitation with oxalic acid and ammonium oxalate at 65 °C; Roasting at 700 °C → Co_2O_3	Fe precipitates as jarosite Metals recoveries: Mn: 85%; Co: 75%; Ni: 78%	[109]
PMN leach liquor → Cu SX → Co, Ni SX from Cu raffinate → Co SX and stripping → Ni SX and stripping	1st stage: Cu SX: 0.3 M Cyanex 923; 2nd stage: Co SX: 0.6 M Cyanex 923; 3rd stage: Ni SX: 0.1M Cyanex 301	Metals recoveries: Co(II), Cu(II), Ni(II): 90%	[58]
PMN leach liquor →Co-extraction of Cu and Ni → selective stripping of Ni → Ni SX-EW	Cu,Ni SX: LIX 64N + kerosene, A/O = 1; scrubbing with ammonia, Ni stripping in 6 stages, Ni EW 12h, 61 °C	Purity of electrocrystallized Ni: 99.82% The effect of organic phase deletorious for the quality of electrowinning product	[110]

* Mn precipitation from the filtrate after the Co, Ni sulfides precipitation.

Mittal and Sen [61] demonstrated the medium scale plant for processing PMN (500 kg/day) by reductive pressure leaching with ammonia and SO_2, combined with SX-EW sections for Cu, Co, and Ni separation. In the leaching stage, Cu, Co, Ni, and Mn are dissolved. Manganese is precipitated and separated as MnO_2 and free ammonia is recovered in a stripper and recycled to the autoclave. Then, Cu is selectively removed from PLS by SX-EW and the raffinate containing Co and Ni is subjected to sulfide precipitation. Precipitated sulfides of Co, Ni, and minor impurities (Cu, Zn, Fe) are dissolved in dilute sulfuric acid. The obtained leach solution is directed to Co-Ni separation by SX, after removal of dissolved impurities. Similar to copper, pure Co and Ni are produced through electro-winning. The recoveries of Cu, Ni, and Co were respectively equal to 85, 90, and 80%.

Solvent extraction of Cu, Co, and Ni from acidic PLS was described in [105]. First, nodules were leached in the mixture of sulfuric acid and activated charcoal at 80–90 °C. Iron and manganese present in the leachate were precipitated before extraction. Commercial acidic extractants (LIX 984, ACORGA M5640) were used for the co-extraction of Cu and Ni. Cobalt remained in raffinate, and the selective recovery of Cu and Ni from the organic phase was based on sequential stripping with an appropriate sulfuric acid solution.

Extraction of Cu, Co, and Ni from HCl leach solution of PMN was studied in [58], together with their mutual separation and separation from accompanying metal ions, such as Ti(IV), Al(III), Fe(III), Mn(II), and Zn(II). The tested extractant compounds (Cyanex 923 and Cyanex 301) were able to efficiently (>90%) extract Cu(II), Co(II), and Ni(II) from highly acidic media (4M). Yet, the residual organic phases still contain some undefined amount of co-existing ions.

Except for primary metals, rare earth elements are often found in seafloor materials [111,112], enhancing their economic value. However, it is critical to separate REE from base metals, and this might be achieved by SX. In [106], the authors used organophosphorus acids to extract REE from a solution after PMN leaching. The main impurity—iron—was precipitated before extraction, leaving in the solution the remaining metals, i.e., Cu, Ni, Co, and Mn. The application of D2EHPA (di -2-Ethylhexyl phosphoric acid) resulted in selective REE extraction with an overall efficiency of around 97%.

5. Conclusions and Future Perspectives

Oceans are known to have particularly rich deposits of various types of minerals (polymetallic nodules, crusts, sulfides). The SMS deposits can be compared to land-based ore deposits and appear to be typically smaller but slightly higher-grade than most sulfide deposits mined today. More similarities may be seen with the volcanogenic massive sulfides (VMS). Additionally, many critical raw materials (CRM) from global crusts and nodule deposits will most likely be relevant as a supplement to land-based mining, rather than a substitute.

A huge effort has been made for exploitation and extraction of polymetallic nodules; however, for other types of marine minerals both mining and processing operations are at very low levels of readiness. Currently, there is no elaborated plan for the processing of SMS and CRC, which is caused by the limited geochemical data and number of mined samples to be tested. In this paper we collected information from peer-reviewed and other technical literature on historical and recent developments in the processing of marine minerals, particularly polymetallic nodules (PMN), polymetallic crusts (CRC), and seafloor massive sulfides (SMS). The proposed processing operations are based on metallurgical approaches applied to terrestrial deposits. However, marine minerals differ significantly from the land resources, and thus the processing will be different as well. Successful methods must consider the characteristics of marine minerals; that is chemistry, mineralogy, porosity, and water content.

Due to the complex mineralogy of polymetallic nodules and polymetallic crusts, separation and enrichment of metals of interest from the other components by physical beneficiation methods would tend to be energy intensive in comparison with the terrestrial resources. Thus, PMN and CRC would be treated either hydrometallurgically or pyrometallurgically, or a combination of both. Currently, the most effective method of polymetallic marine minerals treatment is a combination of pyro- and hydrometallurgical methods. However, the development of new and green methods for metallurgical processing of PMN and CRC is necessary.

The choice of processing routes for seafloor massive sulfides will strongly depend on the metal content and mineralogy, and thus economic causes. Three metallurgical concepts for processing for high-grade, medium-grade and low-grade SMS might be suggested. Rich samples with high concentrations of copper, zinc, and minor elements would be directly processed by pyrometallurgy. The medium-grade ore, which does not meet the pyrometallurgical requirements, would first be enriched by conventional processing, and then either pyro- or hydro- metallurgically treated. Due to complex mineralogy, rapid oxidation of sulfide minerals, and porosity, new flotation reagents would have to be applied here. Finally, a low-grade ore with a complex mineralogy will be directed to hydrometallurgical treatment.

The current processing tests have been performed on the land. However, future preprocessing such as pre-grinding (e.g., comminution) and preconcentration (e.g., sensor-based sorting) might take place on the seabed, disposing of non-toxic associated waste rocks, and only concentrated material could be lifted to the vessel. Such pre-processing might reduce the quantity of waste material lifted from the ocean bed and then reported to the next processing stage. Such a vessel would either transport the ore to the production plant on land or the green production would take place on the vessel. Enrichment by pre-concentration on the seabed and processing on the vessel would significantly decrease the operational costs. The choice of processing plant will strongly depend on economic and technological considerations.

Author Contributions: Conceptualization, K.O.; P.B.K., investigation, K.O., K.A., P.B.K.; writing—original draft preparation, K.O.; writing—review and editing, K.O., P.B.K., K.A.; visualization, K.O., P.B.K., K.A.; supervision, P.B.K.; funding acquisition, K.O. All authors have read and agreed to the published version of the manuscript.

Funding: This research received no external funding.

Acknowledgments: This research was partially financed by statutory activity subsidies from the Polish Ministry of Science and Higher Education for the Faculty of Chemistry of Wroclaw University of Science and Technology.

Conflicts of Interest: The authors declare no conflict of interest.

References

1. Glasby, G.P.; Li, J.; Sun, Z. Deep-Sea Nodules and Co-rich Mn Crusts. *Mar. Georesources Geotechnol.* **2015**, *33*, 72–78. [CrossRef]
2. Molemaker, R.-J.; Gille, J.; Kantor, E.; van Schijndel, M.; Pauer, A.; von Schickfus, M.-T.; van Elswijk, J.; Beerman, R.; Hodgson, S.; Beaudoin, Y.; et al. *Study to Investigate the State of Knowledge of Deep-Sea Mining*; ECORYS: Brussel, Belgium, 2014.
3. Georgiou, D.; Papangelakis, V.G. Behavior of cobalt during sulphuric acid pressure leaching of a limonitic laterite. *Hydrometallurgy* **2009**, *100*, 34–40. [CrossRef]
4. ISA brochure, Seabed Technology. Available online: https://isa.org.jm/files/files/documents/ENG10.pdf (accessed on 14 November 2021).
5. Sharma, R. Environmetal Issues of Deep-Sea Mining. *Proced. Earth Plan. Sc.* **2015**, *11*, 204–211. [CrossRef]
6. Mestre, N.C.; Rocha, T.L.; Canals, M.; Cardoso, C.; Danovaro, R.; Dell'Anno, A.; Gambi, C.; Regoli, F.; Sanchez-Vidal, A.; João Bebianno, M. Environmental hazard assessment of a marine mine tailings deposit site and potential implications for deep-sea mining. *Environ. Pollut.* **2017**, *228*, 169–178. [CrossRef]
7. Abramowski, T. Environment Protection Policy and Monitoring Systems for Polymetallic Nodules Exploitation. *New Trends Prod. Eng.* **2018**, *1*, 523–529. [CrossRef]
8. Monhemius, J. The extractive metallurgy of deep-sea manganese nodules. In *Topics in Non-ferrous Extractive Metallurgy*; Burkin, A.R., Ed.; Blackwell Scientific Publications: Oxford, UK, 1980; pp. 42–69.
9. Spearman, J.; Taylor, J.; Crossouard, N.; Cooper, A.; Turnbull, M.; Manning, A.; Lee, M.; Murton, B. Measurement and modelling of deep sea sediment plumes and implications for deep sea mining. *Sci. Rep.* **2020**, *10*, 5075. [CrossRef]
10. Petersen, S.; Krätschell, A.; Augustin, N.; Jamieson, J.; Hein, J.R.; Hannington, M.D. News from the seabed—Geological characteristics and resource potential of deep-sea mineral resources. *Mar. Policy* **2016**, *70*, 175–187. [CrossRef]
11. Fuerstenau, D.W.; Han, K.N. Metallurgy and Processing of Marine Manganese Nodules. Miner. Process. *Extr. Metall. Rev.* **1983**, *1*, 1–83. [CrossRef]
12. Lusty, P.A.J.; Murton, B.J. Deep-ocean mineral deposits could make a significant contribution to future raw material supply. *Elements* **2018**, *14*, 301–306. [CrossRef]
13. Hein, J.R.; Koschinsky, A. *Deep-Ocean Ferromanganese Crusts and Nodules, in Treatise on Geochemistry*; Elsevier: Amsterdam, The Netherlands, 2014; pp. 273–291. [CrossRef]
14. Navneet, S.; Randhawa, J.; Hait, R.; Kumar, J. A brief overview on manganese nodules processing signifying the detail in the Indian context highlighting the international scenario. *Hydrometallurgy* **2016**, *165 Pt 1*, 166–181.
15. Hein, J.R.; Koschinsky, A.; Kuhn, T. Deep- ocean polymetallic nodules as a resource for critical materials. *Nat. Rev. Earth Environ.* **2020**, *1*, 158–169. [CrossRef]
16. The Manganese Nodule Belt of the Pacific Ocean. In *Geological Environment, Nodule Formation, and Mining Aspects*; Halbach, P.; Friedrich, G.; Von Stackelberg, U. (Eds.) Ferdinand Enke: Stuttgart, Germany, 1988; ISBN 3 432 96381 5.
17. Hoover, M.; Han, K.N.; Fuerstenau, D.W. Segregation roasting of nickel, copper and cobalt from manganese nodules. *Int. J. Miner. Process.* **1975**, *2*, 173–185. [CrossRef]
18. Cruickshank, M.J. Marine mining. In *General Geology. Encyclopedia of Earth Science*; Springer: Boston, MA, USA, 1988. [CrossRef]
19. Haynes, B.W.; Law, S.L.; Barron, D.C.; Kramer, G.W.; Maeda, R.; Magyar, J. Pacific manganese nodules: Characterization and processing. *Bull. United States Bur. Mines* **1985**, *674*, 43.
20. Wegorzewski, A.V.; Köpcke, M.; Kuhn, T.; Sitnikova, M.A.; Wotruba, H. Thermal Pre-Treatment of Polymetallic Nodules to Create Metal (Ni, Cu, Co)-Rich Individual Particles for Further Processing. *Minerals* **2018**, *8*, 523. [CrossRef]
21. Hein, J. Cobalt-rich ferromanganese crusts: Global distribution, composition, origin and research activities. In *ISA Technical Study: No.2 Polymetallic Massive Sulphides and Cobalt-Rich Ferromanganese Crusts: Status and Prospects*; International Seabed Authority: Kingston, Jamaica, 2002.

22. Hein, J.R. Co-Rich Manganese Crusts. In *Encyclopedia of Marine Geosciences*; Springer: Dordrecht, The Netherlands. [CrossRef]
23. SPC. *Deep Sea Minerals: Cobalt-Rch Ferromanganese Crusts, a Physical, Biological, Environmental, and Technical Review*; Baker, E., Beaudoin, Y., Eds.; Secretariat of the Pacific Community: Nouméa, NC, USA, 2013; Volume 1C.
24. Hein, J.R.; Schulz, M.S.; Kang, J.-K. Insular and Submarine Ferromanganese Mineralization of the Tonga-Lau Region. *Mar. Min.* **1990**, *9*, 305–354.
25. Hannington, M.D.; De Ronde, C.E.J.; Petersen, S. Sea-Floor Tectonics and Submarine Hydrothermal Systems. In *One Hundredth Anniversary Volume*; Society of Economic Geologists: Littleton, CO, USA, 2005.
26. Boschen, R.E.; Rowden, A.A.; Clark, M.R.; Gardner, J.P.A. Mining of deep-sea seafloor massive sulfides: A review of the deposits, their benthic communities, impacts from mining, regulatory frameworks and management strategies. *Ocean Coast. Manag.* **2013**, *84*, 54–67. [CrossRef]
27. Hannington, M.; Jamieson, J.; Monecke, T.; Petersen, S.; Beaulieu, S. The abundance of seafloor massive sulfide deposits. *Geology* **2011**, *39*, 1155–1158. [CrossRef]
28. Spagnoli, G.; Jahnb, A.; Halbach, P. First results regarding the influence of mineralogy on the mechanical properties of seafloor massive sulfide samples. *Eng. Geol.* **2016**, *214*, 127–135. [CrossRef]
29. GRID- Arendal. 2014. Available online: https://www.grida.no/resources/8166 (accessed on 17 December 2021).
30. Snook, B.; Drivenes, K.; Rollinson, G.K.; Aasly, K. Characterisation of Mineralised Material from the Loki's Castle Hydrothermal Vent on the Mohn's Ridge. *Minerals* **2018**, *8*, 576. [CrossRef]
31. Cherkashov, G. Prospecting. In *Deep-Sea Mining: Resource Potential, Technical and Environmental Considerations*; Sharma, R., Ed.; Springer International Publishing: Cham, Switzerland, 2017; pp. 143–164. [CrossRef]
32. The Norwegian Petroleum Directorate, Det første funnet i Norskehavet. *Nor. Sokkel* **2020**, *6*. (In Norwegian)
33. Nozaki, T.; Tokumaru, A.; Takaya, Y.; Kato, Y.; Suzuki, K.; Urabe, T. Major and trace element compositions and resource potential of ferromanganese crust at Takuyo Daigo Seamount, Northwestern Pacific Ocean. *Geochem. J.* **2016**, *50*, 527–537. [CrossRef]
34. Hannington, M.; Jamieson, J.; Monecke, T.; Petersen, S. Modern Sea-Floor Massive Sulfides and Base Metal Resources Toward an Estimate of Global Sea-Floor Massive Sulfide Potential. In *The Challenge of Finding New Mineral Resources. Global Metallogeny, Innovative Exploration, and New Discoveries*; Society of Economic Geologists: Littleton, CO, USA, 2010.
35. Su, K.; Ma, X.; Parianos, J.; Zhao, B. Thermodynamic and Experimental Study on Efficient Extraction of Valuable Metals from Polymetallic Nodules. *Minerals* **2020**, *10*, 360. [CrossRef]
36. Drakshayani, D.N.; Sankar, C.; Mallya, R.M. The Reduction Of Manganese Nodules By Hydrogen. *Thermochim. Acta* **1989**, *144*, 313–328. [CrossRef]
37. Zhao, F.; Jiang, X.; Wang, S.; Feng, L.; Li, D. The Recovery of Valuable Metals from Ocean Polymetallic Nodules Using Solid-State Metalized Reduction Technology. *Minerals* **2020**, *10*, 20. [CrossRef]
38. Sommerfeld, M.; Friedmann, D.; Kuhn, T.; Friedrich, B. "Zero-Waste": A Sustainable Approach on Pyrometallurgical Processing of Manganese Nodule Slags. *Minerals* **2018**, *8*, 544. [CrossRef]
39. Kmetova, D.; Stofko, M.; Kmet, S. Ammoniacal Leaching For Extraction Of Non-Ferrous Metals From Deep-Sea Nodules. *Int. J. Miner. Process.* **1985**, *15*, 145–153. [CrossRef]
40. Jana, R.K.; Akerkar, D.D. Studies of the metal–ammonia–carbondioxide–water system in extraction metallurgy of polymetallic sea nodules. *Hydrometallurgy* **1989**, *22*, 363–378. [CrossRef]
41. Wilder, T.C. Two Stage Selective Leaching of Copper and Nickel from Complex Ore. U.S. Patent 3,736,125, 29 May 1973.
42. Mishra, D.; Srivastava, R.R.; Sahu, K.K.; Singh, T.B.; Jana, R.K. Leaching of roast-reduced manganese nodules in NH_3–$(NH_4)_2CO_3$ medium. *Hydrometallurgy* **2011**, *109*, 215–220. [CrossRef]
43. Deng, X.-Y.; He, D.-S.; Chi, R.-A.; Xiao, C.-Q.; Hu, J.-G. The Reduction Behavior of Ocean Manganese Nodules by Pyrolysis Technology Using Sawdust as the Reductant. *Minerals* **2020**, *10*, 850. [CrossRef]
44. Kohga, T.; Imamura, M.; Takahashi, J.; Tanaka, N.; Nishizawa, T. Recovering iron, manganese, copper, cobalt, and high-purity nickel from sea nodules. *JOM* **1995**, *47*, 40–43. [CrossRef]
45. Kane, W.S.; Cardwell, P.H. Recovery of Metal Values from Ocean Floor Nodules. U.S. Patent 3,950,486, 14 August 1973.
46. Sridhar, R.; Jones, W.E.; Warner, J.S. Extraction of copper, nickel and cobalt from sea nodules. *J. Met.* **1976**, 32–37. [CrossRef]
47. Keber, S.; Brückner, L.; Elwert, T.; Kuhn, T. Concept for a Hydrometallurgical Processing of a Copper-Cobalt-Nickel Alloy Made from Manganese Nodules. *Chem. Ing. Tech.* **2020**, *92*, 1–9.
48. UNOET. *Analysis of Processing Technology for Manganese Nodules, Sea Bed Mineral Series*; Graham & Trotman Limited: London, UK, 1968; Volume 3, pp. 31–39.
49. Barik, R.; Sanjay, K.; Mishra, B.K.; Mohapatra, M. Micellar mediated selective leaching of manganese nodule in high temperature sulfuric acid medium. *Hydrometallurgy* **2016**, *165 Pt 1*, 44–50. [CrossRef]
50. Anand, S.; Das, S.C.; Das, R.P.; Jena, P.K. Leaching of Manganese Nodules at Elevated Temperature and Pressure in the Presence of oxygen. *Hydrometallurgy* **1988**, *20*, 155–168. [CrossRef]
51. Han, K.N.; Fuerstenau, D.W. Acid leaching of ocean manganese nodules at elevated temperatures. *Int. J. Miner. Process.* **1975**, *2*, 63–171. [CrossRef]
52. Vua, H.; Jandova, J.; Lisa, K.; Vranka, F. Leaching of manganese deep ocean nodules in $FeSO4$–$H2SO4$–$H2O$ solutions. *Hydrometallurgy* **2005**, *77*, 147–153. [CrossRef]

53. Zhang, Y.; Liu, Q.; Sun, C. Sulfuric acid leaching of ocean manganese nodules using aromatic amines as reducing agents. *Miner. Eng.* **2001**, *14*, 539–542. [CrossRef]

54. Zhang, Y.; Liu, Q.; Sun, C. Sulfuric acid leaching of ocean manganese nodules using phenols as reducing agents. *Miner. Eng.* **2001**, *14*, 525–537. [CrossRef]

55. Pahlman, J.E.; Khalafalla, S.E. Selective Recovery of Nickel, Cobalt, Manganese from Sea Nodules with Sulfurous Acid. U.S. Patent US 4, 138, 465 (US 860249 (771213)), 6 February 1979.

56. Kanungo, S.B.; Jena, P.K. Studies on the dissolution of metal values in manganese nodules of Indian Ocean origin in dilute hydrochloric acid. *Hydrometallurgy* **1988**, *21*, 23–39. [CrossRef]

57. Parhi, P.K.; Park, K.H.; Nam, C.W.; Park, J.T.; Barik, S.P. Extraction of rare earth metals from deep sea nodule using H_2SO_4 solution. *Int. J. Miner. Process.* **2013**, *119*, 89–92. [CrossRef]

58. Gupta, B.; Deep, A.; Singh, V.; Tandon, S.N. Recovery of cobalt, nickel, and copper from sea nodules by their extraction with alkylphosphines. *Hydrometallurgy* **2003**, *70*, 121–129. [CrossRef]

59. Kumari, A.; Natarajan, K.A. Electroleaching of polymetallic ocean nodules to recover copper, nickel and cobalt. *Miner. Eng.* **2001**, *14*, 877–886. [CrossRef]

60. Wang, C.-Y.; Qiu, D.-F.; Chen, Y.-Q. Slurry electrolysis of ocean polymetallic nodule. *Trans. Nonferrous Met. Soc. China* **2010**, *20*, 60–64. [CrossRef]

61. Mittal, N.K.; Sen, P.K. India's first medium scale demonstration plant for treating poly-metallic nodules. *Miner. Eng.* **2003**, *16*, 865–868. [CrossRef]

62. Wang, Y.; Li, Z.; Li, H. A new process for leaching metal values from ocean polymetallic nodules. *Miner. Eng.* **2005**, *18*, 1093–1098. [CrossRef]

63. Asai, S.; Konishi, Y.; Takasaka, Y. Bioleaching of rare metals from manganese nodules by Thiobacillus ferrooxidans. *Miner. Process. Extr. Metall. Rev.* **1995**, *15*, 140–141. [CrossRef]

64. Konishi, Y.; Asai, S.; Sawada, Y. Leaching of marine manganese nodules by acidophilic bacteria growing on elemental sulfur. *Metall. Mater. Trans. B* **1997**, *28*, 25–32. [CrossRef]

65. Li, H.; Feng, Y.; Lu, S.; Du, Z. Bioleaching of valuable metals from marine nodules under anaerobic condition. *Miner. Eng.* **2005**, *18*, 1421–1422. [CrossRef]

66. Mukherjee, A.; Raichur, A.M.; Modak, J.M.; Natarajan, K.A. Bioprocessing of polymetallic Indian Ocean nodules using a marine isolate. *Miner. Process. Extr. Metall. Rev.* **2004**, *25*, 91–127. [CrossRef]

67. Kumari, A.; Natarajan, K.A. Development of a clean bioelectrochemical process for leaching of ocean manganese nodules. *Miner. Eng.* **2002**, *15*, 103–106.

68. Kumari, A.; Natarajan, K.A. Electrochemical aspects of leaching of ocean nodules in the presence and absence of microorganisms. *Int. J. Miner. Process.* **2002**, *66*, 29–47. [CrossRef]

69. Fujimoto, J.; Tanaka, K.; Watanabe, N.; Takahashi, Y. Simultaneous recovery and separation of rare earth elements in ferromanganese nodules by using Shewanella putrefaciens. *Hydrometallurgy* **2016**, *166*, 80–86. [CrossRef]

70. Aishvarya, V.; Mishra, G.; Pradhan, N.; Ghosh, M.K. Bioleaching of Indian Ocean nodules with in situ iron precipitation by anaerobic Mn reducing consortia. *Hydrometallurgy* **2016**, *166*, 130–135. [CrossRef]

71. Mehta, K.D.; Das, C.; Kumar, R.; Pandey, B.D.; Mehrotra, S.P. Effect of mechano-chemical activation on bioleaching of Indian Ocean nodules by a fungus. *Miner. Eng.* **2010**, *23*, 1207–1212. [CrossRef]

72. Mehta, K.D.; Das, C.; Pandey, B.D. Leaching of copper, nickel and cobalt from Indian Ocean manganese nodules by *Aspergillus niger*. *Hydrometallurgy* **2010**, *105*, 89–95. [CrossRef]

73. Beolchini, F.; Becci, A.; Barone, G.; Amato, A.; Hekeu, M.; Danovaro, R.; Dell'Anno, A. High fungal-mediated leaching efficiency of valuable metals from deep-sea polymetallic nodules. *Environ. Technol. Innov.* **2020**, *20*. [CrossRef]

74. Marino, E.; González, F.J.; Kuhn, T.; Madureira, P.; Wegorzewski, A.V.; Mirao, J.; Medialdea, T.; Oeser, M.; Miguel, C.; Reyes, J.; et al. Hydrogenetic, Diagenetic and Hydrothermal Processes Forming Ferromanganese Crusts in the Canary Island Seamounts and Their Influence in the Metal Recovery Rate with Hydrometallurgical Methods. *Minerals* **2019**, *9*, 439. [CrossRef]

75. Lugovskaya, I.G.; Dubinchuk, V.T.; Baturin, G.N. Composition of technological sample of ferromanganese crusts from seamounts and products of its processing. *Lithol. Miner. Resour.* **2007**, *42*, 515–522. [CrossRef]

76. Allen, J.P.; Abercrombie, H.L.; Rice, D.A. Leaching and recovery of metals from cobalt-rich manganese ocean crust. *Mining Metall. Explor.* **1991**, *8*, 97–104. [CrossRef]

77. Kowalczuk, P.B.; Snook, B.; Kleiv, R.A.; Aasly, K. Efficient extraction of copper and zinc from seafloor massive sulphide rock samples from the Loki's Castle area at the Arctic Mid-Ocean Ridge. *Miner. Eng.* **2018**, *115*, 106–116. [CrossRef]

78. Kowalczuk, P.B.; Oliric Manaig, D.; Drivenes, K.; Snook, B.; Aasly, K.; Kleiv, R.A. Galvanic Leaching of Seafloor Massive Sulphides Using MnO2 in H2SO4-NaCl Media. *Minerals* **2018**, *8*, 235. [CrossRef]

79. Kowalczuk, P.B.; Bouzahzah, H.; Kleiv, R.A.; Aasly, K. Simultaneous Leaching of Seafloor Massive Sulfides and Polymetallic Nodules. *Minerals* **2019**, *9*, 482. [CrossRef]

80. Fallon, E.K.; Niehorster, E.; Brooker, R.A.; Scott, T.B. Experimental leaching of massive sulphide from TAG active hydrothermal mound and implications for seafloor mining. *Mar. Pollut. Bull.* **2018**, *126*, 501–515. [CrossRef]

81. Nakajima, Y.; Sato, T.; Thornton, B.; Dodbina, G.; Fujita, T. *Development of Seafloor Mineral Processing for Seafloor Massive Sulfides*; IEEE: Kobe, Japan, 2016; pp. 119–126.

82. Aikawa, K.; Ito, M.; Kusano, A.; Park, I.; Oki, T.; Takahashi, T.; Furuya, H.; Hiroyoshi, N. Flotation of Seafloor Massive Sulfide Ores: Combination of Surface Cleaning and Deactivation of Lead-Activated Sphalerite to Improve the Separation Efficiency of chalcopyrite and Sphalerite. *Metals* **2021**, *11*, 253. [CrossRef]

83. Lipton, I. Mineral Resource Estimate, Solwara project, Bismarck Sea, PNG, Nautilus Minerals Nuigini Limited 2012, SL01-NSG-RPT-7020-001 Rev 1—Golden Resource Report; 2012.

84. Agarwal, J.C.; Beecher, N.; Davies, D.S.; Hubred, G.L.; Kakaria, V.K.; Kust, R.N. Processing of ocean nodules: A technical and economic review. *JOM* **1976**, *28*, 24–31. [CrossRef]

85. Parida, K.M.; Mallick, S.; Mohapatra, B.K.; Misra, V.N. Studies on manganese-nodule leached residues: Physicochemical characterization and its adsorption behavior toward Ni^{2+} in aqueous system. *J. Colloid Interf. Sci.* **2004**, *277*, 48–54. [CrossRef] [PubMed]

86. Mallick, S.; Dash, S.S.; Parida, K.M. Adsorption of hexavalent chromium on manganese nodule leached residue obtained from NH_3–SO_2 leaching. *J. Colloid Interf. Sci.* **2006**, *297*, 419–425. [CrossRef] [PubMed]

87. Parida, K.M.; Dash, S.S.; Mallik, S.; Das, J. Effect of heat treatment on the physico-chemical properties and catalytic activity of manganese nodules leached residue towards decomposition of hydrogen peroxide. *J. Colloid Interf. Sci.* **2005**, *290*, 431–436. [CrossRef] [PubMed]

88. Bhattacharjee, S.; Chakrabarty, S.; Maity, S.; Kar, S.; Thakur, P.; Bhattacharyya, G. Removal of lead from contaminated water bodies using sea nodule as an adsorbent. *Water Res.* **2003**, *37*, 3954–3966. [CrossRef]

89. Pan, L.; Zhang, A.-B.; Sun, J.; Ye, Y.; Chen, X.-G.; Xia, M.-S. Application of ocean manganese nodules for the adsorption of potassium ions from seawater. *Miner. Eng.* **2013**, *49*, 121–127. [CrossRef]

90. Havlik, T.; Laubertova, M.; Miskufova, A.; Kondas, J.; Vranka, F. Extraction of copper, zinc, nickel and cobalt in acid oxidative leaching of chalcopyrite at the presence of deep-sea manganese nodules as oxidant. *Hydrometallurgy* **2005**, *77*, 51–59. [CrossRef]

91. Toro, N.; Pérez, K.; Saldaña, M.; Jeldres, R.I.; Jeldres, M.; Cánovas, M. Dissolution of pure chalcopyrite with manganese nodules and waste water. *J. Mater. Rest. Technol.* **2020**, *9*, 798–805. [CrossRef]

92. Devi, N.B.; Madhuchhanda, M.; Rath, P.C.; Srinivasa Rao, K.; Paramguru, R.K. Simultaneous leaching of a deep-seamanganese nodule and chalcopyrite in hydrochloric acid. *Metall. Mater. Trans. B* **2001**, *32*, 778–784. [CrossRef]

93. Das, R.P.; Anand, S. *Aqueous Reduction of Polymetallic Nodule For Metal Extraction, Proceedings of the Second Ocean Mining Symposium*; ISOPE: Seoul, Korea, 1997; pp. 165–171.

94. Das, R.P.; Anand, S. Metallurgical Processing of Polymetallic Ocean Nodules. In *Deep-Sea Mining*; Sharma, R., Ed.; Springer, Resource Potential, Technical and Environmental Considerations: Cham, Switzerland, 2017. [CrossRef]

95. Charewicz, W.A.; Chaoyin, Z.; Chmielewski, T. The leaching behavior of ocean polymetallic nodules in chloride solutions. *Physicochem. Probl. Miner. Process.* **2001**, *35*, 55–66.

96. Jana, R.K.; Singh, D.D.N.; Roy, S.K. Alcohol modified hydrochloric acid leaching of sea nodules. *Hydrometallurgy* **1995**, *38*, 289–298. [CrossRef]

97. Abramovski, T.; Stefanova, V.; Causse, P.R.; Romanchuk, A. Technologies for the processing of polymetallic nodules from Clarion Clipperton Zone in the Pacific Ocean. *J. Chem. Technol. Metall.* **2017**, *52*, 2.

98. Hirt, W.C.; Rice, D.A.; Shirts, M.B. Flotation of cobalt-rich ferromanganese crust from the pacific ocean. *Miner. Eng.* **1991**, *4*, 535–551.

99. Ito, M.; Tsunekawa, M.; Yamaguchi, E.; Sekimura, K.; Kashiwaya, K.; Hori, K.; Hiroyoshi, N. Estimation of degree of liberation in a coarse crushed product of cobalt-rich ferromanganese crust/nodules and its gravity separation. *Int. J. Miner. Process.* **2008**, *87*, 100–105. [CrossRef]

100. Majima, H.; Peters, E. Electrochemistry of sulphide dissolution in hydrometallurgical systems. *Int. Miner. Processing Congr.* **1968**, *8*, 13–14.

101. Hiskey, J.; Wadsworth, M. Galvanic conversion of chalcopyrite. *Metall. Trans. B* **1975**, *184*, 183–190. [CrossRef]

102. Mehta, A.P.; Murr, L.E. Fundamental studies of the contribution of galvanic interaction to acid-bacterial leaching of mixed metal sulfides. *Hydrometallurgy* **1983**, *9*, 235–256. [CrossRef]

103. Dixon, D.G.; Mayne, D.D.; Baxter, K.G. GALVANOX^{TM}—A novel galvanically-assisted atmospheric leaching technology for copper concentrates. *Can. Metall. Q.* **2008**, *47*, 327–336. [CrossRef]

104. Sommerfeld, M.; Friedmann, D.; Friedrich, B.; Schwarz-Schampera, U. Mineralogical Characterization and Metallurgical Processing of Seafloor Massive Sulphides from The German License Area In The Indian Ocean. In Proceedings of the 58th Annual Conference of Metallurgists (COM) Hosting Copper 2019, Vancouver, BC, Canada, 18–21 August 2019.

105. Sridhar, V.; Verma, J.K. Extraction of copper, nickel and cobalt from the leach liquor of manganese-bearing sea nodules using LIX 984N and ACORGA M5640. *Miner. Eng.* **2011**, *24*, 959–962. [CrossRef]

106. Parhi, P.K.; Park, K.H.; Nam, C.W.; Park, J.T. Liquid-liquid extraction and separation of total rare earth (RE) metals from polymetallic manganese nodule leaching solution. *J. Rare Earths* **2015**, *33*, 207. [CrossRef]

107. Padhan, E.; Sarangi, K.; Subbaiah, T. Recovery of manganese and nickel from polymetallic manganese nodule using commercial extractants. *Int. J. Miner. Process.* **2014**, *126*, 55–61. [CrossRef]

108. Parhi, P.K.; Park, K.-H.; Kim, H.-I.; Park, J.-T. Recovery of molybdenum from the sea nodule leach liquor by solvent extraction using Alamine 304-I. *Hydrometallurgy* **2011**, *105*, 195–200. [CrossRef]

109. Shen, Y.-F.; Xue, W.-Y.; Li, W.; Li, S.-D.; Liu, X.-H. Recovery of Mn^{2+}, Co^{2+} and Ni^{2+} from manganese nodules by redox leaching and solvent extraction. *T. Nonferr. Metal. Soc.* **2007**, *17*, 1105–1111. [CrossRef]

110. Kumar, V.; Pandey, B.D.; Akerkar, D.D. Electrowinning of nickel in the processing of polymetallic sea nodules. *Hydrometallurgy* **1990**, *24*, 189–201. [CrossRef]

111. Zhang, Z.; Du, Y.; Gao, L.; Zhang, Y.; Shi, G.; Liu, C.; Zhang, P.; Duan, X. Enrichment of REEs in polymetallic nodules and crusts and its potential for exploitation. *J. Rare Earths* **2012**, *30*, 621–626. [CrossRef]

112. Pak, S.-J.; Seo, I.; Lee, K.-Y.; Hyeong, K. Rare Earth Elements and Other Critical Metals in Deep Seabed Mineral Deposits: Composition and Implications for Resource Potential. *Minerals* **2019**, *9*, 3. [CrossRef]

 *minerals*

MDPI

Article

The Alkaline Fusion-Hydrothermal Synthesis of Blast Furnace Slag-Based Zeolite (BFSZ): Effect of Crystallization Time

Changxin Li [1,2,*], Xiang Li [1], Qingwu Zhang [1], Li Li [2] and Shuai Wang [3]

[1] College of Safety Science and Engineering, Nanjing Tech University, Nanjing 211816, China; 1905170120@njtech.edu.cn (X.L.); zqw@njtech.edu.cn (Q.Z.)
[2] College of Chemical Engineering and Pharmacy, Jingchu University of Technology, Jingmen 448000, China; lili@jcut.edu.cn
[3] College of Chemistry and Chemical Engineering, Central South University, Changsha 410083, China; wangshuai@csu.edu.cn
* Correspondence: lichangxin2010@njtech.edu.cn or lichangxin20160706@163.com

Abstract: Blast furnace slag (BFS) is usually regarded as a by-product of the steel industry, which can be utilized as raw material for preparing BFS-based zeolite (BFSZ). In this study, BFSZ was successfully prepared from BFS using alkaline fusion-hydrothermal synthesis. Via the analyses by XRD, SEM, EDX, XRF, FT-IR, elemental mapping and BET/BJH methods, BFSZ crystallization was almost complete at 6 h. With a further increase of crystallization time to 8 h, no significant effect on the formation of crystalline phase was found. Meanwhile, the zeolite content Si/Al (Na/Al) molar ratio was highly affected by crystallization time. The main component of BFSZ prepared at 6 h is cubic crystal with developed surface, with particle size around 2 μm. Moreover, further increasing the crystallization time will not significantly influence the size and morphology of BFSZ product.

Keywords: blast furnace slag; conversion; zeolite; characterization; crystallization time

Citation: Li, C.; Li, X.; Zhang, Q.; Li, L.; Wang, S. The Alkaline Fusion-Hydrothermal Synthesis of Blast Furnace Slag-Based Zeolite (BFSZ): Effect of Crystallization Time. *Minerals* **2021**, *11*, 1314. https://doi.org/10.3390/min11121314

Academic Editor: Nikolaos Kantiranis

Received: 5 October 2021
Accepted: 17 November 2021
Published: 25 November 2021

Publisher's Note: MDPI stays neutral with regard to jurisdictional claims in published maps and institutional affiliations.

1. Introduction

According to the data of the National Bureau of Statistics, in 2020, the cumulative output of crude steel in China reached 105.30 million tons, with a cumulative growth of 5.2%. Blast furnace slag (BFS) is usually regarded as a by-product of the steel industry. It is estimated that each ton of iron produced will produce 290 kg BFS [1,2]. There is still a large amount of BFS landfilled in China, because we cannot fully utilize BFS at present [3]. Hence, the landfilled BFS has attracted great attention as regards the pollution of heavy metals and particles in groundwater. Considering this, we urgently need a new way to convert BFS into products of high value-added to meet the requirements of sustainable development in the iron and steel industry.

Recently, the preparation of zeolite from minerals and wastes rich in silicon and aluminum has received great attention, such as ash (fly ash, putty and shell ash) [4–8], clay and slag [9–12] and natural zeolite rock [13,14]. Meanwhile, BFS is rich in silicon and aluminum, which is very suitable for the preparation of zeolite. The zeolite preparation is a heterogeneous reactive crystallization process, and liquid and solid phases generally both exist in the reaction system. The parameters that have the greatest influence on the properties of blast furnace slag-based zeolite (BFSZ) are generally crystallization temperature, crystallization time, initial Si/Al ratio and pressure, etc. In our previous studies, we systematically studied the parameters including crystallization temperature and initial Si/Al ratio in the reaction system on the structure and properties of the BFSZ products. However, further work is needed to clarify all factors controlling the crystal phase and formation of BFSZ. In addition, the crystallization time is known to have a significant effect on the generated crystalline phase and the total crystallinity. Hence, the study of the influence of crystallization time on the synthesis process is very important to clarify and improve the synthesis of BFSZ.

The purpose of this study is to analyze the most effective synthetic route of BFSZ and then to obtain the highest product quality. In order to characterize the synthesized BFSZ, X-ray (powder) diffraction (XRD), scanning electron microscopy (SEM), energy-dispersive X-ray spectroscopy (EDX), X-ray fluorescence (XRF), Fourier transform infrared spectroscopy (FT-IR), elemental mapping and the Brunauer–Emmett–Teller/Barrett–Joyner–Halenda (BET/BJH) methods were utilized to test the BFSZ obtained at different crystallization time.

2. Materials and Methods

2.1. Materials

The BFS samples are collected from multiple positions of a storage yard located in Nanjing, Jiangsu province, China, and then mixed evenly before use. Meanwhile, the same batch of BFS was used to conduct all experiments described in this research. The BFS powder was obtained by passing a 50-mesh sieve after crushing, drying and screening. Table 1 lists XRF analysis results of the BFS used in this study. From Table 1, the main components in the slag were CaO, Al_2O_3, and SiO_2.

Table 1. Determination of chemical composition of blast furnace slag (BFS) and activated BFS (ABFS).

Component	BFS/wt % [a]	ABFS/wt % [a]
CaO	47.08	0.24
SiO_2	29.13	58.59
Al_2O_3	20.59	41.56
MgO	1.11	0.14
Fe_2O_3	1.58	0.58
Na_2O	0.27	0.11
Total	99.76 [b]	99.25 [b]

[a] Measured by X-ray fluorescence (XRF). [b] Sulfur oxides may be present in the remaining components.

2.2. Synthesis of Blast Furnace Slag-Based Zeolite (BFSZ)

Before preparing zeolite, BFS was first activated with nitric acid to obtain activated BFS (ABFS). Namely, 10.0 g of BFS was dissolved in 2.0 mol·L^{-1} HNO_3 solution (100 mL) in a Teflon beaker. After dissolving at 80 °C for 2.0 h, the slurry was filtered to obtain the ABFS (see Table 1 for the chemical components of ABFS). An alkaline fusion-hydrothermal synthesis method was adopted to prepare BFSZ. In a typical program, 1:1.3 (wt %/wt %) ABFS and NaOH was first evenly mixed, after that baked in a muffle furnace at 600 °C for 90 min. Then, 5.0 g of baked products was stirred into 25 mL 1.0 mol/L NaOH (aq) until they were mixed evenly, and 50 mL of aqueous solution containing a certain amount of sodium aluminate or sodium silicate was added to adjust the Si/Al molar ratio to 1:1. The resulting mixture was then transferred to a thermostatic water-bath, stirred at 100 °C for 2 h, and crystallized at the same temperature for a certain crystallization time. Finally, the mixture was filtered, washed with distilled water until neutral and dried for further characterization. To further analyse the impact of crystallization time on the BFSZ synthesis, multiple BFSZ products were produced at different crystallization times.

2.3. Characterization

The XRD pattern of the sample was obtained by D8 discover X-ray diffractometer using Cu Kα radiation in a 5~90° 2θ scan range at 0.02/min scanning rate and operated at 40 mA and 40 kV (D8 discover, Bruker, Berlin, Germany). The samples were chemically analyzed by XRF spectroscopy (Axios, Panalytical, Eindhoven, Holland). The morphological structure of the samples were studied by field-emission scanning electron microscopy (FE-SEM) (SU 8220, Hitachi, Tokyo, Japan) under the following analytical conditions: HV = 20.00 kV, WD = 11.9 mm, HFW = 14.9 μm, mode: SE. ASAP2020 (Micromeritics, Norcross, GA, USA) using N_2 adsorption method was used at 77 K to obtain specific surface area (SSA) through the BET equation. Thermo Scientific Nicolet 6700 FT-IR spectrophotometer (Thermo Fisher Scientific, Waltham, MA, USA) was used to perform Fourier transform infrared spectroscopy (using a KBr method) in the 4000–400 cm^{-1} region.

3. Results and Discussion

3.1. Crystalline Characteristics

In this paper, we have systematically investigated the effect of crystallization time on the crystalline characteristics of obtained samples. The XRD patterns of BFSZ samples obtained at various crystallization time are listed in Figure 1. From Figure 1, after 4 h crystallization, NaA zeolite (JCPDS cards number: 71-0784) and ZK-14 zeolite (JCPDS cards number: 84-0698) is present in the products (see Figure 1a), although the diffraction peak intensity is a little weak, indicating incomplete crystallization. As the crystallization time increased to 6 h, the diffraction intensity and crystallinity increased gradually (see Figure 1b). However, as the crystallization time was further increased to 8 h, the diffraction peak was not further strengthened, indicating that the zeolite crystallization was almost completed within 6 h (see Figure 1c). Moreover, the crystallization products obtained at 6 h of crystallization clearly showed some diffraction peaks at 7.19°, 10.18°, 12.47°, 14.41°, 16.13°, 20.43°, 21.69°, 24.01°, 27.14°, 29.97°, 30.86°, 32.58°, 34.22°, 52.66°, 57.60° and 69.22°, which matched well with the characteristic peaks of NaA zeolite (JCPDS cards number: 71-0784) [15–17]. That is to say, the BFSZ obtained at 6 h is mainly composed of NaA zeolite (see Figure 1b). The results showed that NaA zeolite with good crystallinity was successfully formed.

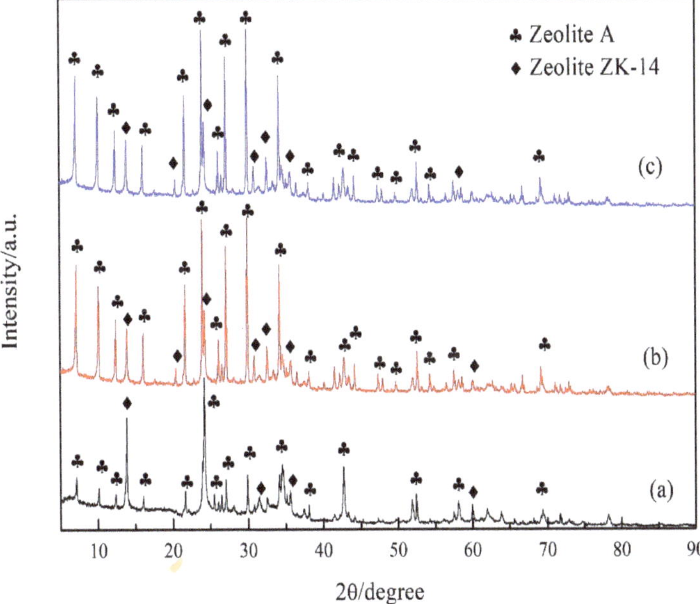

Figure 1. X-ray diffraction (XRD) patterns of the product samples obtained at crystallization time of (a) 4 h (b) 6 h and (c) 8 h.

3.2. Chemical Analysis

The changes of chemical composition and Si/Al (Na/Al) molar ratio are shown in Table 2. Comparing the 0.11% of Na$_2$O content in ABFS, the Na$_2$O content in the BFSZ obtained at various crystallization times were significantly increased. Due to the mechanisms of zeolite growth, the SiO$_4$ tetrahedron can be replaced by Al atoms to form AlO$_4$ tetrahedron. However, the Al atom is trivalent, and the electricity price of one oxygen atom is not neutralized in the AlO_4 tetrahedron, resulting in charge imbalance, which makes the whole AlO_4 tetrahedron negatively charged. In order to remain neutral, there must be positively charged ions to offset, which are generally compensated for by alkali metal and alkaline earth metal ions, such as Na. Hence, the Na_2O content in BFSZ is changed dramatically. In addition, the great change of Na_2O content is also regarded as other evidence of the successful transformation of BFSZ material [18,19]. The zeolite contents in BFSZ were determined to be 90.90%, 95.53% and 97.41% at crystallization time of 4 h, 6 h and 8 h, respectively. The significant increase on zeolite content of BFSZ samples generally clarify that longer crystallization time is favorable for zeolite formation in the reaction system. On the other hand, Si/Al and Na/Al molar ratio both increases with the increase of crystallization time. Namely, under longer crystallization time, more Si and Na ions are involved in the synthesis of BFSZ. Meanwhile, according to the BFSZ preparation process, Si and Al ions content in the filter liquor recovered after BFSZ preparation at 6 h was determined by ICP (Optima 5300DV, Perkin-Elmer, Waltham, MA, USA). The results show that Al ions concentration in the filter liquor was negligible and Si ions concentration in the filter liquor was 248.68 mg/L. Namely, abundant water-soluble silicates ions are not involved in BFSZ synthesis, while almost all Al ions are involved and consumed in BFSZ synthesis. This phenomenon can be attributed to the fact that during the process of zeolite synthesis, Al acts as the controlling substance and the consumption rate of Al is much higher than that of Si in the reaction. Moreover, there was little change in the relevant data such as chemical component, zeolite content and Si/Al (Na/Al) molar ratio of the BFSZ products prepared at 6 or 8 h. Taking the above results into account, an appropriate crystallization time should be controlled at about 6 h.

Table 2. Chemical components and Si/Al (Na/Al) molar ratio of the obtained BFSZ.

Crystallization Time (h)	Chemical Component [a]/wt %						Molar Ratio		Zeolite Content [b] /wt %
	CaO	SiO_2	Al_2O_3	Na_2O	Fe_2O_3	MgO	Si/Al	Na/Al	
4	0.34	35.42	35.66	19.82	0.65	0.17	0.84	0.90	90.90
6	0.43	38.16	36.76	20.61	0.32	0.09	0.88	0.92	95.53
8	0.36	38.72	36.42	22.27	0.38	0.13	0.90	1.01	97.41

[a] Measured by XRF. [b] Defined by $(SiO_2 + Al_2O_3 + Na_2O)/$ (total weight).

3.3. Fourier Transform Infrared (FT-IR) Analysis

The FT-IR spectroscopy of the BFSZ products obtained are displayed in Figure 2. The band at approximately 544 cm^{-1} was attributed to the external vibration of double four-rings (D4R), which is a feature of NaA zeolite [20–22]. The bands at 440 cm^{-1} and 1030 cm^{-1} are attributed to the vibrations of T–O bending and the TO4 asymmetric stretch (T = Si or Al), respectively. The bands at 3336 cm^{-1} and 1646 cm^{-1} were respectively due to intermolecular hydrogen bond and bonding water [23]. The bands at 711 and 764 cm^{-1} were due to the symmetrical stretching of T–O–T, and a band at 958 cm^{-1} was attributed to the asymmetrical stretching of T–O–T. The results of FT-IR for the BFSZ products obtained were consistent with the interpretation of XRD results.

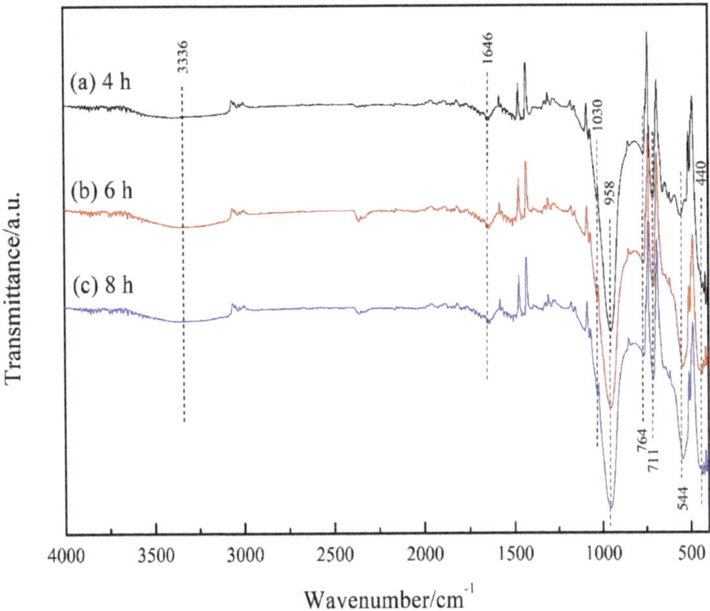

Figure 2. Fourier transform infrared (FT-IR) spectroscopy of the BFSZ prepared at various crystallization time.

3.4. Microstructural Characteristics

The BET/BJH test was carried out to analyze specific surface area (SSA), pore volume and average pore diameter of the BFSZ obtained under various crystallization times. BET/BJH measurements of nitrogen adsorption isotherm and pore size distribution are, respectively, displayed in Figures 3 and 4. Seen from Figure 3, the broad hysteresis loop ranging in 0.2 < P/P0 < 0.8 and steep elevation in the range of $P/P_0 > 0.8$ indicate the existence of mesoporous and microporous structure [24,25]. Moreover, the pore size distribution in Figure 4 again proves that BFSZ has mesoporous and macropore size pores. Meanwhile, the SSA, pore volume and average pore diameter of the BFSZ are demonstrated in Table 3. Results show that the SSA of the BFSZ became greater with the increase of reaction time. The SSA value for the BFSZ obtained at 6 h or 8 h remain largely unchanged. This again shows that the crystallization time of 6 h is sufficient to prepare BFSZ using the alkaline fusion-hydrothermal method. In addition, the average pore size is usually inversely proportional to the SSA value of BFSZ. This may be because relatively small pores can establish a hierarchical connection system of pores in BFSZ, which may lead to the obtained BFSZ samples having high SSA values.

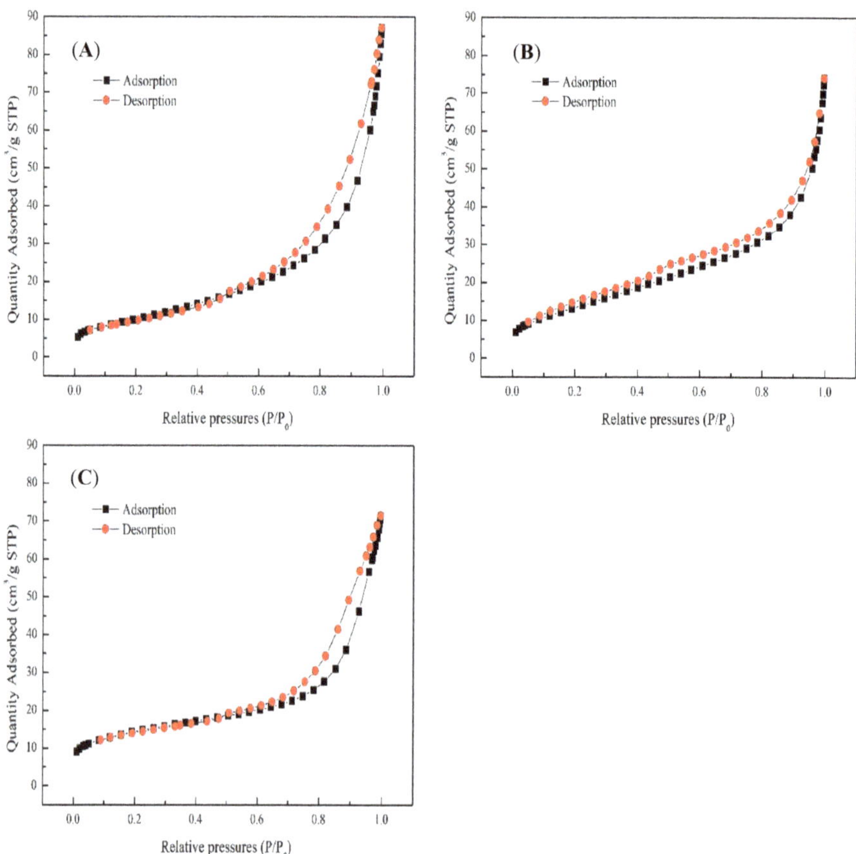

Figure 3. N_2 adsorption-desorption isotherms of BFSZ synthesized at crystallization times of (**A**) 4 h (**B**) 6 h and (**C**) 8 h.

Table 3. Microstructural characteristics results of the BFSZ prepared at various crystallization times.

Crystallization Time (h)	SSA (m²/g) [a]	Pore Volume (cm³/g) [b]	Average Pore Diameter (nm) [b]
4	37.16	0.14	10.95
6	49.64	0.11	7.62
8	49.96	0.11	7.78

[a] Brunauer-Emmett-Teller (BET) surface area. [b] Determined by Barrett-Joyner-Halenda (BJH) desorption data.

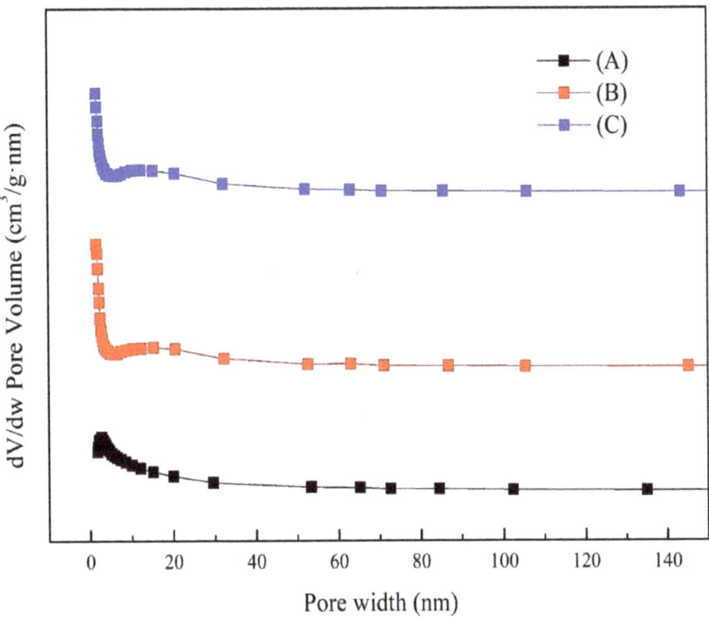

Figure 4. Pore distribution of BFSZ synthesized at crystallization times of (**A**) 4 h (**B**) 6 h and (**C**) 8 h.

3.5. Morphology Analysis

SEM images show that crystallization time has a significant effect on the size and morphology of BFSZ. After 4 h of crystallization, a large number of loosely packed crystals with irregular shapes were discovered, indicating incomplete crystallization. In addition, the morphology listed in Figure 5A again reflected weak peaks displayed in the XRD patterns. As the crystallization time reached up to 6 h, the main component of the product was a much larger cubic crystal with developed surface, and its particle size was about 2 μm (Figure 5B). Moreover, when the crystallization time was further increased to 8 h, all the obtained BFSZ samples were well shaped. In addition, the crystallization time will not significantly affect the crystal size and morphology of the BFSZ (Figure 5C). Moreover, as seen from morphological analyses of BFSZ in Figure 5, the cubic particles with a chamfered shape, along with a small fraction of round crystals can be observed. The cubic and round crystals corresponded to zeolite A and zeolite ZK-14, respectively based on morphology. Moreover, it is worth noting that no aggregates were observed in the SEM pictures which indicate that this shape transformation can provide large surface area with tiny pores. That is to say, 6 h is enough to complete the crystallization process. In addition, with the increase of crystallization time, the surface of BFSZ particles obtained changes from crude to smooth.

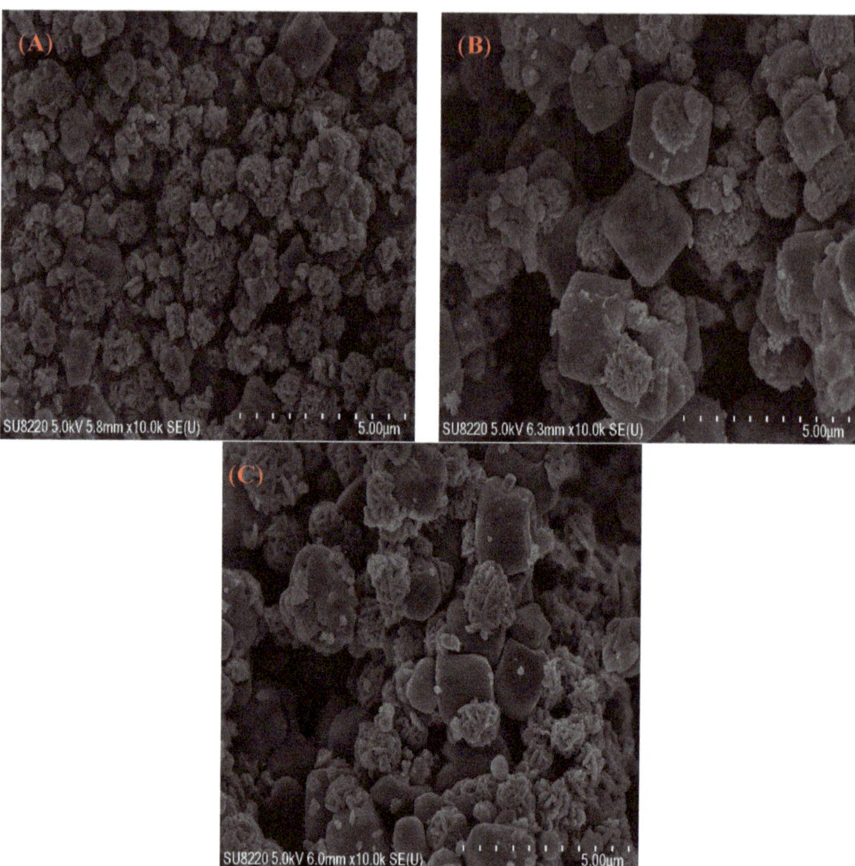

Figure 5. Scanning electron microscope (SEM) images of BFSZ synthesized at crystallization times of (**A**) 4 h (**B**) 6 h and (**C**) 8 h.

3.6. Elemental Mapping Analysis

Elemental mapping analysis was conducted to compare and study the regional (or phase) distribution characteristics of elements in the BFSZ sample obtained under various crystallization times. As seen in Figure 6, the BFSZ sample obtained under various crystallization time contains all BFS-derived metals. As shown in Figure 6A, Si, Al, O and Na-rich phase was formed and the Si, Al and Na were distributed in the same distribution area. From the morphological point of view, this is considered to be zeolite, which once again proves the successful preparation of BFSZ. At the same time, some components such as iron, calcium and magnesium also appear in the mapping image. However, no related crystal structure on iron, calcium and magnesium was discovered in the XRD pattern (see Figure 1). It is likely that those secondary metal ions either cannot form a crystalline phase or may be incorporated into the BFSZ.

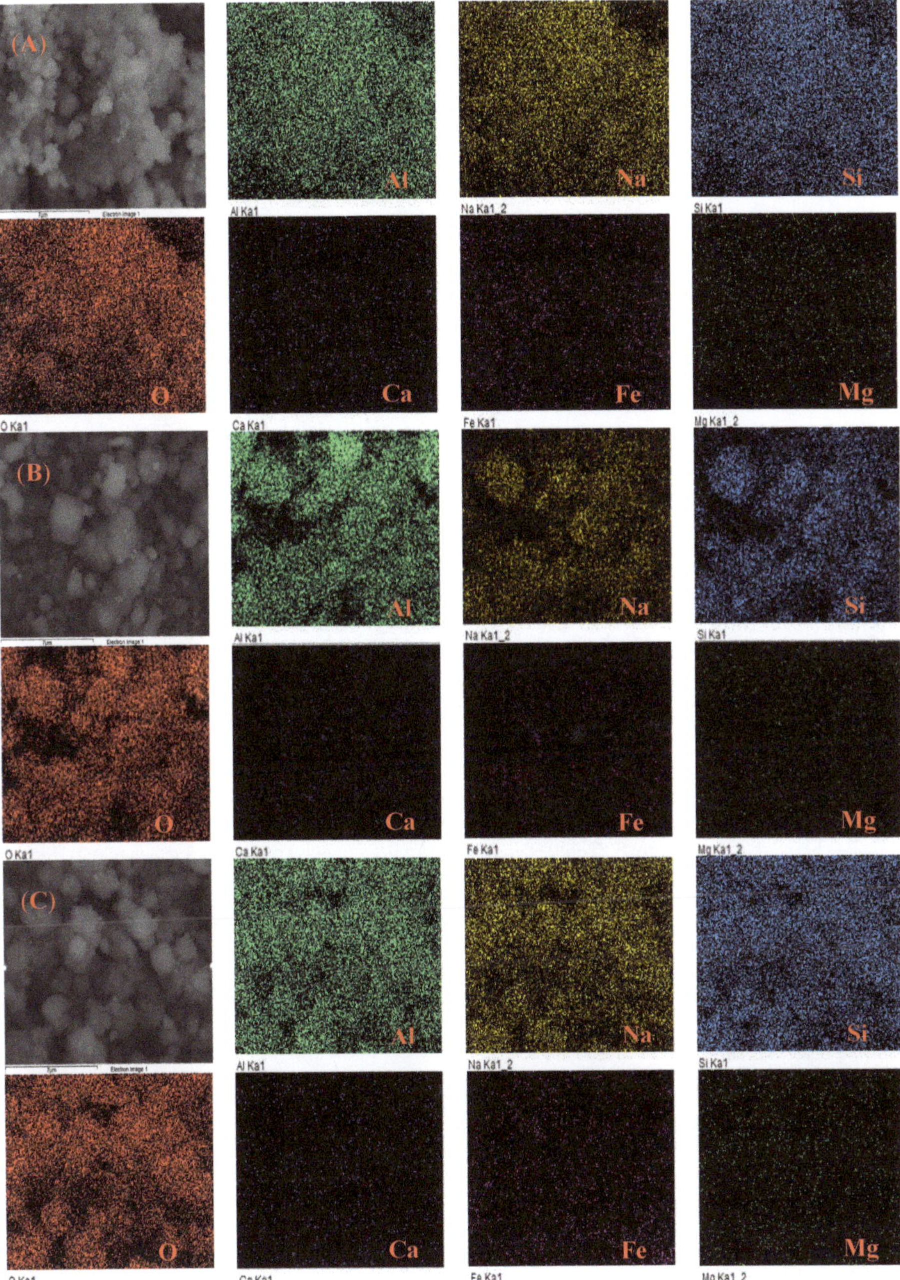

Figure 6. Elemental mapping of BFSZ synthesized at crystallization times of (**A**) 4 h (**B**) 6 h and (**C**) 8 h.

4. Conclusions

The effects of crystallization time on the alkaline fusion-hydrothermal synthesis of BFSZ were investigated. The BFSZ obtained under different crystallization times was measured by XRD, FT-IR, BET/BJH, XRF, FE-SEM and elemental mapping analysis. It was found that BFSZ crystallization was almost complete for 6 h. A further increase of crystallization time did not have a significant effect on the phase formation. Under 6 h aging, the main phases in BFSZ were NaA zeolite with the average SSA of 49.64 m^2 g^{-1}. Additionally, the cubic crystal with a developed surface in BFSZ crystals with particle size of about 2 μm could be clearly observed. Elemental mapping analysis showed that a Si, Al, O and Na-rich phase was formed and the Si, Al and Na were distributed in the same distribution area. Hence, it can be concluded that the optimal crystallization time for the synthesis of BFSZ using alkaline fusion-hydrothermal treatment is around 6 h. Research on the properties of the obtained BFSZ needs to be conducted through further studies.

Author Contributions: Methodology, C.L.; formal analysis, X.L. and L.L.; investigation, Q.Z.; writing-original draft preparation, C.L. and S.W.; writing—review and editing, C.L. and S.W. All authors have read and agreed to the published version of the manuscript.

Funding: The authors acknowledge the support of the Postdoctoral Research Foundation of China (No. 2017M611799) and Basic Research Program of Jiangsu Province (No. BK20190690) for offering the research fund.

Conflicts of Interest: The authors declare no conflict of interest.

References

1. Kuwahara, Y.; Ohmichi, T.; Kamegawa, T.; Mori, K.; Yamashita, H. A novel conversion process for waste slag: Synthesis of a hydrotalcite-like compound and zeolite from blast furnace slag and evaluation of adsorption capacities. *J. Mater. Chem.* **2010**, *20*, 5052–5062. [CrossRef]

2. Ismail, I.; Bernal, S.A.; Provis, J.L.; San Nicolas, R.; Hamdan, S.; Van Deventer, J.S.J. Modification of phase evolution in alkali-activated blast furnace slag by the incorporation of fly ash. *Cem. Concr. Comp.* **2014**, *45*, 125–135. [CrossRef]

3. Tian, P.; Dong, H.W.; Wu, X.M.; Huo, L.Q.; Xue, Y.K. Comprehensive utilization and control measures of iron and steel slag. *Energ. Metall. Ind.* **2020**, *39*, 9–13. (In Chinese)

4. Czuma, N.; Katarzyna, Z.; Motak, M.; Gálvez, M.E.; Costa, P.D. Ni/zeolite X derived from fly ash as catalysts for CO_2 methanation. *Fuel* **2020**, *267*, 117139. [CrossRef]

5. Shawabkeh, R.; Al-Harahsheh, A.; Hami, M. Conversion of oil shale ash into zeolite for cadmium and lead removal from wastewater. *Fuel* **2004**, *83*, 981–985. [CrossRef]

6. Kordatos, K.; Gavela, S.; Ntziouni, A.; Pistiolas, K.N. Synthesis of highly siliceous ZSM-5 zeolite using silica from rice husk ash. *Micropor. Mesopor. Mat.* **2008**, *115*, 189–196. [CrossRef]

7. Belviso, C.; Perchiazzi, N.; Cavalcante, F. Zeolite from fly ash: An investigation on metastable behavior of the newly formed minerals in a medium-high-temperature range. *Ind. Eng. Chem. Res.* **2019**, *58*, 20472. [CrossRef]

8. Belviso, C.; Abdolrahimi, M.; Peddis, D.; Gagliano, E.; Sgroi, M.; Lettino, A.; Roccaro, P.; Vagliasindi, F.G.A.; Falciglia, P.P.; Bella, G.D.; et al. Synthesis of zeolite from volcanic ash: Characterization and application for cesium removal. *Micropor. Mesopor. Mat.* **2021**, *319*, 111045. [CrossRef]

9. Belviso, C.; Piancastelli, A.; Sturini, M.; Belviso, S. Synthesis of composite zeolite-layered double hydroxides using ultrasonic neutralized red mud. *Micropor. Mesopor. Mat.* **2020**, *299*, 110108. [CrossRef]

10. Li, X.Y.; Jian, Y.; Liu, X.Q.; Shi, L.Y.; Zhang, D.Y.; Su, L.B. Direct synthesis of zeolites from a natural clay, attapulgite. *Acs. Sustainable. Chem. Eng.* **2017**, *5*, 6124–6130. [CrossRef]

11. Anuwattana, R.; Balkus, K.J.; Asavapisit, S.; Khummongkol, P. Conventional and microwave hydrothermal synthesis of zeolite ZSM-5 from the cupola slag. *Micropor. Mesopor. Mat.* **2008**, *111*, 260–266. [CrossRef]

12. Li, C.X.; Zhong, H.; Wang, S.; Xue, J.R.; Zhang, Z.Y. A novel conversion process for waste residue: Synthesis of zeolite from electrolytic manganese residue and its application to the removal of heavy metals. *Colloid. Surface. A* **2015**, *470*, 258–267. [CrossRef]

13. Novembre, D.; Sabatino, B.D.; Gimeno, D.; Garcia-Vallès, M.; Martínez-Manent, S. Synthesis of Na–X zeolites from tripolaceous deposits (Crotone, Italy) and volcanic zeolitised rocks (Vico volcano, Italy). *Micropor. Mesopor. Mat.* **2004**, *75*, 1–11. [CrossRef]

14. Lee, C.H.; Lee, M.G. Removal of Cu and Sr ions using adsorbent obtained by immobilizing zeolite synthesized from Jeju volcanic rocks in Polyacrylonitrile. *J. Environ. Sci. Int.* **2018**, *27*, 1215–1226. [CrossRef]

15. Ayele, L.; Pérez-Pariente, J.; Chebude, Y.; Diaz, I. Synthesis of zeolite A using kaolin from Ethiopia and its application in detergents. *New. J. Chem.* **2016**, *40*, 3440–3446. [CrossRef]

16. Mohamed, R.M.; Ismail, A.A.; Kini, G.; Ibrahim, I.A.; Koopman, B. Synthesis of highly ordered cubic zeolite A and its ion-exchange behavior. *Colloid. Surf. A* **2009**, *348*, 87–92. [CrossRef]

17. Liu, H.B.; Peng, S.C.; Shu, L.; Chen, T.H.; Bao, T.; Frost, R.L. Magnetic zeolite NaA: Synthesis, characterization based on metakaolin and its application for the removal of Cu^{2+}, Pb^{2+}. *Chemosphere* **2013**, *91*, 1539–1546. [CrossRef]

18. Li, C.X.; Zhong, H.; Wang, S.; Xue, J.R.; Zhang, Z.Y. Removal of basic dye (methylene blue) from aqueous solution using zeolite synthesized from electrolytic manganese residue. *J. Ind. Eng. Chem.* **2015**, *23*, 344–352. [CrossRef]
19. Kim, S.R.; Lee, J.H.; Kim, Y.T.; Riu, D.H.; Jung, S.J.; Lee, Y.J.; Chung, S.C.; Kim, Y.H. Synthesis of Si, Mg substituted hydroxyapatites and their sintering behaviors. *Biomaterials* **2003**, *24*, 1389–1398. [CrossRef]
20. Kuwahara, Y.; Ohmichi, T.; Mori, K.; Katayama, I.; Yamashita, H. Synthesis of zeolite from steel slag and its application as a support of nano-sized TiO2 photocatalyst. *J. Mater. Sci.* **2008**, *43*, 2407–2410. [CrossRef]
21. Sugano, Y.; Sahara, R.; Murakami, T.; Narushima, T.; Iguchi, Y.; Ouchi, C. Hydrothermal synthesis of zeolite A using blast furnace slag. *ISIJ Int.* **2006**, *45*, 937–945. [CrossRef]
22. Zhang, M.; Zhang, H.; Dan, X.; Lu, H.; Tian, B. Ammonium removal from aqueous solution by zeolites synthesized from low-calcium and high-calcium fly ashes. *Desalination* **2011**, *277*, 46–53. [CrossRef]
23. Wang, W.Q.; Feng, Q.M.; Liu, K.; Zhang, G.F.; Liu, J. A novel magnetic 4A zeolite adsorbent synthesised from kaolinite type pyrite cinder (KTPC). *Solid. State Sci.* **2015**, *39*, 52–58. [CrossRef]
24. Li, C.X.; Li, X.; Yu, Y.; Zhang, Q.W.; Li, L.; Zhong, H.; Wang, S. A novel conversion for blast furnace slag (BFS) to the synthesis of hydroxyapatite-zeolite material and its evaluation of adsorption properties. *J. Ind. Eng. Chem.* **2021**, *105*, 63–73. [CrossRef]
25. Wan, X.K. Synthesis and characterization of composite molecular sieves with mesoporous and microporous structure from ZSM-5 zeolites by heat treatment. *Micropor. Mesopor. Mat.* **2003**, *62*, 157–163.

Review

Titanium: An Overview of Resources and Production Methods

Mohammed El Khalloufi, **Olivier Drevelle** and **Gervais Soucy** *

Département de génie chimique et de génie biotechnologique, Université de Sherbrooke, Sherbrooke, QC J1K 2R1, Canada; Mohammed.El.Khalloufi@USherbrooke.ca (M.E.K.); Olivier.Drevelle@Usherbrooke.ca (O.D.)
* Correspondence: gervais.soucy@USherbrooke.ca; Tel.: +1-819-821-8000 (ext. 62167)

Abstract: For several decades, the metallurgical industry and the research community worldwide have been challenged to develop energy-efficient and low-cost titanium production processes. The expensive and energy-consuming Kroll process produces titanium metal commercially, which is highly matured and optimized. Titanium's strong affinity for oxygen implies that conventional Ti metal production processes are energy-intensive. Over the past several decades, research and development have been focusing on new processes to replace the Kroll process. Two fundamental groups are categorized for these methods: thermochemical and electrochemical. This literature review gives an insight into the titanium industry, including the titanium resources and processes of production. It focuses on ilmenite as a major source of titanium and some effective methods for producing titanium through extractive metallurgy processes and presents a critical view of the opportunities and challenges.

Keywords: titanium; titanium alloys; ilmenite; extractive metallurgy; TiO_2; calciothermic reduction

Citation: El Khalloufi, M.; Drevelle, O.; Soucy, G. Titanium: An Overview of Resources and Production Methods. *Minerals* **2021**, *11*, 1425. https://doi.org/10.3390/min11121425

Academic Editors: Shuai Wang, Xingjie Wang and Jia Yang

Received: 15 November 2021
Accepted: 12 December 2021
Published: 16 December 2021

Publisher's Note: MDPI stays neutral with regard to jurisdictional claims in published maps and institutional affiliations.

1. Introduction

Titanium is a transition metal that is used more in the production of high-strength, corrosion-resistant, and thermally stable metal alloys for the aerospace and shielding industries. The titanium production cost has so far hindered the growth in the use of this metal relative to other base metals on the market, even though titanium is the fourth most abundant structural metal in the earth's crust with 0.6%. It comes after iron, magnesium, and aluminum, but remains exotic due to its prohibitive cost [1,2], which prevents the metal from reaching its full potential in marine and automotive industry applications. Older production technology, high energy losses, and loss of material or metal are some of the problems associated with the production of titanium metal [3]. Further, all base metals are inferior to titanium in terms of specific mechanical and chemical properties, but titanium has not yet been fully exploited [1,2,4]. Titanium exhibits unique properties, some of which are proprietary, which could help it replace common metals and alloys such as steel and aluminum in many applications [5].

The current commercial method of titanium production is the Kroll process, marketed by DuPont Germany in 1948 [4]. It is a discontinuous, energy-, and labor-intensive process whose strict conditions make it expensive; therefore, researchers around the world are investigating new methods for extracting titanium from its precursors. Titanium is mainly produced from minerals such as ilmenite $FeTiO_3$ and rutile (TiO_2) while smaller quantities are produced from perovskite ($CaTiO_3$) and titanite or sphene ($CaTiSiO_5$) [6]. The main ilmenite deposits are located in Australia, China, Norway, Canada, Madagascar, India, South Africa, and Vietnam, while rutile deposits are found in Sierra Leone, the United States, India, and South Africa [7].

Ilmenite is a significant mineral [8], which contains between 40% and 65% titanium dioxide. The other elements are either ferrous oxide or ferric oxide and sometimes small amounts of vanadium, magnesium, and/or manganese. Ilmenite's main sources are the heavy mineral sands (alluvial deposits) but are also commonly distributed in hard rock [9]. Currently, ilmenite accounts for 92% of the world's titanium mineral production. Rutile (TiO_2) has a titanium dioxide content of 93–96% but is difficult to find in natural ilmenite deposits [7]. Thereby, the present review focuses on the alluvial deposit, more precisely the ilmenite, and the extraction of titanium from it.

Titanium metal production consumes a small proportion of total titanium reserves per year [7]. Low density and high tensile strength make titanium attractive for industrial applications, and give titanium-containing alloys the highest strength-to-weight ratio, an important property for metals in the steel industry. In addition, titanium is a valuable metal and can resist corrosion of both seawater and acids. As an alloyed metal, it can also resist corrosion better than copper and nickel alloys and has a low modulus of elasticity that is half that of steel and nickel alloys. The most common titanium alloy is Ti6A-4V (6% aluminum, 4% vanadium, 90% titanium) and is typically used in medical applications such as knee replacement implants. Metal is also one of the main elements in the aerospace industry, architecture, chemical, and automotive applications [10].

The traditional methods of manufacturing Ti follow the same general procedures as steel, including the primary metal production, the melting and casting of alloy ingots, forging and rolling to produce rolling products, and the manufacture of components or structures from rolling products [11]. Many powder production processes are at various stages of development. There are two general approaches for Ti metal production: electrochemical and thermochemical methods. The well-known example of the electrochemical methods is the Fray, Farthing, and Chen (FFC) or Cambridge process [12], which is based on the electrolysis reduction of titanium oxide. The Kroll process [13], is an example of the thermochemical way that is a commercially suitable process for primary Ti metal production today.

This review provides an overview of titanium resources, of which ilmenite is the main source, as well as it focuses on some effective methods for producing titanium powder through extractive metallurgy processes, and highlights a comprehensive view of the opportunities and challenges.

2. Titanium Usage and Market

Titanium, and particularly titanium alloys, have become economically accessible following a drop of nearly 30% in the price of commercially pure titanium over the last five years [14]. Titanium alloys have many physical (lightness, good mechanical properties, resistance to cryogenic temperatures) and chemical (resistance to electrochemical corrosion, biocompatibility) advantages [15,16], which make them an indispensable material for civil and military applications in fields as vast as in energy, transport, medical (MRI magnet for observing the human body), water treatment, and the transport of corrosive liquids and gases (Figure 1) [14].

The uses of titanium have expanded based on its inherent properties as well as the development of new alloys. The main use is still in aerospace and aeronautics applications, such as engines, airframes, missiles, and spacecraft [16]. Aerospace applications are based on the low density (Table 1) and high strength-to-weight ratio of titanium alloyed at high temperatures. Titanium's corrosion resistance makes it a natural material for seawater, marine, and naval applications. In addition, titanium is largely used in seawater-cooled power plant capacitors [16].

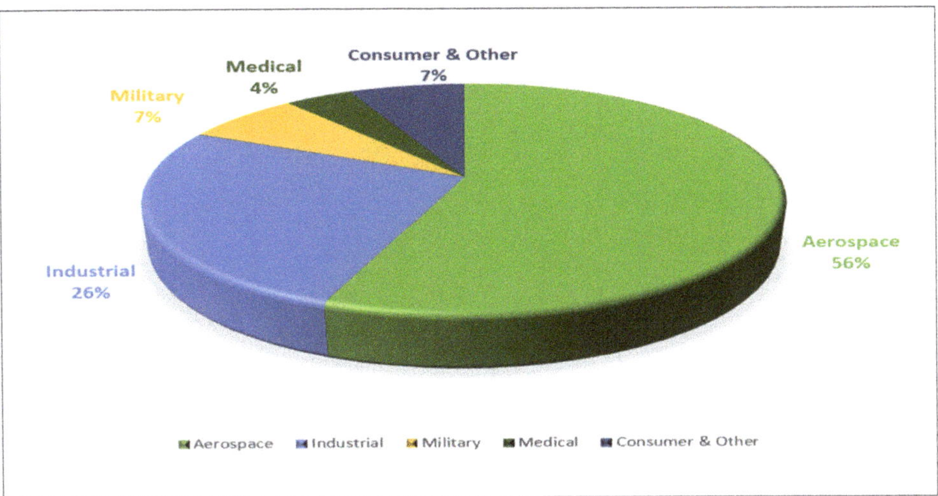

Figure 1. Titanium applications of the Western world by market sector, 2017 [17] (Reprinted with permission from Elsevier, Copyright, (2020)).

Table 1. Some properties of pure titanium [15,16].

Parameter	Value
Atomic number	22
Atomic radius (Å)	1.47
Atomic weight	47.9
Boiling point (K)	3273
Chemical valence	2, 3, 4
Electrical resistivity	42.06×10^{-6}
Ion radius (Å) Ti^{+2}	0.9
Ion radius (Å) Ti^{+4}	0.68
Melting point (K)	2073
Density (g/cm^3)	4.51
Specific heat (J/kg–K)	519
Traction modulus $\times 10^3$ (MPa)	101
Modulus of elasticity $\times 10^3$ (MPa)	103
Hardness (1500 kg load) (HB)	65
Fatigue resistance	0.5–0.6

Titanium can also be utilized in petroleum refineries, paper, and pulp bleaching operations, nitric acid plants, and some organic synthesis production [4]. Moreover, Titanium has found its use in the medical field. In particular, depending on clinical needs, titanium and a multitude of its alloys offer high axial flexibility, good expansion behavior, radio-opacity, and hemocompatibility. Even of its sophisticated bio-applications, the main use of titanium in biomedicine is as a structural prosthesis [1,2,18].

3. Resources

Titanium deposits are huge, with current estimates assuming a global reserve of 650 billion metric tons of titanium oxide. The minable deposits are found in South Africa at Namaqualand and Richards Bay, Australia, Canada, Norway, and Ukraine (Figure 2) [1,7]. The two main minerals being considered for use are ilmenite and rutile and although these are the minerals available for economic mining, TiO_2 is part of almost every mineral, sand, and rock [4].

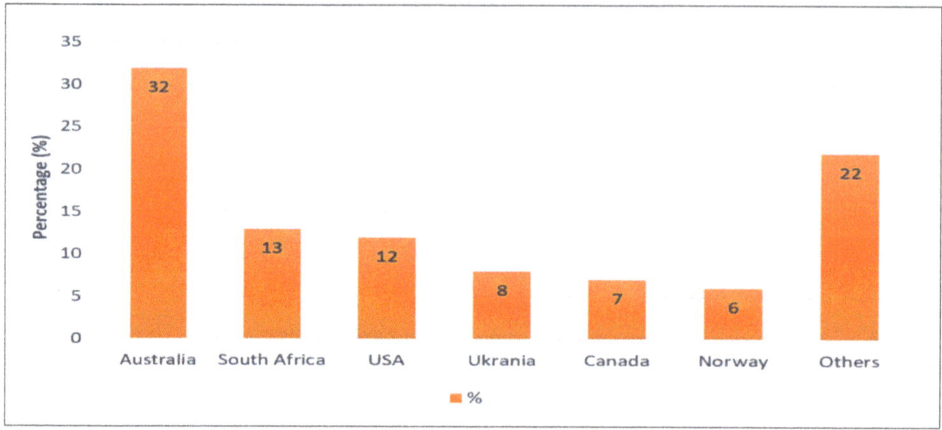

Figure 2. Distribution of the titanium ore reserves in the world [1,4].

The common titanium minerals are anatase, brookite, leucoxene, perovskite, titanite, rutile, and ilmenite. However, only ilmenite, leucoxene, and rutile have crucial commercial value. Ilmenite and rutile are the two main titanium minerals used in industrial applications, mainly for titanium metal production and pigment-grade titanium dioxide [19].

4. Mineral Ilmenite

In 1827, Adolph Theodor Kupffer discovered Ilmenite (titanoferrite) in the Ilmen Mountains of Russia. $FeTiO_3$ is the typical chemical formula of ilmenite, while its chemical composition is (40–65 wt.%) TiO_2 and (35–60 wt.%) Fe_2O_3 [20]. Table 2 shows the properties of pure ilmenite.

Table 2. Basic properties of pure ilmenite ($FeTiO_3$) [21].

Property	Value
Chemical classification	Oxide
Color	Black
Luster	Metallic, sub-metallic
Mohs hardness	5–6
Specific weight	4.7–4.8 g/cm^3
Crystalline structure	Hexagonal
Cleavage	Absent
Unit cell	a = b = 508.854 Å, c = 14.0924 Å

Ilmenite is an economically important mineral, mainly because of its role in the production of titanium oxide pigments. Its magnetic properties and those of ilmenite-hematite solid solutions (Fe_2O_3;) are particularly important in commercial extraction by magnetic separation. The dependence of the ilmenite structure on temperature, pressure, and composition is strongly related to its electronic, magnetic, and optical properties [22]. Due to the coexistence of ilmenite with geikielite ($MgTiO_3$) and pyrophanite ($MnTiO_3$) in these rocks, the typical chemical formula of this mineral in magmatic rocks is (Fe, Mn, Mg) TiO_3 [21]. Ilmenite is often confused with other iron-bearing minerals such as hematite and magnetite because of its high magnetic susceptibility (Figure 3). Ilmenite has a hexagonal crystal structure that is similar to corundum (Al_2O_3) and different from other iron minerals and has lower magnetism compared to hematite and magnetite. The crystal presents an octahedral structure alternating and coordinated with iron and titanium layers [22,23].

Figure 3. Ilmenite sand (**left**) and grinded material (**right**) from the Metchib company, Quebec, Canada.

The Tellnes (Norway) mines produce 550,000 tons of ilmenite per year [24]. The largest Ti producers in the world are China, Australia, and South Africa (Figure 4). China produces ilmenite in significant quantities, while Australia and South Africa have the world's largest natural reserves for ilmenite and rutile [7,21].

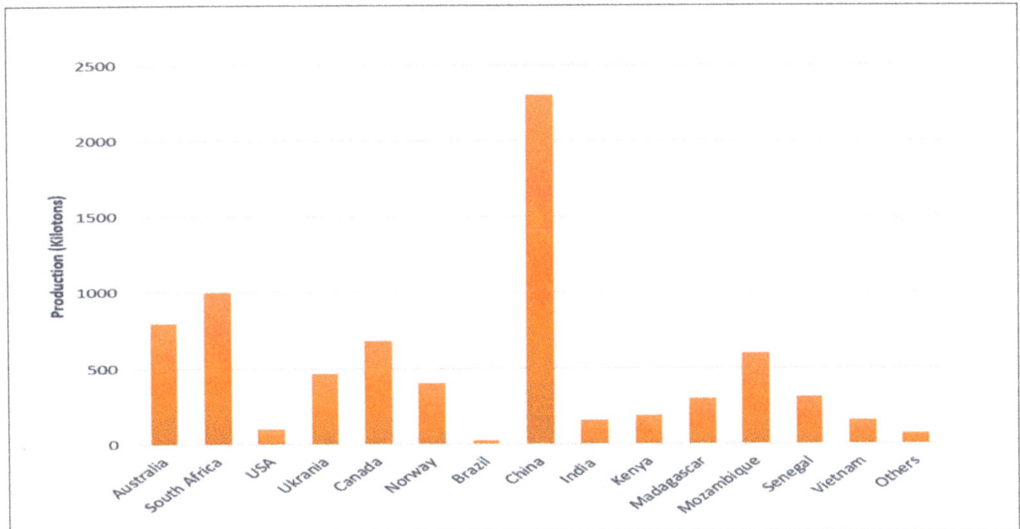

Figure 4. Ilmenite mine production in different countries in 2020 [7].

5. Metallurgical Extraction of Titanium from Its Concentrates

Many methods are used for the production of Ti metal powder. Ti powder is, therefore, the product of extraction processes that produce primary metal by using titanium tetrachloride ($TiCl_4$) or titanium dioxide (TiO_2) as feed material [25,26]. Processes for the manufacture of titanium powder directly as extractive metallurgical products include the manufacture of Ti from $TiCl_4$, purified TiO_2, and/or improved titanium slag (UGS) with a TiO_2 content greater than 90%. Upgraded titanium slag (UGS) is one of the fundamental products of the carbothermal reduction of titanium ore such as ilmenite. Natural rutile and synthetic rutile are also included in this raw material category. These processes can be categorized: (1) thermochemical methods and (2) electrochemical methods [11].

5.1. Thermochemical Processes

5.1.1. Kroll Process

The commercial production of primary Ti metal is generally made either by Kroll [13] or Hunter processes [27]. The standard process against which new technologies are compared is the Kroll process (Figure 5).

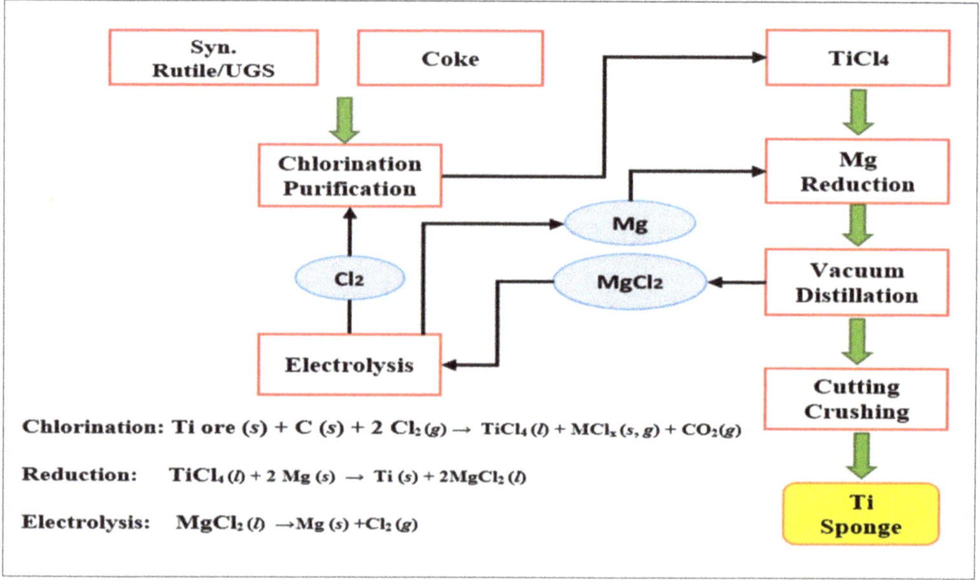

Figure 5. Scheme and reactions of the Kroll process of titanium sponge production [28] (Reprinted with permission from Springer Nature, Copyright, (2020)).

In the process, magnesium metal (the reducing agent) is injected into a retort filled with argon and heated to 800–900 °C. However, the oxides impurity contained in the Ti slag is also chlorinated, so refined $TiCl_4$ has been produced by purifying the crude $TiCl_4$ before the Mg reduction [13,29]. Although most of the by-product $MgCl_2$, and excess magnesium, is drained during reduction, the product sponge contains residual magnesium and $MgCl_2$ in its porosity [30]. Magnesium and $MgCl_2$ are separated by vacuum distillation or helium sweeping followed by leaching. Part of the sponge must be decommissioned due to contamination of the autoclave wall [13,29].

According to one estimate, 70% of the total energy consumption is considered for the distillation to produce the sponge metal. It shows that the cost of metal purification is one of the major cost drivers, in addition to the cost of the precursor and reductant [30].

5.1.2. Hunter Process

The popular process established on $TiCl_4$ reduction using Na is the Hunter process. The Hunter and Kroll processes are quite similar in that they are considered as thermochemical processes based on the reduction of $TiCl_4$ to produce Ti [27]. Economically, the Hunter process is considered non-competitive with the Kroll process. The main difficulty is that to produce one mole of Ti by reducing $TiCl_4$ requires four moles of Na, whereas only two moles of Mg are needed for 1 mole of Ti [11]. In addition, producing Na by electrolysis is at least as expensive as that of Mg. These problems make the processing of Na more expensive than the use of Mg. However, the Hunter process can also produce Ti powder instead of Ti sponge [31].

In the Hunter process, $TiCl_4$ and Na are gradually introduced into the reactor. The process is generally performed over 800 °C [28,32]. Ti is formed at the surface of the molten bath, where the gas $TiCl_4$ is exposed to the Na. Ti crystals then form and are set at the bottom of the liquid bath. According to the operating parameters, some Ti particles can form Ti sponge, while others are deposited as Ti powder. The purity of the powder produced by the Hunter process is often higher (99%) [11,29]. Table 3 provides a comparison between Kroll and Hunter processes.

Table 3. Comparison between Kroll and Hunter processes [31] (Reprinted with permission from John Wiley and Sons, Copyrright, (2013)).

Kroll	Hunter
Batch	Does not last forever
15–50% excess magnesium	A small excess of $TiCl_4$
Few fines	Up to 10% fines
Difficult to rectify	Easy to rectify
Heavy iron contamination from the walls of the autoclave	Little iron contamination from retort walls.
Sponge washed or vacuum distilled	Sponge leached
Retort contains mostly titanium	Retort contains 4 moles of NaCl for each mole of titanium

5.1.3. Armstrong Process

The Armstrong process is considered the most advanced process. It uses the same reactions as the Hunter process [33].

Therefore, the crucial advantage of the Armstrong process [34] is the continuity of the operation, pumping molten sodium in the reactor to react continuously with $TiCl_4$ gas (Table 4). Ti powder and the resulting NaCl are collected from the reactor by the sodium stream. Once the unreacted liquid Na is removed by filtration, as well as Ti powder is purified by washing out the salt. The Armstrong product can be described as mini sponges, i.e., microporous particles [35]. Figure 6 shows the Armstrong process aspects.

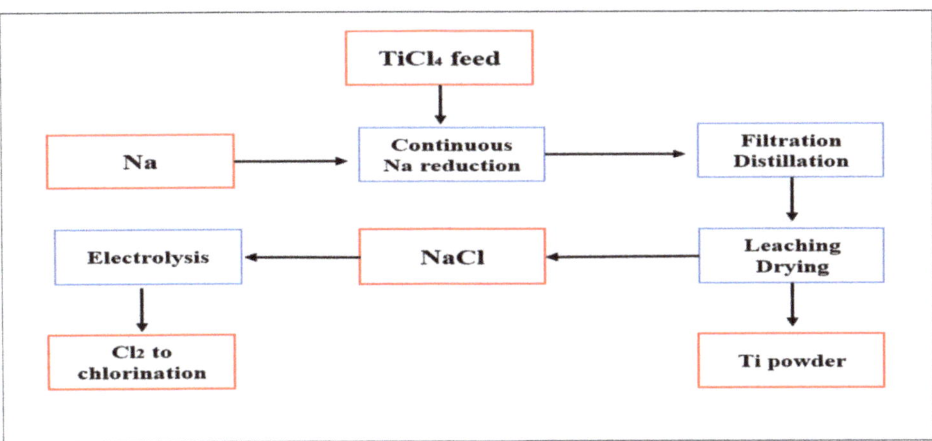

Figure 6. The Armstrong flow diagram [36].

5.1.4. TiRO Process

This process was developed by CSIRO (Commonwealth Scientific and Industrial Research Organization) in Australia and used the same reactions as the Kroll process but in a fluidized bed reactor in which gas–solid fluidization takes place, which considerably increases the reaction rate and reduces both operating and capital costs [37].

The TiRO process consists of two main steps (Figure 7): reduction of $TiCl_4$ in a fluidized bed with Mg powder and vacuum distillation to remove the by-products $MgCl_2$ and Mg [38].

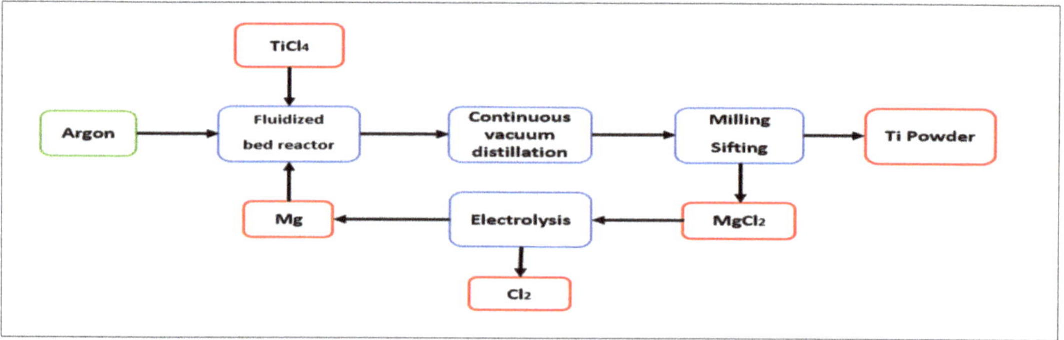

Figure 7. Diagram of the process TiRO™ [39].

According to the information mentioned above, $TiCl_4$ can be reduced by sodium or magnesium. The main characteristics of the powder are well detailed in Table 4 below.

All these processes below use TiO_2 as raw material to produce titanium powder with high purity.

Table 4. Comparison between characteristics of TiCl$_4$ reduction-based processes (adapted from [11] Taylor & Francis, an open-access journal).

Category	Process Identifier	Raw Material	Reducing Agent	Product Size and Morphology	Reported Chemical Product and Composition	Batch or Continuous: Reactions	Salt	Temp (°C)	Ref
	Kroll	TiCl$_4$	Mg Liquid	Sponge	O~0.06%	Batch; 2Mg + TiCl$_4$→Ti + 2MgCl$_2$	No salt	800–1000	[13]
Reducing TiCl$_4$ using Mg	TiRO	TiCl$_4$	Mg Powder	Similar spherical shape but sintered to form large chunks	O ≥ 0.3%, Cl < 0.03%	Continuous; 2Mg + TiCl$_4$→Ti + 2MgCl$_2$(s)	No salt	650–712	[38]
	Vapor Phase Reduction	TiCl$_4$	Mg Vapor	Sub-micrometer, too fine to be captured	Low levels of Mg and Cl, but O ≥ 0.82%.	Continuous; 2Mg + TiCl$_4$→Ti + 2MgCl$_2$	No salt	1000	[40]
	CSIR-Ti	TiCl$_4$	Mg	Irregular shape, size be ranging from 1 to 330 μm	Cl < 50 ppm, N < 50 ppm, O > 0.2%	Continuous; Mg + TiCl$_4$→TiCl$_2$ + MgCl$_2$ TiCl$_2$ + Mg→Ti + MgCl$_2$	No salt	>900	[41]
	Hunter	TiCl$_4$	Na	Sponge and powder	Purer than that produced by the Kroll process (99%)	Batch; 4Na + TiCl$_4$→Ti + 4NaCl	No salt	>800	[27,42]
Reducing TiCl$_4$ using Na	Armstrong	TiCl$_4$	Na	Mini sponge, particulates with micro porosity	O: 0.12–0.23%, N: 0.009–0.026%, C: 0.013%, Fe: 0.012%	Continuous; 4Na + TiCl$_4$→Ti + 4NaCl	No salt	860	[34,35]
	ARC	TiCl$_4$	Na	Powder, small aggregates	Oxide layer contributes a lot to the final high oxygen content	Continuous; 2Na + TiCl$_4$→TiCl$_2$ + 2NaCl TiCl$_2$ + 2Na→Ti + 2NaCl	No salt	>800	[43]

5.1.5. Metal Hydride Reduction (MHR) Process

The MHR process was first introduced in 1945, and the most remarkable work was reported by Borok [44], in 1965 and Froes et al. in 1998 [45]. Calcium hydride was used to directly reduce TiO_2. In Russia, a commercial operation has been reported that uses the same procedure [46].

5.1.6. Electronically Mediated Reduction (EMR) Process

EMR process uses calcium as a reducing agent to produce titanium. The TiO_2 reduction is made with no direct contact with the reducing Ca-Ni alloy, and hence the contamination of titanium can be successfully avoided by using EMR. This approach can be exploited to develop another process to produce titanium powder continuously [47].

5.1.7. Process for Reducing Preforms

Okabe et al. [48] have developed this process where TiO_2 is mixed with either CaO or $CaCl_2$ and sintered in the air. Calcium vapor reduces TiO_2 at temperatures between 800 and 1000 °C, with the calcium oxide dissolving in calcium chloride. A hydrochloric acid solution has been used for leaching the product [46]. Again, these experiments were only conducted in the laboratory [49].

5.1.8. Hydrogen-Assisted Magnesium Reduction (HAMR) Process

The HAMR process is developed to produce a Ti powder with a very low oxygen content, by using a hydrogen atmosphere, molten salt, and deoxygenation step to guarantee that the oxygen amount in titanium powder is sufficiently low (less than 0.15% by weight) [50,51]. Table 5 shows a comparison between the characteristics of these processes based on TiO_2 reduction.

Table 5. Comparison between characteristics of TiO_2 reduction-based processes (reprinted from [11] Taylor and Francis, an open-access journal).

Process Identifier	Raw Material	Reducing Agent	Product Size and Morphology	Reported Chemical Product and Composition	Salt	Temperature (°C)	Ref
MHR	TiO_2	CaH_2	Irregular, sponge	O: 0.19 wt.%, H: 0.34 wt.%, C: 0.03 wt.%, N: 0.06 wt.%	No salt	1100–1200	[52,53]
Calciothermal reduction	TiO_2	Ca	Irregular, sponge	O: <0.2 wt.%, Ca: >0.1 wt.%	$CaCl_2$	900–1200	[54]
Process for reducing preforms	TiO_2	Ca	Irregular, sponge	O: 0.2–0.3 wt.%	$CaCl_2$	>900	[48]
EMR	TiO_2	Ca	Irregular, sponge	O: 0.15–0.2 wt.%	$CaCl_2$ or $CaCl_2$ + CaO	>900	[47,55]
Combustion synthesis	TiO_2	Mg + Ca	Irregular, sponge	O: 0.2–0.3 wt.%	$Ca(OH)_2$	850–1000	[56,57]
HAMR	UGS	Mg + Ca	Dense, globular powder	O: <0.15 wt.%	$MgCl_2$ or $MgCl_2$-KCl	<800	[50,58]

There have been several reports and trials to perform continuous processes based on the same chemistry as the above processes. Lu et al. [59] produced a fine titanium powder from a titanium sponge by the shuttle: the disproportionation reaction and its reverse reaction (proportioning reaction) of titanium ions in molten NaCl-KCl at 750 °C. Moreover, some experiments were performed to synthesize TiH_2 from the reaction between CaH_2 and $TiCl_4$ in the presence of Ni [60].

In another study, porous titanium was obtained in mixtures of molten CaO-$CaCl_2$ salts dissolved in Ca by self-sintering with the exothermic reaction between porous $CaTiO_3$ and calcium vapor at 1000 °C for 6 h under vacuum [61]. Development of a new method based on deoxidation of dissolved O from Ti. The process leads to the formation of YOCl using Mg as a deoxidizer at 1027 °C [62].

Daniel Spreitzer et al. [63] have used a laboratory fluidized bed reactor to cut down the hematite kinetic reaction to produce Fe by hydrogen around 600–800 °C by measuring the change in weight of the sample portion during reduction. Moreover, with combining magnesiothermal reduction of TiO_2 and a leaching purification process, titanium metal powder was obtained with only 2.98% of O [64]. The synthesis of a cermet based on Fe-TiC by carbothermal reduction of ilmenite was successfully produced [65]. This method was done in an atmosphere containing argon in a scope of 850–1350 °C. Another study has led to developing an environmentally friendly pre-treatment process to recycle titanium turning waste and ferrotitanium ingots with low levels of gaseous impurities [66]. The number of gaseous impurities in the titanium scrap before the removal of machining oils from the surface reached 2% [66]. Similarly, Nersisyan et al. [67] have developed a combustion synthesis route for single-phase titanium-based compounds (e.g., FeTi, TiC, TiB_2, and $TiFeSi_2$) from the precursor mixture $FeTiO_3$ (natural ilmenite)-αMg-C (B, Si)-kNaCl. The method allows the process temperature and the phase composition to be controlled by changing the number of moles of Mg and NaCl [67]. Furthermore, new research has been developed to study the effect of the degree of ilmenite reduction on the chemical and phase characteristics of ferrotitanium and slag produced by the SHS aluminothermic process, which is a highly exothermic thermite reaction [68]. Increasing reduction not only reduces the consumption of aluminum and the amount of slag produced in the preparation of ferrotitanium but also reduces the oxygen content and improves the titanium and iron qualities [68]. Further, the Panzhihua ilmenite carbothermal reduction with activated carbon has been studied by using isothermal trials between 1200 °C and 1500 °C [69]. By decreasing the pressure and increasing the temperature, the impurities (Mg, Mn) in the product have been removed [69]. The carbothermal reduction behavior of ilmenite at high temperatures was studied by thermodynamic calculations [70]. FeTi formation is generated at 1650 °C. By increasing the temperature, a clear increase of TiC is observed, which can also encourage the further reduction of ilmenite slag at high temperatures [70].

Otherwise, for patents published in this sense, Mu et al. [71] invented a method that aimed to improve the metallic titanium production with a low-energy titanium-containing material by a molten salt electrolysis process (Table 6).

Similarly, in 2016, Fang et al. [51] presented a research procedure for producing a titanium powder or sponge. For instance, the method may include obtaining a TiO_2-rich material, reducing impurities to produce purified TiO_2, reducing the purified TiO_2, using a metallic reducer at the same temperature and pressure to produce a TiH_2 product [51]. RMI (now part of Arconic) has patented a process that carries out above 900 °C to reduce oxygen in Ti-6Al-4V powder [72].

Table 6. Advantages and disadvantages of some patents in titanium production based on thermochemical methods.

Publication N°	Title	Advantages	Disadvantages	Reference
US 2016/0108497 A1 (2016)	Methods of producing a titanium product	- Hydrogen helps destabilize the Ti-O system. - Titanium hydride is known to be more impermeable to oxidation in the air than α-Ti, making it easier to handle the material after reduction and controlling the oxygen content in the final product - The oxygen content can be reduced to less than 0.2%. - The use of molten salt, especially salt-containing Mg such as $MgCl_2$ to facilitates the reaction and greatly improves the kinetics of the reduction process	- High reduction temperatures - High pressures	[51]
US 9,963,796 B2 (2018)	Method of producing titanium metal with titanium-containing material	- Low energy consumption - Low cost of production - Fewer titanium losses	- Fine powder	[71]
US 10,066,308 B2 (2018)	System and method for extraction and refining of titanium	- Reduce the use of hazardous chemicals - Reduce the production of greenhouse gases, such as CO_2 - High purity up to 90%	- High temperatures - Consumable anode - Relatively high energy consumption	[73]
US 9,067.264 B2 (2015)	Method of manufacturing pure titanium hydride powder and alloyed titanium hydride powders by combined hydrogen-magnesium reduction of metal halides	- To provide cost-effective and highly productive manufacturing of purified titanium hydride powders and alloyed titanium hydride powders from porous titanium compounds with reduced magnesium and hydrogen content - Reduce the thermal reduction reaction time in the horn by 15% to 20% and simultaneously improve the percentage of magnesium used from 40% to 50% to 70% to 80%	- Use of large quantities of hydrogen. - Production of small quantities of titanium powder. - The use of hazardous chemicals such as $TiCl_4$.	[74]
US 8,388,727 B2 (2013)	Continuous and semi-continuous process of manufacturing titanium hydride using titanium chlorides of different valency	- A continuous process for the manufacture of titanium hydride powder - Significantly lower oxygen content in the titanium powder due to the absence of grinding steps and limited contact with atmospheric humidity - Manufacture of pure titanium hydride powder from titanium slag with the recycling of the main amount of hydride and chlorine without an electrolysis process - Use the same equipment as that used in the manufacture of said porous hydrogenated titanium compound - Cost-effective and highly productive production of purified titanium hydride powder	- Carbon monoxide emission - Powder too fine - Formation of intermediate products	[75]
Re. 34,598 (1994)	Highly pure titanium	- A fairly low temperature - Production of high purity titanium - Oxygen content does not exceed 200 ppm	- Contamination problem.	[76]
US 4,923,531 (1990)	Deoxidation of titanium and similar metals using a deoxidant in a molten metal carrier	- The process is effective for most refractory metal alloys - Low-pressure operation - The oxygen level can be reduced to between 10 and 90% of the initial oxygen content, depending on the alloy and conditions	- Deoxidation is limited; it concerns metals and metal alloys containing small quantities of oxygen - Fairly high temperature - Consumable proces	[72]

Table 6. *Cont.*

Publication N°	Title	Advantages	Disadvantages	Reference
US 8,007,562 B2 (2011)	Semi-continuous magnesium-hydrogen reduction process for manufacturing of hydrogenated, purified titanium powder	- Reaction temperatures are quite low - High productivity due to the use of a small mass of $TiCl_4$ - Less electricity consumption - Total powder production time is reduced	- Use of high pressures. - A large amount of hydrogen - Use of large quantities of magnesium - Fine powder - The use of hazardous chemicals such as $TiCl_4$.	[77]
US 10,689,730 B2 (2020)	Methods of producing a titanium product	- Oxygen content can be reduced to less than 0.2%	- Multi-step process	[78]
US 583,492 B2 (2020)	Titanium powder production apparatus and method	- Preventing contamination of titanium powder	- Preparation conditions are difficult	[79]

5.2. Electrochemical Processes

5.2.1. Cambridge FFC Method

For several decades, it was considered difficult to synthesize titanium with a low amount of oxygen. However, in 2000, a new method was introduced [80], which showed that it is possible to reduce TiO_2, completely in the solid-state, to the metal in molten calcium chloride, which is a cheap and non-toxic product. This technique is known as the Cambridge FFC process (Figure 8). It operates in a molten saline environment, typically in the scope of 800 °C to 1100 °C. $CaCl_2$ is used as a salt since $CaCl_2$ can dissolve and transport oxygen ions. The TiO_2 reduction during the FFC process has been reached by the ionization of oxygen from the titanium-containing cathode, which diffuses to the anode and is discharged [46].

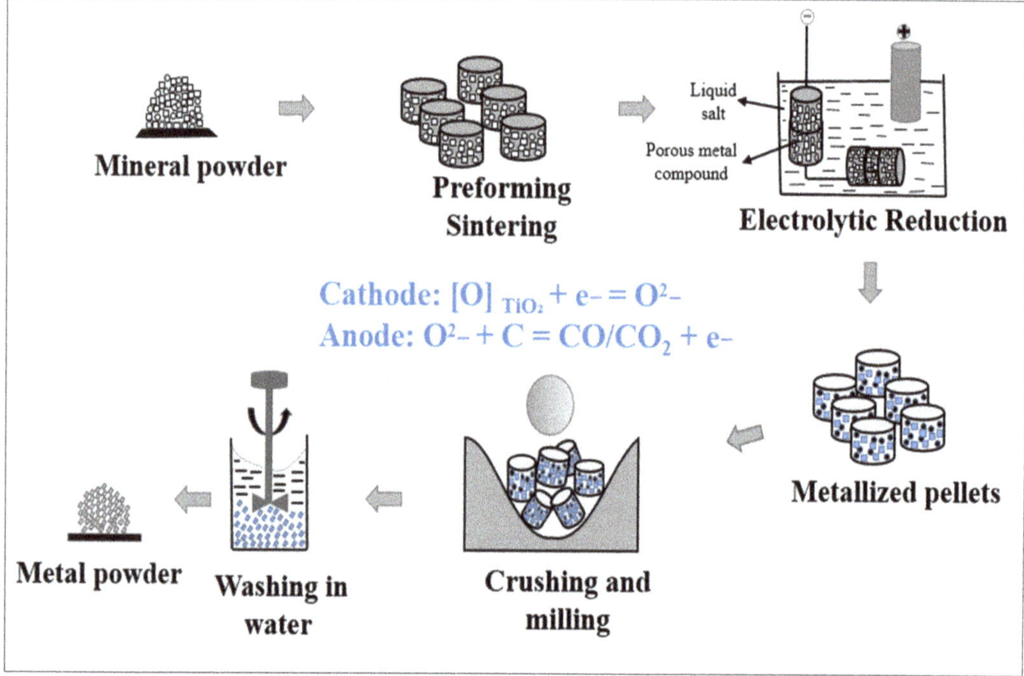

Figure 8. A schematic illustration of the CTF's Cambridge process [81] (Reprinted with permission from Taylor & Francis, Copyright, (2015)).

5.2.2. Ono and Suzuki Process

The (OS) process invented by Ono and Suzuki et al. [54,82,83] is based on a reduction of TiO_2 by Ca. Ca is generated by electrolysis of CaO in molten salt $CaCl_2$. A layer of calcium oxide that inhibits any further reaction was generated by the simple reaction of calcium with titanium dioxide generated (Figure 9); however, the ability of calcium chloride to dissolve significant amounts of calcium oxide may be effective in overcoming this problem [49].

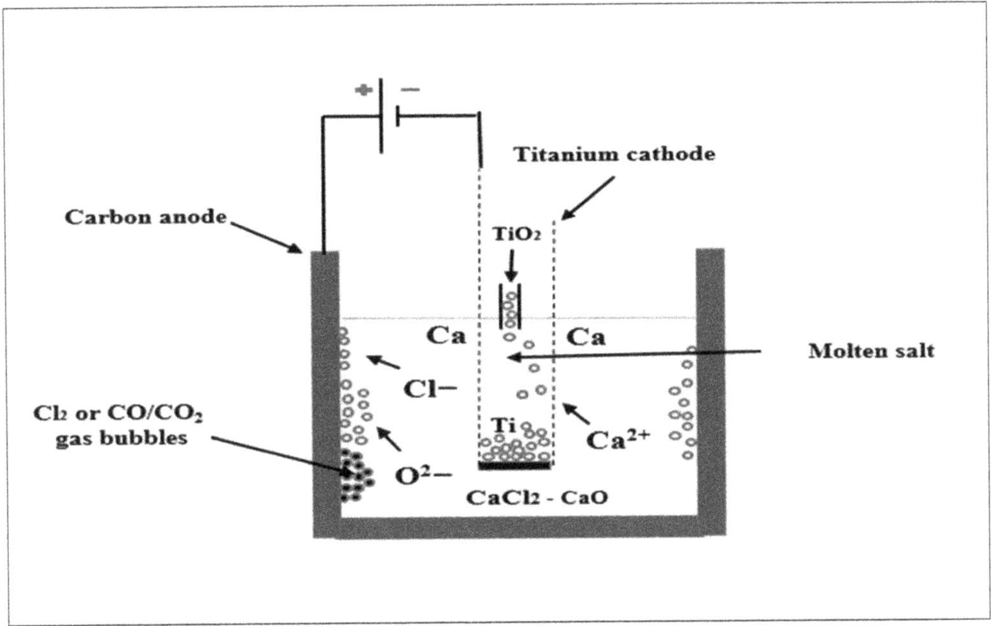

Figure 9. Diagram of the cell used in the OS process [83] (Reprinted with permission from Elsevier, Copyright, (2015)).

5.2.3. Quebec Iron and Titane (QIT) Process

The QIT process, patented by François Cardarelli in 2009 [84], is a process for the electrolytic extraction of Ti metal from compounds containing TiO_2 in a liquid state. Specifically, this process uses a molten compound containing TiO_2 as the cathode on the bottom and a consumable carbon as the anode. A layer of electrolyte, such as molten salts (e.g., CaF_2) or a solid-state ion conductor is used as the O^{-2} ion carrier in the middle. The temperature ranges between 1700 °C and 1900 °C in this process [76].

There have been several reports on the development of continuous processes based on the same chemistry as the above processes. A new process was presented, which integrates carbochlorination and electrolyzation to elaborate metallic titanium in molten $NaCl$-$CaCl_2$ electrolyte [82]. This experiment was made at 577 °C using some specific precursors like carbon-doped TiO_2. Likewise, a pre-treatment is performed to partially reduce TiO_2 to titanium sub-oxides with lower valence. The compacted pre-treated mass is electrolyzed at a temperature of 1000 °C in a molten $CaCl_2$ bath for 1–5 h [85].

An investigation of the $CaCl_2$ effect on the calcium vapor reduction process of Ti_2O_3 has been performed [86]. The compound $CaCl_2$ plays a crucial role in the calciothermic reduction process of Ti_2O_3 to prepare porous titanium [86]. Likewise, the generation of a low oxygen Ti powder was carried out by electrolytic reduction of $CaTiO_3$ in a $CaCl_2$-CaO melt. By using different sizes of raw oxide particles, the concentration of residual oxygen in the cathode imposes a reduction mechanism using calcium from the cast iron [87]. Another method was introduced for the preparation of titanium metal by reduction of $TiCl_4$ in $NaCl$-KCl-NaF eutectic melts [88]. A new method for manufacturing porous titanium by in situ calcium vapor reduction of titanium sesquioxide was presented by Yang et al. [89]. By calcium vapor, Ti_2O_3 and $CaCl_2$ powders have been reduced. The product was leached with hydrochloric acid and deionized water and the porous structure of Ti was obtained [89]. Furthermore, $LiCl$-KCl molten salt was used as an electrolyte due to its relatively low melting point to produce metallic titanium. The oxygen content in the titanium crystal is 1200 ppm [90].

Commercial TiO_2 has been reduced by Mg in a hydrogen atmosphere [91]. A mandatory deoxygenation process was added to ensure that the purity met standard specifications for titanium. The magnesium reduction and deoxygenation combination process is a holistic approach that produces Ti powders meeting ASTM specifications [91].

Otherwise, for patents published in this sense, Zhu et al. [92] invented a method for the electrowinning of titanium metal from a soluble anode molten salt containing titanium and concerns the technical field of non-ferrous metal metallurgy. In 2009, François Cardarelli [84] also invented a process for the electrowinning of Ti metal from compounds containing a liquid TiO_2 (Table 7). This process uses a molten compound containing TiO_2 as the cathode in the bottom. The anode can be a consumable carbon, a stable inert anode, or a gas diffusion anode powered by a combustible gas (e.g., H_2, CO, etc.). The temperature reaches 1700 °C to 1900 °C for this process [76].

As shown above, there are many methods and newly developed technologies to produce titanium powder. Kroll process the most utilized one can produce titanium with less oxygen content and metallic impurities, but its productivity is still very low, magnesium as a reductant has a high cost compared to other metals such as calcium and aluminum, and it is a labor-intensive batch process. Furthermore, the hunter process produces titanium powder purer than Kroll's powder but uses an expensive sodium reductant and heterogeneous exothermic reactions. For other processes, they present advantages at the production level but they are still at the laboratory scale and require more time to be developed in the industry. It is possible to develop a method to directly reduce oxides to efficiently produce Ti with low oxygen amount in the next few years. In principle, the production of high purity metallic Ti by direct reduction of Ti precursors is possible; hence, a method that relies on the melting of these oxides and is based on the principles of the Kroll process should be developed as a new approach to establish a low-cost Ti reduction process.

On the other hand, new technologies are developed as an alternative to the Kroll process. The electrochemical road is the best choice for many industries. The FCC Cambridge provides a product with almost 0.3 mass% oxygen, and it is a semi-continuous operation. However, this method exhibits many drawbacks: low current efficiency (20–40%), slow oxygen diffusion, difficult separation of metal/salt, and is costly for the TiO_2 pellet feed. On the other hand, the OS process can produce titanium directly from TiO_2 reduction in molten salt, but it still has some problem contamination because of the presence of carbon. Furthermore, many electrochemical procedures have been performed recently. They dictate a good enhancement in the cost and production issues, but they are difficult to scale up for the difficulty in facilitating a homogeneous reaction.

Table 7. Advantages and disadvantages of some patents in titanium production based on electrochemical technics.

Publication N°	Title	Advantages	Disadvantages	Reference
US 10, 081, 874 B2 (2018)	Method for electrowinning titanium from titanium-containing soluble anode molten salt	- This invention has the advantages of a short process flow, high carbothermal reduction efficiency, fewer intermediate products, direct availability of high-purity titanium, low purity requirements for anodic raw materials, low energy consumption, environmental friendliness, etc	- Use of toxic products. - High temperatures	[92]
US 7,790,014 B2 (2010)	Removal of substances from metal and semi-metal compounds	- $CaCl_2$ is used as a salt (can dissolve and transport oxygen ion), cheaper and less toxic than $MgCl_2$ - One-step process that does not use Mg as a reducing agent - Can also be used for the direct production of alloys, which could lead to savings in other ways	- Low yield. - Possibility of incomplete or partial reduction of TiO_2	[93]
US 7,504,017 B2 (2009)	Method for electrowinning of titanium metal or alloy from titanium oxide containing compound in the liquid state	- Electrochemical deoxidation is a one-step process - Cheaper raw material. - Use of molten titanium oxide slag as cathode material, preferably as such, without prior treatment and without the introduction of additives - Operation of the electrolysis at a temperature above the liquidus temperature of the titanium oxide slurry and the melting point of the titanium metal, allowing the electrodeposited titanium droplets to be rapidly collected by gravity - Providing a molten material for use as a molten cathode material	- High reaction temperatures - High temperature furnace with consumable carbon anodes - Exhaust of CO_2 gas	[84]
US 7.410,562 B2 (2011)	Thermal and electrochemical process for metal production	- Titanium can be electrolyzed from other titanium compounds that are not oxides - The use of a strong Lewis acid stabilizes and thus increases the activity of the titanium ion. - Extraction of a 99.9% pure titanium sponge - Possibility to produce titanium alloys - Permission to produce titanium with very low oxygen content	- High temperatures and pressures - Problem of oxidation. - Very energy consumable	[94]
EP 2 322 693 B1 (2004)	Electrochemical process for titanium production	- Use of cheaper and less toxic molten salts - Reduction of oxygen content (<400 ppm) - Use of cheaper and non-consumable graphite cathodes	- High temperatures - Difficulty controlling contamination.	[95]
US 2007/0029208A1 (2008)	Thermal and electrochemical process for metal production	- Extraction of the titanium sponge with a purity of 99.9% - Possibility to produce titanium alloys. - Production of titanium with very low oxygen content	- High temperatures and pressures - Oxidation problem - Very energy consumable	[96]

6. Conclusions

Titanium has remained an essential metal because of its wide use in different fields. Its demand in the industry has prompted unprecedented technical progress. Natural ilmenite is the most abundant titanium-bearing mineral in the earth's crust. The mineral ilmenite has been significantly grown since his discovery. Nowadays, it is the most crucial ore of titanium. The presented review shows the presence of enormous techniques for manufacturing titanium powder and titanium metal. Many of these processes are at various stages of development. The incentive for the development of new processes is often firmly rooted in the ambition to achieve a low-cost alternative to the Kroll process for the production of primary Ti metal. Marketed chloride-based thermochemical processes, such as the Kroll and Hunter processes, are batch operations and require high-quality natural rutile or improved synthetic slag, such as feeding and using cost-sensitive chlorination and thermochlorination steps. Many improvements have been made to thermochemical processes, but they offer few opportunities for cost reduction beyond current technology. Several development methods have generated considerable interest and scale-up efforts, including the Armstrong process and the FFC Cambridge process. The former is an example of using a continuous process for the reduction of $TiCl_4$ with Na or Mg while the latter is an example of the various electrochemical methods that convert TiO_2 into Ti metal in molten salt. A recently developed method, named the HAMR process, is based on destabilizing the Ti-O system by using hydrogen as a temporary alloying element during the magnesium thermal reduction of TiO_2. However, compared with the Kroll, these producers are expanding their production to meet the unprecedented demand for titanium. Overall, it is expected that it will take several years before any new process will be in commercial production and compete with the Kroll process.

Author Contributions: Conceptualization, M.E.K. and G.S.; formal analysis, M.E.K.; funding acquisition, G.S.; investigation, M.E.K.; methodology, M.E.K., G.S. and O.D.; resources, G.S.; supervision, G.S.; validation, M.E.K.; writing—original draft preparation, M.E.K.; writing—review and editing, G.S. and O.D. All authors have read and agreed to the published version of the manuscript.

Funding: This study was funded by NSERC (Natural Sciences and Engineering Research Council of Canada) discovery grant (RGPIN-2018-06128) and Metchib company-Quebec, Canada.

Acknowledgments: Metchib Company supported this research. We thank Metchib Company in Canada for providing rock samples and supporting us.

Conflicts of Interest: The authors declare no conflict of interest.

References

1.   Sibum, H. Titanium and titanium alloys—From raw material to semi-finished products. *Adv. Eng. Mater.* **2003**, *5*, 393–398. [CrossRef]
2.   Leyens, C.; Peters, M. *Titanium and Titanium Alloys: Fundamentals and Applications*; John Wiley & Sons: Hoboken, NJ, USA, 2003.
3.   Cui, C.; Hu, B.; Zhao, L.; Liu, S. Titanium alloy production technology, market prospects and industry development. *Mater. Des.* **2011**, *32*, 1684–1691. [CrossRef]
4.   Mutava, T.D. Characterisation of a Titanium Precursor Salt and Study of Some of the Treatment Steps Used for the Extraction Process. Ph.D. Thesis, University of the Witwatersrand, Johannesburg, South Africa, 2009.
5.   Jackson, M.; Dring, K. A review of advances in processing and metallurgy of titanium alloys. *Mater. Sci. Technol.* **2006**, *22*, 881–887. [CrossRef]
6.   Force, E.R. *Geology of Titanium-Mineral Deposits*; Geological Society of America: Boulder, CO, USA, 1991; Volume 259.
7.   U.S. Geological Survey. *Mineral Commodity Summaries 2021*; U.S. Geological Survey: Reston, VA, USA, 2021; 200p.
8.   Force, E.R. *Geology and Resources of Titanium*; Geological Society of America: Boulder, CO, USA, 1976.
9.   Van Gosen, B.S.; Fey, D.L.; Shah, A.K.; Verplanck, P.L.; Hoefen, T.M. *Deposit Model for Heavy-Mineral Sands in Coastal Environments: Chapter L in Mineral Deposit Models for Resource Assessment*; US Geological Survey: Reston, VI, USA, 2014.
10. Fadeel, B. *Handbook of Safety Assessment of Nanomaterials: From Toxicological Testing to Personalized Medicine*; Pan Stanford; CRC Press: Boca Raton, FL, USA, 2014.
11. Fang, Z.Z.; Paramore, J.D.; Sun, P.; Chandran, K.R.; Zhang, Y.; Xia, Y.; Free, M. Powder metallurgy of titanium—Past, present, and future. *Int. Mater. Rev.* **2018**, *63*, 407–459. [CrossRef]

12. Ma, M.; Wang, D.; Wang, W.; Hu, X.; Jin, X.; Chen, G.Z. Extraction of titanium from different titania precursors by the FFC Cambridge process. *J. Alloys Compd.* **2006**, *420*, 37–45. [CrossRef]
13. Kroll, W. The production of ductile titanium. *Trans. Electrochem. Soc.* **1940**, *78*, 35–47. [CrossRef]
14. Berthaud, M. Étude du Comportement de L'alliage de Titane Ti6242S à Haute Température sous Atmosphères Complexes: Applications Aéronautiques. Ph.D. Thesis, Université Bourgogne Franche-Comté, Besançon, France, 2018.
15. Samal, S. *Thermal Plasma Processing of Ilmenite*; Springer: Berlin/Heildelberg, Germany, 2017.
16. Froes, F. *Titanium: Physical Metallurgy, Processing, and Applications*; ASM International: Novelty, OH, USA, 2015.
17. Fang, Z.Z.; Lefler, H.D.; Froes, F.H.; Zhang, Y. Introduction to the development of processes for primary Ti metal production. In *Extractive Metallurgy of Titanium*; Elsevier: Amsterdam, The Netherlands, 2020; pp. 1–10.
18. Biehl, V.; Wack, T.; Winter, S.; Seyfert, U.T.; Breme, J. Evaluation of the haemocompatibility of titanium based biomaterials. *Biomol. Eng.* **2002**, *19*, 97–101. [CrossRef]
19. Bessinger, D.; Geldenhuis, J.M.A.; Pistorius, P.C.; Mulaba, A.; Hearne, G. The decrepitation of solidified high titania slags. *J. Non Cryst. Solids* **2001**, *282*, 132–142. [CrossRef]
20. Nagesh, C.R.; Ramachandran, C.; Subramanyam, R. Methods of titanium sponge production. *Trans. Indian Inst. Met.* **2008**, *61*, 341–348. [CrossRef]
21. Vilakazi, A.Q. Hydrometallurgical Beneficiation of Ilmenite. Ph.D. Thesis, University of the Free State, Bloemfontein, South Africa, 2017.
22. Wilson, N.C.; Muscat, J.; Mkhonto, D.; Ngoepe, P.E.; Harrison, N.M. Structure and properties of ilmenite from first principles. *Phys. Rev. B* **2005**, *71*, 075202. [CrossRef]
23. Charlier, B.; Namur, O.; Bolle, O.; Latypov, R.; Duchesne, J.C. Fe–Ti–V–P ore deposits associated with Proterozoic massif-type anorthosites and related rocks. *Earth-Sci. Rev.* **2015**, *141*, 56–81. [CrossRef]
24. Korneliussen, A.; McENROE, S.A.; Nilsson, L.P.; Schiellerup, H.; Gautneb, H.; Meyer, G.B.; Storseth, L.R. An overview of titanium deposits in Norway. *Norges Geologiske Undersokelse* **2000**, *436*, 27–38.
25. Froes, F. Titanium powder metallurgy: A review-part 1. *Adv. Mater. Process.* **2012**, *170*, 16–22.
26. McCracken, C.G.; Motchenbacher, C.; Barbis, D.P. Review of titanium powder production methods. *Int. J. Powder Metall.* **2010**, *46*, 19–25.
27. Hunter, M. Metallic titanium. *J. Am. Chem. Soc.* **1910**, *32*, 330–336. [CrossRef]
28. Florkiewicz, W.; Malina, D.; Tyliszczak, B.; Sobczak-Kupiec, A. Manufacturing of titanium and its alloys. In *Sustainable Production: Novel Trends in Energy, Environment and Material Systems*; Springer: Berlin, Heidelberg, 2020; pp. 61–74.
29. Hartman, A.D.; Gerdemann, S.J.; Hansen, J.S. Producing lower-cost titanium for automotive applications. *JOM* **1998**, *50*, 16–19. [CrossRef]
30. Kohli, R. *Production of Titanium from Ilmenite: A Review*; Lawrence Berkeley Lab.: Berkeley, CA, USA, 1981.
31. Gambogi, J.; Gerdemann, S. Titanium metal: Extraction to application. In *Review of Extraction, Processing, Properties & Applications of Reactive Metals*; John Wiley & Sons: Hoboken, NJ, USA, 1999; pp. 175–210.
32. Agripa, H.; Botef, I. Modern production methods for titanium alloys: A review. In *Titanium Alloys-Novel Aspects of Their Processing*; IntechOpen: London, UK, 2019.
33. Kraft, E. *Summary of Emerging Titanium Cost Reduction Technologies*; EHK Technologies: Burlington, WA, USA, 2004.
34. Crowley, G. How to extract low-cost titanium. *Adv. Mater. Process.* **2003**, *161*, 25–27.
35. Chen, W.; Yamamoto, Y.; Peter, W.H. Investigation of pressing and sintering processes of CP-Ti powder made by Armstrong Process. In *Key Engineering Materials*; Trans Tech Publications: Stafa-Zurich, Switzerland, 2010.
36. Araci, K.; Mangabhai, D.; Akhtar, K. Production of titanium by the Armstrong Process®. In *Titanium Powder Metallurgy*; Elsevier: Amsterdam, The Netherlands, 2015; pp. 149–162.
37. Brooks, G.; Cooksey, M.; Wellwood, G.; Goodes, C. Challenges in light metals production. *Miner. Process. Extr. Metall.* **2007**, *116*, 25–33. [CrossRef]
38. Doblin, C.; Chryss, A.; Monch, A. Titanium powder from the TiRO™ process. In *Key Engineering Materials*; Trans Tech Publications: Stafa-Zurich, Switzerland, 2012.
39. Van Vuuren, D.S. Direct titanium powder production by metallothermic processes. In *Titanium Powder Metallurgy*; Elsevier: Amsterdam, The Netherlands, 2015; pp. 69–93.
40. Hansen, D.A.; Gerdemann, S.J. Producing titanium powder by continuous vapor-phase reduction. *JOM* **1998**, *50*, 56–58. [CrossRef]
41. Van Vuuren, D.; Oosthuizen, S.; Heydenrych, M.D. Titanium production via metallothermic reduction of TiCl4 in molten salt: Problems and products. *J. S. Afr. Inst. Min. Metall.* **2011**, *111*, 141–147.
42. Mo, W.; Deng, G.; Luo, F. *Titanium Metallurgy*; Metallurgical Industry Press: Beijing, China, 1998.
43. Mishra, B.; Kipouros, G.J. *Titanium Extraction and Processing*; The Minerals, Metals & Materials Society: Warrendale, PA, USA, 1997.
44. Borok, B. Obtaining powders of alloys and steels by a complex reduction of oxide mixtures by CaH2. *Trans. Cent. Res. Inst. Ferr. Met.* **1965**, *43*, 69–80.
45. Froes, F. The production of low-cost titanium powders. *JOM J. Miner. Met. Mater. Soc.* **1998**, *50*, 41–43. [CrossRef]
46. Zhang, Y.; Fang, Z.Z.; Sun, P.; Zheng, S.; Xia, Y.; Free, M. A Perspective on thermochemical and electrochemical processes for titanium metal production. *JOM* **2017**, *69*, 1861–1868. [CrossRef]

47. Park, I.; Abiko, T.; Okabe, T.H. Production of titanium powder directly from TiO_2 in $CaCl_2$ through an electronically mediated reaction (EMR). *J. Phys. Chem. Solids* **2005**, *66*, 410–413. [CrossRef]
48. Okabe, T.H.; Oda, T.; Mitsuda, Y. Titanium powder production by preform reduction process (PRP). *J. Alloys Compd.* **2004**, *364*, 156–163. [CrossRef]
49. Fray, D. Novel methods for the production of titanium. *Int. Mater. Rev.* **2008**, *53*, 317–325. [CrossRef]
50. Zhang, Y.; Fang, Z.Z.; Xia, Y.; Huang, Z.; Lefler, H.; Zhang, T.; Guo, J. A novel chemical pathway for energy efficient production of Ti metal from upgraded titanium slag. *Chem. Eng. J.* **2016**, *286*, 517–527. [CrossRef]
51. Fang, Z.Z.; Zhang, Y.; Xia, Y.; Sun, P. Methods of Producing a Titanium Product. U.S. Patent 2016/0108497 A1, 23 June 2016.
52. Froes, F.; Eylon, D. Powder metallurgy of titanium alloys. *Int. Mater. Rev.* **1990**, *35*, 162–184. [CrossRef]
53. Alexander, P.P. Production of Titanium Hydride. U.S. Patent 2,427,338, 16 September 1947.
54. Suzuki, R.O.; Ono, K.; Teranuma, K. Calciothermic reduction of titanium oxide and in-situ electrolysis in molten $CaCl_2$. *Metall. Mater. Trans. B* **2003**, *34*, 287–295. [CrossRef]
55. Abiko, T.; Park, I.; Okabe, T.H. Reduction of titanium oxide in molten salt medium. In Proceedings of the 10th World Conference on Titanium, Ti-2003, Hamburg, Germany, 13–18 July 2003.
56. Won, C.; Nersisyan, H.; Won, H. Titanium powder prepared by a rapid exothermic reaction. *Chem. Eng. J.* **2010**, *157*, 270–275. [CrossRef]
57. Nersisyan, H.; Lee, J.; Won, C. Combustion of TiO_2–Mg and TiO_2–Mg–C systems in the presence of NaCl to synthesize nanocrystalline Ti and TiC powders. *Mater. Res. Bull.* **2003**, *38*, 1135–1146. [CrossRef]
58. Zhang, Y.; Fang, Z.Z.; Sun, P.; Zhang, T.; Xia, Y.; Zhou, C.; Huang, Z. Thermodynamic destabilization of Ti-O solid solution by H_2 and deoxygenation of Ti using Mg. *J. Am. Chem. Soc.* **2016**, *138*, 6916–6919. [CrossRef] [PubMed]
59. Lu, X.; Ono, T.; Takeda, O.; Zhu, H. Production of fine titanium powder from titanium sponge by the shuttle of the disproportionation reaction in molten NaCl–KCl. *Mater. Trans.* **2019**, *60*, 405–410. [CrossRef]
60. Ardani, M.R.; Al Janabi, A.S.M.; Udayakumar, S.; Hamid, S.A.R.S.A.; Mohamed, A.R.; Lee, H.L.; Ibrahim, I. Reduction of $TiCl_4$ to TiH_2 with CaH_2 in presence of Ni powder. In *Rare Metal Technology 2019*; Springer: Berlin/Heidelberg, Germany, 2019; pp. 131–144.
61. Lei, X.; Xu, B.; Yang, G.; Shi, T.; Liu, D.; Yang, B. Direct calciothermic reduction of porous calcium titanate to porous titanium. *Mater. Sci. Eng. C* **2018**, *91*, 125–134. [CrossRef]
62. Zheng, C.; Ouchi, T.; Iizuka, A.; Taninouchi, Y.K.; Okabe, T.H. Deoxidation of titanium using Mg as deoxidant in $MgCl_2$-YCl_3 flux. *Metall. Mater. Trans. B* **2019**, *50*, 622–631. [CrossRef]
63. Spreitzer, D.; Schenk, J. Iron ore reduction by hydrogen using a laboratory scale fluidized bed reactor: Kinetic investigation—Experimental setup and method for determination. *Metall. Mater. Trans. B* **2019**, *50*, 2471–2484. [CrossRef]
64. Bolivar, R.; Friedrich, B. Magnesiothermic reduction from titanium dioxide to produce titanium powder. *J. Sustain. Metall.* **2019**, *5*, 219–229. [CrossRef]
65. Mojisola, T.; Ramakokovhu, M.M.; Raethel, J.; Olubambi, P.A.; Matizamhuka, W.R. In-situ synthesis and characterization of Fe–TiC based cermet produced from enhanced carbothermally reduced ilmenite. *Int. J. Refract. Met. Hard Mater.* **2019**, *78*, 92–99. [CrossRef]
66. Chae, J.; Oh, J.M.; Yoo, S.; Lim, J.W. Eco-friendly pretreatment of titanium turning scraps and subsequent preparation of ferro-titanium ingots. *Korean J. Met. Mater.* **2019**, *57*, 569–574. [CrossRef]
67. Ri, V.; Nersisyan, H.; Kwon, S.C.; Lee, J.H.; Suh, H.; Kim, J.G. A thermochemical and experimental study for the conversion of ilmenite sand into fine powders of titanium compounds. *Mater. Chem. Phys.* **2019**, *221*, 1–10. [CrossRef]
68. Gao, Z.; Cheng, G.; Yang, H.; Xue, X.; Ri, J. Preparation of ferrotitanium using ilmenite with different reduction degrees. *Metals* **2019**, *9*, 962. [CrossRef]
69. Zhang, G.; Gou, H.; Wu, K.; Chou, K. Carbothermic reduction of Panzhihua ilmenite in vacuum. *Vacuum* **2017**, *143*, 199–208. [CrossRef]
70. Kim, D.H.; Kim, T.S.; Heo, J.H.; Park, H.S.; Park, J.H. Influence of temperature on reaction mechanism of ilmenite ore smelting for titanium production. *Metall. Mater. Trans. B* **2019**, *50*, 1830–1840. [CrossRef]
71. Mu, H.; Mu, T.; Zhao, S.; Zhu, F.; Deng, B.; Peng, W.; Yan, B. Method of Producing Titanium Metal With Titanium-Containing Material. U.S. Patent 9,963,796, 8 May 2018.
72. Fisher, R.L. Deoxidation of Titanium and Similar Metals Using a Deoxidant in a Molten Metal Carrier. U.S. Patent 4,923,531, 8 May 1990.
73. Cox, J.R.; DeALWIS, C.L.; Kohler, B.A.; Lewis, M.G. System and Method for Extraction and Refining of Titanium. U.S. Patent 10,066,308, 4 September 2018.
74. Moxson, V.S.; Duz, V.A.; Klevtsov, A.G.; Sukhoplyuyev, V.D.; Sopka, M.D.; Shuvalov, Y.V.; Matviychuk, M. Method of Manufacturing Pure Titanium Hydride Powder and Alloyed Titanium Hydride Powders by Combined Hydrogen-Magnesium Reduction of Metal Halides. U.S. Patent 9,067,264, 30 June 2015.
75. Klevtsov, A.; Nikishin, A.; Shuvalov, J.; Moxson, V.; Duz, V. Continuous and Semi-Continuous Process of Manufacturing Titanium Hydride Using Titanium Chlorides of Different Valency. U.S. Patent 8,388,727, 5 March 2013.
76. Shimotori, K.; Ochi, Y.; Ishihara, H.; Umeki, T.; Ishigami, T. Highly Pure Titanium. U.S. Patent RE34,598, 3 May 1994.

77. Kasparov, S.A.; Klevtsov, A.G.; Cheprasov, A.I.; Moxson, V.S.; Duz, V.A. Semi-Continuous Magnesium-Hydrogen Reduction Process for Manufacturing of Hydrogenated, Purified Titanium Powder. U.S. Patent 8,007,562, 30 August 2011.

78. Fang, Z.Z.; Zhang, Y.; Xia, Y.; Sun, P. Methods of Producing a Titanium Product. U.S. Patent 10,689,730, 23 June 2020.

79. Hanusiak, W.M.; McBride, D.R. Titanium Powder Production Apparatus and Method. U.S. Patent 10,583,492, 10 March 2020.

80. Chen, G.Z.; Fray, D.J.; Farthing, T.W. Direct electrochemical reduction of titanium dioxide to titanium in molten calcium chloride. *Nature* **2000**, *407*, 361. [CrossRef]

81. Chen, G.Z. The FFC Cambridge process and its relevance to valorisation of ilmenite and titanium-rich slag. *Miner. Process. Extr. Metall.* **2015**, *124*, 96–105. [CrossRef]

82. Zhao, K.; Wang, Y.; Gao, F. Electrochemical extraction of titanium from carbon-doped titanium dioxide precursors by electrolysis in chloride molten salt. *Ionics* **2019**, *12*, 6107–6114. [CrossRef]

83. Furuta, T.; Nishino, K.; Hwang, J.; Yamada, A.; Ito, K.; Osawa, S.; Kuramoto, S.; Suzuki, N.; Chen, R.; Saito, T. Development of multi functional titanium alloy, "Gum Metal". In *Ti-2003 Science and Technology*; Lütjering, G., Albrecht, J., Eds.; Wiley-VCH: Weinheim, Germany, 2004.

84. Cardarelli, F. Method for Electrowinning of Titanium Metal or Alloy from Titanium Oxide Containing Compound in the Liquid State. U.S. Patent 7,504,017, 17 March 2009.

85. Mohanty, J.; Behera, P.K. Use of pre-treated TiO_2 as cathode material to produce Ti metal through molten salt electrolysis. *Trans. Indian Inst. Met.* **2019**, *72*, 859–865. [CrossRef]

86. Yang, G.; Xu, B.; Wan, H.; Wang, F.; Yang, B.; Wang, Z. Effect of $CaCl_2$ on microstructure of calciothermic reduction products of Ti_2O_3 to prepare porous titanium. *Metals* **2018**, *8*, 698. [CrossRef]

87. Noguchi, H.; Natsui, S.; Kikuchi, T.; Suzuki, R.O. Reduction of $CaTiO_3$ by electrolysis in the molten salt $CaCl_2$-CaO. *Electrochemistry* **2018**, *86*, 82–87. [CrossRef]

88. Zhu, F.; Qiu, K.; Sun, Z. Preparation of titanium from $TiCl_4$ in a molten fluoride-chloride salt. *Electrochemistry* **2017**, *85*, 715–720. [CrossRef]

89. Yang, G.; Xu, B.; Lei, X.; Wan, H.; Yang, B.; Liu, D.; Wang, Z. Preparation of porous titanium by direct in-situ reduction of titanium sesquioxide. *Vacuum* **2018**, *157*, 453–457. [CrossRef]

90. Li, C.; Song, J.; Li, S.; Li, X.; Shu, Y.; He, J. An investigation on electrodeposition of titanium in molten LiCl-KCl. In *Materials Processing Fundamentals 2019*; Springer: Berlin/Heidelberg, Germany, 2019; pp. 193–202.

91. Xia, Y.; Fang, Z.Z.; Zhang, Y.; Lefler, H.; Zhang, T.; Sun, P.; Huang, Z. Hydrogen assisted magnesiothermic reduction (HAMR) of commercial TiO_2 to produce titanium powder with controlled morphology and particle size. *Mater. Trans.* **2017**, *58*, 355–360. [CrossRef]

92. Zhu, H.; Wang, Q.; Jiao, S. Method for Electrowinning Titanium from Titanium-Containing Soluble Anode Molten Salt. U.S. Patent 10,081,874, 25 September 2018.

93. Fray, D.J.; Farthing, T.W.; Chen, Z. Removal of Substances from Metal and Semi-Metal Compounds. U.S. Patent 7,790,014, 7 September 2010.

94. Withers, J.C.; Loutfy, R.O. Thermal and Electrochemical Process for Metal Production. U.S. Patent 7,985,326, 26 July 2011.

95. Withers, J.C.; Loutfy, R.O. Electrochemical Process for Titanium Production. U.S. Patent EP 2 322 693 B1, 18 August 2004.

96. Withers, J.C.; Loutfy, R.O. Thermal and Electrochemical Process for Metal Production. U.S. Patent 2007/0029208A1, 8 February 2008.

MDPI
St. Alban-Anlage 66
4052 Basel
Switzerland
Tel. +41 61 683 77 34
Fax +41 61 302 89 18
www.mdpi.com

Minerals Editorial Office
E-mail: minerals@mdpi.com
www.mdpi.com/journal/minerals